Lecture Notes in Mechanical Engineering

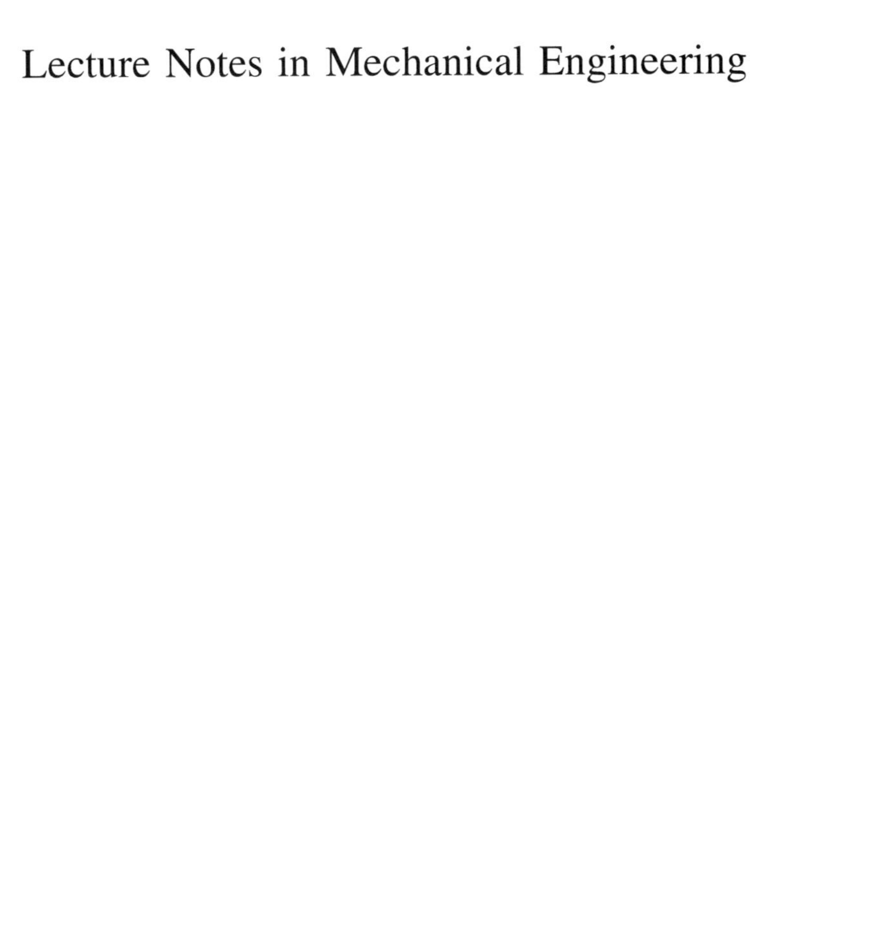

For further volumes:
http://www.springer.com/series/11236

Amaresh Chakrabarti · Raghu V. Prakash
Editors

ICoRD'13

Global Product Development

Part II

Editors
Amaresh Chakrabarti
Centre for Product Design
and Manufacturing
Indian Institute of Science
Bangalore, Karnataka
India

Raghu V. Prakash
Mechanical Engineering
Indian Institute of Technology Madras
Chennai, Tamil Nadu
India

ISSN 2195-4356 ISSN 2195-4364 (electronic)
ISBN 978-81-322-1049-8 ISBN 978-81-322-1050-4 (eBook)
DOI 10.1007/978-81-322-1050-4
Springer New Delhi Heidelberg New York Dordrecht London

Library of Congress Control Number: 2012954861

© Springer India 2013
This work is subject to copyright. All rights are reserved by the Publisher, whether the whole or part of the material is concerned, specifically the rights of translation, reprinting, reuse of illustrations, recitation, broadcasting, reproduction on microfilms or in any other physical way, and transmission or information storage and retrieval, electronic adaptation, computer software, or by similar or dissimilar methodology now known or hereafter developed. Exempted from this legal reservation are brief excerpts in connection with reviews or scholarly analysis or material supplied specifically for the purpose of being entered and executed on a computer system, for exclusive use by the purchaser of the work. Duplication of this publication or parts thereof is permitted only under the provisions of the Copyright Law of the Publisher's location, in its current version, and permission for use must always be obtained from Springer. Permissions for use may be obtained through RightsLink at the Copyright Clearance Center. Violations are liable to prosecution under the respective Copyright Law.
The use of general descriptive names, registered names, trademarks, service marks, etc. in this publication does not imply, even in the absence of a specific statement, that such names are exempt from the relevant protective laws and regulations and therefore free for general use.
While the advice and information in this book are believed to be true and accurate at the date of publication, neither the authors nor the editors nor the publisher can accept any legal responsibility for any errors or omissions that may be made. The publisher makes no warranty, express or implied, with respect to the material contained herein.

Printed on acid-free paper

Springer is part of Springer Science+Business Media (www.springer.com)

Preface

Design is ubiquitous, yet universal; it pervades all spheres of life, and has been around ever since life has been engaged in, purposefully changing the world around it. While some designs that matured many centuries ago still remain in vogue, there are areas in which new designs are being evolved almost every day, if not, every hour, globally. Research into design and the emergence of a research community in this area has been relatively new, its development influenced by the multiple facets of design (human, artefact, process, organisation, and the micro- and macro-economy by which design is shaped) and the associated diversification of the community into those focusing on various aspects of these individual facets, in various applications. Design is complex, balancing the needs of multiple stakeholders, and requiring a multitude of areas of knowledge to be utilised, with resources spread across space and time.

The collection of papers in this book constitutes the Proceedings of the 4th International Conference on Research into Design (ICoRD'13) held at the Indian Institute of Technology Madras in the city of Chennai, India during 7–9 January 2013. ICoRD'13 is the fourth in a series of biennial conferences held in India to bring together the international community from diverse areas of design practice, teaching and research. The goal is to share cutting edge research about design among its stakeholders; aid the ongoing process of developing a collective vision through emerging research challenges and questions; and provide a platform for interaction, collaboration and development of the community in order for it to address the challenges and realise the collective vision. The conference is intended for all stakeholders of design, and in particular for its practitioners, researchers, teachers and students.

Out of the 201 abstracts submitted to ICoRD'13, 175 were selected for full paper submission. One hundred and thirty-two full papers were submitted, which were reviewed by experts from the ICoRD'13 International Programme Committee comprising 163 members from 127 institutions or organisations from 32 countries spanning five continents. Finally, 114 full papers, authored by over 200 researchers from 91 institutions and organisations from 23 countries spanning five continents, have been selected for presentation at the conference and for publication as chapters in this book. ICoRD has steadily grown over the last three editions, from a humble beginning in 2006 with 30 papers and 60 participants, through 75 papers and 100 participants in ICoRD'09, and 100 papers and 150 participants in ICoRD'11. This is also the first time that ICoRD has taken place outside Bangalore, with the Indian Institute of Technology Madras and Indian Institute of Science Bangalore jointly sharing the responsibility for its organisation.

The chapters in this book together cover all three major areas of products and processes: functionality, form and human factors. The spectrum of topics range from those focusing on early stages such as creativity and synthesis, through those that are primarily considered in specific stages of the product life cycle, such as safety, reliability or manufacturability, to those that are relevant across the whole product life cycle, such as collaboration, communication, design management, knowledge management, cost, environment and product life cycle management. Issues of delivery of research into design, in terms of its two major arms: design education and practice, are both highlighted in the chapters in this book. Foundational topics such as the nature of design theory and research methodology are also major areas of focus. It is particularly encouraging to see in the chapters the variety of areas of application of research into design—aerospace, healthcare, automotive and white goods sectors are but a few of those explored. The theme of this year's conference is Global Product Development. The large number of chapters that impinge on this theme reflects the importance of this theme within design research.

On behalf of the Patron, Steering Committee, Advisory Committee, Local Organising Committee and Co-Chairs, we thank all the authors, reviewers, institutions and organisations that participated in the conference and the Conference Programme Committee for their support in organising ICoRD'13 and putting this book together. We are thankful to the Design Society and Design Research Society for their kind endorsement of ICoRD'13. We thank the Indian Institute of Technology Madras and the Indian Institute of Science, Bangalore for their support of this event. We also wish to place on record and acknowledge the enormous support provided by Mr. Ranjan B. S. C., Ms. Kumari M. C. and Ms. Chaitra of IISc in managing the review process, and in preparation of the conference programme and this book, and the group of student-volunteers of Indian Institute of Technology Madras led by Swostik, Suraj and Sahaj in the organisation and running of the conference. Finally, we thank Springer India for its support in the publication of this book.

Amaresh Chakrabarti
Raghu V. Prakash

Contents

Part II Design Creativity, Synthesis, Evaluation and Optimization

Part III Design Aesthetics, Semiotics, Semantics

Part IV Human Factors in Design

Part V Eco-Design, Sustainable Manufacturing, Design for Sustainability

Part VI Design Collaboration and Communication

Part VII Design Management, Knowledge Management and Product Life Cycle Management

Organizing Committee

Patron

Ramamurthi, Bhaskar, Director, Indian Institute of Technology Madras, Chennai, India

Chairs

Prakash, Raghu, Indian Institute of Technology Madras, Chennai, India
Chakrabarti, Amaresh, Indian Institute of Science, India

Co-Chairs

Blessing, Lucienne, University of Luxembourg, Luxembourg
Cugini, Umberto, Politecnico di Milano, Italy
Culley, Steve, University of Bath, UK
McAloone, Tim, Technical University of Denmark, Denmark
Ray, Gaur H., Indian Institute of Technology Bombay, India
Taura, Toshiharu, Kobe University, Japan

Steering Committee

Athavankar, Uday, Indian Institute of Technology Bombay, India
Gero, John, George Mason University, USA
Gurumoorthy, B., Indian Institute of Science, India
Harinarayana, Kota, National Aeronautical Laboratories, India (Chair)
Lindemann, Udo, Technical University of Munich, Germany

Advisory Committee

Arunachalam, V. S., Center for Study of Science, Technology and Policy, India
Das, Anjan, Confederation of Indian Industry, India
Dhande, Sanjay G., Indian Institute of Technology Kanpur, India
Forster, Richard, Airbus Industrie, Toulouse, France
Gnanamoorthy, R., IIIT D&M, Kancheepuram, India
Horvath, Imre, Delft University of Technology, The Netherlands
Jaura, Arun, Eaton India Engineering Center, Pune, India
Jhunjhunwala, Ashok, Indian Institute of Technology Madras, Chennai, India
Leifer, Larry, Stanford University, USA
Pathy, Jayshree, CII National Committee on Design, New Delhi, India
Mitra, Arabinda, International Division, Department of Science and Technology, India
Mohanram, P. J., Indian Machine Tool Manufacturers Association, India
Mruthyunjaya, T. S., Chairman Emeritus, CPDM, Indian Institute of Science, India
Nair, P. S., ISRO Satellite Centre, Indian Space Research Organisation, India
Pitroda, Sam, National Innovation Council, India
Saxena, Raman, USID Foundation, India
Subrahmanyam, P. S., Aeronautical Development Agency, India
Sumantran, V., Hinduja Automotive, UK
Lt. Gen. Sundaram, V. J., National Design and Research Forum, India

International Programme Committee

Albers, Albert, Karlsruhe Institute of Technology, Germany
Allen, Janet K., University of Oklahoma, USA
Anderl, Reiner, Technical University of Darmstadt, Germany
Arai, Eiji, Osaka University, Japan
Aurisicchio, Marco, Imperial College London, UK
Badke-Schaub, Petra, Delft University of Technology, The Netherlands
Bernard, Alain, Ecole Centrale, Nantes, France
Bhattacharya, Bishakh, Indian Institute of Technology Kanpur, India
Blanco, Eric, Institut Polytechnique de Grenoble, France
Bohemia, Erik, Northumbria University, UK
Boks, Casper, Norwegian University of Science and Technology, Norway
Bolton, Simon, Cranfield University, UK
Bonnardel, Nathalie, University of Provence, France
Bordegoni, Monica, Politecnico di Milano, Italy
Borg, Jonathan C., University of Malta, Malta
Braha, Dan, University of Massachusetts, USA
Brazier, Frances, Delft University of Technology, The Netherlands
Brissaud, Daniel, Institut Polytechnique de Grenoble, France
Bruder, Ralph, TU Darmstadt, Germany
Burvill, Colin, University of Melbourne, Australia

C. Amarnath, Indian Institute of Technology Bombay, India
Caillaud, Emmanuel, Universite de Strasbourg, France
Cascini, Gaetano, Politecnico di Milano, Italy
Cavallucci, Denis, INSA Strasbourg, France
Chakrabarti, Debkumar, Indian Institute of Technology Guwahati, India
Childs, Peter, Imperial College London, UK
Clarkson, John P., University of Cambridge, UK
Das, Amarendra K., Indian Institute of Technology Guwahati, India
Deb, Anindya, Indian Institute of Science, India
Dong, Andy, University of Sydney, Australia
Dorst, Kees, University of Technology Sydney, Australia
Duffy, Alex, University of Strathclyde, UK
Duflou, Joost, Katholieke Universiteit Leuven, Belgium
Duhovnik, Jožef, University of Ljubljana, Slovenia
Echempati, Raghu, Kettering University, USA
Eckart, Claudia, Open University, UK
Eckhardt, Claus-Christian, Lund University, Sweden
Eynard, Benoit, Universite de Technologie de Compiegne, France
Fadel, Georges, Clemson University, USA
Fargnoli, Mario, University of Rome "La Sapienza", Italy
Friedman, Ken, Swinburne University of Technology, Australia
G. K. Ananthasuresh, Indian Institute of Science, India
Ghosal, Ashitava, Indian Institute of Science, India
Girard, Phillippe, University of Bordeaux, France
Goel, Ashok, Georgia Institute of Technology, USA
Goldschmidt, Gabriela, Technion, Israel
Gooch, Shayne, University of Canterbury, New Zealand
Grimhelden, Martin, Royal University of Technology, Sweden
Gu, Peihua, University of Calgary, Canada
Hanna, Sean, University College London, UK
Hekkert, Paul, Delft University of Technology, The Netherlands
Helander, Martin, Koln International School of Design, Germany
Heskett, John, Hong Kong Polytechnic University, China
Hicks, Ben, University of Bath, UK
Horne, Ralph, RMIT University, Australia
Hosnedl, Stanislav, University of West Bohemia, Czech Republic
Howard, Thomas, Technical University of Denmark, Denmark
Ijomah, Winifred, University of Strathclyde, UK
Ion, William, Strathclyde University, UK
Iyer, Ashok Ganapathy, Manipal University—Dubai Campus, UAE
Jagtap, Santosh, Lund University, Sweden
Johnson, Aylmer, University of Cambridge, UK
Kailas, SatishVasu, Indian Institute of Science, India
Karlsson, Lennart, Lulea University of Technology, Sweden
Kasturirangan, Rajesh, National Institute of Advanced Studies, India

Keinonen, Turkka, University of Art and Design Helsinki, Finland
Kim, Jongdeok, Hongik University, Korea
Kota, Srinivas, Institut Polytechnique de Grenoble, France
Krishnan, S. S., Center for Study of Science, Technology and Policy (CSTEP), India
Kumar, Kris L., University of Botswana, Botswana
Larsson, Tobias C., Blekinge Institute of Technology, Sweden
Lee, Soon-Jong, Seoul National University, Korea
Leifer, Larry, Stanford University, USA
Lewis, Kemper, The State University of New York, Buffalo, USA
Lin, Rung-Tai, National Taiwan University of Arts, Taiwan
Linsey, Julie, Texas A&M University, USA
Lloyd, Peter, Open University, UK
MacGregor, Steven P., University of Girona, Spain
Macia, Joaquim Lloveras, Polytechnic University of Catalunya, Spain
McMahon, Chris, University of Bristol, UK
Magee, Christopher, Massachusetts Institute of Technology, USA
Malmqvist, Johan, Chalmers University of Technology, Sweden
Mani, Monto, Indian Institute of Science, India
Manivannan, M., Indian Institute of Technology Madras, India
Marjanovic, Dorian, University of Zagreb, Croatia
Matsuoka, Yoshiyuki, Keio University, Japan
Matthew, Mary, Indian Institute of Science, India
Mckay, Alison, University of Leeds, UK
Meerkamm, Harald, Friedrich-Alexander-Universität Erlangen-Nürnberg, Germany
Ming, Henry X. G., Shanghai Jiao Tong University, China
Mistree, Farrokh, University of Oklahoma, USA
Montagna, Francesca, Politecnico di Torino, Italy
Mulet, Elena, Universitat Jaume I, Spain
Mullineux, Glen, University of Bath, UK
Murakami, Tamotsu, University of Tokyo, Japan
Nagai, Yukari, Japan Advanced Institute of Science and Technology, Japan
Papalambros, Panos, University of Michigan, USA
Petrie, Helen, University of York, UK
Poovaiah, Ravi, Indian Institute of Technology Bombay, India
Popovic, Vesna, Queensland University of Technology, Australia
Prasad, Sathya M., Ashok Leyland, India
Radhakrishnan, P., PSG Institute of Advanced Studies, India
Rahimifard, Shahin, Loughborough University, UK
Rao, N. V. C., Indian Institute of Science, India
Rao, P. V. M., Indian Institute of Technology Delhi, India
Ravi, B., Indian Institute of Technology Bombay, India
Riitahuhta, Asko, Tampere University of Technology, Finland
Rodgers, P., Northumbria University, UK

Rohmer, Serge, Universite de Technologiede Troyes, France
Roozenburg, Norbert, Delft University of Technology, The Netherlands
Rosen, David W., Georgia Institute of Technology, USA
Roy, Rajkumar, Cranfield University, UK
Roy, Satyaki, Indian Institute of Technology Kanpur, India
Saha, Subir Kumar, Indian Institute of Technology Delhi, India
Salustri, Filippo A., Ryerson University Canada
Sarkar, Prabir, Indian Institute of Technology Ropar, India
Sato, Keichi, Illinois Institute of Technology, USA
Seliger, Guenther, Technical University of Berlin, Germany
Sen, Dibakar, Indian Institute of Science, India
Shah, Jami, Arizona State University, USA
Shimomura, Yoshiki, Tokyo Metropolitan University, Japan
Shu, Li, University of Toronto, Canada
Sikdar, Subhas, Environmental Protection Agency, USA
Singh, Mandeep, School of Planning and Architecture New Delhi, India
Singh, Vishal, Aalto University, Finland
Sotamaa, Yrjo, University of Art and Design Helsinki, Finland
Srinivasan, V., Technical University of Munich, Germany
Srinivasan, Vijay, National Institute of Standards and Technology, USA
Sriram, Ram, National Institute of Standards and Technology, USA
Storga, Mario, University of Zagreb, Croatia
Subrahmanian, Eswaran, Carnegie Mellon University, USA
Sudarsan, Rachuri, National Institute of Standards and Technology, USA
Sudhakar, K., Indian Institute of Technology Bombay, India
Thallemer, Axel, Universitaetfürindustrielle und kuenstlerische Gestaltung, Austria
Thompson, Mary Kathryn, Technical University of Denmark, Denmark
Tiwari, Ashutosh, Cranfield University, UK
Tomiyama, Tetsuo, Delft University of Technology, The Netherlands
Torlind, Peter, Lulea University of Technology, Sweden
Tripathy, Anshuman, Indian Institute of Management Bangalore, India
Tseng, Mitchell, The Hong Kong University of Science & Technology, China
Umeda, Yasushi, Osaka University, Japan
Vancza, Jozsef, MTA SZTAKI, Hungary
Vaneker, Tom, Univeristy of Twente, The Netherlands
Vasa, Nilesh, Indian Institute of Technology Madras, India
Verma, Alok K., Old Dominion University, USA
Vermass, Pieter, Delft University of Technology, The Netherlands
Vidal, Rosario, UniversitatJaume I, Spain
Vidwans, Vinod, FLAME School of Communication, India
Vijaykumar, A. V. Gokula, University of Strathclyde, UK
Voort, M. C. van der, University of Twente, The Netherlands
Wiegers, Tjamme, Delft University of Technology, The Netherlands
Yammiyavar, Pradeep, Indian Institute of Technology Guwahati, India
Yannou, Bernard, Ecole Centrale Paris, France

Zavbi, Roman, University of Ljubljana, Slovenia
Zeiler, Wim, Technical University Eindhoven, The Netherlands
Zwolinski, Peggy, Institut Polytechnique de Grenoble, France

Local Organising Committee

Swaminathan, Narasimhan, Indian Institute of Technology Madras, India
Siva Prasad, N., Indian Institute of Technology Madras, India
Krishnapillai, Shankar, Indian Institute of Technology Madras, India
Seshadri Sekhar, A., Indian Institute of Technology Madras, India
Balasubramaniam, Krishnan, Indian Institute of Technology Madras, India
Varghese, Susy, Indian Institute of Technology Madras, India
Ganesh, L. S., Indian Institute of Technology Madras, India
Ramkumar, Indian Institute of Technology Madras, India
Srinivasan, Sujatha, Indian Institute of Technology Madras, India
Sivaprakasam, Mohan Sankar, Indian Institute of Technology Madras, India

Student Organising Committee

Das, Swostik, Indian Institute of Technology Madras, India
Gullapalli, Suraj, Indian Institute of Technology Madras, India
Anand, Sahaj Parikh, Indian Institute of Technology Madras, India
Dhinakaran, S., Indian Institute of Technology Madras, India
Arun Kumar, S., Indian Institute of Technology Madras, India
Sarkar, Biplab, Indian Institute of Science, India
Devadula, Suman, Indian Institute of Science, India
Kumari, M. C., Indian Institute of Science, India
Madhusudanan, N., Indian Institute of Science, India
Ranjan, B. S. C., Indian Institute of Science, India
Uchil, Praveen T., Indian Institute of Science, India

Organised By

Indian Institute of Technology Madras, Chennai, and Indian Institute of Science, Bangalore, India

Endorsed By

The Design Society and The Design Research Society

Part VII
Design Management, Knowledge Management and Product Life Cycle Management

Modeling and Analyzing Systems in Application

Maik Maurer and Sebastian Maisenbacher

Abstract Technical products and processes do not represent complex but complicated systems. Complexity gets implemented into such systems by including users and use cases. Hereby, technical systems can be interpreted as enablers, which fulfill functions for the user. We define the combination of users and enablers as a "system in application" and propose applying methods from structural complexity management for its modeling and analysis. Therefore, we introduce two structural characteristics and their interpretation. Based on modeling, analysis and interpretations we present procedures for system improvement and evaluation in terms of increased system usability. The practical application of the new approach on the check-in process for air travel shows achievable benefits from systematic improvement and evaluation strategies. Future work will cover the extension of applicable structure analyses and methods of multi-domain analyses.

Keywords Structural complexity · Enabler · User · Design structure matrix

1 Introduction

In a modern society, people often deal with networked and complex issues [1–3]. Today, new technologies and innovative products enable users to manage many of those issues, which were difficult to solve in the past. Therefore it is a prerequisite

M. Maurer (✉) · S. Maisenbacher
Product Development, Technische Universität München, Garching, Germany
e-mail: maik.maurer@pe.mw.tum.de

S. Maisenbacher
e-mail: maisenbacher.seidl@t-online.de

A. Chakrabarti and R. V. Prakash (eds.), *ICoRD'13*, Lecture Notes in Mechanical Engineering, DOI: 10.1007/978-81-322-1050-4_56, © Springer India 2013

that users are familiar with those supporting technologies and products. This can be illustrated in the context of mobile communication: Many people desire permanent internet connection and worldwide communication even while traveling. Linking their mobile devices to different networks asked for significant effort until a few years ago. Today, smartphones can establish connections to many different networks without any manual user interaction. Thus, permanent internet connection can be easily realized today. The precondition for this simplification is that the user knows how to interact with the smartphone.

"Complexity increases" is a common statement, which often focuses on technical products and processes (e.g. smartphones). In contrast to that we state that processes and products can be complicated but not complex. Application of products and processes requires the user to possess sufficient capabilities. These capabilities differ from one user to the other. A system can become complex when products or processes are linked with a user. In such a "system in application", complexity reduction means shifting complexity from the area of user interactions to the product or process and to simplify the interface between user and the product or process.

This contribution introduces a method for modeling, analyzing and interpreting a complex system as a combined view of the user context and the technical system. This can help adjusting products and processes to user capabilities and making products and processes more user-friendly.

In Sect. 2 we discuss the difference between complicated and complex system and define the terms, which are relevant for our approach. Then we describe the state of the art concerning inclusion of users into complex system consideration. In Sect. 3, we explain how to use possibilities of structural complexity management for modeling and analyzing systems in application. The application process then gets detailed in Sect. 4. Section 5 describes a case study of the mobile check-in process for air travel.

2 Definitions

The introduction of the new approach toward complex systems modeling requires some initial definitions. Based on these definitions we explain why products and processes can be complicated but not complex. And we introduce the "system in application", which allows including the subjectivity of users into complex system consideration.

2.1 Simple, Complicated and Complex Systems

Cotsftis [4] quotes that systems belong to one of the following three states: simple, complicated and complex. Simple systems consist of a small number and weakly coupled elements, which are acting according to well understood laws [5]. The behavior of simple systems is predictable.

Complicated systems consist of lots of elements and relations between them. These elements and relations are constant and the behavior of the system is deterministic [6]. According to [4] complicated systems can be decoupled and elements can be identified. The system elements are connected to external elements, which can influence the system. Complicated systems have a limited range of response to environmental changes [5]. A defect or change on critical parts brings the entire system to a halt. Complicated systems can easily generate complex dynamics, if they are influenced by external elements.

Complex systems comprise of a large number of elements, which are strongly interconnected [7]. Interconnection and interaction between the elements prevents the system from decomposition, it cannot be divided into subsystems [6]. Complex systems are also strongly connected to their environment, which results in a high amount of possible system conditions. Thus, complex systems possess internal dynamics [4] with changing system behavior [6]. The dynamics can arise from the system and from its environment (in contrast to that, dynamics in simple or complicated systems only arises from the environment). Cotsftis and Richardson [4, 6] define this as self-organization. This enables a complex system to respond to changes in the environment [5]. All aspects mentioned can be summarized in the following definitions for the three system states:

A simple system consists of a small number of weakly connected elements. The structure is easily manageable and different system states result from external sources only.

Complicated systems possess a hardly manageable structure with lots of elements and relations. As in simple systems, different system states can only result from external sources. A complicated system becomes static, if it gets isolated from its environment.

Complex systems possess a high number of elements and relations. The crucial characteristics are the existence of internal dynamics and self-organization, which make complex systems hardly predictable and uncertain.

2.2 *Internal and External Complexity*

Several authors distinguish internal and external complexity in an industry context (e.g. [8]). Internal complexity describes the variety and interdependencies of elements, which are required to realize the product portfolio offered by the company. External complexity results from the market, customers and competitors, who represent the (also interlinked) requirements, which have to be met by the product offer. Eppinger and Browning [9] mention that external complexity is the input from the environment to the system.

In the context of our approach, internal and external complexity means complexity inside or outside of the system borders (thus definitions are not limited to a company view). Consequently, only complex systems possess internal complexity (see definitions above). Complicated systems do not contain internal complexity, because they only show static behavior when separated from their environment.

2.3 System Classification of Products and Processes

Technical products and processes can be classified as complicated systems in the context of our definitions. A product consists of a specified number of components, which are partly interconnected. A process consists of a specified number of process steps, which are ordered in a sequence. Technical products and processes do not contain internal complexity. They can only adopt one static condition, if considered without input from their environment.

However, products and processes are influenced by the external environment, if they get applied. Such application can change the condition of products and processes by implementing external complexity to the system. For example, changing network connection of a smartphone can result in interruptions in communication; new downloads, updates or malware can lead to system improvement or deterioration and mistreatment by the user can cause malfunctions or destruction of the smartphone.

2.4 System in Application

The explanations on system classification suggest not limiting complexity analyses to technical system descriptions only, because external complexity needs to be considered as well. The inclusion of users and use cases into system models allows representing the application of technical systems and processes and therefore integrates external complexity into the system descriptions. We define a "system in application" as consisting of a subjective user, a technical product or process and different use cases (or environmental conditions). The technical system or process is also called enabler, as it fulfills functions for the user.

The subjective user represents the source of external complexity for the enabler. UML use case diagrams are often applied for modeling the connection between users or use cases and technical systems [10]. Such diagrams allow describing a specific (external) influence to a technical system. However, this influence does not get correlated to the internal structure of the technical system [11]. But the integrated consideration of an internal complicated product structure and the external complexity is required for analyzing the behavior of a system in application.

Design Structure Matrices (DSM) represent a powerful method for modeling technical products and processes [12], but aspects of external complexity are rarely integrated.

So far the combined consideration of enablers and users in a structural system model is not covered sufficiently. Investigations on system complexity require a comprehensive description, which also provides insights into the interfaces between users and products or processes.

3 Modeling and Analyzing a Systems in Application

In this chapter we present a model scheme, followed by analysis techniques for successfully manage systems in application. Interpretations of analysis results allow drawing conclusions on the system behavior, which then can be optimized by adequate measures. Such measures can be the redesign of the interface between users and enabler and the beneficial positioning of system complexity in the user or enabler area.

3.1 Modeling a System in Application

A system in application can be split up into two parts describing an enabler (technical product or process) and a user. The enabler permits the user to execute an intended task or to solve a problem. The structure of an enabler is formed by its elements (e.g. product components or process steps) and dependencies between them.

Users apply the enabler. Differences between users (subjectivity of users) can be important aspects of systems in application [13]. And this can increase systems' complexity. Also non-human, unpredictable or dynamic processes (e.g. weather conditions) can be modeled as system users. Elements (e.g. processes executed by the user) and their mutual dependencies form the structure of a user. As the system user and his connections to the enabler area can be complex, this system part can hardly be modeled completely. This is not a drawback of the new approach, because the intention is to build only a useful view of the real system [12].

Dependencies between user and enabler represent the mutual influence between both subsets and can implement complexity into the system. The example of an applied smartphone can clarify this: Users interact with their smartphone mainly by input devices like the keypad or touchpad. Input devices change over time, e.g. from keypads to voice recognition. User groups, e.g. elderly people, show the subjectivity of users, which influences the design of input devices. For example, larger keys and reduced menu navigation are offered for elderly people.

The structure of a system in application can be modeled using the Multiple-Domain Matrix (MDM) [12]. The general layout is shown in Figure 1. This matrix is applied in the method of structural complexity management [12]. This method includes approaches for information acquisition, definition of the objective as well as system boundaries and also provides analysis procedures for systems in application. User and enabler represent subsystems, which are modeled as DSM. The interface between both subsets is modeled by two Domain Mapping Matrices (DMM). One indicates inputs from the user to the enabler; the other DMM contains responses or outputs from the enabler to the user.

The general MDM layout from Fig. 1 needs to be further detailed for system representation. Therefore, relevant domains have to be defined for specifying users as well as the enabler. A domain means a set of similar system elements, e.g. components or process steps. The selection of appropriate domains requires basic system knowledge.

For all domains system elements and dependencies between them need to be collected. It is advisable not only to focus on acquiring elements and dependencies in the single subsets. Rather modelers should separate the acquisition of elements and dependencies from their subsequent assignment to users or enablers. When allocating elements to the user and enabler area the interfaces between both areas result automatically in the upper right and lower left sector of the matrix. Figure 2 shows the two steps by using a generalized example, which consists of three domains (persons, components and process steps).

3.2 Analysis Approach

Structural complexity management provides possibilities for identifying characteristics in system structures. This can also be used for analyzing structural models of systems in application. The analysis of the entire matrix (complete system in application including all four subsets) allows interpreting the overall system interdependencies. A graph representation of the system can be helpful, if an appropriate layout algorithm visually indicates structural characteristics like clusters or articulation nodes [12]. And metrics like the level of connectivity or the amount of clusters are useful for comparing two or more systems (e.g. variants or alternative designs) on a general level.

	User	Enabler
User	Interlinked activities of the subjective user: *„User activity impacts another user activity“*	User activities as input for the enabler: *„User activity impacts enabler element“*
Enabler	Enabler elements as output to user activities: *„Enabler element impacts user activity“*	Dependencies within the technical system: *„Enabler element impacts another enabler element“*

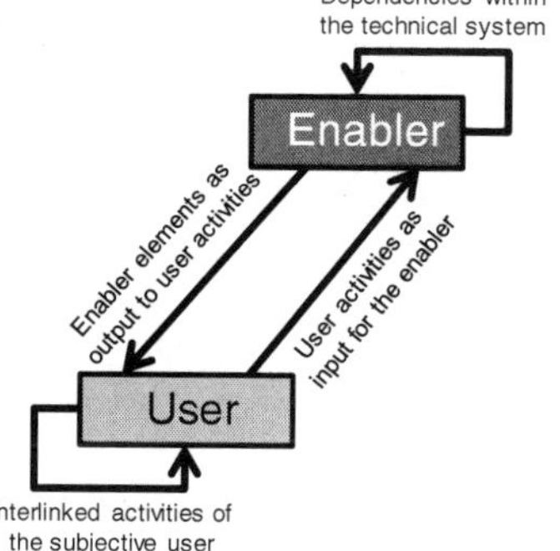

Fig. 1 System in application modeled as multiple-domain matrix (*left*) and graph (*right*)

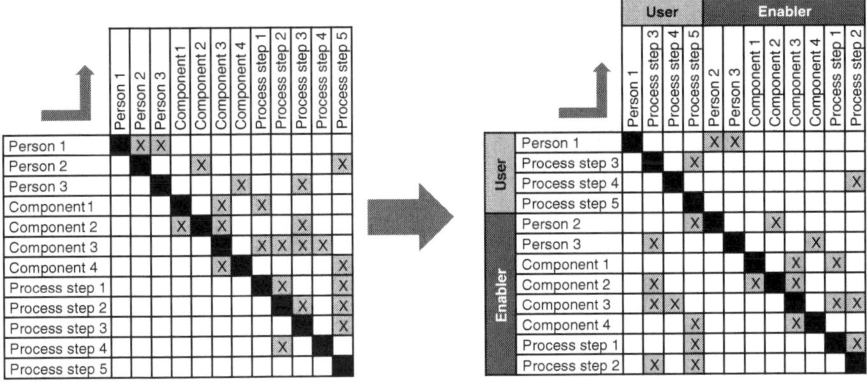

Fig. 2 Acquisition of elements and dependencies (*left*) and their classification (*right*)

Detailed structure analysis can be executed for all four subsets and cover different objectives. One has to consider that the two DSM (isolated views on the technical system and the user) partly allow other analyses than the two DMM (interfaces between users and technical system) [12]. E.g. feedback loops can only occur in DSM.

In the following paragraphs we will describe the application of two analysis methods taken from structural complexity management and explain their interpretation in the context of the four subsets of a system in application.

3.2.1 Clustering in Systems in Application

A cluster is a group of system elements, which possess many internal and only few external dependencies [12]. Clusters can exist in all four areas of the system model, as indicated by Fig. 3 (left). A cluster in the two DSM can be interpreted as strongly interrelated user activities (1) or technical components (2) suitable for being implemented as a module. Clusters in DMM combine elements of two domains and have to be interpreted differently. A cluster, which links user activities to enabler elements (3) describes a set of activities, which are all required as input for the same group of technical system components. In extreme cases one missing user input could make all other inputs obsolete and could block fulfillment by the enabler. The cluster indicated as (4) in Fig. 3 can be interpreted as a set of enabler elements which all provide output to a defined set of user activities. A missing output results in lacking information for all involved user activities.

When considering specific systems in application, beneficial analyses need to concretize cluster interpretations. However, the general identification of clusters without context specific interpretation allows drawing useful conclusions. For example, clusters of user inputs to enablers can indicate the existence of a systematically structured interface.

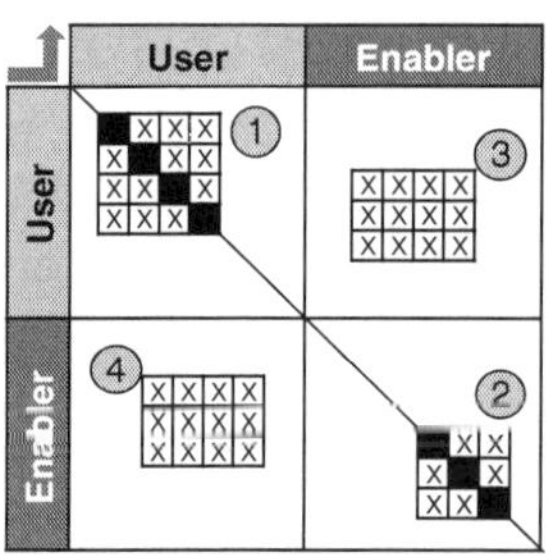

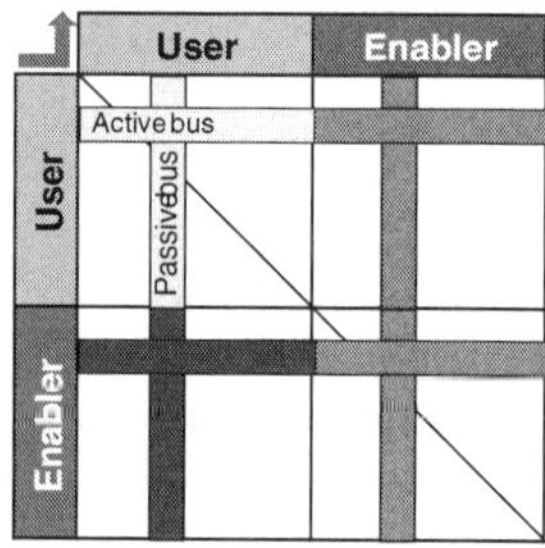

Fig. 3 Clusters and bus elements in a system in application

3.2.2 Bus Elements in Systems in Application

A bus element is characterized by many dependencies to other system elements. Busses can be active or passive. This means they influence the entire system or get influenced by the system [12]. In the model of a system in application a bus can appear in all four subsets (Fig. 3, right). The position of a bus element impacts its interpretation.

If located within the user domain (upper left area), an active bus means a user activity influencing many other activities, e.g. the starting of a phone call on the mobile phone. A passive bus then means a user reaction to several other decisions made before. Interpretations of bus elements within the enabler domain (lover right area) can be deduced similarly. Bus elements in the upper right and lower left area represent important interfaces, which connect users and enablers. Such interfaces contain many technical components or many user activities. The initiation of an internet connection via a smartphone was such an interface, as long as many manual inputs (user activities) were required for this enabler function. Moving bus elements from interface areas to internal areas of users or enablers can be advantageous for the system's usability and is described in Sect. 4.

3.2.3 Interpreting Further Structural Characteristics

The new approach models systems in application according to the methods of structural complexity management. More structural characteristics from this management approach should be transferable to systems in application. For example, consideration of feedback loops, hierarchies and articulation nodes seem to be beneficial for interpretation and optimization of systems in application. A systematic transfer and evaluation of structural characteristics from structural complexity management will be part of the future work.

4 Application Process

Analysis and interpretation of systems in application can serve two general objectives. Firstly, identification of possibilities for system improvement can be desired. Improvements mean reducing complexity for users when interacting with the enabler. This does not mean that products or processes need to be simplified. They can even be complicated (see definitions in Sect. 2), but their application should be as simple as possible. In the depiction of a system in application this implies the transfer of nodes and edges from the user area to the enabler area and the facilitation of interfaces between user and enabler.

Secondly, one could ask for evaluating the impacts of ongoing design activities to the system. Therefore, two states of the system structure need to be compared (delta analysis) and rated by appropriate indicators. Both objectives are indicated in Fig. 4 and explained in the following paragraphs.

4.1 Identification of Possible Improvements

From a user perspective, possibilities of improvement can be found in the user area and the two DMM, which connect users and enabler. The intention is to transfer complexity from user's responsibility to the technical system. In the matrix that means transferring system elements from the user to the enabler domain. This can be implemented, if those elements can also be executed by the technical system (enabler).

An example is the initiation of internet connections via mobile phones. Some years ago users were responsible for implementing all settings for connecting to the internet. If users were not familiar with technical vocabulary and network specifications, internet connections were hard to establish. Modern smartphones do not require any user specifications and provide an internet connection almost automatically. This shows that moving elements from the user to the enabler does also move interdependencies within the system. Whereas the manual internet

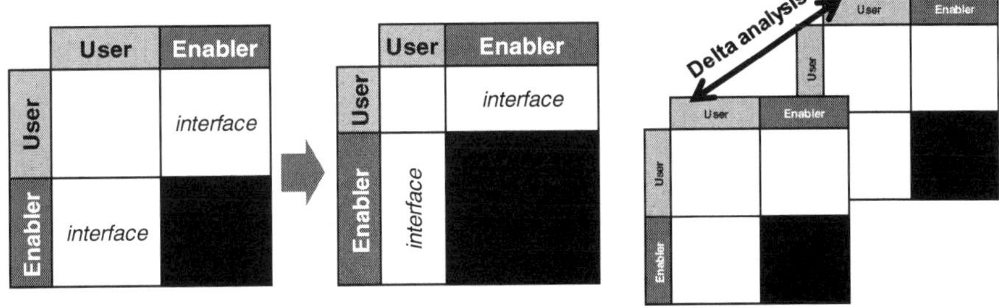

Fig. 4 Objectives for system interaction, system improvement and design evaluation

connection required an extensive interface between user and enabler, automatic internet connection simplified this interface. Such a measure does not change the overall system complexity. But elements and dependencies become aggregated in the technical system and user interactions are decreased (Fig. 4).

4.2 Evaluation of System Design Activities

It can be helpful comparing different versions of a system in application for evaluating the technical progress. For example, many user-involved service processes (e.g. check-in process for air-travel) have changed in the last few years by integrating mobile media devices. In this case, one could ask if the process has been improved for users.

When comparing two system versions, some indicators can be applied for evaluation: The amount of elements in both areas (users and enabler) allows estimating the degree of system automation. The amount of dependencies in the user DSM and the two DMM (interfaces) indicate if the degree of user interactions could be decreased. Such delta analyses between system states are also applicable for evaluating design measures concerning their effectiveness in terms of increased system usability. As well, design measures can be initiated by identifying system improvements according to Sect. 4.1.

5 Case Study: Check-in Process for Air Travel

Possibilities for passengers to check-in for air travel have changed over the last few years. In the past, the process was characterized by printed documents (receipt, ticket, boarding pass) and related user activities. Since some years the check-in can be executed by using a smart phone. Users do not have to interact with documents (e-tickets) and can directly proceed to the gate, when using the mobile check-in. The integration of new technologies should facilitate the entire process. However, many people do still not use the mobile check-in. They declare the new process to be more difficult than the prior one.

Figure 5 shows the model of the mobile check-in with two domains for the user and the enabler (element names start with "U_" and "E_"). The number of elements in the user domain decreased in comparison to the model of the prior (paper-based) check-in process (not shown here) from 35 to 25. In the enabler domain the number of elements increased from 12 to 17. The degree of connectivity in users' input interface (upper right subset) dropped from 6.7 to 5.2 %. These changes can be interpreted as simplification of users' need for interaction and their input interface.

Whereas the model indicates a system improvement, still many users refuse mobile check-in and describe it as being too difficult. In Fig. 5 one can see that

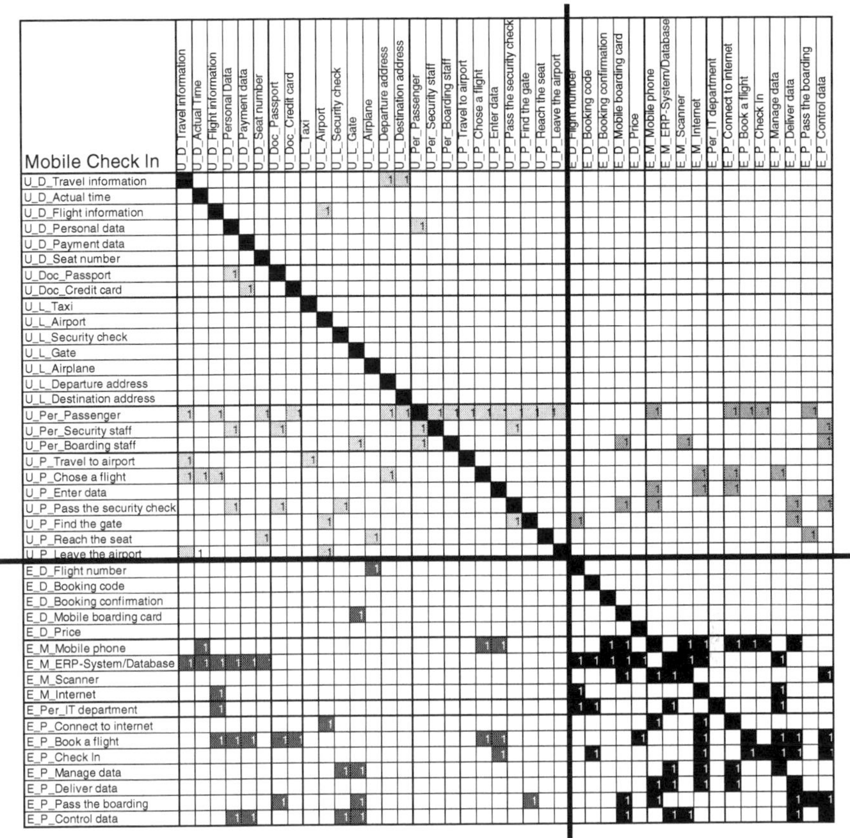

Fig. 5 System model of the mobile check-in process for air travel

passengers have to interact intensely with the smartphone and have to enter data. If users are not familiar with these tasks, mobile check-in can not be executed.

The model of the mobile check-in process provides possibilities of improved system usability. E.g. the process "U_P_Pass the security check" is an element of the user domain with an intensive interface to the enabler area. The interface would get simplified, if this step could be moved into the enabler domain in future check-in processes. In fact, the concept of "trusted passengers" is currently under consideration. Hereby, passengers would not have to pass security checks any more. In this concept the enabler executes background security checks of passengers. If successful, passengers obtain their status as "trusted passenger" and enter airport security zones without further checks.

6 Conclusion and Future Work

We argue that technical products and processes are not complex but complicated. Complexity arises when a system is formed by subjective users and use cases in combination with products or processes. Such a combination we call a "system in application". We propose applying methods from structural complexity management for modeling and analysis of such systems. This is the basis for system interpretation and implementation of optimization measures. We showed two structural characteristics and their interpretation in the context of user related and enabler related system elements. And we introduced procedures for system improvement and evaluation in terms of usability. Practical application on the check-in process for air-travel indicated suitability of the modeling approach and achievable benefits from systematic improvement and evaluation strategies.

So far, only few possibilities of analysis and interpretation have been adapted from structural complexity management. Several others could be helpful for better understanding and adapting systems in applications. In future work we will evaluate the applicability of structural characteristics like feedback loops, hierarchies and articulation nodes.

Systems in applications are formed by the user and enabler domain. And both domains typically contain further sub-domains (e.g. processes and components). If the entire system should be investigated, possibilities of multi-domain analyses are required—but not available so far. Therefore, we will also focus on expanding structure analysis to multi-domain contexts in future work.

References

1. Calvano C, John P (2004) Systems engineering in an age of complexity. J Syst Eng 7(1):25–34
2. Chung L, Cooper K (2004) Defining goals in a COTS-aware requirements engineering approach. J Syst Eng 7(1):61–83
3. Kohn A, Lindemann U (2009) Combination of algorithms and visualization techniques considering user requirements—a case study. 11th international design structure matrix conference
4. Cotsftis M (2009) What makes a system complex?—an approach to self organization and emergence. In: Aziz-Alaoui M et al (ed) From system complexity to emergent properties. Springer, Berlin, pp 49–99
5. Amaral L, Ottino J (2004) Complex networks. Eur Phys J B 38:147–162
6. Richardson K (2001) On the status of natural boundaries: a complex systems perspective. Systems in management 7th annual ANZSYS conference, pp 229–238
7. Anderson P (1999) Complexity theory and organization science. Organ Sci 10(3):216–232
8. Marti M (2007) Complexity management: optimizing product architecture of industrial products. Dissertation University St. Gallen
9. Eppinger SD, Browning TR (2012) Design structure matrix methods and applications. MIT Press, Cambridge

10. Von der Maßen T, Lichter H (2002) Modeling variability by UML use case diagrams. International workshop on requirements engineering for product lines, Essen
11. Koch N, Kraus A (2002) The expressive power of UML-based web engineering. Second international workshop on web oriented software technology CYTED, pp 105–119
12. Lindemann U, Maurer M, Braun T (2009) Structural complexity management. Springer, Berlin
13. Gulliksen J (1996) Designing for usability—domain specific Human-computer interfaces in working life. PhD thesis, University of Uppsala

A Categorization of Innovation Funnels of Companies as a Way to Better Make Conscious Agility and Permeability of Innovation Processes

Gwenola Bertoluci, Bernard Yannou, Danielle Attias and Emilie Vallet

Abstract It is common in the Management Science and Design Engineering communities to represent the processes contributing to innovation in companies as a funnel or similar variants. It is assumed it is possible to represent an analogy to the stages of planning and idea generation (the so-called fuzzy front-end), conception generation, as well as idea and concept selection to end up with the very few emerging developed and launched products and services on the market. First, this analogy may feature different innovation process layers, each of them independently as well as the entire set of these innovation process layers. After a review of literature on this funnel representation, we show that this analogy may be meaningful to globally represent and discuss about some properties of the innovation capability of a company at different locations: the R&D process as well as a given NPD process. We further describe a survey carried out within 28 large European technological companies through 48 detailed face-to-face interviews. Our questionnaire has allowed us to observe some characteristic patterns in the

---

G. Bertoluci · B. Yannou (✉) · D. Attias
Laboratoire Genie Industriel, Ecole Centrale Paris, Grande Voie des Vignes 92290 Chatenay-Malabry, France
e-mail: bernard.yannou@ecp.fr

G. Bertoluci
e-mail: gwenola.bertoluci@ecp.fr

D. Attias
e-mail: danielle.attias@ecp.fr

G. Bertoluci
AgroParisTech, 1 Avenue des Olympiades 91744 Massy, France

E. Vallet
Logica Business Consulting, Immeuble CB16, 17 Place des Reflets 92400 Courbevoie, France
e-mail: emilie.vallet@logica.com

A. Chakrabarti and R. V. Prakash (eds.), *ICoRD'13*, Lecture Notes in Mechanical Engineering, DOI: 10.1007/978-81-322-1050-4_57, © Springer India 2013

innovation funnels. We finally propose a model of five innovation funnels varying by their shape, permeability of emerging ideas and agility in terms of innovation management. We also hypothesize that these 5 funnels evolve in a sequential and cyclic way and that our cyclic model may be used as a questioning tool for the continuous improvement of the innovation management.

Keywords Innovation funnel · Development funnel · Product development · Design process · Innovation management

1 Introduction to Funnel Representations

Looking for ideas for an individual or group to document the resolution of a problem implements a cognitive process [1] of divergence and convergence. This is in this divergence part of ideation that creativity tools are solicited. This second convergence part is often represented as a synthesis funnel (see [2] and Fig. 1) in which ideas are selected, recombined or eliminated to result in the very few selected ideas that merit to be further detailed.

Design exploration process is an important part of design creativity and novelty [3]. The design activities in conceptual design are contained in two kinds of steps: divergent and convergent [4–6]. Cross [4] thought of the conceptual design process as mostly being convergent with the necessity to contain a deliberate divergence in the search for novel ideas (see Fig. 2).

Pugh's model [5] for conceptual design and the definition of the convergence and divergence activities can be represented as on Fig. 3. Pugh mentioned that it is essential to carry out concept generation and evaluation in a progressive and disciplined manner so as to generate better designs. This progressive and disciplined manner is illustrated as an iterative, repeated divergent and convergent process with the number of solutions gradually decreased (see [7]).

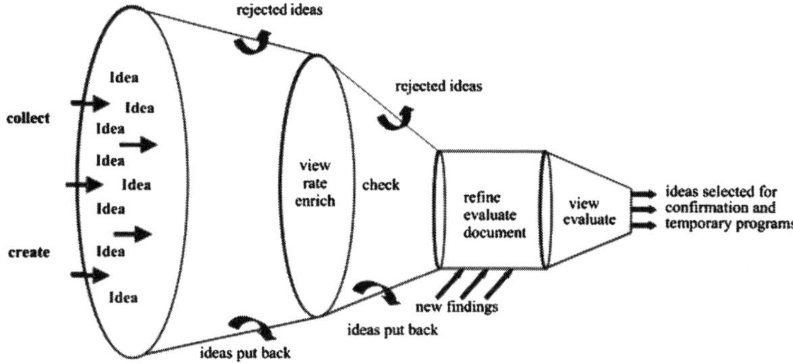

Fig. 1 Idea funnel [2]

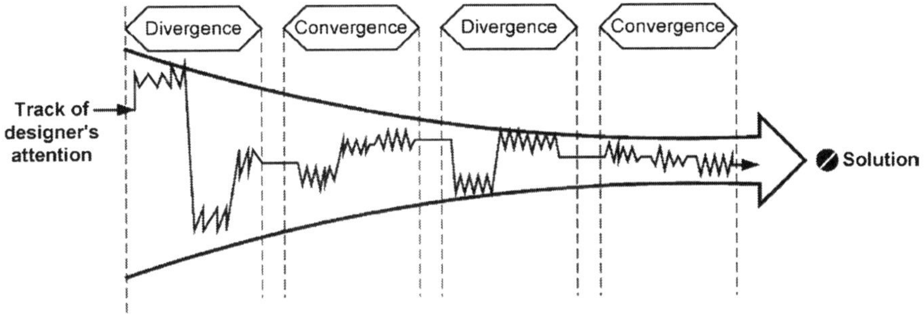

Fig. 2 The conceptual design process defined by cross [4]

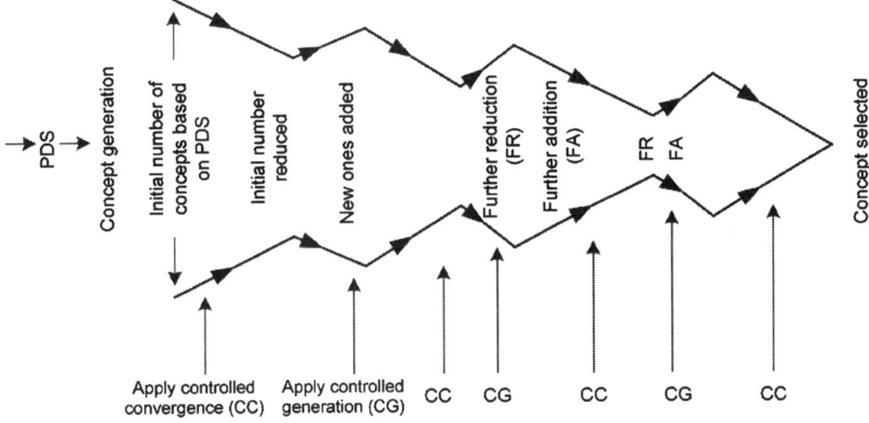

Fig. 3 Pugh's conceptual design process [5]

This representation of the progressive reduction of the possible design space has been extended by various authors [8, 9] to take into account the development of new technologies whenever a new technology is a key lever of innovation. Wheelwright et al. note that the companies that supply these funnels by R&D (Fig. 4a) on technologies, processes and products also extend the scope of the funnel to bottom-up uses (Fig. 4b) to identify and filter out new ideas from a strategic market analysis.

The two situations of convergent part of an ideation process (innovative process) and technology push or market pull representations (strategic innovation management) are very different. Whether one refers to very general models such as the Chain Linked of Kline and Rosenberg [10] and the Stage and Gates of Cooper [11], or more detailed and structuring processes proposed in design engineering [9, 12–15], this analogy of forward selection and progressive shrinking of the design space appears to be a highly interpretative task. However, it is very well illustrated by Wheelwright and Clark [9] who clearly associate it with the product planning stage preceding that of product development (Fig. 5).

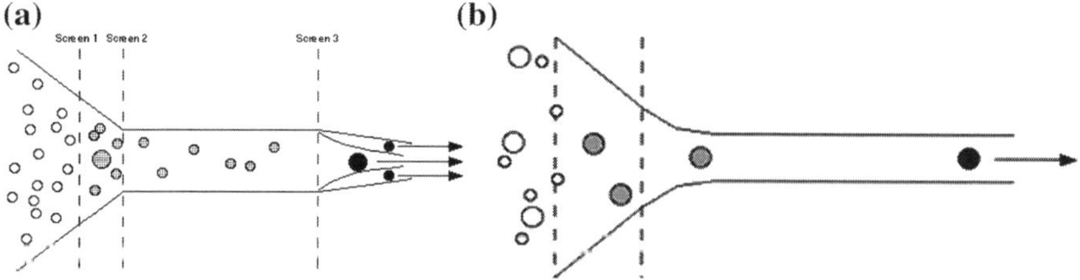

Fig. 4 Funnel representations of Wheelwright [9] **a** Funnel of technology push. **b** Funnel of market bottom up

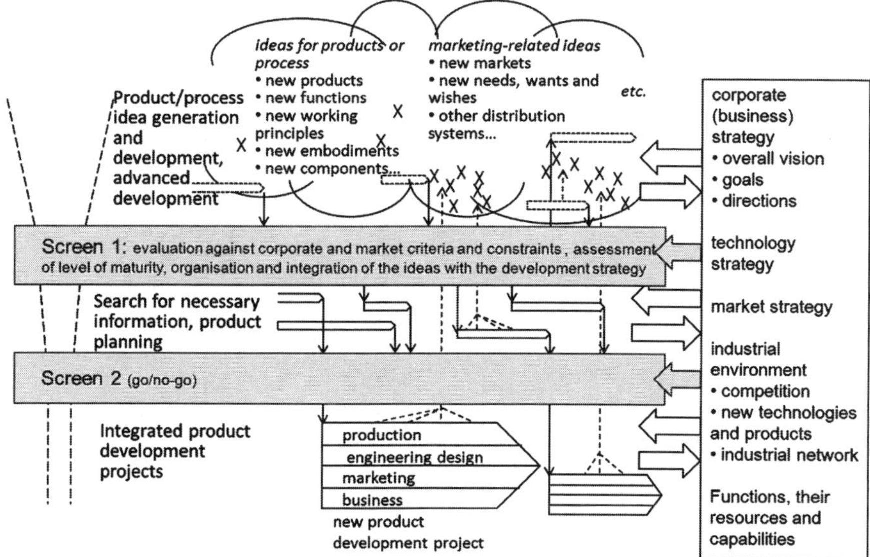

Fig. 5 The funnel development process of Motte et al. [19], adapted from Wheelwright and Clark [9]

These representations in funnels do not aim to facilitate the structuring of a process of innovation but to illustrate how an organization has the means to open and close its creative space and thus its products space. In this sense, to interrogate the businesses associated with the innovation process on how they actually process opens or closes to integration of new ideas and new opportunities, throughout a development process, is a way to identify how this funnel is actually generated. It is a question asked on the innovative organization behind that could lead to a continuous improvement of the innovation management. The objective of this paper is to propose a tool for the continuous improvement of the innovation management in companies based on a funnel categorization obtained after an innovation survey in 28 large companies.

2 Survey on Innovation in Large Companies and Lessons Learned

In a recent survey of 28 large European companies in the industry, we investigated the state of practice in innovation and innovative organizational models in large companies. This survey has provided qualitative and quantitative results which have been reported in a book [16]. We interviewed 48 R&D or innovation directors by asking them to self-diagnose their business practices according to the five management areas that contribute to value creation: strategy and business intelligence, organization of R&D, management of innovation processes, innovation culture and management of human resources and R&D, measurement of innovation performance. Our learning are numerous and sometimes surprising. We report hereafter just a few of them to contribute to this paper.

Two thirds of respondents reported having profoundly transformed or reorganized their R&D over the past three years. For reasons to support the international expansion, pooling and centralization of research upstream, location in the business unit of applied research and development or reorganization of the development process and resource allocation of R&D. These reorganizations are made in trial and error mode (no apparent method) and reflect a search for greater innovation performance. Indeed, companies face a real difficulty in measuring performance and the benefits of innovation, investment in research that results often lately by the market launch of innovative products and services. Management indicators of innovation or value creation are often of "rear-view mirror" type like the number of patents rather than of "looking ahead" type (able to monitor the value increasing).

Finally, the companies say that the upstream processes of ideas management are poorly organized. Indeed, 47 % of companies do not use a methodology for generating ideas. The methods used are the idea boxes and idea contests without, in most cases, budget for the exploitation of good ideas that emerge. Also appalling, the only methods of generating ideas and driving innovation that are sometimes referred by high-level managers are TRIZ and Design to Cost and Objectives.

The companies surveyed acknowledge the fact that the so-called "innovation process" of a company is actually a series of strata or four processes (see Fig. 6 which is our own representation) with interconnections but acting at different times, with specific strategies, roadmaps and different but interdependent budgets. These four processes are:

1. The process of ideas generating on products, technologies, processes or organization,
2. That of research or technology management
3. That of product lines or programs management or planning

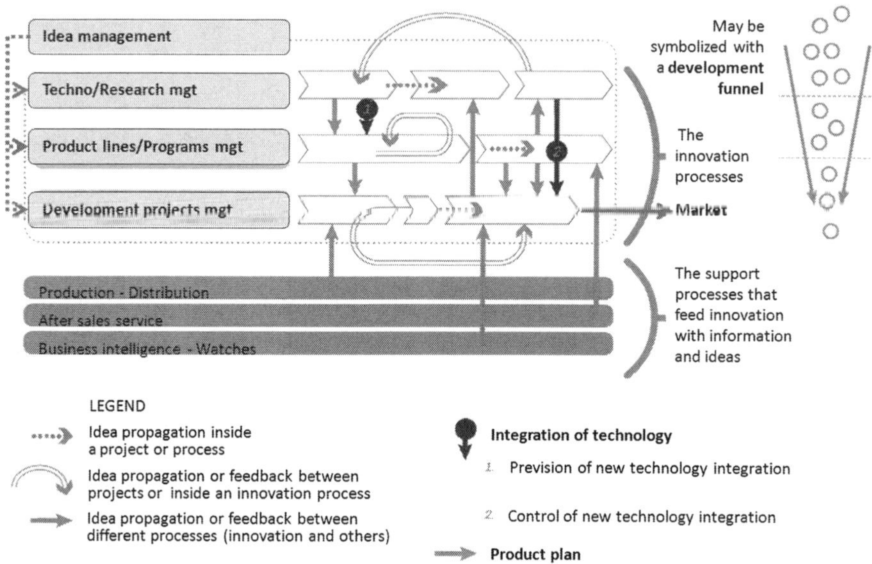

Fig. 6 Modeling innovation or value creation processes in companies

4. The very activity of project management of New Product Development (NPD), which supplies the market with new offers and contributes largely to the creation of business value.

It may be noted that a number of business support processes such as marketing, customer relations, after sales service, purchasing and competitive intelligence also contribute more indirectly to the increased degree of effective innovation of the company's products. Note also that according to the ideas (nature, size, maturity), process #1 of Ideas Management can feed processes #2, #3 and #4. All this must necessarily be organized within the company with the collection process of ideas and transfer of ideas within mature roadmap that invest and plan their development and their deployment in research, product lines or project development. However, these transfer processes are currently poorly organized and coordinated so that the process of generating and collecting ideas as we have seen. But there are organized and standardized processes within the company such as product development. These are step by step processes with intermediate outcomes expected, so-called "stage and gate". It turns out that these processes are as well necessary to ensure a minimum quality and coordinate development activity on a complex project, as sometimes too rigid and not very permeable to new ideas and opportunities that would upset too much a strategic positioning or that would appear during the project.

In our book [16] we have thus classified the 28 surveyed companies into 5 categories of innovation funnels (there are 6 companies per category in average), considering the characteristics of the whole 4-layer innovation process of Fig. 6

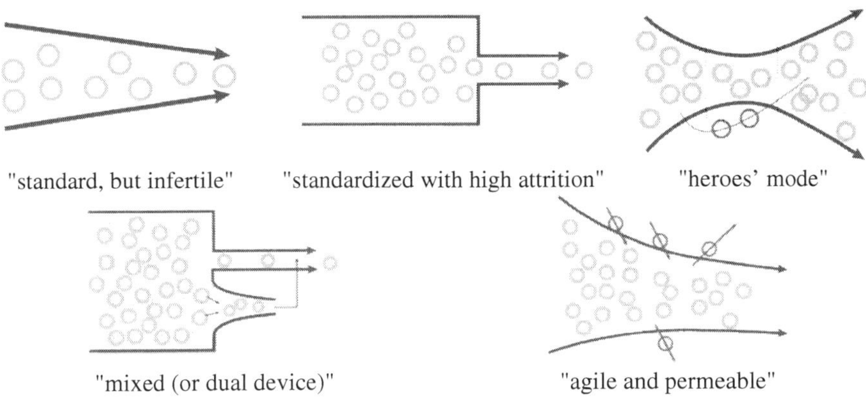

Fig. 7 Our model of five innovation funnels

after the three general properties of *shape*, *permeability* to emerging ideas along the process and process *agility*. This classification has been made by us, the 10 researchers having contributed to interview the company managers.

We have baptized them: the funnel "standard, but infertile", the funnel "standardized with high attrition", the "heroes' mode" funnel, the funnel "mixed (or dual device)", the funnel "agile and permeable". We have also graphically represented them in simple and expressive pictorials with small circles for ideas and contours for the shape (see Fig. 7).

3 Definition of 5-Funnel Model and Correlations with Company Features

3.1 The "Standard, but Infertile" Funnel

For this funnel, few ideas and innovative proposals are selected upstream. The process R&D is classical and relatively heavy. It usually results in a limited number of development projects that generally go until the completion; it is not very selective. This model is found in sectors such as aeronautics, submitted to severe constraints of technical reliability in operation and intense capital investments. It can also be found in the energy sector, a sector where a R&D Director said: *Today the results will kill creativity*; *R&D is today is a super design department*.

The main characteristic of firms adopting this funnel is to have a limited number of ideas in the sequence generation/collection. However, the steps of selection/maturation and launch/development are well standardized and instrumented (Stage and Gate, TRL-Technology Readiness Levels—project mode). Despite a cumbersome process, the dropout rate of development projects is low, once the concept validated and the decision to develop the product made.

These companies are mainly technology companies where the product renewal period is long (more than 10 years) and where the type of innovation is "product" dominant. However, some companies doing predominantly "product" innovation may refer to other funnel models.

The competitive intensity corresponds to relatively oligopolistic markets which, even if they evolve with new entrants, do not change they belong to the exclusive club of leaders in their field.

3.2 The "Standardized with High Attrition" Funnel

Here, the process Research and Development (TRL, Stage and Gate) is unique, classic and relatively heavy, with progressive selection and strong attrition of the proposals. In this model, it is difficult to stop the projects launched once they have passed the stage—gate—of product development. This funnel is found in sectors such as pharmacy or mass retailing where the selection of ideas for innovation is strong before starting the process of development itself.

In this model, many ideas are killed as they pass upstream phase (vision/selection/feasibility of innovation concepts), sometimes despite an abundance of leads and ideas for innovation in this phase. For the transition phase in product development phase, the company takes the minimum possible risk and projects that pass this stage are then under control. But as explained by a Director of Research, Technology and Innovation, *Risk reduction is the anti-innovation. The goal is not to reduce risks, but to manage them. The difficulty comes from people who want to reduce the risks.*

However, the development projects initiated are not arrested or very little even when they no longer meet the original expectations: *I have a project in mind. When I arrived ten years ago, people said It's been ten years since we talk of it*, it's still there!.

These companies face a double problem. First, how to reduce the number of tracks to explore and to evaluate in upstream stage—which has a cost—to focus on the most promising? Then, how to accept anyway to take risks, knowing that establishing sales forecasts of innovation on the market is both a delicate task and a choke point? In other words, how not to "sterilize" automatically waiving any idea if its business case is not solid in terms of apparent ROI, given that the expected revenue is often not accurate?

3.3 The "Heroes' Mode" Funnel

The "heroes' mode" funnel makes coexisting a standardized funnel with a heroic mode. It is a R&D process (Stage and Gate) classic that can turn out to be cumbersome with a progressive selection of ideas and difficulty to kill the projects once they started. It is characterized by a weak support to potential innovations proposed by the field and which have failed to take the step of selecting initial ideas. The innovations are defended by tenacious individuals with strong personalities that sometimes succeed to make their ideas acknowledged by others and to put them into an official project portfolio.

Companies in this scheme feed well the funnel in ideas in the upstream phase. But there may be two reasons for not ensuring a good transfer from upstream phase to downstream phase: (1) A R&D process too cumbersome, judged too bureaucratic and procedural or (2) conversely, a lack of a real ideas selection process.

Innovations that pass this barrier are supported for some literally "overreach" by the holders of ideas, which we called the "heroes" and who, with tenacity and by using their internal network, particularly among the bosses of business units, manage to insert their idea into the mainstream product development.

3.4 The "Mixed (or Dual Device)" Funnel

As its name suggests, this funnel combines two processes. On the one hand, a conventional process of Research and Development (TRL type, stage and gate) works, but it can be cumbersome. It is often keyed to the yearly budget. Once the candidate ideas issued, it works more in "top down" mode. Its main features are a progressive selection and difficulty to kill projects launched. On the other hand, the process is light, but institutionalized, organized and resourced to capture the ideas and proposals. It is "bottom up", which gives real resources to support potential innovations, giving the possibility to impact the official portfolio of research and development projects.

This funnel consists in putting another competing device in parallel of the standard R&D process (standardized funnel). This parallel process encourages the expression and development of ideas, if possible "out of the box". It is light, rather "bottom up", but well processed and sequenced. It is generally funded fairly "light".

Conditions to help ideas holders are implemented effectively, albeit on a modest scale in comparison with the official process. They can take different forms: a department dedicated to breakthrough projects and exploration of new growth territories, small committees with financial resources which support the evolution and development of the idea, fund dedicated to disruptive innovation. In the best case, an idea supported and emerging from this parallel device may reinstate the traditional process, or even take the place of a program or development project already in the portfolio.

These models are very new in the organizations we met. Some have produced tangible results, that is to say that ideas have been supported and reintegrated into the process of a traditional project. However, none has yet succeeded in bringing an innovation project standalone, ready to integrate the portfolio of projects issued from the official planning

3.5 The "Agile and Permeable" Funnel

Here, the innovation process is instrumented and institutionalized to generate and capture ideas constantly. The whole organization—not only the functions dedicated to Research and Development—and its culture are adapted to support the ideas and give them means. The permeability of the funnel acts both for "input" and "output": stopping of the project, outsourcing of a part of the cycle, selling the idea and the first results obtained for continuing the development by a third party, etc. The engagement process of the projects is disconnected from the annual budget process for more flexibility while roughly maintaining the budgets allocated to them.

This funnel that one can qualify of "ideal" is pretty close to the previous. When the "dual device" opens a new avenue for innovation, this funnel seeks to open innovation to all disciplines (research, engineering, marketing, design, sociology, risk management, regulatory watch) and to external partners, from the upstream stage of the process (exploration of innovation fields).

4 Let us Hypothesize an Evolution Law in a 5-Funnel Cycle

4.1 An Evolution Law of Companies in a 5-Funnel Cycle

We have also observed that these different forms of funnel tend to be on a sequential series which is even cyclic. Indeed, the surveyed companies described their 3–5 years past and were entitled too to describe their idealized future in terms of innovation management. Therefore, we experimentally hypothesize that there is a natural law of evolution like it exists in TRIZ theory for technical systems [17, 18].

We define these five stages of evolution of a company in terms of innovation management maturity as:

- The start-up, spontaneously permeable and agile
- The growing company that tends to standardize processes

- The healthy reaction to the normalization: the heroes that defend innovations
- The awareness of the need to establish mechanisms that promote innovation "out of the box"
- The organization of agility

In more details, the 5 stages are:

1. **Stage 1**: At the beginning of a company, at the stage of start-up, the involvement of the founders, the small size of teams and the low volume of activities make that an oral culture is adequate. The whole company is spontaneously committed, led by leaders heavily involved at the operational level, to ensure the output of innovations. The funnel is naturally agile and permeable (or "ideal") without the need for formal processes.
2. **Stage 2**: The growing company creates specialized functions and provide them with dedicated resources. The implementation of processes, operating rules, management tools is needed to control the volume of activity and resource use. The company becomes "technocratic" and sometimes "bureaucratic". A standardized funnel is then implemented. Depending on the types of innovation practiced and sectors concerned, the standardized funnel can be "infertile" (number of ideas for innovation low, often linked to highly technical products and processes and to their renewal duration) or, on the contrary, can be of "high attrition".
3. **Stage 3**: To fight against the negative effects of this form of selection, heroes appear, which personally defend innovative ideas that have failed to pass the formal selection process and eventually turn into real success stories.
4. **Stage 4**: Companies capitalizing on these successes and eventually recognizing the value of one or more parallel channels to the formal and standardized selection process, finally put in place a dual system by putting them in competition and, at best, in making them cooperate and exploiting synergies.
5. **Stage 5**: At the ultimate stage, companies are able to merge all the channels of innovation in addressing innovation transversely to functions, approaching it from a multidisciplinary perspective, identifying and exploiting at best external sources of innovation or the value of their Research and Development. They manage to implement, on an industrial scale, an ideal, organized and equipped, agile and permeable funnel, returning to the virtues of roots. In practice, companies in our study [16] are almost all in stages 1–4; only a few have some characteristics of the ultimate stage 5. This clearly reveals how much progress still to implement a process for ideas and projects management that is not too rigid while being sufficiently organized. The challenge, in our opinion, for the company that wants to be fully innovative is to get to stage 5, at least for innovations that are not strictly incremental. To be truly innovative as it grows, the company must recover the dynamics and virtues of its beginning as illustrated by Fig. 8 of the evolutionary cycle of the five funnels.

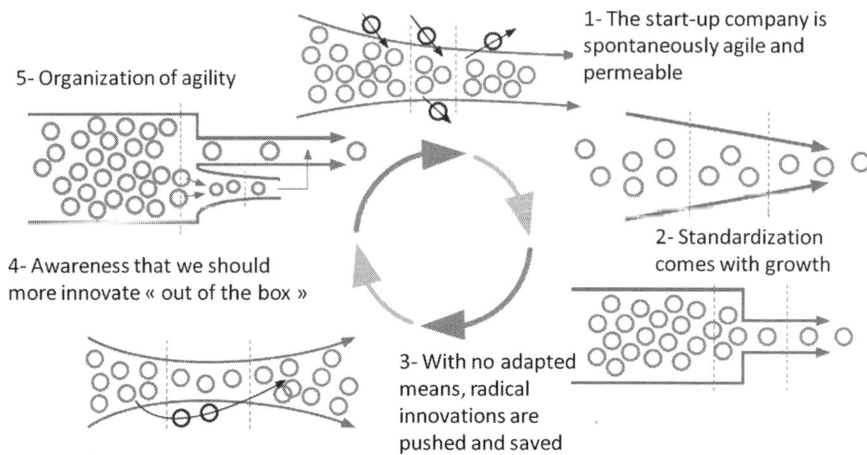

Fig. 8 The evolutionary cycle of the five funnels

4.2 How to Find the Agility and Creativity of the Origins?

It is not mandatory at all to evolve towards the next stage of innovation funnel for a company. But one believes that, like the TRIZ evolution laws of technical systems [17, 18], making people conscious of the innovation funnel their company is likely to be close to is already fruitful for them to visualize their company process and start discussions about what today contributes to the funnel *shape*, *permeability* and *agility*. Each people can also try to imagine how to personally interpret this innovation funnel.

In the case where the company is not satisfied with innovation management and performance, we propose to use the previous evolution law as a tool for continuous improvement. A creative workshop can be organized as follows:

- Gather a short-list of some players representing the ideas and projects management processes in the company (generation, selection, development, managers).
- Make them positioning their business on the 5-funnel cycle. Make them express the reasons to be at a given location and try to result in a consensus for the whole company.
- Then, lead a brainstorming session on how to move from the present company funnel to the next stage.
- Work to a strategic and operational roadmap.

5 Conclusion

It is common in the Management Science and Design Engineering communities to represent the processes contributing to innovation in companies as a funnel pictorial. We noticed that it could be used in several situations: representing the ideation process, as well as more strategic innovation processes. We propose to use the funnel representation as a holistic representation of the 4-layer processes that well model innovation or value creation processes in companies (see Fig. 6).

Our survey on innovation practices in 28 large companies [16] has led to the observation that the innovation funnels of these companies can be categorized into 5 funnel types differentiated by their shape, permeability to ideas and process agility, and that they could be easily sketched as meaningful pictorials (see Fig. 7). Our surveyed companies also learnt us that the past or the expected future of the companies let us think that the 5 funnel types follow a sequential and cyclic evolution law, as in TRIZ theory.

We finally propose to use the 5-funnel model as an inspiring tool for making people conscious of the innovation processes of their company and discussing of it. In addition, they can use it in a creativity workshop to imagine a strategic and operational roadmap for making their company evolving from one funnel stage to the next. These tools are being tested in companies and they truly appear insightful.

References

1. De Brabandere L, Mikolajczak A (2002) Le plaisir des idées. Libérer, gérer et entrainer la créativité au sein des organisations", Dunod, ISBN 210040847, Paris
2. Deschamps JP, Nayak PR (1995) Product juggernauts—how companies mobilize to generate a stream of market winners. Harvard Business School Press, Arthur D. Little, Inc., ISBN-10: 0875843417, Boston
3. Gero JS, Kazakov V (1996) An exploration based evolutionary model of a generative design process. Microcomput Civil Eng 11:209–216
4. Cross N (1994) Engineering design methods—strategies for product design. Wiley, Chichester
5. Pugh S (1990) Total design: integrated methods for successful product engineers. Addison Wesley, New York
6. Roozenburg NF, Eekels J (1995) Product design: fundamentals and methods. Wiley, Chichester
7. Liu YC, Chakrabarti A, Bligh T (2003) Towards and 'ideal' approach for concept generation. Des Stud 24(4):341–355
8. Brem A, Voigt K-I (2009) Integration of market pull and technology push in the corporate front end and innovation management—insights from German software industry. Technovation 29(5):351–367
9. Wheelwright SC, Clark KB (1992) Revolutionizing Product Development. Free Press, New York
10. Kline SJ, Rosenberg N (1986) An overview on innovation, in the positive sum strategy.In: Landau R, Rosenberg N (eds) National Academy Press, Washington, DC

11. Cooper RG (1990) Stage-gate system: a new tool for managing new products. Business Horizons, May–June, pp 44–55
12. Cooper RG (1983) A process model for industrial new product development. IEEE Trans Eng Manag 30:2–11
13. Cooper RG (1988) Predevelopment activities determine new product success. Ind Mark Manage 17:237–248
14. Pahl G, Beitz W (1984) Engineering design. The design council. Springer, London
15. Ulrich KT, Eppinger SD (1995) Product design and development, McGraw-Hill, New York
16. Cuisinier C, Vallet E, Bertoluci G, Attias D, Yannou B (2012) Un nouveau regard sur l'innovation—Un état des pratiques et des modèles organisationnels dans les grandes entreprises. Techniques de l'Ingénieur, ISBN 978-2-85059-130-3, Paris
17. Petrov V (2002) The laws of system evolutions. The TRIZ J, Mar 2002, paper #2, see http://www.triz-journal.com/archives/2002/03/b/index.htm.
18. Savransky SD (2000) Engineering of creativity—introduction to TRIZ methodology of inventive problem solving. CRC Press, Boca Raton
19. Motte D, Bjärnemo R, Yannou B (2011) On the interaction between the engineering design and development process models—Part I: elaborate elaborations on the generally accepted process models. In: Proceedings of ICoRD: 3rd international conference on research into design, Jan 10–12, Bangalore

A Methodology for Assessing Leanness in NPD Process

B. A. Patil, M. S. Kulkarni and P. V. M. Rao

Abstract Lean concepts are routinely used in manufacturing but are still relatively new to New Product Development (NPD) process. If manufacturing can be modeled as flow of material, an analogous model for NPD process would be flow of information. This paper addresses the issue of eliminating wastes in new product development process by bringing lean concepts through an assessment system. The paper first defines wastes in the context of NPD process. A large number of waste drivers which add to wastes in any NPD process are identified. Similarly lean enablers and lean tools which if implemented can reduce wastes are also identified. A model connecting wastes and lean enables is arrived which gives a holistic picture of relationship to address and assess wastes in any NPD process. In order to measure the effectiveness of NPD process five lean performance measures are used which include design cost overrun, product cost overrun, schedule overrun, knowledge capture and customer satisfaction. Industrial inputs are used to arrive at assessment system proposed in this work and a further feedback from industry is used to enrich the assessment system after it is developed. Though the proposed system is developed for machine tool development process, it is generic in nature and can be used to assess any new product development activity.

Keywords New product development · Lean product development · Leanness assessment

B. A. Patil · M. S. Kulkarni · P. V. M. Rao (✉)
Mechanical Engineering Department, IIT Delhi, New Delhi, India
e-mail: pvmrao@mech.iitd.ac.in

A. Chakrabarti and R. V. Prakash (eds.), *ICoRD'13*, Lecture Notes in Mechanical Engineering, DOI: 10.1007/978-81-322-1050-4_58, © Springer India 2013

1 Introduction

Greater competition and more demanding customers are forcing industries to continuously responds to their demands and develop new products. As a result of this organizations are bringing new products at frequent intervals to keep customers excited about their products. In other words to survive and succeed, industries should now be able to design new products as per customer requirements in a short period of time. Failure to do so can be very costly, not only in terms of lost market share, but also in terms of investment made to develop a new product.

New Product Development (NPD) consists of a set of activities beginning with the perception of a market opportunity and ending in the production, sale, and delivery of a product [1]. The major steps included in an NPD process are product planning, concept development, system level design, detail design, testing and refinement and production ramp-up. According to Cooper and Kleinschmidt [2], a new product or process is closely related to the stages of its development process the success of which depends on the manner in which these stages are executed, the way in which these stages complement one another, and the effectiveness with which persons responsible for these stages interact with each other. The communication among suppliers, customers, designers, shop-floor people are important and any miscommunication between these members severely affects the smooth execution of NPD process [3, 4]. One way to be effective in this process is by adopting lean practices.

The term lean which is synonymous with manufacturing was popularized by Womack et al. [5] to describe the concepts and philosophy behind Japanese auto manufacturers. Since then the concept of lean has spread to cover other industries, other parts of the world and other product life-cycle activities. Lean concepts are now integrated and practiced extensively in new product development process as well. Lean when implemented translates as less human effort, less resources, less time and even less space while simultaneously delivering products that customer really wants to pay for. In this way, it facilitates increasing value while decreasing wastes at the same time [6]. The underlying premise is that any feature or an activity, which does not add value to the product, is waste and customer is not willing to pay for it.

There have been many initiatives in the past to adopt lean practices in an NPD process. MIT's Lean Aerospace Initiative has pointed out that improvements in manufacturing alone with the help of lean principles will make only a marginal difference in ultimate system costs, because production and service of a quality product by manufacturing, usually seen as the delivered value, are not valuable if the product itself does not please the customer [7]. In other words, planned value by design process only ensures the delivered value satisfying the end user. In this manner product development (PD) has a great deal of leverage on both the creation of the right product, and the enabling of lean production through appropriate design [8].

It has been emphasized that lean NPD requires the right information at the right place at the right time [9], bringing focus to the value of information. Lean NPD helps companies to develop a seamlessly flowing PD value stream with minimal waste, defined and pulled by the customer [10]. According to Oehmen and Rebentisch [11] the lean product development focuses on defining and creating successful and profitable product value streams which facilitates global need to develop products and systems faster, cheaper and better. Though there have been efforts to adopt lean practices in an NPD process there are no significant efforts to measure the effectiveness of lean implementations. It is needless to state that in order to improve the performance of a process, it must be measured. In other words measurement of leanness in an NPD process can facilitate in effective adoption and implementation of lean practices. The present work proposes a methodology for the same which is termed as lean function deployment (LFD).

2 Previous Work

Though there have been some efforts to quantify leanness in the domain of manufacturing, the same is not true in the area of new product development. Following paragraphs discuss a few efforts made in this direction and some insights given by researchers to understand leanness.

McManus et al. [12] tried to differentiate between lean NPD process and lean manufacturing. Their effort to study application of five basic lean principles applied to two domains gives a good perspective as summarized in Table 1.

It is clear that an assessing system developed for manufacturing can not work or be extended to NPD process because the two are essentially different. In NPD process value is harder to see, and the definition of value-added is more complex in this case [9, 13]. Due to uncertainties or interdependencies, branching or iterative flows may be beneficial because designers learn from iterations that what is worked and what did not, which is not true in case of manufacturing. The iterations help to improve the tacit knowledge as well as the explicit knowledge of the designers. The flow in manufacturing is of material which is being processed whereas PD processes deals with the flow of information which is generated at each step of the process, and value of information is increasing at each step which helps to reduce the risk of not meeting the product performance. Pull in manufacturing is more related to the market demand and the capacity of the plant, whereas in PD process pull is more dependent on the timely requirements of the downstream processes from the upstream processes of design [14]. Finally, perfection is even harder to reach in case

Table 1 Lean manufacturing versus lean NPD [12]

Lean principles	Manufacturing	Product development
Value	Value at each step, defined goal	Harder to see, emergent goals
Value stream	Parts and material	Information and knowledge
Flow	Iterations are waste	Planned iterations are must
Pull	Driven by take time	Driven by needs of enterprise
Perfection	Process repeatable without errors	Process enables enterprise improvement

of product development as simply doing the process very fast and perfectly with minimal resource used is not the final goal. The goal is to ensure the efficiency of the design process (in terms of product output with resources consumed, ensuring right method of developing a product) as well as effectiveness of the PD process ensuring the development of the right product.

According to Murman et al. [15] lean concepts and approaches should be used to create and deliver value for all stakeholders, and not just end users. Hines et al. [16] have developed a framework for lean product development which is driven by use of lean tools by industries. Anand and Kodali [17], have proposed a framework to make existing NPD process leaner by classifying activities into value added and non value added.

While developing a framework for creation of value in PD process, Chase [13] has discovered that value is facilitated using four tools namely, right tasks, resources, environment, and management approach. For measurement of value addition by an activity, ten different attributes of value are considered. In yet another attempt, McManus and Team [14] used a questionnaire based approach to know the occurrence of flow and pull in PD process.

Stanke and Murman [18] developed a framework for achieving a lifecycle value consisting of three phases as value identification, value proposition, and value delivery. Six value creation attributes covering holistic perspective, organizational factors, requirements metrics, tools and methods, enterprise relationship and leadership management were used to assess effectiveness of PD process. In a survey carried out by Hoppman et al. [19], leanness assessment was done based on three supporting measures (goals defined, human resources for introduction to LPD process, use of value stream mapping). The goal of the survey was to investigate the process of introducing a lean PD system.

Some important observations can be drawn from the above studies. Most of the works assess leanness based on whether or not certain lean tools are in practice. They do not consider the inherent relationship between lean tools and waste drivers. These approaches miss identification of value and quantification of waste in PD process. Some of the methods end with assessment of leanness without proposing any remedial measures. As one of the goals of lean implementations is continuous improvement, the leanness systems proposed are not able to address this aspect. Lastly the goals of adopting lean concepts in an NPD process differ from one organization to another. The systems proposed are not able to address this aspect and assume a common goal for all lean implementations.

To overcome these limitations, a more effective and a holistic approach is proposed in the present work. Though the proposed work is developed for assessment of leanness in new machine tool development process, the framework is generic in nature and can be easily applicable to NPD process of other products and organizations as well. The term new in an NPD process can mean any of the following four cases:

1. An entirely new product based on new technology or new concept
2. A new product lines or family which allows entry into newer markets

3. Making additions to existing product lines
4. Improvements in existing product.

A preliminary study of machine tool industry, particularly that of Indian machine tool industry has shown that most of NPD activity belongs to improvement of an existing a product, product line or product family. Keeping this in mind, the present research focuses on classification of new products as defined under Sects. 2 and 3.

3 Lean Function Deployment

The methodology proposed in the present work called as *lean function deployment* (LFD) is both qualitative and quantitative in nature. It can be used to benchmark one organization with respect to another or can be used for assessing continuous improvement of any organization from one state to another. It not only identifies problem areas but also gives remedial measures to correct the same. The proposed methodology is based on constructing many houses of leanness each with a definite objective. The process has similarities with Quality Function Deployments (QFD) often used to capture customer requirements. The proposed methodology consists of following steps.

3.1 Identification of Performance Measures

Organizations which implement lean often do with certain specific objectives in mind. These are referred as *lean performance measures* in the present work. The reason for referring them as performance measures is that these measures should improve or reach an improved state at the end lean implementation. The performance measures identified include: Project cost overrun, schedule overrun, product cost overrun, customer satisfaction and knowledge capture. It is true that for any organization all these measures are important. However, organizations would like to concentrate on improving one or more from existing state to an improved state. For example an organization doing well in terms of project cost over run and customer satisfaction would like to implement lean to improve its delivery schedules. First step in the proposed methodology is to identify, prioritize and if necessary assign weights to these performance measures.

3.2 Identification of Wastes

Any activity or process which does not add value to the product and customer is not willing to pay for it can be called as waste. Identification of wastes is key to any lean activity. The eight wastes commonly defined in manufacturing domain can not be borrowed in the case of product development. Some are retained and some are newly defined for NPD process. The five wastes identified for NPD process in the present work are untapped human potential, waiting time, wasting time, overdesign and rework. As wastes are difficult to quantify, sub-wastes are defined which are specific manifestation of wastes and are somewhat quantifiable. For example lack of a system to address suggestions from team members is a sub-waste under the category of waste untapped human potential. Waste driver is even one level below which tries to find the root cause(s) because of which a particular waste or sub-waste happens. In this work five wastes are classified into 19 sub-wastes which are further sub divided into 49 waste drivers. In other words the probability and magnitude of presence of a type of a waste is evaluated by knowing the waste drives are extent to which it happens in any given organization.

3.3 Identification of Lean Enablers and Lean Tools

A systematic application of lean and its principles ensures the delivery of value to all stakeholders in a process. *Lean enablers* are the attributes/characteristics of NPD Process which enhance the lean component in the process and thus ensure the creation of value. Lean enablers are not only derived from the industry best practices but also are extracted from the core lean thinking principles. 21 lean enablers are identified in this work.

Lean tools are those which promote leanness in a system by strengthening lean enablers. The different lean tools identified and used for assessment include customer relationship management (CRM), quality function deployment (QFD), Design for X (DFX), Failure modes and effects analysis (FMEA), cross functional teams (CFT), set based concurrent engineering (SBCE), design structure matrix (DSM), theory of inventive problem solving (TRIZ), knowledge management, product life-cycle management (PLM) etc.

3.4 Relating Lean Performance Measures to Wastes

There exist a strong relation between lean performance measures and wastes. Establishing this relationship is vital for the assessment system. Given the performance measures which organization wants to improve, it is possible to know from this relationship what wastes or waste drivers need to be addressed for achieving the same. This relationship is in the form of a matrix or house of leanness where rows are lean performance measures and columns are waste drivers. The relative importance of performance measures is first established. This can be done by pair-wise comparison using tools such as AHP. This helps in addressing waste drivers which are more important than the others. This step not only considers presence of a specific waste in an organization but also considers magnitude and probability of its occurrence.

3.5 Relating Wastes to Lean Enablers

In this step wastes measured in terms of sub-wastes are related to lean enablers. This step is important as it gives an insight into what lean enablers influence which waste and vice versa. In other words it gives enough information to identify lean enablers to overcome certain wastes predominant in an organization being assessed. Together with step 4 it also quantifies the readiness of organization for implementing lean.

3.6 Relating Lean Tools to Lean Enablers

Previous five steps identify a set of lean enablers which need to be addressed to overcome weaknesses in an organization being assessed. Lean enablers can be strengthened by implementing lean tools as discuss previously. The relation between lean tools and lean enablers is not straightforward as one lean tool can influence many lean enablers and vice versa. One needs to know the relative influences of each lean tool on lean enablers. Building this relation is the sixth step in lean function deployment (LFD).

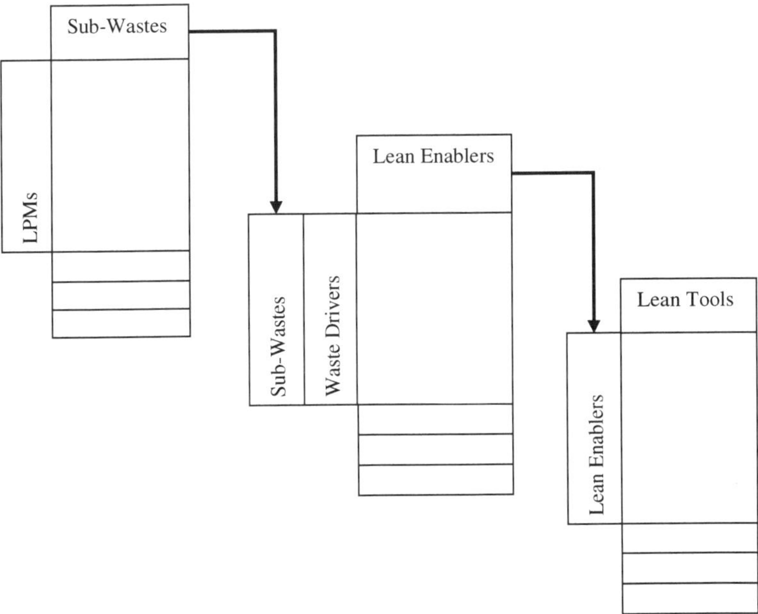

Fig. 1 Lean function deployment framework

Sub-wastes → Lean Performance Measures	People waiting for information	Overdesign	…….	Unutilized tacit knowledge
schedule overrun				
project cost overrun				
product cost overrun				
customer satisfaction				
knowledge capture				

Fig. 2 Relating lean performance measures with sub-wastes

Lean Enablers → Waste Drivers	Organization of Knowledge	Reuse of components	…….	Standardize work processes
Not allowing/motivating employees to put forward their ideas and views				
Unevenness in work distribution				
………….				
Over commitment of resources				

Fig. 3 Relating waste drivers to lean enablers

Lean Tools → Lean Enablers	QFD	DFX	…….	PLM
Organization of knowledge				
Reuse of components				
………….				
Standardize work processes				

Fig. 4 Relating lean enablers to lean tools

4 Implementation

Figure 1 gives pictorial representation of all six steps discussed above. Once all the relations have been built and validated, the proposed framework can be used in multiple ways to assess leanness as well as to identify tools necessary for continuous improvement. For example one can assess an organization by extent to which it uses lean tools. It is important to mention here that mere notional adoption of lean tool is not a true assessment. For effective results, these have to be assessed at various levels as is done in a typical capability maturity models (CMMs). As per the proposed framework, implementation of lean tools is effective if it finally improves the performance measures as desired.

Secondly one can start with performance measures to be improved and using the frame work, identify lean tools which need to be implemented or strengthened. At each stage of assessment, one can visualize both qualitative and quantitative information about performance measures, wastes and lean enablers. The detailed quantification measures are not discussed here and are a subject of another work.

Figure 2 shows how lean performance measures can be related to wastes. In this first level of assessment either all lean performance measures can be given equal weightage or they can be prioritized by assigning different weights. This would depend on what the organizations goals are in implementing lean programme. Figure 3 establishes relation between wastes and lean enablers. This is one of the larger matrices in the entire assessment system which requires industrial input and experience to structure it. Last stage of assessment as shown in Fig. 4 is optional. If one is interested in only assessing leanness in NPD, this stage can be avoided. However most of the organizations would be not only interested in knowing weakness but also to overcome them. In this context relating lean enablers with lean tools become an important task.

5 Conclusions and Future Work

The proposed framework has been arrived at after studying many of the exiting assessments system in practice and those proposed in the literature. Inputs from industrial experts have gone in coming up with proposed framework. This has been done by discussion with experts from multiple industries. A workshop of machine tool experts too has been organized to capture required inputs.

However, the proposed methodology is still a qualitative framework. It needs to be fully validated quantitatively in multiple industrial scenarios. The efforts are presently on in this direction the results of which will be communicated later.

Acknowledgments The authors acknowledge the financial support received from Department of Science & Technology (DST), International Division in carrying out this work.

References

1. Ulrich KT, Eppinger SD (2004) Product design and development. McGraw Hill, New York
2. Cooper RG, Kleinschmidt T (1986) An investigation into the new product process: steps, deficiencies, and impact. J Prod Innov Manage 3:71–85
3. Brown SL, Eisenhardt KM (1995) Product development: past research, present findings, and future directions. Acad Manag Rev 20(2):343–378

4. Olson E, Walker O, Ruerket R, Bonner J (2001) Patterns of cooperation during new product development among marketing, operations and R&D: implications for project performance. J Prod Innov 8:258–271
5. Womack J, Jones D, Roos D (1990) The machine that changed the world. Rawson Associates, New York
6. Womack JP, Jones D (1996) Lean thinking: banish waste and create wealth in your corporation. Simon and Schuster, London
7. Murman E, Thomas A, Bozdogan T, Cutcher-Gershenfeld J, McManus H, Nightingale D, Rebentisch E, Shields T, Stahl F, Walton M, Warmkessel J, Weiss S, Widnall S, (2002) Lean enterprise value: insights from MIT's lean aerospace initiative, Palgrave, New York
8. Mcmanus HL (2005) Product development value stream mapping (PDVSM) manual, lean aerospace initiative. MIT Cambridge, MA
9. Browning TR (2000) Value-based product development: refocusing lean. In: Proceedings of the 2000 IEEE, Engineering Management Society
10. McManus HL, Millard RL (2002) Value stream analysis and mapping for product development. Int Counc Aeronaut Sci 1–5
11. Oehmen J, Rebentisch E (2010) Waste in lean product development. LAI paper series lean product development for practitioners. Lean advancement initiative (LAI), Massachusetts Institute of Technology, July 2010
12. McManus H (2005) Lean engineering: doing the right thing right. In: Proceeding of the 1st international conference on innovation and integration in aerospace sciences. Queen's University Belfast, Northern Ireland,1–10 2005
13. Chase JP (2001) Value creation in the product development process. M.Sc. thesis, Massachusetts Institute of Technology
14. McManus HL, Team LP (2000) Flow and pull in PD. Working paper series, the lean aerospace initiative, Massachusetts Institute of Technology
15. Murman E, Stanke A (2005) Virtual assessment of lean user experience (VALUE) survey (v. 3.4.2), MIT
16. Hines P, Francis M, Found P (2006) Towards lean product lifecycle management: a framework for new product development. J Manuf Techno Manag 17(7):866–887
17. Anand G, Kodali R (2008) Development of a conceptual framework for lean new product development process. Int J Prod Dev 6(2):190–224
18. Stanke A, Murman E (2002) A framework for achieving lifecycle value in aerospace product development. International Council of Aeronautical Sciences Congress, 1–10 2002
19. Hoppmann J (2009) The lean innovation roadmap—a systematic approach to introducing lean in product development processes and establishing a learning organization. Diploma thesis, Institute of Automotive Management and Industrial Production, Technical University of Braunschweig, June 2009

PREMΛP: Exploring the Design and Materials Space for Gears

Nagesh Kulkarni, Pramod R. Zagade, B. P. Gautham, Jitesh H. Panchal, Janet K. Allen and Farrokh Mistree

Abstract Design of a gear that meets specified requirements is a challenging task. Competition from other power transmission components as well as increasing demands from industry such as increased power density, low noise etc., are forcing gear designers to design gears using novel methods (which are beyond the traditional standards based design methods). We, at Tata Consultancy Services, are developing a Platform for Realization of Engineered Materials and Products (PREMΛP), which helps a designer exploit the synergy between component design, material design and manufacturing. One of the key features of PREMΛP is its decision support capability, which we are demonstrating using design of spur gear as an example. The compromise Decision Support Problem (cDSP) construct

N. Kulkarni · P. R. Zagade · B. P. Gautham (✉)
Tata Consultancy Services, Pune 411013, India
e-mail: bp.gautham@tcs.com

N. Kulkarni
e-mail: nagesh.kulkarni@tcs.com

P. R. Zagade
e-mail: pramod.zagade@tcs.com

J. H. Panchal
School of Mechanical Engineering, Purdue University, West Lafayette, Indiana 47907, USA
e-mail: panchal@purdue.edu

J. K. Allen
School of Industrial and Systems Engineering, University of Oklahoma, Norman, Oklahoma 73019, USA
e-mail: janet.allen@ou.edu

F. Mistree
School of Aerospace and Mechanical Engineering, University of Oklahoma, Norman, Oklahoma 73019, USA
e-mail: farrokh.mistree@ou.edu

A. Chakrabarti and R. V. Prakash (eds.), *ICoRD'13*, Lecture Notes in Mechanical Engineering, DOI: 10.1007/978-81-322-1050-4_59, © Springer India 2013

is used to formulate the problem and the software DSIDES to solve it. Results obtained are well in agreement with existing knowledge of gear design. Ternary contour plots are created which show the compromise between various goals and associated standard deviation.

Keywords Gear design · Compromise decision support problem

1 Introduction

1.1 PREMΛP

PREMΛP—Platform for Realization of Engineered Materials and Products is conceived at Tata Consultancy Services. It is based on integrated computational materials engineering (ICME) [1] approach that is expected to (a) reduce the time and cost of discovery and development of materials and their manufacturing processes, and (b) enable faster development of products augmented with richer material information, is described in four parts:

- PREMΛP—A Platform for the Realization of Engineered Materials and Products [2],
- PREMΛP—Exploring the Design Space for Continuous Casting of Steel [3],
- PREMΛP—Exploring the Design and Materials Space for Gears (this paper),
- PREMΛP—Knowledge Driven Design of Materials and Engineering Processes, [4].

It hosts various tools for modeling and simulation supported by tools for informatics, knowledge engineering, robust design and decision support along with appropriate databases and knowledge bases. It is discussed in detail in Gautham et al. [2]. One of the key features of the platform is its decision support capability. In the second [3] and this, third paper of the series, application of compromise Decision Support Problem (cDSP) [5] to two problems, manufacturing process design and component design, are discussed. In the second paper [3] we discuss two features of PREMΛP, namely, robust design and the exploration of the design space through the use cDSP in the context of the manufacturing process, continuous casting of steel. In this the third paper, we illustrate a method for exploration of design space for the material and gear geometry simultaneously using the design of a gear as an example.

1.2 Designing a Gear—Challenges

Gears are extremely important components of transmission systems of many kinds and steel gears are the most widely used ones. According to the vision report of the gear industry [6], the global gear market exceeds US$45 billion per annum. Gears are mostly used in automobiles (which constitute nearly three forth of the entire gear market) along with industrial equipment, airplanes, helicopters, marine vessels and other applications. The vision document [6] lists various goals for the gear industry in areas such as profits and market share, gear performance, gear manufacturing techniques etc. It also lists key technological challenges and innovations required to achieve these goals. One of the challenges is innovative gear design and development.

Apart from the historical demand for high life and reliability, new demands have emerged such as low noise, high efficiency, low weight, high quality, etc., which are driven by demands from customers as well as by legislation. Traditionally gear design has been based on gear design standards from organizations such as the American Gear Manufacturers Association (AGMA), the International Organization for Standardization (ISO), and the German Institute for Standardization (DIN). These standards provide methods to select suitable materials and design the geometry of gears. They use different factors such as the overload factor, the load distribution factor, the geometry factor, and the size factor to correct for basic assumptions in analysis and to adjust for the uncertainties in load estimation, load distribution, stress concentration, etc. Similarly, a number of factors are considered such as the reliability factor and the safety factor to account for variability in material and manufacturing processes based on historical data and experimentation. The resulting design is safe but it may be bulky, hence the gear design process needs to be improved. It should take advantage of recent developments in numerical modeling of gears and materials and manufacturing processes, in order to make it amenable to competition and to survive. The new process should be such that it can take into account the effects of material and manufacturing processes such as residual stresses, detailed distribution of properties and microstructure, etc. [2, 7]. Defects introduced during steel mill production gives rise to problems during further manufacturing process of these mill products. For example, segregation influences distortion during forging of gear blanks; the fatigue life is impacted by the residual inclusion content during the continuous casting stage. This leads to the complex problem of making design decisions across multiple domains—mechanical design, materials processing and manufacturing. The integrated computational platform PREMΛP described in Gautham et al. [2] is envisaged to provide means for such development.

The decisions on selection of gear geometry, material and manufacturing parameters are riddled with compromises among performance, weight and cost, for example, high reliability design in general is costly. The cDSP is one of the possible constructs for decision support for such situations. In this work, we present cDSP for gear geometric design and material selection, a subset of the larger problem described earlier, to explore the compromises one has to make between the compactness, cost and reliability of gears. We use DSIDES[1] for solving the cDSP. The present work is seen as a first stage high throughput

[1] Mistree F., "DSIDES-Decision Support in the Design of Engineering Systems", User manual.

screening using AGMA standards and will be extended through the incorporation of a detailed analysis of gear performance and manufacturing processes in an integrated computational framework in the future.

2 Gear Design in PREMΛP Using the cDSP Construct

2.1 PREMΛP Based Gear Design

An integrated design of a gear envisages concurrent design of geometry, material selection and manufacturing process design starting from a round bar of steel sourced from a steel supplier. This needs a preliminary gear design using AGMA standards for setting the initial design space, followed by detailed analysis of gear performance utilizing the material information from the manufacturing process simulation. This in turn requires detailed simulation of manufacturing processes that provide the material property evolution as well as estimates of manufacturing costs. Armed with this information, the designer can arrive at an acceptable solution that meets the targets on cost, performance and constraints. In Sect. 4.2 of Ref. [2] the authors discuss such aspects in detail. Further to an integrated design, evolving specifications of mill products or developing new alloys starting from the chemistry requires further integration with upstream processes of the steel mill. For example, the inclusion content that remains at the casting stage influences final fatigue properties. Designing the continuous casting process parameters under cDSP construct is discussed in Ref. [3]. The integrated design is envisaged to be carried out on the ICME platform PREMΛP being developed at Tata Consultancy Services [2]. In this paper we deal only a small part of the chain—AGMA based geometric design and material selection—in a cDSP framework. This initial geometry and material is taken as initial guess and appropriate design space is built around it for integrated design discussed in Ref. [2].

2.2 The Compromise Decision Support Problem

The cDSP differs from standard optimization formulations in that it is a hybrid formulation based on mathematical programming and goal programming. It enables the construction of different practical scenarios in a multi-objective formulation by giving appropriate weightage to different goals and exploring the compromise among them. It works by minimizing the difference between the desired (the target G_i) and the achieved ($A_i(x)$) value of a goal. The difference between these values is the deviation value, d_i^+ and d_i^-, which represents overachievement and under-achievement of each goal respectively. A cDSP is constructed such that constraints and deviations are always positive and no simultaneous over or under achievement is allowed. The details of cDSP can be found in Refs. [8, 9].

2.3 The cDSP for Gear Design

In traditional gear design based on AGMA, the designers make use of guidelines and procedures given by AGMA to make decisions on the design variables such as module, number of teeth, etc. At this stage, the designer has a number of choices and would eventually pick one of them that satisfies the constraints and one or more of targets. This design process can be put in a formal optimization loop to search a design which satisfies the constraints and provides the best possible value of the objective function. On the contrary, robust design using cDSP involves exploring the design space for flat regions while trying to satisfy the constraints. The additional information given by cDSP allows the designer to add/modify constraints and goals systematically to achieve better designs.

2.3.1 Problem Statement

The current problem deals with design of a pinion for the first gear reduction for a compact sized automobile. The problem statement is as follows

> Design a gear set (pinion and gear) to transfer maximum torque of 113 Nm @ 4,500 rpm with a speed reduction ratio of 3.5. The gear set has standard full depth teeth 20° pressure angle. The teeth are generated with a rack cutter. Design compact, light weight, cost effective and reliable spur precision (AGMA quality no 8) gearing (moderate shock in driving engine, moderate shock in driven machinery). A minimum reliability of 99.99 % is expected. The minimum expected fatigue life is 10^{9} cycles. The maximum allowable center distance is 300 mm. Materials for the gear are available in the range of 800 to 1,600 MPa. Space available for gear and pinion is 600 mm. Standard gear geometry as per AGMA standard is desired.

2.3.2 Formulation

Following the example in Ref. [8], the cDSP for the design of a gear is formulated as given below.

Given

It is required to design a pinion of a commercial spur gear system of AGMA precision no. 8 for torque of 113 Nm @ 4,500 rpm with speed reduction of 3.5 ($G = 3.5$) having a minimum of 99.99 % reliability (though helical gear are preferred for such applications, we restrict this study to spur gears for easier illustration). The pressure angle of gear teeth is given to be 20° and cut using rack cutter. The gear pair is to be designed to have fatigue life of 10^9 cycles.

The materials for gear are available in the range of tensile strength of 800–1,600 MPa. Space available for gear and pinion is 600 mm.

Design variables: The standard gear geometry is expressed using AGMA standards [10] in terms of: module (m) in mm, number of teeth (N), and face width (b) in mm. We use these geometry variables along with material strength (UTS) in MPa and reliability (R) as design variables. Reliability is considered part of design variable, as the design space will be explored with different levels of reliability.

Bounds on design and deviation variables

- B1: $4 \leq m \leq 8$ (mm)
- B2: $18 \leq N \leq 40$
- B3: $40 \leq b \leq 80$ (mm)
- B4: $800 \leq UTS \leq 1{,}600$ (MPa)
- B5: $0.95 \leq R \leq 0.9999$, though reliability of 99.99 % is required, we are exploring this range of reliability to study its influence in a wider range.
- B6 to B8: restriction on deviation variables, defined in Sect. 2.2, $d_i^+ \cdot d_i^- = 0$ for $i = 1\ldots3$ and $d_i^+, d_i^- \geq 0$ for $i = 1\ldots3$

Design constraints

- C1: Minimum face-width*: $b \geq 3m$ (1)

- C2: Maximum face-width*: $b \leq 5m$ (2)

- C3: Maximum limit on center distance#: $d = m(1 + G)N/2 \leq 300$ mm (3)

- C4: Bending stress induced *#: $\sigma_{allowable}^{bending} - \sigma_{induced}^{bending} \geq 0$ (4)

- C5: Contact stress induced*#: $\sigma_{allowable}^{contact} - \sigma_{induced}^{contact} \geq 0$ (5)

- C6: Minimum contact ratio*: $R_c \geq 1.4$ (6)

- C7: Maximum contact ratio*: $R_c \leq 1.8$ (7)

- C8: Number of teeth are integer value†: $(N - INT(N)) \leq 0.01$ (8)

* *as per AGMA guidelines* [10, 11],
\# *as per problem statement,*
† *cDSP requires all variables to be of type real, N is also treated as real variable and an additional constraint is added to make it an integer.*

Goals:

- G1: Maximize reliability R

  In AGMA based design, the influence of reliability is brought in through a correction factor (Y_z) in the allowable stresses during gear design through constraints C4 and C5.

- G2: Minimize center distance d (make the design compact)
- G3: Minimize cost C

  Total cost consists of material cost and manufacturing cost and calculated as

$$C = W\left(a_0 + b_0\left(1 + \left(\frac{S_{ut} - (S_{ut})_{\min}}{(S_{ut})_{\max} - (S_{ut})_{\min}}\right)^{1.5}\right)\right) \quad (9)$$

where, W is weight of component, a_0 is the manufacturing cost (taken as INR. 5/Kg), b_0 is the material cost (taken as INR. 40/Kg for 800 MPa UTS steel). The increase in material cost with UTS is as per the form given in Eq. (9). The above equation for cost is valid from $(S_{ut})_{\max}$ and $(S_{ut})_{\min}$, which are maximum and minimum tensile strengths.

These system constraints and goals are normalized as suggested in Mistree et al. [9]. The deviation function is constructed as shown below

$$Z = \{(d_1^- + d_1^+), (d_2^- + d_2^+), (d_3^- + d_3^+)\} \quad (10)$$

As we are minimizing for multiple goals [maximization of reliability (R) is considered as minimization of ($1-R$)], the negative deviation is always 0 and Eq. 10 simplifies as

$$Z = \{d_1^+ + d_2^+ + d_3^+\} \quad (11)$$

Archimedean formulation [9] is used to construct the deviation function and it is solved using the software DSIDES.

$$Z(d^-, d^+) = \sum W_i^+ d_i^+ \quad i = 1\ldots3$$
$$\sum_{i=1}^{3} W_i = 1 \text{ and } W_i \geq 0 \text{ for all } i \quad (12)$$

Table 1 Gear design scenarios

Scenario No	Goal	Requirements	Conflicts
1	Maximum reliability design	Larger gear geometry and high strength material	Compact design, minimum cost design
2	Compact design	Smaller m and N, and high strength material	High reliability design, possible weak conflict with cost
3	Minimum cost design	Smaller gear geometry and low cost material	High reliability design

2.3.3 Gear Design Scenarios

A designer often encounters conflicting demands while designing a gear, such as a reliable design and compact design, or reliable design and minimum cost design. In this gear design problem, we have 3 goals as described earlier. In Table 1 we list the requirements and conflicts of a given goal with other goals.

We explore the above three goals with different weights using the Archimedean approach as discussed earlier. Figure 1 represents different practical scenarios explored.

3 Results and Discussion

DSIDES has many control options, namely, adaptation to control exploration of the design space, specification of maximum number of iterations and criteria for convergence of design and deviation variables. We set the maximum number of iterations to 100, stationarity of design and deviation variables to 2 %.

3.1 Validation

We obtained various designs for different scenarios using DSIDES and few representative gear designs obtained are selected for validation. As DSIDES uses adaptive linear programming, which linearizes constraints and goals during exploration, there could be a scope for error creeping in. In order to verify this, we have plugged in m, N, R and UTS (obtained from gear design by the DSIDES) in the AGMA gear design formulae and computed the face width for selected five cases. The face width computed using AGMA and that given by the DSIDES are given in Table 2.

From Table 2, we can see that the results obtained using DSIDES are well in agreement with that of AGMA. The difference in the value of face width observed is due to the existence of active constraint, i.e., allowable contact stress, which has

Table 2 Validation of results obtained using DSIDES

	Design 1	Design 2	Design 3	Design 4	Design 5
b (mm) using AGMA	77.81	67.72	39.23	72.85	58.40
b (mm) using DSIDES	78.73	57.32	40.00	68.57	61.34
% Error	1.18	−15.36	1.96	−5.88	5.03

been violated within acceptable limits. In the worst case, i.e., Design 2, we observe that the actual factor of safety in contact is 1.3, violating the specified factor of safety of 1.4 used in AGMA calculations (and this is acceptable situation as far as factor of safety is concerned).

3.2 Results and Discussion

The seven different scenarios (S1 through S7) illustrated in Fig. 1 are analyzed in detail using DSIDES and discussed in Sect. 2.3.3. While analyzing results of different scenarios we made the following observations. In scenarios such as compact design (S1) and high reliability design (S3) there are oscillations in convergence behavior of the design variables and deviation variables. In Fig. 2a we show the convergence behavior of design variables for scenario S1. We see that the variables that define compact design (i.e., minimum m and minimum N) converge within the first few iterations. The other three design variables oscillate between three possible states (contact stress constraint is active and violated within acceptable limit for these three designs). This is expected as a single S1 goal leads to an under constrained system as it does not put any constraint on face width, reliability and tensile strength and these three parameters can take theoretically infinite feasible combinations. A similar trend is observed for other scenarios like S3 where same reliability can be achieved through different combinations of geometry and material. For the scenario S6, which is a combination of scenarios

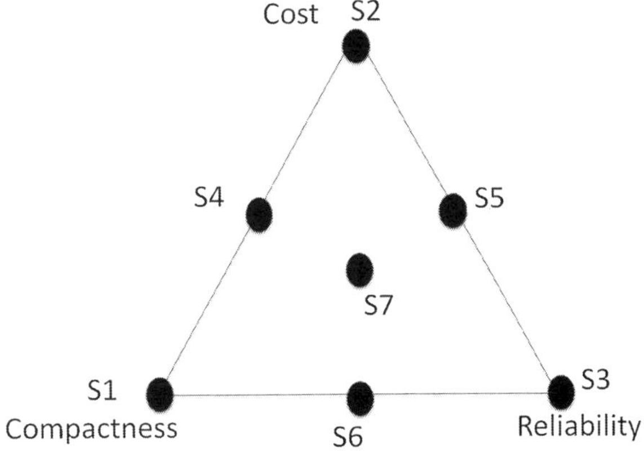

Fig. 1 Design exploration scenarios

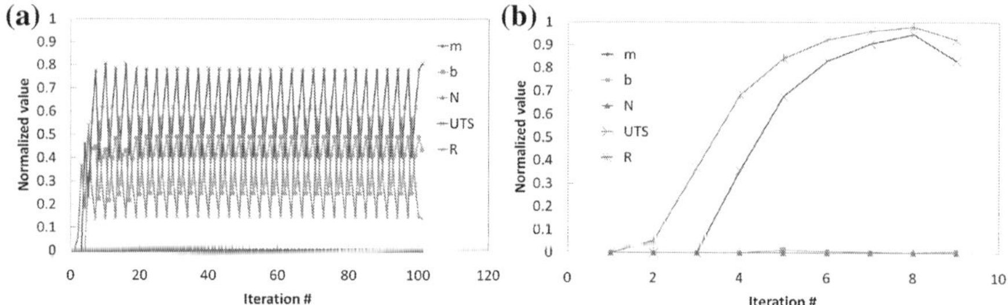

Fig. 2 Convergence behavior of design variables for **a** compact design (S1), **b** minimum cost design (S2)

S1 and S3, oscillation is observed in *UTS* and *b* in initial few iterations and finally a converged solution is obtained.

On the contrary, no oscillations are observed (Fig. 2b) in the convergence behavior of design variables for scenarios S2, S4, S5 and S7 (where cost goal is involved) as estimation of cost involves all the design variables and hence is well-constrained.

In Table 3 we list the best design and respective goals achieved for each scenario. In Fig. 3 we show the comparison of normalized goals achieved for different scenarios. These results are in line with known aspects of gear design such as:

- when compact design is required, the material strength needs to be high, and
- high reliability requirement needs increase in the size of gear as well as strength of the material.

We observe from the results that the minimum cost design is simultaneously a compact design and a minimum face width design. Compact design is not necessarily a minimum cost design (as it is a design with larger face width compared to the minimum cost design). The requirement for high reliability conflicts with compactness or cost as can be expected. Scenarios S5 and S7 involve a compromise between various goals. In order to observe the achievability of goals visually, we have collated all feasible designs obtained during the analysis and made ternary plots of average values of goals and their standard deviation as shown in Fig. 4.

Table 3 Results: design variables and goals for different scenarios

	m (mm)	b (mm)	N	UTS (MPa)	R (%)	d (mm)	C (INR)
S1	4.00	57.32	18	1,444	99.30	162.00	134.55
S2	4.00	40.26	18	1,464	95.40	162.00	96.14
S3	6.59	67.29	18	1,221	99.99	266.94	349.48
S4	4.00	40.26	18	1,464	95.40	162.00	96.14
S5	5.69	57.59	18	1,550	99.99	230.53	300.96
S6	4.99	76.72	18	1,550	99.99	201.95	307.71
S7	4.99	76.70	18	1,550	99.99	210.96	307.67

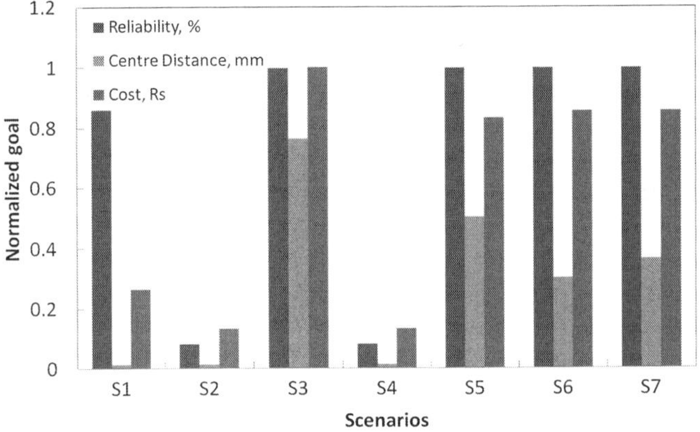

Fig. 3 Comparison of goals for different scenarios

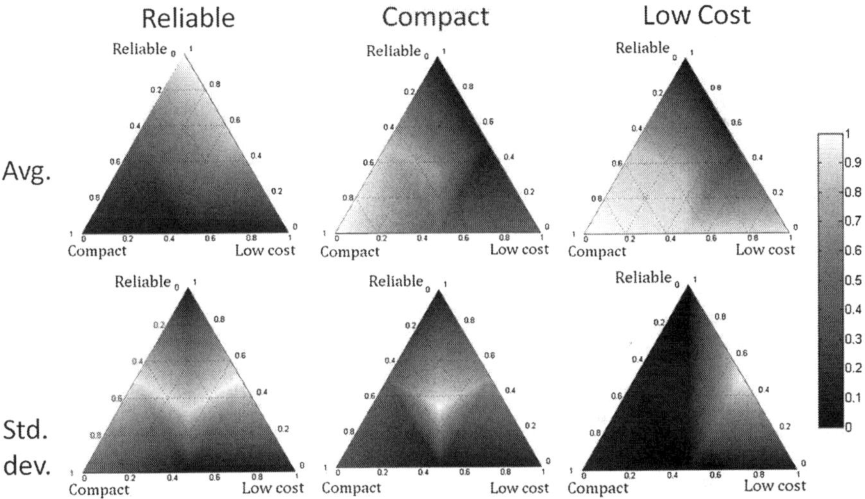

Fig. 4 Visual representation of average and standard deviation of achieved goals with different weights

As can be expected, when a specific goal has highest weight, it achieves the desired target fully. Highly reliable designs have least admissible space (large dark areas), as this requires larger gears and materials with high strength. Low cost gears can be achieved if we compromise on reliability. We further observe that the center region provides compromise between various goals. However, the center region also shows higher standard deviation. The center region represents the design space where equal weightage is given to all the goals. High standard deviation in this region represents same level of achievement in goals as a weighted sum can be obtained through different combinations of individual goals (diverse design choices are available). In such a situation one can formulate preemptive cDSP and explore the compromise in a narrowed down region; see Ref. [9]. This ternary plot (Fig. 4) can be used to make initial decision on the weights to be given so as to exploit the compromise between reliable, compact and low cost design. This is illustrated through the following example. For automotive gears, reliability and compactness are important considerations. It is very likely that the designer may settle for equal weight, i.e., 0.5 to these two goals. However, Fig. 4 suggests that the region corresponding to equal weightage to each of above goals is region of high variability, which may not be the ideal choice. In order to get a less sensitive design, one may select weights for compactness and reliability as 0.4 and 0.6 respectively.

4 Closing Remarks

Designing a gear that meets specified functional, non-functional as well as performance requirements is a challenging task. PREMΛP provides integrated systems engineering framework to design such complex components. In this work, we illustrate a method for exploration of the design and the material space simultaneously using the cDSP and through the design of a gear as an example. Various practical gear design scenarios such as compactness, reliability, cost and compromise among them are considered. DSIDES is used to solve the cDSP and results obtained are validated against AGMA standards. Results (geometry and material strength requirement) obtained are well in agreement with existing knowledge regarding gear design and the methods used provide means for systematic exploration of the design space.

We have collated all feasible designs obtained during the analysis and made ternary plots of average values of goals and their standard deviation. These charts show the regions of compromise. These charts can act as guideline, which should be helpful to the gear designer in making an informed choice among the various design objectives. This work also demonstrates two of key features of PREMΛP, viz., the robust design and exploration of geometric design and material space (through UTS) using cDSP for design of gears as an example.

Acknowledgments Authors acknowledge the support provided by management of TCS in pursuing this research. Janet Allen gratefully acknowledges financial support from the John and Mary Moore Chair at the University of Oklahoma. Farrokh Mistree gratefully acknowledges financial support from the L.A. Comp Chair at the University of Oklahoma.

References

1. Pollock TM, Allison J (2008) Integrated computational materials engineering: a transformational discipline for improved competitiveness and national security. The National Academies Press, Washington
2. Gautham BP, Singh AK, Ghaisas SS, Reddy SS, Mistree F (2013) PREMΛP: a platform for the realization of engineered materials and products. In: Proceedings of the 4th international conference on research into design, IIT Madras, Chennai, India, Paper Number: 63
3. Kumar P, Goyal S, Singh AK, Allen JK, Panchal JH, Mistree F (2013) PREMΛP: exploring the design space for continuous casting of steel. In: Proceedings of the 4th international conference on research into design, IIT Madras, Chennai, India, Paper Number: 65
4. Bhat M, Das P, Kumar P, Kulkarni N, Ghaisas SS, Reddy SS (2013) PREMΛP: knowledge driven design of materials and engineering processes. In: Proceedings of the 4th international conference on research into design, IIT Madras, Chennai, India, Paper Number: 66
5. Allen JK, Seepersad C, Choi H-J, Mistree F (2006) Robust design for multiscale and multidisciplinary applications. J Mech Des 128:832–843
6. Gear industry vision: a vision for the gear industry in 2025 (2004) Developed by the gear community, (http://agma.server294.com/images/uploads/gearvision.pdf: as on 27 May 2012)
7. Cavallaro GP (1995) Fatigue properties of carburized gear steels, Ph D Thesis, Gartrell School of Mining, Metallurgy and Applied Geology, University of South Australia, Adelaide
8. Rolander N, Rambo J, Joshi Y, Allen JK, Mistree F (2006) An approach to Robust design of turbulent convective systems. J Mech Des 128:844–855
9. Mistree F, Hughes OF, Bras BA (1993) The compromise decision support problem and adaptive linear programming algorithm. In: Kamat MP (ed) Structural optimization: status and promise, AIAA, pp 247–286
10. Design guide for vehicle spur and helical gears, ANSI/AGMA 6002-B93
11. Fundamental rating factors and calculation methods for involute spur and helical gear teeth, ANSI/AGMA 2001-D04

PREMAP: Exploring the Design Space for Continuous Casting of Steel

Prabhash Kumar, Sharad Goyal, Amarendra K. Singh, Janet K. Allen, Jitesh H. Panchal and Farrokh Mistree

Abstract Continuous casting is a crucial step in the production of a variety of steel products. Its performance is measured in terms of productivity, yield, quality and production costs, which are conflicting. In this paper an integrated design framework has been developed based on metamodels and the compromise Decision Support Problem (cDSP) for determining a robust solution. Further, the design space for continuous casting has been explored to determine robust solutions for different requirements. Moreover, the utility of the framework has been illustrated for providing decision support when an existing configuration for continuous casting is unable to meet the requirements. This approach can be easily instantiated for other unit operations involved in steel manufacturing and then can be used to

P. Kumar · S. Goyal · A. K. Singh (✉)
Tata Consultancy Services, Pune 411013, India
e-mail: amarendra.singh@tcs.com

P. Kumar
e-mail: kumar.prabhash@tcs.com

S. Goyal
e-mail: sharad.goyal@tcs.com

J. K. Allen
School of Industrial and Systems Engineering, University of Oklahoma, Norman, OK 73019, USA
e-mail: janet.allen@ou.edu

J. H. Panchal
School of Mechanical Engineering, Purdue University, West Lafayette, IN 47907, USA
e-mail: panchal@wsu.edu

F. Mistree
School of Aerospace and Mechanical Engineering, University of Oklahoma, Norman, OK 73019, USA
e-mail: farrokh.mistree@ou.edu

A. Chakrabarti and R. V. Prakash (eds.), *ICoRD'13*, Lecture Notes in Mechanical Engineering, DOI: 10.1007/978-81-322-1050-4_60, © Springer India 2013

integrate the host of operations for the development of materials with specific properties and the combined design of products and materials. This enables an integrated simulation based design framework, PREMΛP, and will lead to a paradigm shift in the manufacturing industry.

Keywords Robust design · Compromise Decision Support Problem · Continuous casting of steel · Metamodels

1 Introduction

Global market trends are leading to continuously increasing demand for high quality and cost effective steel in order to meet the requirements and challenges posed by other advanced materials and increasingly demanding industries such as the automobile manufacturing industry. To remain competitive and to survive, steel manufacturers need to address the challenges of increasing productivity and quality as well as reducing production cost. This requires careful control of individual process parameters as well as improvement in the whole steel making process route [1].

In order to address such complex problems, efforts are going on at Tata Consultancy Services to develop a platform, "PREMΛP-Platform for the Realization of Engineered Materials and Products", based on an integrated systems engineering approach. It is expected to: (a) reduce the time and cost of discovery and development of materials and their manufacturing processes and (b) enable faster development of products augmented with richer material information. Detailed discussion on the development, platform architecture and components of PREMΛP is presented in Refs. [2] and [3].The efficacy of using the cDSP is illustrated using two foundational problems: (a) development of steel mill products and (b) integrated design of steel gears. Details of integrated design of steel gear are discussed in Ref. [4] and in this paper we discuss on development of steel mill products with a focus on illustrating two features of PREMΛP, namely, robust design [5] and the exploration of the design space using the cDSP for continuous casting of steel. The cDSP has been shown to be an effective construct for robust design [6–8].

Steel manufacturing involves a host of complex processing steps such as secondary steel making, continuous casting, rolling, heat treatment, etc. Among these, the continuous casting operation is a crucial step involved in production of a variety of steel products from liquid steel, for example slabs and billets, which are subsequently used for sheet and gear manufacturing respectively. The performance of the slab casting process is generally assessed using parameters such as productivity, yield, quality of slab and cost of production. All of these parameters need to satisfy very stringent requirement norms. But, achieving this is very difficult as these requirements are often conflicting. For example, if design variables are selected so as to maximize the productivity, then quality goes down and vice versa. Thus, we need to determine the design variables that balance these conflicting requirements. In principle, one could explore the effects of design variables through experiments and plant trials. However, the slab quality is governed by complex physics of heat transfer, fluid flow, mass transfer, stress evolution, etc. and the highly non-linear nature of these interacting phenomena. Hence obtaining these parameters through experimentation is very costly, time consuming and nearly impossible. It is quite common to sacrifice productivity to meet the quality specifications and run the process sub-optimally.

Considering its importance, continuous casting of slab has been widely studied and the latest edition of "Making, Shaping and Treating of Steel" [9] provides the current advances in technologies as well as present understating of various phenomena governing this process. There have been numerous studies on mathematical modeling of various phenomena related to continuous casting of steel and these are aimed at understanding the complex nature of the process. Very few studies, e.g., Ref. [10], use this information to get better control of the design variables by optimization. Even then, the solutions thus obtained are of limited scope owing to the fact that these constructs seek optimum rather than robust solutions. Due to the unsteady nature of the process, optimum solutions can become unsatisfactory with small changes in the inputs. Hence, there is a need to designing robust solutions.

In this paper, we describe how we explore the design space for this important intermediate operation of slab casting. Comprehensive mathematical models of continuous casting operations have been used to develop metamodels which are used to predict shell thickness, segregation index, temperature at critical locations, oscillation mark depth, etc., as a function of different design variables. These metamodels are then integrated with the cDSP to develop an integrated design framework. This formulation uses performance and robustness related goals in terms of productivity and quality and explores the design space bounded by constraints and limits for a set of design variables such as superheat, casting speed, mold oscillation frequency, cooling conditions in different segments of strand, etc.

Our aim is to demonstrate the potential of mathematical modeling and a multi-objective robust formulation in an integrated design framework supported by PREMΛP. We discuss the potential of the developed integrated design framework to provide insight and answer some of the key questions with respect to the casting process, such as: (a) What values of design variables should be used for a casting process to meet quality and productivity requirements, (b) Can meet these requirements with the existing configuration? If not, what configuration changes are required?

The significance of the above work is that it can be easily instantiated for other unit operations involved in steel manufacturing and can be used to integrate the host of unit operations, supported by PREMΛP.

2 Problem Description and Solution Strategy

2.1 Overview of Continuous Casting Process

A schematic diagram of continuous casting process is shown in Fig. 1 [9]. The process is briefly described here. As hot melt from tundish enters the continuous casting mold through the submerged entry nozzle,it gets cooled by the heat extracted through the mold (primary cooling) and starts to solidify at the metal-mold interface. The mold oscillates in order to avoid sticking (as shown in the inset (a) of Fig. 1 [9]). As superheat of the melt from tundish varies with time, the casting speed is dynamically adjusted. The thickness of solidified layer (shell thickness) keeps on increasing as it moves down the mold. Typically, the mold height is of the order of 1 m and the shell thickness at mold exit should be such that it can withstand the ferrostatic head of the liquid metal. Beyond the mold exit, the movement of the slab is guided and supported by rollers and it is cooled by water with the help of spray nozzles. There are several segments of rollers, varying in roll pitch and diameter. In this paper, we have considered only four segments of rolls, which are clearly depicted in Fig. 1, where the spray cooling by water, the secondary cooling, is done. In this area, a small amount of cooling also takes place by air (natural convection) and support rolls (conduction).As the slab moves further, solidification progresses only due to natural convection and roll

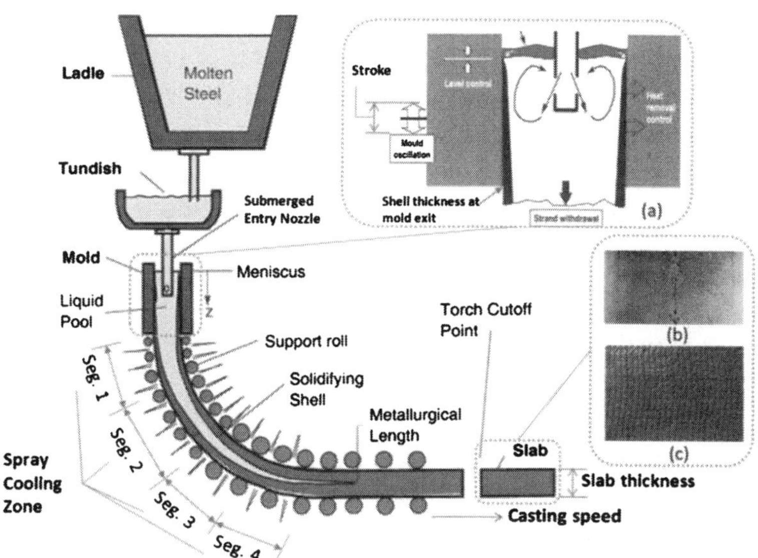

Fig. 1 Schematic diagram of continuous casting process with inset (a) mold oscillation, (b) segregation and (c) oscillation mark [9]

conduction. At the end of the process, the solidified steel is cut to predefined lengths into slabs. Details of the process are reported in Ref. [9].

As mentioned earlier, the performance of the caster is generally assessed in terms of productivity and quality. The productivity of the caster is dependent on casting speed and width and thickness of the slab. Casting speed, in turn, is dependent on the superheat of the melt received from tundish and the required grade of steel. The overall quality of the slab depends on both qualitative and quantitative parameters. In order to define slab quality in a way which can be incorporated in a mathematical framework, we have focused on estimating key quantitative parameters such as segregation index and oscillation mark depth. These parameters are shown in inset (b) and (c) of Fig. 1. Some of the qualitative aspects of quality have been incorporated as constraints and bounds. These are obtained by experience and from the literature [10]. For example, a constraint on shell thickness at the mold exit ensures that there are no breakouts or excessive bulging, which, in turn, ensures better surface and internal properties of the slab.

2.2 Problem Statement

Our objective is to obtain a robust set of design variables, namely, superheat, casting speed, slab width, cooling conditions off our consecutive segments in the spray cooling zone (as shown in Fig. 1) and the mold oscillation frequency for slab casting, to meet the conflicting requirement of maximizing both productivity and quality. Minimum productivity and quality required are 3,000 t/day and 0.6 respectively (refer to Eqs. 18 and 19 presented later). In addition to this, some other constraints also must be satisfied such as segregation index should be less than 2.5, the metallurgical length should be less than 18.84 m, the shell thickness at mold should be greater than 10 mm, reheating (refer to Fig. 3) in each segment should be less than 100 °C, temperature at the unbending point should avoid the ductility trough, oscillation mark depth should be less than 0.25 mm and cooling condition in successive segments of the strand should be in decreasing order.

2.3 Solution Strategy

In this context, where we need productivity and quality and simultaneously minimize their variation for robust solution while satisfying a set of constraints, a mathematical construct capable of handling multiple objectives and constraints is required. For this purpose, the cDSP construct is used. The steps of the solution strategy are given in Fig. 2. Detailed simulation models have been developed for various problems and sets of requirements. Based on these models, Response Surface Models (RSM) have been developed. These response surface models have been integrated with the cDSP to develop an integrated design framework. Later

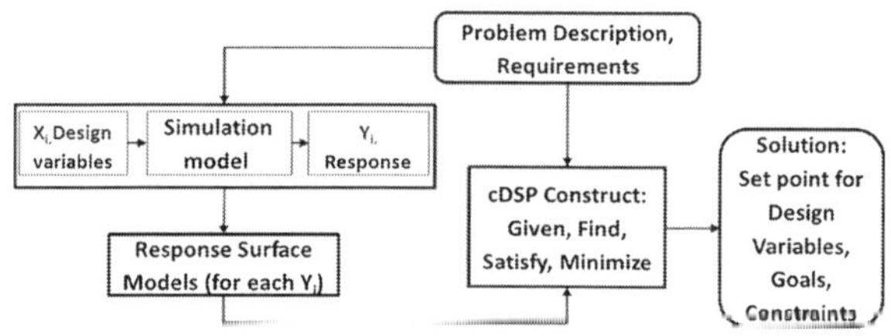

Fig. 2 Flow diagram showing the solution strategy

these are used to explore the design space for continuous casting and arrive at design variables to meet the conflicting requirements of both productivity and quality while satisfying the constraints.

3 Models

3.1 Mathematical Model for Heat Transfer

Mathematical models have been used to model the underlying physical phenomena and identify several additional parameters apart from process and design parameters that play an important role in quality control of the slab. A transient, 2-D finite-difference based heat transfer model has been developed to get the temperature evolution profile, shell thickness at mold exit and segregation index. The formulation is based on the fundamental heat transport equation [9] and modified Scheil's equation [11]. It is assumed that heat flow by conduction in the axial direction is low compared to the heat flow by bulk movement of slab in the axial direction. This reduces the problem to 2-D. Also, due to symmetry, only a quarter of the full cross-section of the slab has been modeled. Appropriate boundary conditions have been used in each zone. The trends observed in results are in good agreement with the data published in literature [10]. A typical temperature profile on surface of slab along the casting direction is shown in Fig. 3.

3.2 Meta Models

Ideally comprehensive and detailed mathematical models should be used for design to get high levels of accuracy. But, this may require intensively high computational power and time for analysis. Compared to these, metamodel approximations are significantly cheaper to run and can be easily linked to optimization algorithms for fast design exploration. Although, by doing this, some information is lost and accuracy decreases, metamodels can be used in the preliminary stages of design to reduce the design search space and later more comprehensive and detailed mathematical models can be used in the reduced design space to obtain accurate results.

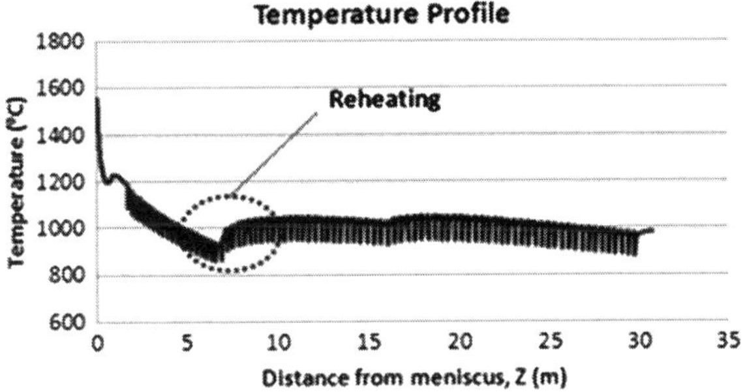

Fig. 3 Typical temperature profile on surface of slab along casting direction

Polynomial response surfaces are the most widely used approximating functions for constructing metamodels. We use a second-order polynomial function as shown in Eq. 1 for the response surface models for all the responses generated from the heat transfer model. The constants in the Eq. 1 are calculated using least square regression analysis of the data generated using the design of experiments. Some of the constant terms in a response surface model have been neglected because they are small as compared to other constant terms. The details of constructing response surface models and the design of experiments are described in Ref. [12]. The ranges of design variables for which these metamodels are developed are shown in Table 1. A summary of all the response surface models is presented in Table 2. The responses studied include segregation index (SI), shell thickness at mold exit (ST_{ME}), metallurgical length (ML), i.e., distance from meniscus to the point of complete solidification (see Fig. 1), temperature at unbending point (T_{UP}), reheating in segment 1 ($T_{R,1}$), reheating in segment 2 ($T_{R,2}$), reheating in segment 3 ($T_{R,3}$) and reheating in segment 4 ($T_{R,4}$). All the design variables are used in normalized form from 0 to 1 in the response surface models. For this example, a low carbon steel with composition (wt%) of C-0.055, S-0.005, P-0.011, Si-0.012, Mn-0.45 and Al-0.053 has been considered.

$$f(x_i) = a_0 + \sum_{i=1}^{n} a_i x_i + \sum_{i}^{n} a_{ii} x_i^2 + \sum_{i=1}^{n} \sum_{j>i}^{n} a_{ij} x_i x_j \quad (1)$$

Table 1 Range of design variables

Sr.No.	Design variables	Symbol	Range
1	Super heat	X_1	10–45 °C
2	Casting speed	X_2	0.8–2.5 m/min
3	Slab width	X_3	1,000–1,550 mm
4	Cooling condition in segment 1	X_4	236–720 W/m^2 K
5	Cooling condition in segment 2	X_5	333–450 W/m^2 K
6	Cooling condition in segment 3	X_6	331–420 W/m^2 K
7	Cooling condition in segment 4	X_7	172–235 W/m^2 K
8	Mold oscillation frequency	X_8	95–250 cycles/min

Table 2 List of response surface models

Response surface models	
SI ($R^2 = 0.89$)	$=(0.01142 + 0.00098\ X_1 + 0.00129\ X_2 - 0.00077\ X_3 + 0.00094\ X_4 - 0.00046\ X_1^2 - 0.00091\ X_2^2 + 0.00026\ X_3^2 - 0.00018\ X_4^2 - 0.00026\ X_1X_2 + 0.00019\ X_1X_4 - 0.00048\ X_2X_4 + 0.00024\ X_2X_5 + 0.0001\ X_2X_6)/0.005$
ST_{ME} ($R^2 = 0.99$)	$=(0.02241 - 0.00436\ X_1 - 0.01641\ X_2 + 0.00558\ X_2^2 + 0.0024\ X_1X_2)$
ML ($R^2 = 0.99$)	$=(9.775 + 0.613\ X_1 + 18.412\ X_2 - 1.132\ X_4 - 0.4\ X_5 + 4.663\ X_2^2 + 0.248\ X_4^2 + 1.474\ X_1X_2 - 0.767\ X_2X_4 - 0.629\ X_2X_5 - 0.66\ X_2X_6 - 0.651\ X_2X_7)$
T_{UP} ($R^2 = 0.99$)	$=(886.68 + 13.55\ X_1 + 502.79\ X_2 - 29.29\ X_4 - 27.74\ X_7 - 261.82\ X_2^2 + 14.62\ X_2X_4)$
$T_{R,1}$ ($R^2 = 0.98$)	$=(163.42 + 3.5\ X_1 - 14.5\ X_2 - 190\ X_4 + 0.61\ X_1^2 + 43.58\ X_2^2 + 81.3\ X_4^2 + 0.19\ X_1X_2 - 2.50\ X_1X_4 - 36.18\ X_2X_4)$
$T_{R,2}$ ($R^2 = 0.99$)	$=(104.44 + 2.11\ X_1 - 84.66\ X_2 + 60.32\ X_4 - 30.35\ X_5 + 2.42\ X_1^2 + 30.93\ X_2^2 + 18.9\ X_4^2 + 6.3\ X_5^2 - 0.85\ X_1X_2 - 4.44\ X_1X_4 + 5.93\ X_2X_4 + 6.17\ X_2X_5 - 24.7\ X_4X_5)$
$T_{R,3}$ ($R^2 = 0.87$)	$=(0.7557 - 0.10625\ X_1 - 1.0887\ X_2 + 0.15201\ X_4 + 8.93837\ X_5 - 12.372\ X_6 + 1.1766\ X_1^2 + 0.63\ X_2^2 - 0.1175\ X_4^2 + 9.2787\ X_5^2 + 6.9296\ X_6^2 + 0.01658\ X_1X_2 - 1.9403\ X_1X_5 - 0.08636\ X_2X_4 - 1.637\ X_2X_5 + 3.0596\ X_2X_6 + 6.2287\ X_4X_5 - 17.5402\ X_5X_6)$
$T_{R,4}$ ($R^2 = 0.97$)	$=(125.82 + 2.39\ X_1 - 32.51\ X_2 - 3.82\ X_4 + 0.324\ X_5 + 27.87\ X_6 - 46.64\ X_7 + 2.196\ X_2^2 + 0.166\ X_4^2 - 1.119\ X_5^2 + 0.4\ X_6^2 + 8.219\ X_7^2 - 3.841\ X_1X_2 + 8.499\ X_2X_4 + 4.68\ X_2X_5 + 7.993\ X_2X_6 + 1.185\ X_4X_5 + 1.958\ X_5X_6 - 5.848\ X_6X_7)$

In addition to the response surface models, an empirical equation for oscillation mark depth (d) has been utilized, which is shown in Eqs. 2 and 3 [13]. Here, t_N, v_{cast}, f, S stand for negative strip time (s), casting speed (m/min), mold oscillation frequency (cycles/min) and mold stroke (mm) respectively. The oscillation mark depth has a significant effect on surface quality as it can act as a nucleation site for surface cracking and transverse cracks.

$$d = 0.065 \times 1.145^s \times (200 \times 0.9^s)^{t_N} \tag{2}$$

$$t_N = \frac{60}{\pi f} \cos^{-1} \frac{1,000\, v_{cast}}{\pi f s} \tag{3}$$

4 The cDSP for Exploration of Design Space of Continuous Casting of Slab

The detailed mathematical formulation of the cDSP for the problem discussed is given in Table 3. Some of the variables such as slab thickness, downtime for casting, density of steel (ρ), mold stroke and variance (σ^2) for all the eight design variables

Table 3 The cDSP for continuous casting of slab

Given:

Fixed parameters: Slab thickness = 210 mm, $\rho = 7.8$ g/cc, mold stroke = 6 mm, caster downtime = 1 h., $\sigma^2_{X_1} = 0.04$, $\sigma^2_{X_2} = 0.01$, $\sigma^2_{X_3} = 0.03$, $\sigma^2_{X_4} = \sigma^2_{X_5} = \sigma^2_{X_6} = \sigma^2_{X_7} = 0.01$ and $\sigma^2_{X_8} = 0.002$

Response surface models (Table 2) and equations (Eq. 2, 4, 18, 19)

Maximum and minimum value of each goal (calculated based on range of design variables), $P_{max} = 8{,}800$ t/day, $P_{min} = 1{,}800$ t/day, $\sigma^2_{P,max} = 650{,}000$ (t/day)^2, $\sigma^2_{P,min} = 170{,}000$ (t/day)^2, $Q_{max} = 1$, $Q_{min} = 0.08$, $\sigma^2_{Q,max} = 0.7$, $\sigma^2_{Q,min} = 0$

Target for each goal (normalized based on maximum and minimum value, 0–1): $P_{Target} = Q_{Target} = 1$ (maximize), $\sigma^2_{P,\,Target} = \sigma^2_{Q,\,Target} = 0$ (minimize)

Number of design variables = 8, Goals = 4, Constraints = 14

Find:

The value of design variables: X_i, i = 1,….,8

The value of deviation variables d_i^+, d_i^-, i = 1,….,4

Satisfy:

Constraints:

Shell thickness (m): $ST_{ME} - 0.01 \geq 0$ (4)

Oscillation mark depth (mm): $d - 0.25 \leq 0$ (5)

Metallurgical length (m): $ML - 18.84 \leq 0$ (6)

Temperature at unbending (°C): $(T_{UP} - 800) \times (T_{UP} - 1{,}000) \geq 0$ (7)

Reheating (°C): $T_R - 100 \leq 0,$ *for all four segments* (8)

Segregation index: $(SI - 2.0) \times (SI - 2.5) \leq 0$ (9)

Cooling condition (W/m^{2}K): $(X_i - X_{i+1}) \geq 1, \quad for\, i = 4, 5, 6$ (10)

Productivity (tons/day): $(P - 3{,}000) \geq 0$ (11)

Quality: $(Q - 0.6) \geq 0$ (12)

Goals:

$[P(x_i) - P_{min}]/[P_{max} - P_{min}] + d_1^- - d_1^+ = 1$ (13)

$\left[\sum_i \left[\left(\frac{\partial P(x_i)}{\partial x_i}\right)^2 \times \sigma^2_{x_i}\right] - \sigma^2_{P_{min}}\right] / \left[\sigma^2_{P,max} - \sigma^2_{P,min}\right] + d_2^- - d_2^+ = 0$ (14)

$[Q(x_i) - Q_{min}]/[Q_{max} - Q_{min}] + d_3^- - d_3^+ = 1$ (15)

$\left[\sum_i \left[\left(\frac{\partial Q(x_i)}{\partial x_i}\right)^2 \times \sigma^2_{x_i}\right] - \sigma^2_{Q_{min}}\right] / \left[\sigma^2_{Q,max} - \sigma^2_{Q,min}\right] + d_4^- - d_4^+ = 0$ (16)

Bounds:

$0 \leq X_i \leq 1$, i=1,….,8 (Table 1)

$d_i^-.d_i^+ = 0$, $d_i^+,d_i^- \geq 0$, i = 1,….,4

Minimize:

The deviation function (Z): Archimedean formulation

$Z = \sum_{i=1}^{m} W_i(d_i^-, d_i^+); \sum_{i=1}^{4} W_i = 1, \; W_i \geq 0 \quad for\; i = 1, \ldots, 4$ (17)

have been fixed. The values for these have been determined based on literature review and experience. The metamodels have been used to formulate the constraints and goals. As discussed in Sect. 2.1, we have considered only two important performance goals, namely, productivity and quality for casting operation. Now, we discuss the mathematical formulation of these goals. Productivity [$P(x_i)$, tons/day] has been defined in terms of casting speed, thickness and width of slab, density of steel and caster downtime as shown in Eq. 18.

$$P(x_i) = x_2 \times x_3 \times \rho \times slab\,thickness \times (24 - caster\,down\,time) \times \frac{60}{10^6} \quad (18)$$

The Quality ($Q(x_i)$) of the slab has been formulated in terms of the normalized segregation index and oscillation mark depth from 0 to 1 as shown in Eq. 19. The weights for each of these parameters (W_{SI} and W_d) can be determined based on the application of the slab, i.e., if slab is being produced for an application which is highly sensitive to the segregation index, then a greater weight can be assigned to it. Further, other suitable quality indicator parameters such as the crack index can also be incorporated in this formulation, as appropriate. In this illustration, equal weights ($W_{SI} = W_d = 0.5$) have been given to both the segregation index and oscillation mark depth.

$$Q(x_i) = 1 - (W_{SI} \times SI(x_i) + W_d \times d(x_i)), \sum W_j = 1 \quad (19)$$

For the robustness of the solution, two other goals, variance in productivity (σ_P^2, Eq. 14) and variance in quality (σ_Q^2, Eq. 16) have also been incorporated and must be minimized. For all four goals, maximum and minimum limits have been calculated based on the range of design variables and these have been used to normalize the goals from 0 to 1.

The constraints are mainly metallurgical constraints (Eqs. 4–8). The details of these metallurgical constraints are reported in Ref. [10]. Considering economic aspects, plant practices and other requirements, some additional constraints are added (Eqs. 9–12). All the constraints must be satisfied to obtain a feasible solution.

Depending on the requirements for productivity, quality and robustness, different scenarios can be addressed by varying the weights given to each goal. Table 4 shows the different scenarios and the form of the corresponding deviation function (Z). Next we discuss the physical significance of these scenarios.

Scenario 1 represents a case where only productivity and quality goals are important. The process is controlled, so that there is very little variation in the design variables, and there is no concern for robustness. An equal trade-off between productivity and quality has been considered with constraints on them (Eqs. 11 and 12).

Table 4 Design scenarios for continuous casting of slab

Scenario no.	Scenario description	Deviation function
1	Overall, without robustness	$Z = (d_1^- + d_3^-)/2$
2	Productivity and its robustness	$Z = (d_1^- + d_2^+)/2$
3	Quality and its robustness	$Z = (d_3^- + d_4^+)/2$
4	Overall, with robustness	$Z = (d_1^- + d_2^+ + d_3^- + d_4^+)/4$
5	Ladle changeover	$Z = (d_3^- + d_4^+)/2$

Scenario 2 represents a case where there is an equal importance of productivity and its variance. Quality and its variance are either not needed or they are already in an acceptable range. For this, a constraint has been added on quality (Eq. 12).

In Scenario 3 quality and its variance are equally weighted. Productivity and its variance are not considered, but a constraint on productivity (Eq. 11) has been added to maintain a critical level of production.

Scenario 4 represents an equal trade-off among all goals. Productivity and quality has been constrained in order to satisfy minimum requirements (Eqs. 11 and 12).

Scenario 5 deals with ladle changeover during the casting operation where an empty ladle is being replaced by a filled ladle. During ladle changeover, maintaining the quality of slab is important. For this, an equal trade-off between quality and its variance have been considered with a constraint on quality (Eq. 12). It should be noted that this scenario is different from Scenario 3 because the bounds on the design variables have changed. Casting speed is lowered (0.8–1.31 m/min) to ensure proper transition and the superheat also decreases (10–35 °C).

5 Results and Discussion

The values of design variables and goals achieved are shown in Table 5. These are in close agreement with existing plant practices. Moreover, we observe a different solution in terms of design variables as well as goals, for all the five scenarios discussed earlier. This shows the importance of goals and helps in understanding the trade-offs among them. It is clear that slab casting process performed using design variables which are in agreement with the requirements, i.e., design variables should be fixed for operation depending on the requirements of productivity and quality.

In Fig. 4a and b, we show the convergence in the values of deviation variables and goals achieved respectively with successive iterations for Scenario 1. Similar trends are observed for other scenarios also. As expected, in Scenario 1, we see a decrease in value of the deviation function and increase in the value of goals (productivity and quality) achieved with successive iterations. It is important to note that as desired goals for productivity and quality are being maximized, their variances are also increasing. This means that even though the productivity and quality is high, the solution obtained is not robust because the variance is also high. In order to obtain robust solution greater weights must be given to the variance in productivity and variance in quality. This is explored in Scenarios 2, 3 and 4. On comparing the results of Scenario 1 with that of Scenario 4, we see more robust solution in Scenario 4 where variance is less but productivity and quality decreases, again this is as expected. Thus it is important to assign appropriate weights to each goal as per the requirements.

In Scenario 5 where ladle changeover has been considered, we observe that the quality reduced from 0.82 (Scenario 3) to 0.74. This indicates that with the existing

Table 5 Design exploration results for continuous casting of slab

	Scenario 1	Scenario 2	Scenario 3	Scenario 4	Scenario 5
Design variables					
X_1 (^{0}C)	10	10	10	10	26
X_2 (m/min)	1.68	1.75	1.34	1.62	1.31
X_3(mm)	1,550	1,000	1,550	1,270	1,550
X_4 (W/m^{2}K)	420	694	411	418	419
X_5 (W/m^{2}K)	419	450	350	333	361
X_6 (W/m^{2}K)	417	412	331	331	331
X_7 (W/m^{2}K)	235	235	195	235	196
X_8 (cycles/min)	95	95	250	250	250
Goals					
Productivity (tons/day)	5,897	3,960	4,705	4,637	4,590
Standard deviation in productivity (tons/day)	700	540	662	600	659
Quality	0.84	0.64	0.82	0.77	0.74
Quality variance	0.1574	0.1678	0.0029	0.0029	0.0013
Deviation function (Z)	0.2955	0.4707	0.0972	0.3119	0.1448

plant configuration, it is not possible to meet higher quality (more than 0.74) requirements during ladle changeover. Even to maintain a low quality, the set point for superheat goes up from 10 °C (Scenario 3) to 26 °C, with other parameters remaining more or less the same, which is economically not suitable. To understand the modifications in the plant configuration needed to meet the quality requirements, we looked at the constraint values. We observed that constraints on reheating and temperature at the unbending point are just satisfied. These all indicate that to maintain the same level of quality during ladle changeover as in normal operating condition (Scenario 3), the bounds for cooling conditions (X_4 to X_7) need to be decreased. Alternatively this can also be achieved

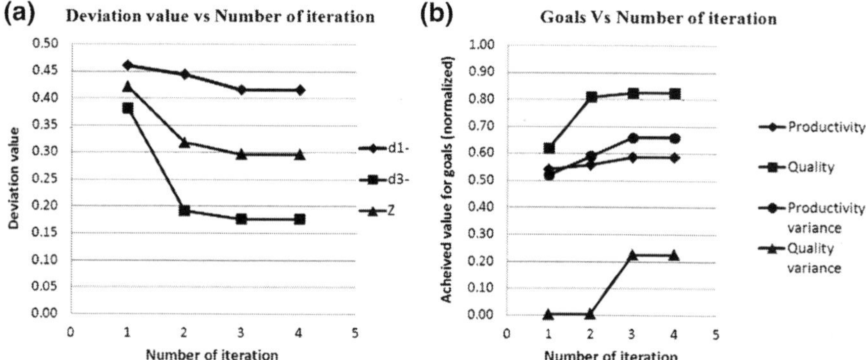

Fig. 4 Results of scenario 1 for variation in **a** deviation value and **b** goals with iterations

by incorporating some heating arrangements in the tundish to increase superheat. This leads to a guided change in existing plant configuration for meeting the requirements.

Further, it is important to observe that the superheat in all the cases (except Scenario 5) is close to its lower bound which is consistent with the basics. From a fundamental perspective, superheat should be as low as possible since higher superheat promotes segregation which affects the quality adversely. Also, high superheat limits the casting speed which adversely affects productivity. This has been observed in Scenario 2 where quality has not been considered yet superheat is low because it allows casting at a higher casting speed which increases productivity (goal for Scenario 2). Thus, it can be concluded that superheat should be low irrespective of the requirements.

In this paper, we have only dealt with a part of the process chain for the development of steel mill products [2]. Same approach can be used for other processes in the process chain and these can be used to integrate the processes for the development of materials with specific properties and the combined design of product and material. In future we will explore these issues. Along with these, we will also extend the present work to billet casting operation in order to address the challenges in gear manufacturing [2, 4].

6 Closing Remarks

An integrated design framework comprising of metamodels and a heuristic based cDSP has been developed. It requires various components such as simulation models, knowledge bases, optimization schemes etc., which are supported by PREMΛP. The design framework has been successfully employed to explore the design space of continuous casting of slab. The trade-offs between competing requirements has been clearly demonstrated. Using the framework, different scenarios based on different requirements have been analyzed and set points for design variables have been obtained. It is clear that the continuous casting operation for slab should have different sets of design variables for different set of requirements although superheat should be maintained at a lower value in each case. During ladle changeover, in order to maintain quality similar to that during normal operations, either the cooling pattern of the existing set up or configuration of tundish must go under a guided change. The cDSP can be further improved by improving upon models used and imposing more accurate constraints. More importantly, PREMΛP can be augmented for exploring the design space and determining the robust design of other unit operations involved in steel making.

Acknowledgments The authors thank TRDDC, Tata consultancy services, Pune for supporting this work. Janet Allen gratefully acknowledges financial support from the John and Mary Moore Chair at the University of Oklahoma. Farrokh Mistree gratefully acknowledges financial support from the L.A. Comp Chair at the University of Oklahoma.

References

1. Singh AK, Pardeshi R, Goyal S (2011) Integrated modeling of tundish and continuous caster to meet quality requirements of cast steels. In: 1st world congress on integrated computational materials engineering, vol 1, pp 81–85
2. Gautham BP, Singh AK, Ghaisas SS, Reddy SS, Mistree F (2013) PREMAP—a platform for the realization of engineered materials and products. In: Proceedings of the 4th international conference on research into design, IIT Madras, Chennai, India. Paper no.: 63
3. Bhat M, Shah S, Das P, Kumar P, Kulkarni N, Ghaisas SS, Reddy SS (2013) PREMAP—knowledge driven design of materials and engineering processes. In: Proceedings of the 4th international conference on research into design, IIT Madras, Chennai, India, Paper no: 66
4. Kulkarni N, Zagade PR, Gautham BP, Panchal JH, Allen JK, Mistree F (2013) PREMAP—exploring the design and materials space for gears. In: Proceedings of the 4th international conference on research into design, IIT Madras, Chennai, India. Paper no.: 64
5. Allen JK, Seepersad CC, Mistree F (2006) A survey of robust design with applications to multidisciplinary and multiscale systems. J Mech Des 128(4):832–843
6. Mistree F, Hughes OF, Bras BA (1993) The compromise decision support problem and the adaptive linear programming algorithm. In: Kamat MP (ed) Structural optimization: status and promise, Washington, D.C., AIAA, pp 247–286
7. Simpson TW, Chen W, Allen JK, Mistree F (1999) Use of the robust concept exploration method to facilitate the design of a family of products. In: Roy U, Usher JM, Parsaei HR (eds) Simultaneous engineering: methodologies and applications. Chapman-Hall, New York, pp 247–278
8. Choi H-J, McDowell DL, Rosen D, Allen JK, Mistree F (2008) An inductive design exploration method for robust multiscale materials design. J Mech Des 130(3):031402-1/13
9. Cramb AW (2010) The making, shaping and treating of steel-casting volume, 11th edn, AIST
10. Santos CA, Cheung N, Garcia A, Spim JA (2005) Application of solidification mathematical model and a genetic algorithm in the optimization of strand thermal profile along the continuous casting of steel. Mater Manuf Proces 20:1–14
11. Ghosh A (1990) Principles of secondary processing and casting of liquid steel. Oxford and IBH publishing Co. Pvt. Ltd
12. Myers RH, Montgomery DC, Anderson-Cook CM (2009) Response surface methodology: process and product optimization using designed experiments. Wiley, New York (Wiley series in probability and statistics)
13. http://www.ewp.rpi.edu/hartford/users/papers/engr/ernesto/lankep/THESIS/Supplemental/CC_UserGuide.pdf

Requirements for Computer-Aided Product-Service Systems Modeling and Simulation

Gokula Vasantha, Romana Hussain, Rajkumar Roy and Jonathan Corney

Abstract The design of product-service systems (PSS) is a co-production process which involves the manufacturer, customer and suppliers as well as any other stakeholders. This multi-organizational, collaborative environment along with the unique tangible and intangible characteristics of PSS demands a novel computer-based design platform. To identify the requirements for modeling and simulating PSS, existing academic software for PSS, namely, Service CAD integrated with a life cycle simulator (ISCL), Service Explorer and the commercially available system modeling software OPCAT™ are analyzed using a truck provider PSS problem. These platforms are used to represent PSS, model the factors and simulate the system in order to appreciate and compare the impact of different PSS designs. The evaluation reveals the scope of these platforms and proposes overall requirements for modeling and simulating PSS.

Keywords Product-service system · Modeling · Simulation

G. Vasantha · J. Corney
Design Manufacture and Engineering Management, University of Strathclyde, Glasgow, UK
e-mail: gokula.annamalai-vasantha@strath.ac.uk

J. Corney
e-mail: jonathan.corney@strath.ac.uk

R. Hussain · R. Roy (✉)
Manufacturing and Materials Department, Cranfield University, Bedfordshire, UK
e-mail: r.roy@cranfield.ac.uk

R. Hussain
e-mail: r.s.hussain@cranfield.ac.uk

A. Chakrabarti and R. V. Prakash (eds.), *ICoRD'13*, Lecture Notes in Mechanical Engineering, DOI: 10.1007/978-81-322-1050-4_61, © Springer India 2013

1 Introduction

Product-Service Systems (PSS) is defined as an integrated product and service offering that delivers value in use [1]. Goedkoop et al. [2] define a product-service system as—a system of products, services, networks of players and supporting infrastructure that continuously strives to be competitive, satisfy customer needs and have a lower environmental impact than traditional business models. A commonly sighted PSS example is the CorporateCare™ offering from Rolls-Royce Plc, UK [3]. CorporateCare™ is a comprehensive *cost-per-flight-hour* service, designed to deliver a highly competitive engine maintenance program to corporate customers. The key merits to the customers of this offering are low risk, predictable maintenance costs, increased aircraft availability and 24 h per day/365 days per year service support. However, Neely's [4] findings from the analysis of a large industrial database were that designing, implementing and managing this kind of offering is a huge challenge to the manufacturer as there is a distinct possibility of economic. However, in this challenging, globalized economy, PSS can help manufacturers to lock the customer into a long-term relationship, inhibit any replication of the customized offering by competitors, provide greater insights into how products are used and ultimately, if suitably delivered, also increase revenue. However, currently in industry, PSS conceptual design, in practice, is ad-hoc and it lacks a systematic approach. In order to support industry and designers in designing PSS, there has been an increasing interest with regards to developing theories, methodologies, tools and techniques.

A review of current PSS literature [5] reveals that the theories and methodologies to aid the design of PSS are still in their initial stages of development and substantial research is required to develop a practical PSS design methodology along with supporting tools. The interaction of the stakeholders in the design process and the unique characteristics of products, services, networks of players as well as the supporting infrastructure which is involved in the design of PSS, all demand a novel computer-aided modeling and simulation platform.

In this paper, we aim to enumerate the requirements for computer-aided conceptual modeling and the simulation of PSS designs. In order to identify the requirements for modeling and simulating PSS, existing academic software for PSS namely Service CAD integrated with a life cycle simulator (ISCL) [6], Service Explorer [7] and the commercially available system modeling software, OPCAT™ [8], are analyzed using a truck provider's PSS issue. ISCL and Service Explorer are greatly referred to within the PSS literature and OPCAT™ is an academic based system modeling platform which is based on the robust Object-Process Methodology (OPM). These platforms are used to represent PSS, model the factors and simulate the system in order to understand and compare the impact of different PSS designs. The evaluation reveals the scope of these platforms and proposes the overall requirements for modeling and simulating PSS.

2 PSS Design and Modeling Techniques

In the PSS literature, PSS design is not yet fully defined or commonly accepted. The research question 'what constitutes PSS design?' is yet to be answered and agreed upon by the PSS community. It is our opinion that PSS design can be defined as a process to synthesize and create sustained functional behavior through tangible products and intangible services [9]. Sustained functional behavior represents the degree to which a system can continuously achieve its purpose. PSS design involves the design of business models, the design of products and services, the design of processes and the interactions of all of the elements involved within the system. The primary motivation in PSS modeling is to co-create conceptual models that can be systematically shared and commonly interpreted by stakeholders.

Komoto and Tomiyama [6] have proposed that in PSS design processes, designers define an activity to meet a specified goal and quality and also define the environment as being the circumstance within which that activity is realized. We also argue that activity based mapping is an efficient representation for describing PSS design. By extending Komoto's and Tomiyama's PSS design process, we propose a representation for PSS design to be the mapping of each activity into inputs, outcomes, resources, competences, responsibilities, environmental variables and customer needs. These parameters are based upon the approach of capability-based PSS design [9]. A capability can be defined as the continuing ability to generate a desired operational outcome. Capabilities should be mapped depicting activities, outcomes and the reasons for deficiency. The core principles of this approach are:

- Fulfilling the customer's goal more closely by adjusting the customer's system to decrease system deficiencies [10].
- Understanding the capability shift between tangible and intangible elements and also between stakeholders.

Figure 1 details the activity based representation of PSS design. The sequence of activities is linked based on the satisfaction of preceding activity's outcomes. This representation highlights the necessary PSS parameters and the interactions amongst them.

To map and represent the parameters depicted in Fig. 1, a wide variety of representational techniques have been proposed in the literature. Representational techniques have been used in PSS design methodologies to define different processes involving products and services [11, 12]. They include Unified Modeling Language (UML), Structured Analysis and Design Technique (SADT), Functional Analysis, Service Blueprinting, Business Process Model and Notation (BPMN), Integrated DEFinition method of modeling functionality and information modeling (IDEF0), Data Flow Diagrams (DFD), Graphes à Résultats et Activités Interreliés (GRAI), Functional Flow Block Diagrams (FFBD), N2 charts, Behavior Diagrams, Design Structure Matrix (DSM) and Systems Modeling Language (SysML).

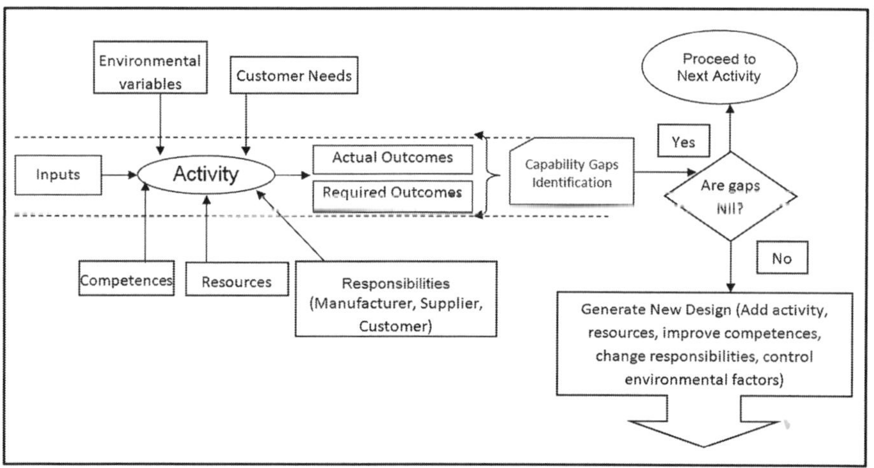

Fig. 1 Activity based representation of PSS Design

Detailed research studies are required to analyze and identify the suitability of these representational techniques in mapping the various parameters to define and present PSS design. Detailed elaboration and assessment of these techniques are out of the scope for this paper. In this paper, we aim to assess the representation used in three software platforms which could be potentially used for designing PSS: Service CAD integrated with a life cycle simulator (ISCL), Service Explorer, and OPCAT^{TM}. Along with these representational techniques, the assessment also aims to highlight the features required for designing and simulating PSS designs.

3 Truck Industry PSS Problem

In this competitive environment, Reliability, Availability, Maintainability and Safety (RAMS) are the four key performance indicators for the business-to-business activity for most large technical system manufacturers. The major goal of the truck company in question is to satisfy their customer's needs which are primarily based upon increasing truck uptime and lowering life cycle operational costs. Although the truck manufacturer has a large market share, they would like to become market leaders which they could attain by providing value-adding PSS offerings to their customers. An analysis of the existing capability gaps that their customers have and the identification of the issues with the trucks such as frequently occurring defects and the frequency and cost of maintenance (by analyzing historical maintenance data of the trucks) reveals that:

Table 1 PSS design solution for the truck problem

PSS design 1	PSS design 2	PSS design 3
Customer owns truck	Customer leases truck	Manufacturer owns truck
Guaranteed truck uptime and MOT pass with constant pricing throughout contract duration	Various services such as MOT and maintenance record management are provided in 'pay-as-you-go' mode	Customer pays for the amount of mileage they incur on the monthly basis

- The driver's skills (to drive correctly, that is, not commit too many driving faults such as harsh braking that could damage the truck) greatly influence truck performance, and hence services and life cycle cost.
- There is a requirement for an efficient IT system for maintaining truck records dealing with concerns such as inspection, maintenance periods and first time MOT pass.
- There is a lack of skills in performing regular maintenance checks and preliminary maintenance tasks.
- The cost of maintenance for frequent defects is very low, but the downtime due to frequent defects is very high.
- Different systems and components in trucks fail at different times creating difficulties in scheduling maintenance activities.
- If a preventive maintenance schedule is implemented focused primarily around the few critical systems, it could also lead to the replacement of several other components before they adversely affect the truck.
- There is difficulty in building relationships between how the vehicle is driven and the repairs that are incurred.

In order to add value to their offering to customers by eliminating or reducing these issues, various PSS designs have been generated. Three PSS designs have been chosen here to assess the aforementioned software are detailed in Table 1. Along with these designs, the parameters mentioned in Fig. 1 are used to describe software in the next section.

4 Illustrations of Chosen Platforms

4.1 Service CAD Integrated with a Life Cycle Simulator

ISCL [6] is a Python based computer-aided tool primarily developed to support the design of PSS. Its process modeling is based on service modeling and the simulation of process models with discrete event simulation are the main features of ISCL. ISCL consists of elements classified into Entity (which represents products and stakeholders), Attribute (which defines the state of entities), Activity (which represents any type of process), Scene (used to classify entity instances), and

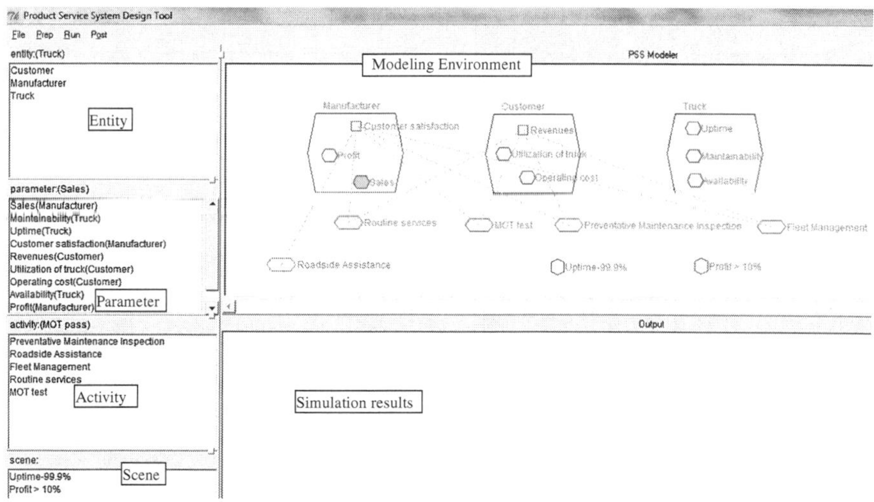

Fig. 2 The modeling environment of service CAD integrated with a life cycle simulator

Specification (which defines the requirements). Furthermore, ISCL structures the relations between these model elements. The elements and relationships are based on the concept that within PSS design, designers define the activity to meet a specified goal and quality and also define environment as being the circumstance within which that activity is realized. The "grammar" for modeling PSS design is programed into this platform and assists the designer in modeling and reasoning about the relationships between the elements.

Designers can create these elements on the canvas of ISCL and also move, inspect, and delete these elements on the canvas. Conditions and consequences are detailed for each activity to aid structure during discrete event simulation. Also activities can be sequenced to specify the simulation configuration. Activities treat scenes as the execution conditions and change the value of attributes and realize specifications. Simulation results are plotted in two-dimensional plots. Also, all simulation results can be downloaded in an excel sheet format (.csv). This software is currently in the public domain, downloadable and is actively supported by the developers. Figure 2 details the application of ISCL to the truck problem.

4.2 Service Explorer

Service explorer [7] was firstly developed for service engineering. Service explorer is a computer-aided tool which can be used to evaluate existing services and to design new services by allowing the visualization of the relationships from the customer requirements to the service delivery process. The elements of service are defined by the provider, receiver, contents and channel. A 'service' is defined as "an activity that a provider causes a receiver, usually with consideration, to change from an existing state to a new state that the receiver desires, where both contents and a channel are means to realize the service" [13]. It is based on the notion that service contents are provided by a service provider and delivered through a service channel. Depending on the business model, physical products are either the service contents or the service channel. The parameters are the Receiver State Parameter (RSP—this expresses the receiver's state), the Contents Parameter (CoP–is directly associated with the contents and factors in the RSP change), and the Channel Parameter (ChP—indirectly contributes to the RSP change with an action through the channel).

Service explorer structures a service model which consists of four sub-models: the 'flow' model (the "*who*"—the chain of agents who take roles of service providing and receiving), the 'scope' model (the "*what*"—the intended area of a service), the 'scenario' model (the "*why*"—the reasons for value)' and the 'view' model (the "*how*"—the functional structure for an RSP). An extended blueprint with Business Process Model and Notation (BPMN) is used to represent the view model. The design process is structured in the following steps: define the service target, represent the service target, describe the realization structures, evaluate a service, modify a service and create a service. Service explorer supports the development of the service case base, the design rule base and supports the reasoning for designers to construct service models. It also supports various techniques such as Quality Function Development (QFD), Analytical Hierarchy Process (AHP), DEMATEL, Activity Based Costing (ABC) and the Petri Net based technique (external simulator) for customer requirements and solution analysis as well as simulating the designed service process. As Service Explorer contains many features, the overall conceptual scheme underlying Service Explorer is detailed in Fig. 3.

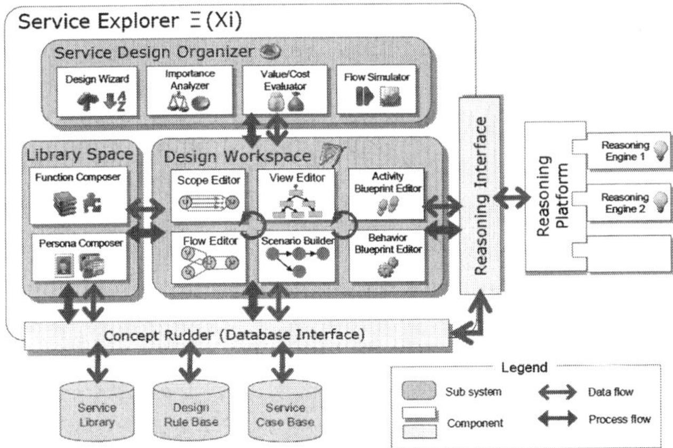

Fig. 3 The conceptual scheme underlying service explorer [2]

4.3 OPCAT^{TM}

The Object-Process Methodology (OPM) [8] is an underlying language and modeling approach for OPCAT software. OPM is a holistic approach for the conceptual modeling of complex systems. OPM is a modeling language based on a paradigm that views processes and objects as being equally important in the system model. OPM is modeled through three different types of entities: objects (things in the system that exist), processes (the things in the system that transform objects), and states. These elements are connected through structural (static time-independent relations) or procedural links (to describe the behavior of a system). The System Diagram (SD) is the topmost diagram in a model. It presents the most abstract view of the system, typically showing a single process as the main function of the system, along with the most significant objects that enable it (preconditions) and the ones that are transformed by it (results/effect/value). For each process the preconditions and the post-conditions of that process are described.

Features of in-zooming, unfolding and state expressing are used to show the various levels of detail which are represented of the system. These features aid scalability to system modeling and increase the information processing ability of designers. Also, different system views (a view is an assembly of system elements which specifies a certain aspect of the system) could be generated to explain the system to different stakeholders. The abstraction-refinement mechanism ensures that the context of a thing at any level of detail is never lost and the "big picture" is maintained at all times. The key benefit is that the OPM model integrates the functional, structural, and behavioral aspects of a system into a single, unified view, expressed bi-modally in equivalent graphics and text with a built-in refinement-abstraction mechanism. The OPM syntax specifies consistent ways by which entities can be connected via structural and procedural links, such that each legal entity-link entity combination bears specific, unambiguous semantics. Importantly it provides options for adding user text to define unidirectional and bidirectional between objects or processes. The OPCAT platform supports model repository, reuse engineering, simulation and validation of the model, and automatic document generation and management. Figure 4 illustrates application of the OPCAT in this truck industry problem.

5 A Comparison of PSS Modeling Environments

Table 2 compares the three platforms detailed in the previous section. This comparison indicates that operand, operator and the relationship definitions vary largely within the PSS domain itself. In order to avoid these problems with semantics it could be more fitting to use simple and precise terminologies as used in OPCAT^{TM} or to develop PSS ontology which should be commonly accepted

Table 2 Comparison of three platforms under consideration

Aspects	ISCL	Service explorer	OPCAT™
Operand	Specification and entities	Receiver state, content, channel parameters	Object and state
Operator	Activity	Service channel	Process
Relationship	Conditions and consequences	Four sub-models	Arrows—structural and procedural links
Representation	No formal visualization	BPML, blueprint	OPM
Application domains	PSS	PSS	System description/Simulation
Evaluation tool	Discrete event simulation	QFD, AHP, dematel, petri nets	Life span graph, time limits, GUI presentation with animation

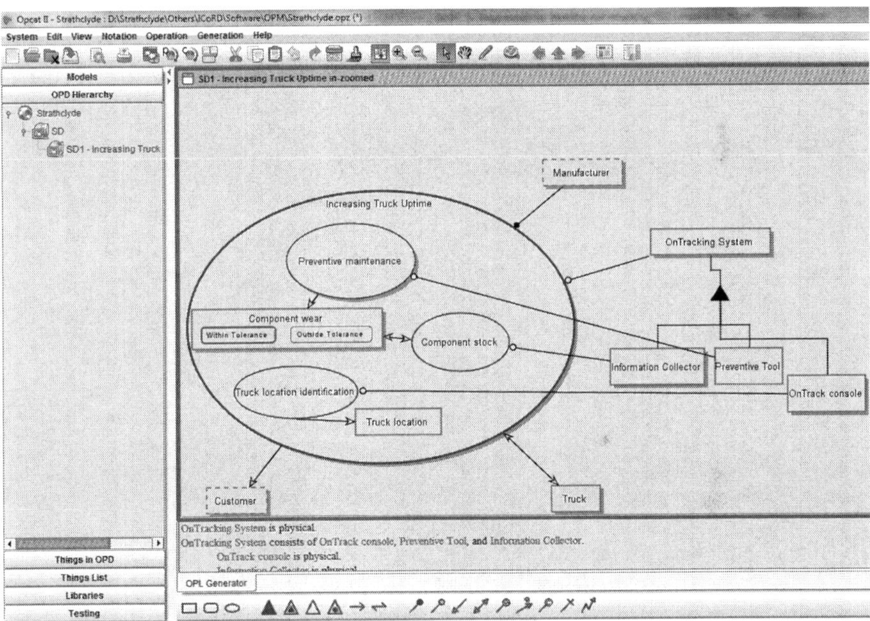

Fig. 4 The modeling environment of the OPCAT™ platform

within academics and practitioners. Even though all three platforms provide a rigid structure for defining the elements within a system, OPCAT™ facilitates users to define relationships between objects or processes. Information cognition loading for designers is suitably handled in the OPCAT™ platform which is facilitated by in-zooming, unfolding and state expressing features which aid to limit defining 3–6 processes in each SD. Also, the system representation which is expressed bi-modally in equivalent graphics and text facilitates comprehension and

Table 3 Requirements for computer-aided PSS modeling and simulation

Modeling requirements	ISCL	Service explorer	OPCAT^{TM}
Multiple views of the PSS design	Available with only single view point	Available with only single view point	Multiple views which specify different aspects of the system
Illustration of the 'need behind the need' of the customer	Links between the goals are not described	State change of the customer is described with RSP	Illustrated through overall functional need
Responsibility sharing and Preferences of the stakeholders	Could be mapped through conditions specification	Scope model aids these descriptions	Procedural links could define these parameters
Several PSS designs configurations	Each configuration should be individually modeled	Each configuration should be individually modeled	Each configuration should be individually modeled
Mapping of PSS design functions and processes	Goals, activities, parameters and scenes are structured	RSP, CoP, ChP, activities are mapped	Objects, processes and states are structured
Co-production process	Single user platform	Single user platform	Single user platform
Assistance to designer to generate PSS design	Assist in syntax checking	Service case base could be used	Assist in syntax checking
Activity precedence and sequences (sequential, concurrent, coupled or conditional)	Issues with sequential, and coupled representation	BPMN supports these processes	Issues with sequential, and coupled representation
Consistent notation and common interpretation	Common interpretation could be an issue	Common interpretation could be an issue	Graphics and text based support common interpretation
Representation of differences between the current and the designed system	Need to be compared separately	Need to be compared separately	Supports comparison through multiple frames
Simple, flexible and easy to maintain	Becomes complex with more elements	Becomes complex with more elements	Handles complexity with in-zooming
Redundancy and inefficiency identification	Not supported	Not supported	Provides checking mechanism
Predictability of PSS design outcomes	Predictability not guaranteed	Predictability not guaranteed	Predictability not guaranteed
Change impact analysis and risk analysis	Simulates through discrete event simulation	Simulates through Petri Nets	Simulates through life span graph

(continued)

Table 3 (continued)

Modeling requirements	ISCL	Service explorer	OPCAT™
Iteration loop for PSS design optimization	Not supported	Not supported	Not supported
Integrated quantitative and qualitative analysis	Qualitative need to be supported	Qualitative need to be supported	Qualitative need to be supported
Product and service improvements	Not supported	Not supported	Not supported

interpretation between the stakeholders. With regards to ISCL, this provides a very useful feature in linking discrete event simulation to evaluate each PSS design.

A detailed list of requirements for computer-aided PSS modeling and simulation is enumerated based on the application of a truck industry problem to the three software platforms, our understanding of PSS design methodology [9], the proposed representation structure (Fig. 1) and a thorough review of PSS literature. Table 3 reveals that the computer-aided platforms, to assist designers in developing PSS, could be improved with the following features: multiple views of the PSS design, PSS design configurations, the co-production process, activity precedence and sequences, gap analysis, complexity management, redundancy and inefficiency identification, predictability of PSS design outcomes, an iterative loop for PSS design optimization, support for qualitative analysis.

6 Conclusion

The analysis of three software platforms (ISCL, Service Explorer and OPCAT™) by their application to a truck industry issue has revealed the scope of these platforms; these findings are used as the basis for the overall requirements which have been proposed for a software platform which aids the modeling and simulation of PSS. A list of the improvements which are required has been analyzed and presented. Although this list could be added to by other researchers in the field, even in this initial form, it still provides a current benchmark against which future PSS software platforms can be evaluated. Furthermore, PSS design configurations and the predictability of PSS design outcomes are issues which require particular attention.

Acknowledgments The authors would like to express their gratitude to Professor Dov Dori, Professor Yoshiki Shimomura and Dr. Hitoshi Komoto for providing access to the software referenced within this paper.

References

1. Meier H, Roy R, Seliger G (2010) Industrial product–service systems–IPS2. CIRP Ann Manuf Technol 59(2):607–627
2. Goedkoop M et al (1999) Product service-systems, ecological and economic basics. Report for Dutch Ministries of environment (VROM) and economic affairs (EZ)
3. http://www.rolls-royce.com/civil/services/corporatecare/. Accessed on 17 July 2012
4. Neely A (2008) Exploring the financial consequences of the servitization of manufacturing. Oper Manage Res 1:103–118
5. Annamalai GV et al (2011) A review of product–service systems design methodologies. J Eng Des. doi:10.1080/09544828.2011.639712
6. Komoto H, Tomiyama T (2008) Integration of a service CAD and a life cycle simulator. CIRP Ann Manuf Technol 57(1):9–12
7. Sakao T et al (2009) Modeling design objects in CAD system for service/product engineering. Comput Aided Des 41:197–213
8. Dori D et al (2008) From conceptual models to schemata: an object-process-based data warehouse construction method. Inf Syst 33(6):567–593
9. Vasantha A, Vijaykumar G et al (2011) A framework for designing product-service systems. International conference on engineering design '11, Denmark
10. Hussain R et al (2012) A framework to inform PSS conceptual design by using system-in-use data. Comput Ind 63:319–327
11. Durugbo C, Tiwari A, Alcock JR (2010) A review of information flow diagrammatic models for product–service systems. Int J Adv Manuf Technol 52(9–12):1193–1208
12. Becker J et al (2010) The challenge of conceptual modeling for product–service systems: status-quo and perspectives for reference models and modeling languages. IseB 8(1):33–66
13. Tomiyama T (2001) Service engineering to intensify service contents in product life cycles. Second international symposium on environmentally conscious design and inverse manufacturing. IEEE Computer Society, Tokyo, pp 613–618

Designers' Perception on Information Processes

Gokula Annamalai Vasantha and Amaresh Chakrabarti

Abstract Understanding information needs and managing organizational information resources are vital to face competitive globalized industrial environment. Support development to aid these processes is often failed in real-time environments due to lack of in-depth awareness about organizational interactions. In this paper a systematic approach to understand the information processes and information sources available in an aerospace organization was studied through a questionnaire survey. The analysis reveals that designers perceive that they get the right information at the right time in only 4 or more out of 10 for most of the times. Also they are unable to differentiate among the types of interactions they perform during their daily activities, due to which the information processes occurring within the interactions are not perceptible to them. This perception illustrates there is substantial need for the development of support to create awareness and satisfy the information needs of designers.

Keywords Information processes · Resources · Questionnaire

1 Introduction

Presently organizations are heavily pressurized to develop products which should be novel and useful in order to meet customers' demands and wishes. There are various barriers which influence the companies from achieving these targets,

---

G. A. Vasantha (✉)
Design Manufacture and Engineering Management, University of Strathclyde, Glasgow, UK
e-mail: gokula.annamalai-vasantha@strath.ac.uk

A. Chakrabarti
Center for Product Design and Manufacturing, Indian Institute of Science, Bangalore, India
e-mail: ac123@cpdm.iisc.ernet.in

A. Chakrabarti and R. V. Prakash (eds.), *ICoRD'13*, Lecture Notes in Mechanical Engineering, DOI: 10.1007/978-81-322-1050-4_62, © Springer India 2013

notably the increasing attrition rate of employees and the increasing amount of time spent by employees in information acquisition and dissemination. In this globalizing era, information has been seen as the foremost resource in an organization. It has been argued that in order to maximize business-specific working information, all sources and forms of organizational information have to come into play [1]. Information will be potentially utilized, if it is properly managed. Managing information will potentially retain the competence held by the organization if information generated across its projects and units is captured, structured and reused. It is noted that capture, storage, transfer and exploitation of information play a critical role throughout the information cycle and the cost of not retrieving information is indeed high [2].

Besides retaining competence of an organization, information capture and reuse practises will be substantially helpful in communication between agents, in understanding design, in training novices and in avoiding repetition. It could also help satisfy information requirements of designers at the right time and in the right format. Various tools are proposed in the literature to support information needs of the designers but the acceptance rate is low. These may be due to 'technology push' from vendors without adequate understanding of the unique nature of the information processes that each organization possesses. These issues, especially those related to understanding of the information processes of organizations are need to be understand in the industrial context.

Since studies involving organizational environment are time- and effort-consuming activity and also its outcomes substantially impact the information management strategy of the organization and technology to be developed to support information processes, it is necessary to define precisely and clearly the objectives of the study. There is little practical guidance on defining the scope of this study considering the complexity and scale of the undertaking, and how to tailor it to individual circumstances and goals [3]. In this paper, a systematic approach for conducting informational study in an aerospace organization is detailed through a questionnaire survey along with earlier results obtained through personal observation method.

2 Related Literature

In their pioneering work on identifying information needs, Kuffner and Ullman [4] define design history as a representation of the evolution of a product from its initial specifications. They argue that in order to develop a usable design history, it is necessary to determine the types of information needed by designers when they attempt to understand a design. The two most significant findings from this work are: 51 % of the questions and conjectures were about old topics, and a high percentage of questions and conjectures were about the construction of features and components. Khadilkar and Stauffer [5] have shown that, for generating new product concepts using information from previous design effort, the designers used both conceptual and detail level information almost in equal proportions. Also, the number of queries about product construction and description accounted for almost half of the total queries, and the subject-class 'component level' received 43 % of the queries.

Information requirements of designers as found from existing literature [4, 6] are summarized as follows: test or analysis data, explanation of structure and function information, feature, location, operation of components, purpose of assemblies of components, customer preferences, cost, projected sales, projected manufacturing runs, maintenance constraints, materials, processes, tools, standards, patents, cultural trends, competition details, terminology, values, contacts and market information. Aurisicchio [7] investigates the nature of requests formed by designers during design processes and their associated searches. The categories found in the information 'request' group are: *objective, subject, response process, response type, directions of reasoning, and behavior type*. The main findings are the following: in the total number of requests recorded, in the group of *response process* the percentage of *retrieval-recognition* (finding and returning information) (50 %) is higher than *reasoning* (making an inference) (30 %) and *deliberation* (following paths of inference and weighing arguments) (20 %); and 70 % of the requests were sourced through interactions with colleagues.

Marsh [8] observes that the majority of the information is obtained from personal contacts rather than formal sources. Also he noted that designers spend on average 24 % of their time in information acquisition and dissemination. Crabtree et al. [9] point out that project delays are mainly due to time spent in information acquisition and information access. The associated delays range from a single day to a year. MacGregor et al. [10] observe that engineers use company systems and colleagues in the same office to get information, and engineers perceive that 34 % of their time is taken in sourcing and locating relevant information. The amount of time spent for these processes is substantially high, and need to be reduced. Capture and retrieval could play a significant role in reducing designers' time spent in information acquisition and dissemination.

Even though many results have been produced in the literature, the approaches used, number of subjects studied, subjects' background, organizational context, projects involved and its design stages and location lead to prevent developing a unified methodology for informational study and transferring results from an organization to another. Also, the number of research papers published in informational study (from engineering design context) from India is almost none. To fill this gap, this paper aims to study the nature of information processes and resources in an aerospace organization in India.

3 Research Questions and Approach

This study was conducted in a research and development organization focused primarily on design and development of special purpose aircraft. It is a thirty years old company with around 240 employees in a centre studied in this work. The designation of these employees is noted with scientist grades from 'A'–'H' (highest grade) and early starters with 'Research Fellow'. In order to understand the information processes and sources, personal observation method was initially planned. Framing and using research methods in an organization set-up is challenging and time consuming. Even though personal observation method provides in-depth information in real-time situations for this research, it has been heavily restricted by the organization concerned, and obtained little cooperation from its designers. The study has been limited to seven designers instead of the 18 designers originally planned. In order to understand the perspectives of other designers and to further validate the results obtained from the personal observations of seven designers, a questionnaire has been developed. The detailed results obtained from personal observation are presented in [11]. This questionnaire survey was carried out in the organization to validate the results obtained from the seven designers who were observed through the personal observation method. The questionnaire is developed in order to understand the perception, of a broader group of designers, about the following:

- The types, processes and characteristics of information handled by the designers,
- Types of documents and their characteristics handled by the designers and
- Interactions carried out by the designers leading to information activities and their characteristics.

The questionnaire also provides insights into some of the issues such as importance and satisfaction of purpose of the interactions, reasons for not using the documents, and usage and usability of the software, which cannot be observed from the personal observations. Since the usage of the term 'knowledge' annoyed the designers in the organization, we used the term 'information' throughout the questionnaire. Since this questionnaire is a successor to the observations made in the organization for one and half years, the types of information for this questionnaire were chosen based on the observation results and also the subjects are given the option to fill in any other information which they felt is missing in the questionnaire. The designers taking part in the actual survey were informed that the questionnaire takes about 30 min to complete.

Out of the 39 circulated, 21 filled questionnaires were returned. The response rate is 54 %. The response rate is high due to continuous push from the researcher and due to the adoption of various methods to improve response rate, such as sitting with the subjects, filling the response by the researcher while asking the questions to the subjects, as well as allowing subjects to fill the questionnaire alone. In this survey, the subjects left the columns unfilled if they thought these were inapplicable to their work. The number of years of experience of the subjects participated range from 1 to 23 years. The average number of years of experience is 12. The subjects are widely involved in original, adaptive and variant designs. Also, the subjects together cover various stages in the product development starting from conceptual to testing phases. In the analysis, the unfilled columns are separated in the name of 'not answered'. Subsequent sections elaborate results and analyzes obtained from the data collected through the questionnaire, and comparison of these results with the results obtained through personal observations.

4 Results

The results from analyzing the returned questionnaires are discussed in this section.

4.1 Types of Information Versus Information Processes

Figures 1, 2 show that the subjects agree about the importance of capturing and reusing all types of information. For capture, the least perceived importance are for cost of the product, and for installation, maintenance and service data. Frequency analysis reveals that information about behavior/operation, requirements, dimensions and performance of the product are the most frequently captured, while that for function, structure, materials, cost, assembly and, installation, maintenance and service data are the least frequently captured. Personal observations had revealed that property, component, feature, value and method were considerably needed by the designers. These results clearly show that there is mismatch between the information required and information captured by the designers.

The correlation coefficient between what is important for capture and reuse is moderate (0.58, $p < 0.01$). Some of the least preferred information for reuse are behavior/operation and cost of the product. Information about function/purpose, structure, components, assembly and performance of the product lead in the frequency of reuse. The correlation between the frequency of capture and reuse is very poor (0.19). Substantial support needs to be given in order to increase this

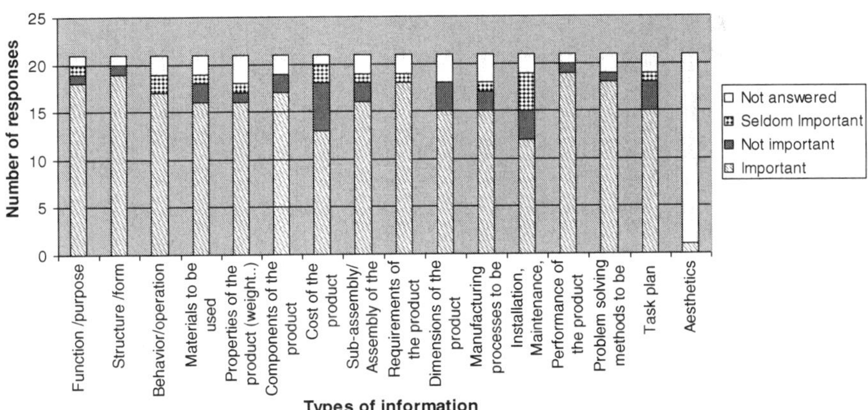

Fig. 1 Importance of capturing various types of information

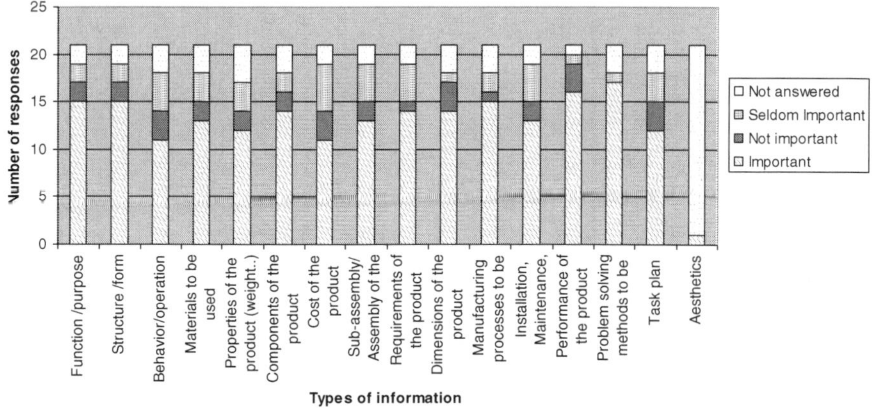

Fig. 2 Importance of reusing various types of information

coefficient of correlation; for instance, the designers' perception towards capture and reuse of information has to be changed. This is in line with the personal observations where on average only 10 % of the questions asked by designers are answered in the documents. Compared to the variation observed in personal observations for information needs, the replies in the questionnaire reveals that the designers perceive all types of information as important.

4.2 Interactions Versus Information Processes

The subjects perceived it important to interact with people having scientist grades for generating and sharing information. Even interactions with research fellows and contract employee are highly rated. Frequency of interaction with scientist grade equal to 'D' or below is highly rated for both generation and sharing of information. Communication mechanism is mostly mixed (verbal and written), for Scientists having above or below grade 'D' designation, within their project as well as outside the project. In contrast, interactions with research fellows and contract employees are mostly verbal. The written communications are very few in number for information generation and sharing, in all the designations. The frequency of written communication is higher with other directorates and communications outside the organization. Even though designers perceived that oral and written communications are mixed equally in the interactions, the personal observations reveal that oral communication dominates in various interactions among designers with all designations.

The subjects rated interactions within their group and directorate as important and frequent for both their project and for other projects, for generating and sharing information in the organization. This result is in line with the personal observations where most interactions occurred within the group and directorate.

4.3 Types of Interactions Versus Information Processes

Figures 3, 4 and 5 show that subjects rated all types of interactions except a few as important and frequent for information generation and sharing. The interaction 'One + Computer + Document' should be read as: 'designers interacting with computer and document'. The few interactions excluded are online chat, documents through post and video conferencing. The correlation between the importance and frequency of interactions for generating information is high (0.9, $p < 0.01$). The correlation between the importance and frequency of interactions for sharing information is very high (0.97, $p < 0.01$). These results are contradictory to the personal observations in which only interactions 'designer working with computer' and 'two designers working with a computer' frequently occurred. This possibly indicates that designers are unable to differentiate among the types of interactions they perform during their daily activities, due to which the information processes occurring within the interactions are not perceptible to them.

4.4 Time Consumed Versus Types of Interaction

The main interactions in which designers spend more than one hour in each day for generating information are 'alone', 'designer with a document', 'designer with a computer', and 'designer with a computer and document'. This result is in line with the personal observations, where 'designer working with computer' and 'two designers working with a computer' occupied most of the time during designing. The major interactions in which less than an hour is spent for sharing information are: 'designer interacting with another', 'designer interacting with another and notebook or documents', 'designer interacting with another in front of computer and/or document', and, meetings. This result is also in line with the personal observations, in which 'designer interacting with another' and 'designer interacting with another in front of computer' occupied most of the time for sharing

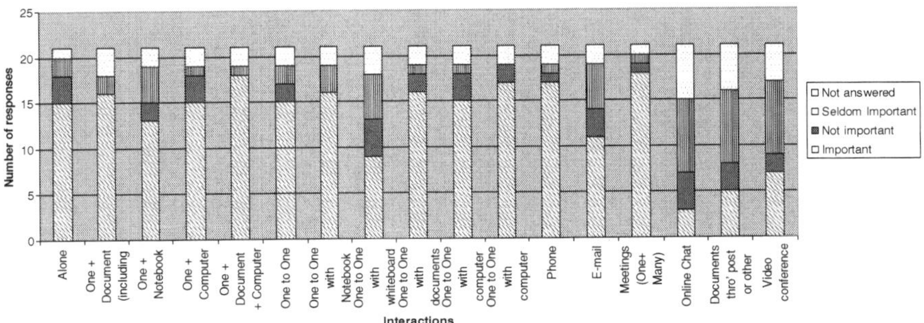

Fig. 3 Importance of types of interactions for generating information

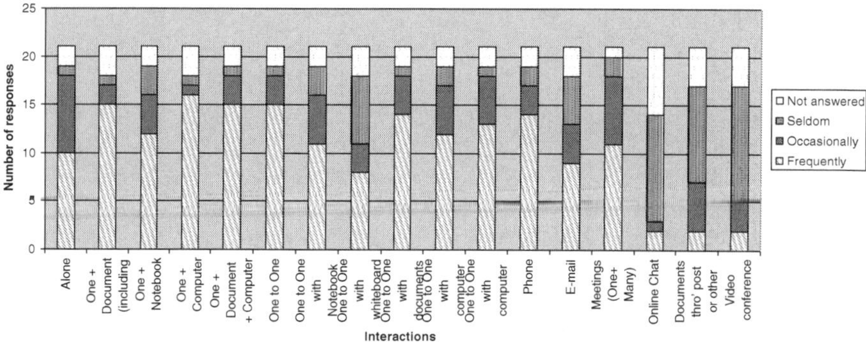

Fig. 4 Frequency of types of interactions for generating information

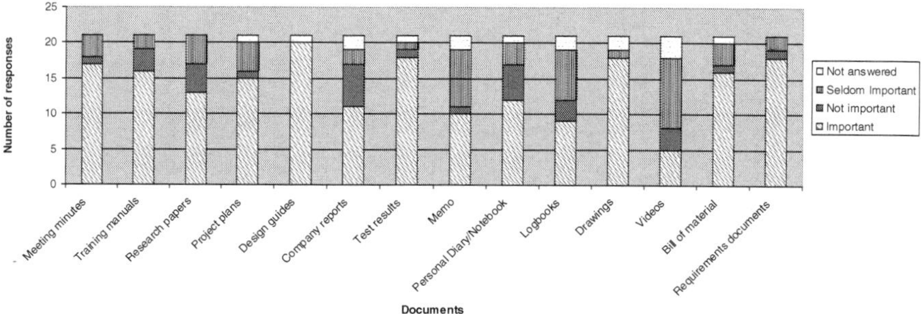

Fig. 5 Importance of capturing information in various documents

information. Even though designers perceive that they spend more than an hour alone, the personal observations reveal that the interaction 'alone' is rarely performed by the designers.

The subjects perceive that the interactions which mostly satisfy the purpose for generating information are 'designer interacting with a computer and/or document' and 'designer interacting another'. It is noted that the all collaborative (i.e. designer interacting with another or many, and with or without tools) interactions are rated less satisfactory, especially meetings. The interactions which most satisfy the purpose for sharing information are 'designer interacting with another including or excluding notebook'. It is interesting to observe that designers' perceive all the interactions as important, with differing levels of satisfaction. This shows that each interaction has its own purpose and one interaction may not be able to replace or override another, even though satisfaction through that type of interaction is less than through another.

4.5 Capturing and Searching Information Versus Types of Interaction

The subjects are conscious about and give importance to capturing information, in all types of interactions. This shows that designers are aware of the various interactions for capturing information. The interactions 'designer interacting with computer and/or document' and 'designer interacting with another in front of computer and/or document' dominate the time for capturing information. The interactions frequently used for capturing information are 'designer interacting with notebook and/or computer and/or document'. Even though designers' perceive that they capture information frequently, the personal observations reveal designers only occasionally capture information in their daily activities.

The subjects rated that they spent considerable amounts of time in each interaction per day for searching information, and most of the interactions are rated as important for finding required information. The interactions 'designer interacting with computer' and 'designer interacting with internet' are used frequently to obtain required information. This is in line with the personal observations where the interaction 'designer interacting with a computer' dominates in searching for information. It is curious to observe that designers prefer to use external media (Internet) to search for required information rather than internal systems and documents.

The subjects rated the capturing information as frequently used for understanding design, communicating with others, for design of similar products, and aiding in redesign. It has been rated low for training novices and for repeatability avoidance.

4.6 Capturing and Reusing Information Versus Documents

Figure 5 shows that the subjects rated all the documents as important to capture except videos and memos. The documents frequently generated are design guides, test results, drawings, bill of materials and requirements document. The documents most seldom generated are logbooks and videos. The correlation between importance and frequency of capturing information is high (0.84, $p < 0.01$). The subjects required between 1 to 5 days for preparing most of the documents. More than 5 days are required for training manuals and research papers. These results indicate that designers are aware of the importance of capturing information in various documents.

Documents that are rated as important for reuse are design guides, project plans, test results, drawings, requirements documents, bill of materials, books, standards and data handbooks. Important reused documents are frequently reused. The correlation between importance of capturing and reusing documents is high (0.84, $p < 0.01$). The correlation between frequency of capturing and reusing documents is also high (0.90, $p < 0.01$). This presents a good trend of capturing and reusing of documents. Figure 6 shows that all the documents could be searched within one hour. But time spent for searching being less than a day is considerable for the following documents: meeting minutes, training manuals, memos and drawings. Also, time spent for searching being less than a week is considerable for the following documents: research papers, test results, drawings, company reports and design guidelines. These results show that considerable attention has to be given to identify the reason for taking more time to find relevant documents.

Personal diary or notebook is mainly available with the designers themselves. It has been rated that most of the documents are available within the respondents' group or directorate. This could be interpreted as: designers primarily look for information within their group and directorate. Even though designers perceive that documents which are not available are negligible, considerable thought has to be given on the validity of this perception. The documents considerably rated and not available are design guidelines, personnel diary and videos. Among these documents, particular attention should be given to design guidelines. The documents which are rated as considerably difficult to access are training manuals, research papers, project plans, company reports, logbook, videos and standards.

The subjects rated indexing mechanisms of most of the documents as good. Research papers are rated as badly indexed. Some of the documents which are mentioned as seldom indexed are test results, logbook, videos, catalogs and requirements documents. Subjects' rated that the accuracy of most of the documents as high. The documents in which accuracy is substantially low are project plans and company reports. The subjects rated the recency of most of the documents as high. The documents in which recency is substantially low are training manuals, company reports, logbooks, videos and bill of materials. The subjects rated the clarity of most of the documents as high. The documents in which clarity is substantially low are meeting minutes, project plans, company reports and logbooks. The subjects rated the completeness of most of the documents as high. The documents in which completeness is substantially low are meeting minutes, training manuals, test results, logbooks and bill of materials.

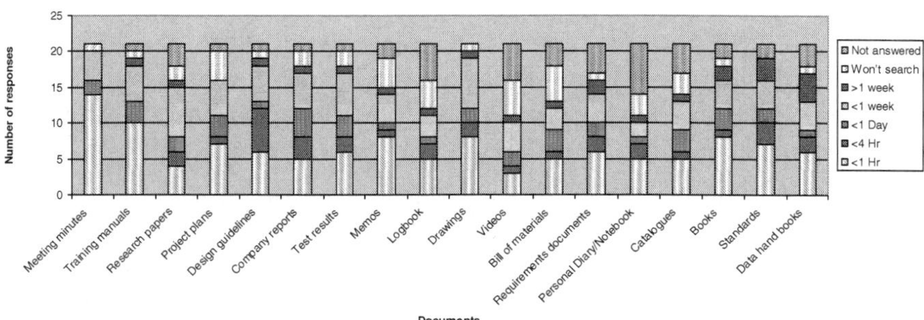

Fig. 6 Amount of time spent for searching various documents

4.7 Behavior of the Designers

The subjects tend to frequently search for more information if required information for taking decision is not available. This behavior would prolong the amount of time spent by subjects on information acquisition and shows that designers do not want to compromise on the required information. The subjects perceived that they get the right information at the right time in 4 or more out of 10 for most of the times. This perception illustrates that there is substantial need for development of support to satisfy the information needs of the subjects. The designers perceive the impact of required information on all important design criteria as very high.

5 Discussion and Conclusions

The primary outcome illustrated through the results discussed in the last section is that triangulation of results is foremost important in the informational studies. Using personal observation and questionnaire methods demonstrate that usage of methods to collect data influence the results generated. Even though personal observation method is time- and effort-consuming and also heavily restricted in the organizational environment, it provides realistic understanding and occurrences of information processes which could leads to develop enhanced information management strategies. The questionnaire brings out the perception of designers which is unrealistic in many results but also needs to take into account while formulating and developing support for information processes. This study provides the current organizational status on information processes. The questionnaire approach presented in this paper could be used by other organization to understand their processes and benchmark with reference to other industries. The core recommendation for the organization involved in this study is that substantial effort should be spend to bring interaction as the core in the information processes so that capture and reuse can be significantly improved. Each interaction has its own purpose and one interaction may not be able to replace or override another, even though satisfaction through that type of interaction is less than through another. The results highlight that the core question to be answered to enrich information processes is 'how to minimize mismatch between information needs, capture and reuse'. Currently substantial amount of time have been invested in capturing processes due to standards regulation rather than understanding reuse value of captured content. Awareness of available resources within the organization and easy retrieval of well-structured resources are necessary to improve poor ratio of 40 % to get right information at the right time.

References

1. Handzic M, Hasan H (2003) The search for an integrated KM framework. In: Handzic M, Hasan H (eds) Australian studies in knowledge management. UOW Press, Wollongong, pp 3–34
2. Gantz JF (2012) Expanding digital universe: a forecast of worldwide information growth through 2010. IDC white paper. Available at: http://www.cmc.com/collateral/analyst-reports/expanding-digital-idc-white-paper.pdf. Accessed on 09/07/2012
3. Buchanan S, Gibb F (2007) The information audit: role and scope. Int J Inf Manage 27:159–172
4. Kuffner TA, Ullman DG (1991) The information requests of mechanical design engineers. Des Stud 12(1)
5. Khadilkar D, Stauffer L (1995) Evaluating the reuse of design information: an experimental case study. In: Proceedings of ICED '95, vol 2, pp 484–489
6. Ahmed S (2001) Understanding the use and reuse of experience in engineering design. PhD Dissertation, Cambridge University
7. Aurisicchio M (2005) Characterising information acquisition in engineering design. Ph.D. thesis, Cambridge University
8. Marsh JR (1997) The capture and utilisation of experience in engineering design. Ph.D. thesis, Department of Engineering, University of Cambridge
9. Crabtree RA et al (1997) Case studies of coordination activities and problems in collaborative design. Res Eng Design 9(2):70–84
10. MacGregor SP, Thomson AI, Juster NP (2001) Information sharing within a distributed collaborative design process: a case study. In: Proceedings of international design engineering technical conferences (DETC)
11. Annamalai Vasantha GV, Chakrabarti A (2011) Understanding collaborative activities for knowledge processes in Indian industry. In: MacGregor SP, Carleton T (eds) Sustaining innovation: collaboration models for a complex world. Part of the Innovation, technology and knowledge management series, Springer

Assessing the Performance of Product Development Processes in a Multi-Project Environment in SME

Katharina G. M. Kirner and Udo Lindemann

Abstract Competiveness of small and middle-sized enterprises (SME) correlates with recognizing customer needs and being able to efficiently react to it. At the same time, SME face a limitation of their resources. Therefore, it is crucial to monitor the strength of product variety management (PVM) and its impact on the performance product development processes (PDP) in a multi-project environment. This paper aims at creating the means for SME to gain awareness of the interrelations of PVM and PDP. The latter are described using a dependency model, which has been derived from literature and empirical data gathered in case studies with six SME. The model serves as basis of a self-assessment-tool for SME to support the identification of the need to improve their PDP and the prioritization of improvement methods.

Keywords Small and middle-sized enterprises · Product variety · Product development process performance

1 Introduction

Small and middle-sized enterprises (SME) need to recognize and react to customer needs efficiently to maintain their competitiveness. This results in a high number of subsequent and overlapping product development projects—i.e. a multi-project

K. G. M. Kirner (✉) · U. Lindemann
Institute of Product Development, Technische Universität München, Boltzmannstrasse 15, 85748 Garching, Germany
e-mail: katharina.kirner@pe.mw.tum.de

U. Lindemann
e-mail: udo.lindemann@pe.mw.tum.de

A. Chakrabarti and R. V. Prakash (eds.), *ICoRD'13*, Lecture Notes in Mechanical Engineering, DOI: 10.1007/978-81-322-1050-4_63, © Springer India 2013

environment. Thus, their business and product strategy regarding the necessary product variety should be aligned with the available resources. This is often not the case, due to a lack of attention for marketing and financial planning.

It is important for a company to determine the appropriate level of variety, in order to maintain its position in the market and to be able to compensate for the increasing operational effort induced by additional product variants.

Consequently, it is crucial to monitor the strength of product variety management (PVM) and its impact on the performance product development processes (PDP) in a multi-project environment. Literature in this area focuses on the impact of product architectures either on PDP, e.g. the composition of design teams [1], or on downstream processes, e.g. supply chain management [2]. Neither the specific circumstances of SME nor a multi-project environment are taken into account sufficiently.

This paper aims at creating the means for SME to gain awareness of the interlinked performance of PVM and PDP. Therefore, it elaborates on how to monitor the impact of PVM on the performance of PDP in order to identify and prioritize the need to improve PDP. With the aim to enable SME to assess the performance of PDP in regard of PVM self-reliantly, the dependency between PDP and PVM will be described and serve as a basis for a self-assessment-tool (SAT) to monitor the performance of PDP in regard of PVM.

1.1 Related Work

SME play an important role in the market, but there are disadvantages they have to face. Not only are their resources limited, but also their environment is rather uncertain [3], as markets and customer wishes are rapidly changing. To maintain or improve their competitive market position, the strategy of a SME needs to be planned carefully. Thus, business and product strategy regarding customer wishes and necessary product variety should be aligned with the available resources. But in reality, the strategy is often not planned, as resources are scarce and e.g. no staff position for strategy planning can be afforded [4]. This is also emphasized by Vossen [5] who describes the lack of attention for marketing and financial planning and Schmidt-Kretschmer et al. [6] who criticize the lack of a holistic approach of requirements management.

Da Silveira [7] emphasizes the strategic importance of product variety, which faces the two aspects of meeting market requirements and maintain an operational performance in regards to manufacturing processes and supply chain management. Thus, he adds that product and part variety should be managed with adaptive and flexibility strategies. The impact of product variety on manufacturing and supply chains [2] is mentioned by several authors.

Although there is research on the interdependencies between product architectures—as an important aspect in PVM—and organisational structures, i.e. product development processes, they only focus on the linkage of process and product

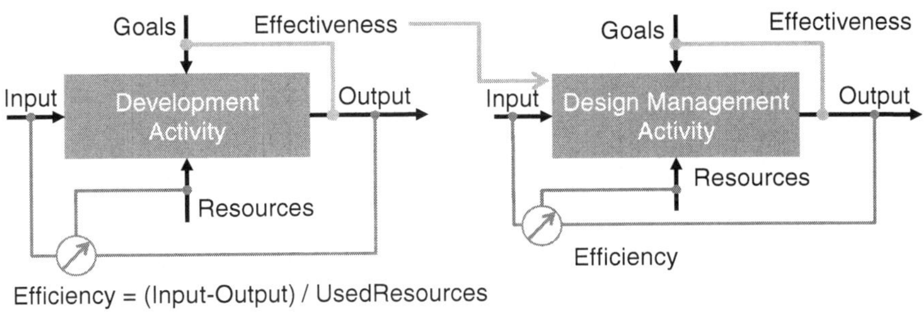

Fig. 1 Model of design performance according to [10]

component networks with a scope of single projects (e.g. [1]). The identification of mismatches between product architecture and organizational structure moreover serves as means to examine the impact of a coordination deficit in product quality [8] of singular projects. Moreover, although the impact of the strategy of PVM on manufacturing and the supply chain has been well researched, the impact of the strategy on the design productivity—for example whether a modular architecture is feasible in certain circumstances—lacks attention in literature [9].

Many authors have contributed to the assessment of performance in product development. Within this paper the model of the performance of product development created by O'Donnel and Duffy [10] serves as a basis for the assessment of the impact of PVM on the performance of PDPs. O'Donnel and Duffy [10] describe design performance using the terms efficiency and effectiveness of development and design management activities (see Fig. 1). The former stand for the concretisation of the product, while design management activities represent tasks, e.g. related to project management, necessary to coordinate the development activities. Within the model by O'Donnel and Duffy [10] efficiency is assessed by comparing the amount of knowledge gained within development activity—by comparing the input and output knowledge—and the used resources. Thus, an activity can only be effective if enough knowledge could be created, while an appropriate amount of resources has been used. Effectiveness is reached if the output of knowledge fulfills the goals that have been set for an activity. The effectiveness of development activities serves as input data for conducting design management activities, as within those development activities are coordinated. An overall efficiency and effectiveness can only be reached if it is strived for a high performance in both types of activities.

Although there is much research on process performance, only few authors have contributed to the question of how the strategy of product variety impacts the performance of the development process.

Lean Development (LD) focuses on the efficiency of processes and its main directive is value orientation [11–13]. The definition of value is essential in order to be able to guide improvement processes [12]. Processes are improved by eliminating waste—i.e. unnecessary activities or time loss—and thus shortening cycle time of process steps and the lead time of the overall process [12]. Methods—so called lean enablers—are used to eliminate waste [14] and structured along the lean principles of value, value stream, flow, pull and perfection [11]. According to Oehmen and Rebentisch [14] different categories of waste (waste sources) are highly linked, impact each other and cause other wastes. Thus, waste sources cause other waste symptoms. In order to successfully eliminate the waste, it is important to identify their root causes [15] and apply the lean tools to them.

Within this research the Lean concept of waste is interpreted using the design performance model by O'Donnell and Duffy [10]. On the one hand waste is understood as a lack in efficiency, i.e. too many resources are consumed during a development activity. Thus, Lean tools are used to identify and analyze inefficiency and ineffectiveness in PDPs. On the other hand waste is understood as a lack in effectiveness, i.e. that the goals for development and design management activities cannot be met due to an insufficient PVM strategy. This equals with the interpretation of value as effectiveness, i.e. setting and reaching the right goals.

Thus, the interrelation between PVM and PDP are described using the concept of symptoms of inefficiency and ineffectiveness and their causes. In order to optimize the interrelations between PVM and PDP—i.e. avoiding the occurrence of symptoms—the causes of inefficiency and ineffectiveness need to be eliminated.

1.2 Implications from Related Work

PVM is well researched, as an abundance of approaches exists as product platforms, different approaches of modularity, product families or customized products. But investigations on how PVM impacts on the development processes are scarce. Literature on the alignment of product architectures and PDPs does not take into account a multi-project environment in SME, and only focus the impact of product architectures on PDPs in single projects and the influence of the latter on product quality. They do not take into account PVM or a multi-project environment. Thus an understanding the interrelations of PVM and PDPs in a multi-project environment needs to be gained. Moreover, the means to assess the performance of both PVM and PDPs in a multi-project environment have to be developed, in order to enable the identification of the improvement of, taking into account the limited resources of SME. These gaps are addressed by in the subsequent sections, as the interrelations between PDP and PVM are be described and a self-assessment-tool (SAT) to monitor the performance of PDP in regard of PVM is presented.

1.3 Research Methodology

The paper describes the interrelations between specific symptoms and causes of inefficiency and ineffectiveness. These specific factors and their interrelations have been gathered by literature review [16], a web-based survey [17] on the understanding of the interrelations of methods to enhance value and waste causes in PDP, and from the results of distinct case studies in six SME [18]. Within these case studies the companies' performance PDP in a multi-project environment in the companies has been analyzed by examining existing documentation, and conducting interviews and workshops with representatives of different hierarchical levels in the enterprise. Thereby, the impact of PVM on the performance of PDP has been investigated in these companies. Based on the analyzes results criteria have been deduced to describe the interrelations between PVM and PDP in the form of a dependency model of symptoms as well as causes of inefficiency and ineffectiveness. These criteria have been incorporated to a procedure to identify symptoms of inefficiency and ineffectiveness in a specific company using a set of deductive questions. A set of inductive questions based to the interrelations between the symptoms and causes of inefficiency and ineffectiveness serves for the identification of the causes in a specific company. This served as basis to develop a self-assessment-tool (SAT) for SME, to facilitate the identification and prioritization of the need to change. By applying the SAT the user is suggested a sequence of steps to improve the performance of PDP in a multi-project environment. The SAT has been evaluated by industrial application.

2 Interrelation of PVM and Product Development Processes

Based on literature review, the results of a web-based survey [17] and empirical data from six case studies [18] a dependency model of criteria has been deduced describing the interrelations between PVM and PDP. The dependency model has been created using matrix-based methodologies as the design structure matrices (DSM) and domain mapping matrices (DMM), which can be depicted in a multiple-domain matrix (MDM) [19]. The dependency model consists of three MDMs—one based on the findings from literature, one based on the results of the survey and one based on the case studies (the latter is shown in Fig. 2). Each of the MDMs contains a set of symptoms of inefficiency and ineffectiveness in the PDP and a set of corresponding causes due to PVM. In a MDM the interrelations between symptoms and causes are modeled in a DMM (dark gray shaded part of Fig. 2, a dependency is indicated if a cause induces a symptom in one of the categories) and interrelations between several causes in an additional DSM (a dependency is modeled if one cause occurred with another cause in at least three of the six companies).

As depicted in Fig. 2, the part of the dependency model, which has been deduced from the six case studies, comprises the following categories of causes of inefficiency and in effectiveness: goals, processes, information flow, documentation, employees and tools. Figure 2 indicates specific causes within each category. The causes induce symptoms of inefficiency and ineffectiveness in the following categories:

		Causes																		Symptoms					
		Planning of strategies and priorities - company perspective	Elicitation of requirements and definition of goals and priorities - project	Lack in use of resources (also competences of employees) and	Lack of process modelling and lack of execution of processes	Lack of process analysis, process monitoring and adaption of processes	Lack of awareness of decisions (responsibilities, criteria, communication	Lack of information flow (execution)	Wrong handling of organisational interfaces	Lack of process standardisation	Lack of project management	lack of modelling of information and artefact flows (Knowledge)	Lack of data management	Lack of documentation	Specialisation of employees in many areas, multitude of tasks	Lack of self-responsibility	Lack of process comprehension and sensivity	Lack of awareness of own resposibilities and tasks	Lack of discipline (not following rules and guidelines such as the process)	Lack of hardware and software tools	Unnecessary Waiting / Delays	Unnecessary Movement	Over-production	Over-processing	Defects	Lack of reuse of existing information
		G1	G2	P1	P2	P3	P4	P5	P6	P7	P8	I1	D1	D2	E1	E2	E3	E4	E5	T1	S1	S2	S3	S4	S5	S6
Causes	G1		●			●			●												●		●	●		
	G2	●																			●	●	●	●		
	P1							●	●					●		●	●		●		●	●				
	P2					●		●	●	●				●	●	●		●		●	●	●		●	●	●
	P3	●			●			●	●					●		●			●		●	●	●			●
	P4																									
	P5			●	●	●			●					●						●	●	●			●	
	P6	●		●	●	●		●									●		●	●	●	●				
	P7				●									●	●			●	●	●	●	●			●	●
	P8																				●					
	I1																				●					
	D1													●					●		●					●
	D2			●	●	●		●		●			●			●	●		●	●	●	●				●
	E1				●					●							●		●		●	●				●
	E2			●	●	●								●			●	●		●	●	●	●			
	E3			●					●					●	●	●		●		●	●	●	●			
	E4				●					●						●	●			●	●	●				●
	E5			●		●			●	●			●	●	●						●	●	●			
	T1				●			●	●	●				●		●	●	●			●	●				●

Fig. 2 Dependency model of the interrelations of product variety management (PVM) and product development processes (PDP)—excerpt based on data from case studies [18]

- **S1—Unnecessary Waiting/Delays**—e.g. delays due to a lack of knowledge of project status, task responsibilities or status of documents
- **S2—Unnecessary Movement**—being unable to fulfill planned tasks e.g. in projects due to unplanned tasks in day-to-day business
- **S3—Over-processing**—product oriented, generation of too much detail in documents, the final product, too many or too strict requirements
- **S4—Over-production**—process oriented, executing unnecessary tasks, e.g. process of generating documents that is already exist, unnecessary testing
- **S5—Defects**—defects in drawings, the product etc.
- **S6—Lack of reuse of existing information**—Information existing in a company is not reused

3 Self-Assessment-Tool for SME: Monitoring the Performance of PDP in Regard of PVM

In this section a self-assessment-tool (SAT) is presented (see Fig. 3), which is aimed at enabling SME to gain an understanding of the dependency between PVM and the performance of PDP and to identify and prioritize the need to improve

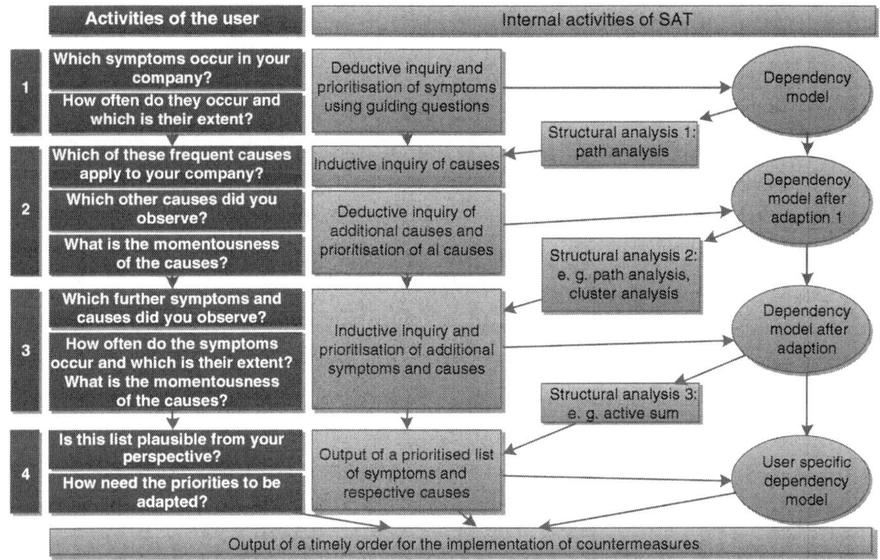

Fig. 3 Application procedure of SAT to monitor the performance of PDP in regard of PVM

specific aspects of PDP. The SAT is based on a comprehensive dependency model of causes and symptoms of inefficiency and ineffectiveness of PDP in regard on PVM in a multi-project environment as described in Sect. 2.

3.1 Applying the SAT

The SAT application procedure is decomposed in four stages as described below. Within these stages the user is iteratively asked to indicate and rate the importance of symptoms as well as causes of inefficiency and ineffectiveness of PDP. This is facilitated by alternating deductive and inductive inquiries that guide the user based on the dependency model. The possibility is maintained to name symptoms and causes that have not been covered in the underlying model up to that point. In order to guide the latter step, the users are asked to indicate a category for the supplementary symptoms and causes, which are suggested based on the dependency model. The iterative as well as inductive and deductive application procedure aims at ensuring a high quality of the obtained results.

On **stage 1** the users are asked to specify any symptoms that occur in their company. Further the user needs to indicate how often and up to which extend these symptoms occur. In order to rate the frequency and extent a scale is available with four levels from very high, high, medium to low. A verbal description for each level is suggested, but the users may change the description or replace it by a numerical scale depending on the circumstances in the specific company.

The deductive inquiry is realized by presenting the users a set of guiding questions corresponding to the symptoms of inefficiency and ineffectiveness in the dependency model. The users do not have to work with the dependency model itself. Based on the users' answers a path analysis [19] is performed in the dependency model. The dependency paths to possible causes are identified for each indicated symptom. Thereby, a list of possible causes is created. The rating of frequency and extent is used to assess the criticality of a symptom and to prioritize the symptoms on stage 4. The higher frequency and extent of a symptom are, e.g. the more delays are caused in PDP. Therefore, this symptom is of a high economic importance.

On **stage 2** the users are presented the list of possible causes, in order to indicate those that apply for the specific company. If none of these causes occurs, the users can specify other causes. The users need to indicate the degree of its momentousness for each of the causes, i.e. how many persons or departments it affects. The rating of the momentousness uses a four level scale from very high to low. A verbal description is available for each level, which again the users can change according to a company's circumstances.

The results of the inductive inquiry of causes—using the list of possible causes—as well as those of the deductive inquiry of further causes are used to adapt the dependency model to the users' companies. Those dependencies, which do not occur according to the users, are removed from the model. Subsequently, path and cluster analyzes are executed for all occurring causes, in order to identify a list of further possible symptoms and causes. The rating of the momentousness of causes is used to prioritize the causes on stage 4. The higher the momentousness, the more people and departments are affected by a cause of inefficiency and in effectiveness and thus the implementation of counter measures might be more complex.

Stage 3 uses the list of further possible symptoms and causes for a final inductive inquiry. Additionally, the users have another possibility to list further symptoms and causes they might not have been aware before or are not part of the underlying dependency model. Thereby, it can be ensured that all relevant symptoms and causes of the specific company are covered.

All user answers are used to adapt the dependency model by only maintaining those symptoms, causes and interrelations, which occur in the specific company, and by adding all symptoms and causes indicated additionally. Based on the resulting company specific model a short structural analysis is carried out in order to identify the most critical symptoms and causes. For example the active sum is calculated for each item by counting the number of other causes and symptoms it is related to. The higher the active sum the higher is the influence of the item in the dependency model and also its criticality. The results of these analyzes are combined with the user rating of extent and frequency of symptoms as well as momentousness of causes. Thereby, a prioritized list of symptoms and causes is generated, in order to indicate a timely order for the implementation of countermeasures ordered from the most to the least critical symptoms based on the criticality of their causes.

On **stage 4** the users inspect the prioritized list regarding its plausibility. They have the possibility to change the prioritization in order to obtain a guideline for the timely order of implementing countermeasures for specific symptoms and their causes.

Finally, all symptoms, causes and their dependencies, which the users listed but have not been part of the dependency model so far, are incorporated in the comprehensive underlying dependency model. Hence, the dependency model grows continuously with every application, in order to assure a continuous improvement of the SAT.

3.2 Example for an Application of the SAT

This section presents the results of an exemplary application of the SAT. In this case stages 1, 2 and 4 are executed, as only criteria from the dependency model are taken into account.

The deductive inquiry of symptoms of inefficiency and ineffectiveness using guiding questions results in the following symptoms, while for each of the symptoms extent and frequency are indicated: on the one hand a symptom occurs in the category S4 *Over-processing*. Employees produce unnecessary documents, which do not add any value in the PDP and are not used in further development processes. This happens rarely, approximately once every month, with an extent of a few hours of delay. On the other hand a symptom of the category S5 *Defects* occurs. Defective documents are generated frequently—once per week—bringing along several days of delay.

Subsequently, cluster and path analyzes of the underlying dependency model result in the following inductive suggestion of possible causes for these symptoms (Table 1).

In this example only part of the suggested causes apply. For symptom S4 *Over-processing* the causes of the category goals G1 "Planning of strategies and

Table 1 Exemplary application of SAT—suggestion of causes for symptoms S4 and S5

Causes for symptom S4—*Over-processing*	Causes for symptom S5—*Defects*
G1—Planning of strategies and priorities—company perspective	P1—Lack in use of resources and workload leveling
G2—Elicitation of requirements and definition of goals and priorities—project perspective	P2—Lack of process modeling and lack of execution of processes
P2—Lack of process modeling and lack of execution of processes	P5—Lack of information flow (execution)
P6—Wrong handling of organizational interfaces between departments	P6—Wrong handling of organizational interfaces between departments
	P7—Lack of process standardization
	D2—Lack of documentation
	T1—Lack of hardware and software tools

			Causes				Symptoms	
			G1	G2	P5	D2	S4	S5
Causes	G1	Planning of strategies and priorities - company perspective		•			•	
	G2	Elicitation of requirements and definition of goals and priorities - project perspective	•				•	
	P5	Lack of information flow (execution)				•		•
	D2	Lack of documentation			•			•
Symptoms	S4	Over-processing	•	•				
	S5	Defects			•	•		

Fig. 4 Company specific dependency model for the assessment of performance of PDP

priorities—company perspective" and G2 "Elicitation of requirements and definition of goals and priorities—project perspective" apply. Symptom S5 *Defects* is caused by one cause in the category process and one in documentation: P5 "Lack of information flow (execution)" and D2 "Lack of documentation".

In a next step the dependency model can be reduced to a company specific one as depicted in Fig. 4. In order to prioritize the symptoms and causes, their active sums are calculated based on the company specific dependency model. Both seem equally critical. They have the same active sum of 2, as they are both caused by two causes. The same applies for their causes, which only contribute to one symptom of inefficiency or ineffectiveness. At this point, the additional rating of the extent and frequency of symptoms and momentousness of causes by the user allows for a prioritization. The user rated the extent and frequency of S5 higher than those of S4. Thus, the causes of S5 need to be be eliminated first. In order to eliminate the causes P5 and D2 the execution of necessary information flow and documentation needs to be ensured.

4 Discussion and Conclusions

This paper aims at creating the means for SME to gain awareness of the interlinked performance of product variety management (PVM) and product development processes (PDP) in a multi-project environment. Therefore, the interdependency between PVM and PDP is described in a dependency model. The model is derived from literature as well as from the results of a web-based survey and six industrial case studies.

It is the goal to enable SME to gain awareness about the dependency between PVM and PDP and thus of the possible negative impact of PVM on the performance of PDP. Consequently, a self-assessment-tool (SAT) for SME is presented, which uses the dependency model as underlying database. The tools allows for monitoring the performance of PDP in regard of PVM, in order to identify and prioritize the need to improve PDP in a multi-project environment and to support the choice of improvement methods. An example for the application of the tool is presented, which results in a company specific dependency model and the amendment of the initial dependency model.

So far the dependency model is based on literature, survey results and six industrial case studies. Thus, the main limitation of the SAT is the fact that only a small number of companies contributed to the dependency model. Therefore, future work will focus on enhancing the empirical database for the SAT. Moreover, the SAT will be enhanced in regard of supporting the use of the tool, in order to ensure the quality of the obtained results by systematically supporting the user to perform a valid assessment.

References

1. Sosa ME, Eppinger SD, Rowles CR (2004) The misalignment of product architecture and organizational structure in complex product development. Manage Sci 50(12):1674–1689
2. Jiao J, Simpson T, Siddique Z (2007) Product family design and platform-based product development: a state-of-the-art review. J Intell Manuf 18(1):5–29
3. Löfqvist L (2009) Design processes and novelty in small companies: a multiple case study. International conference on engineering design (ICED09), Stanford, USA, pp 265–277
4. Braun T, Gausemeier J, Lindemann U, Orlik L, Vierenkötter A (2004) Design support by improving method transfer—a procedural model and guidelines for strategic product planning in small and medium-sized enterprises. 9th international design conference (DESIGN 2004), Dubrovnik, Croatia, pp 143–148
5. Vossen RW (1998) Relative strengths and weaknesses of small firms in innovation. Int Small Bus J 16(3):88–94
6. Schmidt-Kretschmer M, Gericke K, Blessing L (2007) Managing requirements or being managed by requirements—results of an empirical study. International conference on engineering design (ICED07), Paris, France
7. Da Silveira G (1998) A framework for the management of product variety. Int J Oper Prod Manag 18(3):271–285
8. Gokpinar B, Hopp WJ, Iravani SMR (2010) The impact of misalignment of organizational structure and product architecture on quality in complex product development. Manage Sci 56(3):468–484
9. Ramdas K (2003) Managing product variety: an integrative review and research directions. Prod Oper Manag 12(1):79–101
10. O'Donnell FJ, Duffy AHB (2005) Design performance. Springer, London
11. Haque B, James-Moore M (2004) Applying lean thinking to new product introduction. J Eng Des 15(1):1–31
12. McManus H (2005) Product development value stream mapping (PDVSM). Manual, Lean Advancement Initiative (LAI), Massachusetts Institute of Technology, Cambridge, USA
13. Ward AC (2007) Lean product and process development. The Lean Enterprise Institute, Cambridge
14. Oehmen J, Rebentisch E (2010) Waste in lean product development. LAI paper series lean product development for practitioners. Lean Advancement Initiative (LAI) Massachusetts Institute of Technology, Cambridge, USA
15. Kato J (2005) Development of a process for continuous creation of lean value in product development organizations. Thesis (PhD), Massachusetts Institute of Technology, Cambridge, MA, USA

16. Siyam GI, Kirner KGM, Wynn DC, Lindemann U, Wynn DC, Clarkson PJ (2012) Value and waste dependencies and guidelines. 14th international DSM conference (DSM2012), Kyoto, Japan
17. Kirner KGM, Siyam GI, Lindemann U, Wynn DC, Clarkson PJ (2013) Information in lean product development—assessment of value and waste. International conference on Research into Design (ICoRD13), Chennai, India
18. Eben KGM, Helten K, Lindemann U (2011) Product development processes in small and middle-sized enterprises—identification and elimination of inefficiency caused by product variety. ICED international conference on engineering design (ICED11), Copenhagen, Denmark
19. Lindemann U, Maurer M, Braun T (2009) Structural complexity management—an approach for the field of product design. Springer, Berlin

Information in Lean Product Development: Assessment of Value and Waste

Katharina G. M. Kirner, Ghadir I. Siyam, Udo Lindemann, David C. Wynn and P. John Clarkson

Abstract The value stream in product development (PD) is information flow. Therefore, the value of information needs to be increased, while waste of information needs to be eliminated in lean product development (LPD). Although the concepts value and waste in PD have both been the object of research activities, the dependency between them has been only addressed by a few authors. Therefore, this paper aims to examine the understanding of value and waste of information and their dependency in industrial practice. A web-based survey has been conducted in industry, which resulted in insight in industrial practice regarding definitions of value and waste, approaches to manage and assess value and waste, as well as companies' reasons for and challenges in the implementation of LPD. Finally, the need for further research from an industrial perspective was revealed.

Keywords Lean product development · Value and waste of information

K. G. M. Kirner (✉) · U. Lindemann
Institute of Product Development, Technische Universität München, Boltzmannstrasse 15 85748 Garching, Germany
e-mail: katharina.kirner@pe.mw.tum.de

U. Lindemann
e-mail: udo.lindemann@pe.mw.tum.de

G. I. Siyam · D. C. Wynn · P. John Clarkson
The University of Cambridge, Department of Engineering, Trumpington Street, Cambridge CB2 1PZ, UK
e-mail: gs417@eng.cam.ac.uk

D. C. Wynn
e-mail: dcw24@cam.ac.uk

P. John Clarkson
e-mail: pjc10@eng.cam.ac.uk

A. Chakrabarti and R. V. Prakash (eds.), *ICoRD'13*, Lecture Notes in Mechanical Engineering, DOI: 10.1007/978-81-322-1050-4_64, © Springer India 2013

1 Introduction

Lean in product development (LPD) has been deployed in order to improve processes and their output in industrial practice [1]. LPD is oriented towards the fulfillment of customer requirements by enhancing value and minimizing waste. The question of how LPD—with its principles, guidelines and techniques—can be better adopted in product development (PD) is attracting increased attention.

Value in product development (VPD) can be defined as 'a capability provided to a customer at the right time at an appropriate price, as defined in each case by the customer' [2]. A wide range of tools is associated with Lean Thinking. These tools, such as standardization of processes, effective communication, and a continuous flow of information, aim to increase the value (e.g. [3]).

Womack [2] defines waste as "any human activity which absorbs resources but creates no value". Engineering design can be regarded as an information creation and transformation process that aims to deliver a 'recipe' that satisfies the customer requirements [3]. Thus, in PD the value stream is represented by the flow of information produced within the product development process (PDP) [4].

Therefore, it is crucial to regard waste in PD in terms of information. Waste of information is therefore considered to occur when information is created, transformed and/or transferred without adding any value regarding the fulfillment of customer requirements [5]. Several authors have studied different types of waste in LPD, such as over-production, over-processing, waiting, transportation, unnecessary movement, defective product, and inventory (e. g. [6, 7]).

As stated in previous work [5], we propose that LPD can be used in the PDP more effectively if it focuses on value and waste of information. Further, we propose that the implementation of LPD is more successful, if a company gains a comprehensive understanding of value and waste of information as part of lean thinking. Research in LPD has developed various approaches to enhance value generation and to improve process aspects such as flow by applying lean methods (e. g. take time [8]). Although there has been research on waste types in PD, only few authors focused on waste of information [4, 7]. Moreover, the linkage between value methods and waste types has not been described thoroughly in literature [5].

This paper focuses on the understanding of value and waste in information and the linkage of both aspects. A web-based survey has been conducted in order to investigate the understanding of value and waste of information in practice as well as the dependency between techniques to foster the generation of value and waste types as observed in industry. Moreover, current practices to assess the extent of value and waste and to manage value and waste in practice are gathered.

2 Survey Set-up

The web-based questionnaire on value and waste of information in PD was sent to 450 companies in Germany and the UK. There have been 55 participants in total. Their roles in the company span from design engineer, project or product manager to head of department or combinations of all roles. The industry sectors of their companies comprise aerospace, automotive, consumer goods, furniture industry, heavy equipment, medical engineering, microscopy, plant engineering and construction, ship building, software, telecommunications and the operation in various sectors.

The questionnaire has been developed based on literature review. The results of this literature review are described in detail by Siyam et al. [5]. The survey consists of 32 questions, of which five are open questions and 27 comprise a combination of fixed answers and the option to add a supplementary comments and descriptions.

The 32 questions are grouped in five sections. Section one focuses on the application of lean thinking. The survey participants are asked whether their company makes use of lean in product development and why they adopted it. Moreover, their definition of value and waste, their assessment of common practices to enhance value and of the criticality of common waste types, as well as challenges faced during the implementation of lean practices are gathered. The topic of the second section is assessment of value and waste. Participants can name procedures for and the frequency of value and/or waste measurement, as well as metrics and attributes that are used. The third section of the survey covers the dependency between value and waste. The participants rate the dependency between specific best practices to enhance value and five waste types that were identified as most critical in the literature review. Further questions concern the participants' opinion regarding the need for research to provide support in different research areas of lean. Part four covers general information about the participant, e. g. the person's position in the company and experience in lean, and the company. Finally, the fifth part of the survey covers the possibility to give general comments and to volunteer for further inquiries.

3 Survey Results

In this section we elaborate on the results of the web-based survey on information in LPD. The results are classified into the following topics:

- Understanding of value and waste of information
- Management of value and waste
- Assessment of value and waste
- Dependency between value and waste of information
- General information on implementation of lean in practice
- Perceived need for further research.

Not all of the 51 survey participants answered all questions. The following sections indicate the number of answers for each question.

3.1 Understanding of Value and Waste of Information

Value is understood as receiving the right information at the right time, in the right format by 43.6 % of the participants (number of answers n = 55). At the same time 41.8 % of the answers stated that—additionally o the first definition—value of information is information or data that contributes to features in the product based on customer requirements.

The latter definition is also emphasized by several comments given by the participants. Not only should information be generated based on the actual need for information, but it is also important to fulfill the customers' needs. Moreover, information is considered to be valuable, if it enables the creation of innovative products or services, which exceed the customers' expectations. The majority (37.7 %) of the participants (n = 53) consider waste to be on the one hand information or data that does not contribute to the creation or enhancement of value in order to satisfy customer requirements (definition no. 1). At the same time they define wasteful information as information that is not available for the right person, at the right time and in the appropriate amount (definition no. 2). The reminder of the participants decided almost evenly between the two definitions (30.2 % for no. 1 and 32.1 % for no. 2).

Further, the participants expressed that information impeding value-adding activities are considered as waste. Information overflow, like unnecessary e-mails and test reports without sufficient summary, is one example of such waste.

The participants indicate the importance of differentiating between non-value-added information and activities—which are necessary to complete value-added activities or generate valuable information—and completely wasteful information and activities.

The participants have been asked to rate the criticality of different waste types. The latter have been structured into three dimensions, the transformation of information, the transmission of information and the content of a piece of information [5]. The left part of Fig. 1 shows the distribution of answers over the total scale from no effect (value 1) to very critical effect (value 5). The right part of the figure shows the arithmetical average of all answers.

As depicted in Fig. 1 the main wastes during the transformation or processing of information are rework—rated as critical or very critical by 66 % of the participants—and a poor synchronization of tasks—rated as critical or very critical by 64 %. Regarding the waste of the content of a piece of information defective content (rated as critical or very critical by 62 %) and incompleteness of information (rated as critical or very critical by 66 %) are the most critical wastes. During the transmission or delivery of information the most crucial waste is

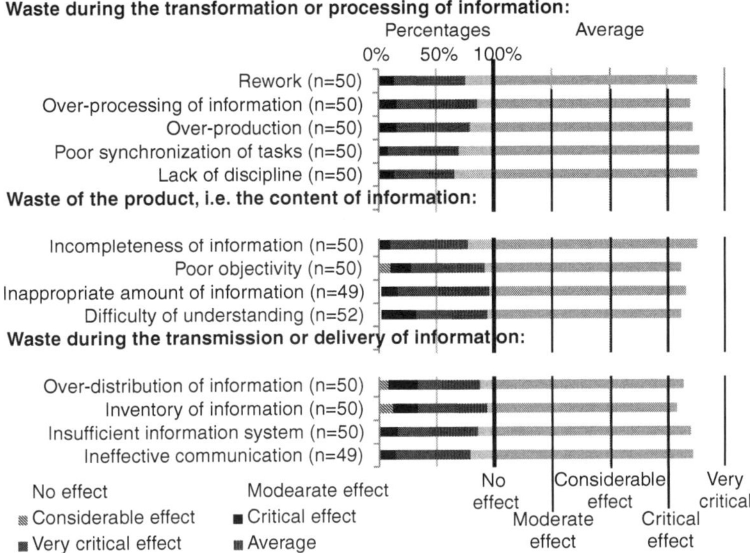

Fig. 1 Results of rating the criticality of waste types

waiting for information, which has been rated as critical or very critical by 64 % of the participants.

Other critical wastes named in the questionnaires are the existence and use of different versions of information by different people involved in the process, unclear responsibilities for pieces of information, a lack of trust in expert estimations, a lack of need-based information deployment and waste due to outdated information systems and inappropriate modeling (e.g. BOM systematics).

3.2 Management of Value and Waste

A majority of the participants regard the practices of standardization (67.3 %), integration of supplier and customer (63.8 %), simultaneous engineering (58 %) and communication (63.8 %) as effective and applicable, although it might need a considerable amount of training (Fig. 2, left part: distribution of answers, right part arithmetical average of all answers). The least applied approaches are set-based engineering (38.3 %), pull and flow of information (36.2 %) and adaptability (31.9 %).

The participants state that they use other tools with lean methods. These methods comprise Complexity Management, Design Structure Matrix, Module Oriented Value Engineering, Design-to-Cost, flow charts, swim-lane-diagrams, industry standards, ISO 9000, Total Quality Management, Quality Function Deployment, Failure Mode and Effect Analysis, the morphological chart, Kanban,

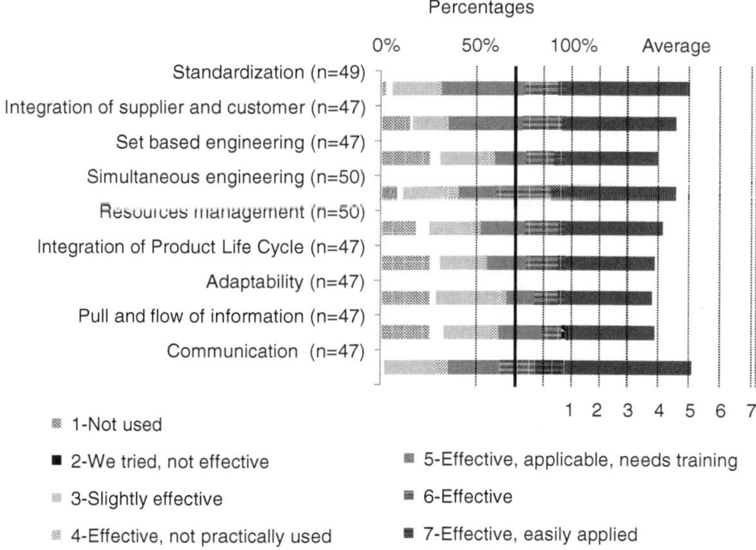

Fig. 2 Rating results of efficiency of practice to enhance value of information

moderation by process attendants from central administration, project management, rapid development cycles (e.g. agile development, SCRUM), knowledge capturing techniques, root cause analysis and best-practice-templates.

3.3 Assessment of Value and Waste

A majority of 54.8 % of the participants stated that they do not assess value or waste of information. While 28.6 % of the participants assess value or waste by frequently measuring specific attributes, such as cost, time, quality and risk. Only 16.7 % have established their own measurement (n = 42).

Further it has been stated that the enterprises' general process monitoring—not focused on lean in PD—is used for the monitoring and identification of challenges. This includes the use of general key performance indicators such as milestone adherence and engineering change order throughput. While in some companies the assessment of value and waste is modeled in workflow systems, other enterprises only apply the assessment in certain areas.

According to 13.4 % of the questionnaires (n = 22) the assessment is carried out monthly, while 86.4 % of the participants state that the measurements are conducted at pre-defined stages, e.g. project milestones.

Figure 3 shows which attributes or measurements are used according to the participants. The metrics used and considered to be easy to measure are cost

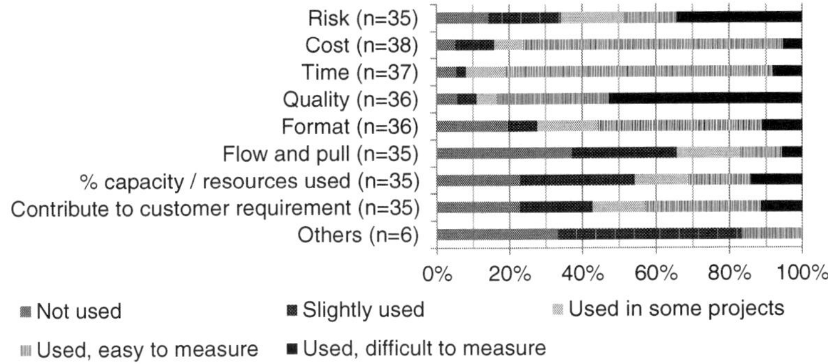

Fig. 3 Rating results of attributes most used to quantify value and/or waste

(71 %), time (73 %), format (44.4 %) and to contribution to customer requirements (31.4 %). Also used but difficult to measure are risk (34.2 %) and quality (52.8 %).

From the additional comments can be concluded that these measures, i.e. cost, time and used resources, are often monitored in terms of project management, but not exclusively used to assess value and/or waste.

Being asked which attributes had an impact on value and waste in PD the participants considered uncertainty (76.2 %, n = 42), a lack of measurement culture (44.2 %, n = 43) and the dynamics of PD processes (51.2 %, n = 43) to have a high impact on value and waste in PD.

Further the participants stated the impact on value of the following other aspects: a lack of decisions, an unclear decision structure, a change of requirements within the process, a lack of transparency of responsibilities, information status and over-all strategy, a high variety of technologies in one product, a lack of a mature marketing strategy before product definition and competition between separate departments.

3.4 Dependency Between Value and Waste of Information

The participants have been asked to rate the dependency between the five waste types rework, over-processing, over-distribution, defective content and waiting, which have been identified as the most critical in the literature review [5] and the most effective value methods standardization, management of resources and effective communication methods [5]. In 49.33 % of the questionnaires fostering an efficient and effective communication is considered to have a high positive impact regarding the elimination of waste. Also the comments given by the participants reflect the importance of communication as the basic requirement for a lean enterprise.

In more detail communication has the highest positive impact on the waste occurrence (73.3 % for waiting, 46.7 % for rework, 38.7 % for over-processing, 43.3 % for over-distribution of information) except for the defective content of

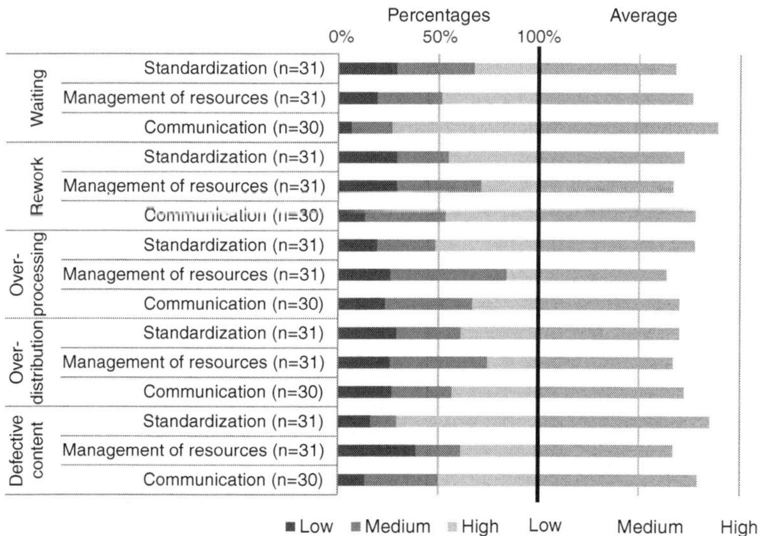

Fig. 4 Rating results of the impact of value methods on waste types

information (50 %). The latter is supposed to be eliminated best by standardization (71 % in Fig. 4). Management of resources is regarded as the approach with the least impact on the elimination of waste. Nevertheless, it is considered by a majority of all participants to have at least a medium impact on all wastes.

3.5 General Information on Implementation of Lean in the Company

When asked about whether Lean methods have been implemented in the participants' organizations 5.2 % (n = 58) stated that they had not implemented Lean in their product development processes. A percentage of 32.8 stated that they had implemented LPD, but were still trying to define lean (e.g. appropriate methods to be applied, involved teams). Further, a fraction of 20.7 % answered that they had adopted general lean tools with no specific strategy (e.g. Value Stream Mapping, Lean Principles). 15.5 % stated that they had implemented LPD with a specific focus, while 25.9 % indicated that several of these aspects apply.

The aims of the implementation of LPD differed as follows: 20.9 % named processes and activity analysis (e.g. visualization, set-based engineering), 4.7 % people (i.e. culture, leadership, and training), 2.3 % information (i.e. quality, flow, format), 2.3 % customer (i.e. requirements) and 69.8 % a combination of two or more of these aspects as focus of the implementation of LPD. In further comments the companies' orientation on value generation have been also indicated as a focus of LPD implementation.

The reasons for adopting lean have been gathered in an open question and can be summarized as follows. A majority stated that they implemented LPD as part of a holistic company strategy aiming at process improvement, increasing efficiency and effectiveness, e. g. resources management, process orientation, visualization of progress, reducing time to market, development cost and risk, increasing the reliability of delivery time and quality. Furthermore, the need to change regarding efficiency and competence development evolved due to planned strategic decisions for the companies' future. Moreover, previous success of Lean Production serves as a main driver for LPD. Finally, the urge to increase the ability to innovate, to enhance competitiveness and to involve customers in development are important factors for the implementation of LPD.

The companies faced a variety of problems during the implementation of LPD. The statement that tools, which have been developed for lean implementation (e.g. Value Stream Mapping, standardization) are difficult to apply and/or we have gained limited benefit has been named by 39.5 % of the participants (n = 38). Moreover, the dynamic and iterative process of product development was considered as limiting by 42.1 %. Finally, both challenges have been faced by 18.4 %.

The participants have been asked to list further challenges they encountered during the implementation of LPD. These challenges can be summarized as follows:

- Challenges due to application of Lean Production approaches to PD
- Difficulty to apply the general principles of LPD
- Shortage of resources
- Challenge to define the appropriate level of standardization
- Lack of acceptance of the value of standardization
- Challenge of coordination of standardized processes between customer and development organization.
- Cultural barriers (company culture differs depending on organization, departments, sites, countries).
- Challenge of clear understanding of descriptions (e.g. of waste types).
- Lack of commitment or management support
- Difficulty to achieve success quickly—conflict of high effort and low benefit in the beginning
- Lack of means to visualize the improvements due to LPD (e.g. contribution to companies' goals).

3.6 *Perceived Need for Further Research*

We asked the participants to state whether they consider it as important to further explore the following aspects using a scale from a low (value of 1) to a high (value of 10) importance. A number of 33 participants answered this question. For each aspect the average rating is as shown in Table 1.

Table 1 Perceived need for further research

Proposed research topic	Average rating of importance
Further analysis of waste causes	8
Reshaping lean principles for a better application in PD	7
Further analysis of waste types	6
Determining the impact of methods on the occurrence of waste	6
Categorizing and weighting value	6
(Further) Development of lean modeling tools (e.g. value stream mapping)	6
(Further) Development of measurements	5

Thus, the participants consider as most important research on waste causes and further development of lean principles to facilitate their application in industrial practice. The latter is also stated frequently as the most important issue, which needs to be addressed by researchers aiming to develop LPD. Moreover, the need to evaluate and validate developed LPD methods in industrial practice is expressed in order to ensure the availability of easy manageable principles and methods. Explicitly the development of methods allowing for quick analyzes and improvement approaches in LPD ensuring quick wins and thus a high acceptance of the LPD implementation.

Additionally, the prerequisites for a successful application of LPD methods regarding organization, processes and persons in industrial practice should gain higher attention, as well as the development of objectively measurable criteria e.g. for value, waste or the success of LPD. Finally, psychological aspects such as collaboration in teams, conflict resolution and emotional competency should be explored.

4 Conclusions

Value in PD is widely understood on the one hand as receiving the right information at the right time, in the right format and on the other hand as information that contributes to the fulfillment of customer requirements. Waste in PD is perceived as the opposite, while further information impeding value-adding activities is considered as waste. Thus, the understanding of value and waste in PD in practice aligns to the academic perspective. Notwithstanding from the latter, the dependency between value and waste in PD can be observed in industry and is rated according to its strength.

Regarding the management of value and waste in PD, the participants evaluated the applicability of lean methods to enhance value and eliminate waste. Further a list of aspects has been deduced, which are affecting value and waste negatively. The results covering the assessment of value and waste show that both aspects are mostly measured indirectly in the course of the enterprises' general process monitoring using general key performance indicators. Lastly, the need for the development or improvement of lean methods and tools for LPD ensuring for a better applicability in practice as current approaches is deduced.

This paper is part of ongoing research aiming to gain a comprehensive understanding of the dependency between value and waste of information in LPD and to develop support to improve PD based on the integrated assessment of value and waste of information in PD. Therefore, the results presented in this paper will be evaluated in workshops and interviews with representatives from industry and serve as a basis for the development of an approach to assess in further understanding and managing value and waste of information in PD.

References

1. Browning TR (2003) On customer value and improvement in product development process. Syst Eng 6(1):49–61
2. Womack JP, Jones DT (1996) Lean thinking. Simon Schuster, New York
3. Browning TR, Deyst JJ, Eppinger SD, Whitney DE (2000) Complex system product development: adding value by creating information and reducing risk. In: Tenth annual international symposium of INCOSE, Minneapolis
4. Graebsch M, Seering WP, Lindemann U (2007) Assessing information waste in lean product development. International conference on engineering design (ICED07), Paris
5. Siyam G, Kirner K, Wynn DC, Lindemann U, Clarkson PJ (2012) Relating waste types to value methods in lean product development. International design conference (DESIGN 2012), Dubrovnik
6. McManus H (2005) Product development value stream mapping (PDVSM), manual, lean advancement initiative (LAI). Massachusetts Institute of Technology, Cambridge
7. Oehmen J, Rebentisch E (2010) Waste in lean product development, LAI Paper Series "lean product development for practitioners," lean advancement initiative (LAI). Massachusetts Institute of Technology, Cambridge
8. Oppenheim BW (2004) Lean product development flow. Syst Eng 7(4):352–376

A Method to Understand and Improve Your Engineering Processes Using Value Stream Mapping

Mikael Ström, Göran Gustafsson, Ingrid Fritzell and Gustav Göransson

Abstract This paper describes two ways of mapping engineering processes in product development—Value Stream Mapping (VSM) and a simplified variant of VSM—which are compared with Process Mapping (PM). PM is closely related to VSM but applied differently although the goal—to identify possible process improvements—is often the same. The results of the study indicate that simplified ways of doing VSM are the most feasible. They are easier to get started with, they have a higher potential for improvement of the process and one gets an instant overview of the mapped process. Further, it is more likely that the improvement will be implemented when the users are committed through their involvement in the mapping process.

Keywords Value stream mapping · Process mapping · Lean

M. Ström (✉)
Product Realization, Swerea IVF AB, 104 SE-431 22 Mölndal, Sweden
e-mail: mikael.strom@swerea.se

G. Gustafsson · G. Göransson
Department of Product and Production, Development Chalmers University of Technology, SE-412 96 Gothenburg, Sweden
e-mail: gorang@chalmers.se

G. Göransson
e-mail: goranssg@alumni.chalmers.se

I. Fritzell
Volvo Car Corporation, SE-405 31 Gothenburg, Sweden
e-mail: ingrid.fritzell@volvocars.com

A. Chakrabarti and R. V. Prakash (eds.), *ICoRD'13*, Lecture Notes in Mechanical Engineering, DOI: 10.1007/978-81-322-1050-4_65, © Springer India 2013

1 Introduction

The increased global competition during the last decades, when quality and short time to market has been the key to survival, has forced companies to streamline their processes [1]. Since the discovery of Toyota's superiority in quality and lead time in the nineties, many companies have turned to the lean philosophy as a potential solution to their needs [2]. From having had the initial focus on manufacturing, the lean movement has more recently spread to other functions like product development (PD) [3].

One lean method, Value Stream Mapping (VSM), see Sect. 3.2, has successfully been used to revamp manufacturing processes. By visualizing the production flow, VSM helps to map processes and identify wasteful activities and serve as an input for continuous improvement [4]. Wasteful activities and other problems detected when using VSM often fit into different categories of waste. Examples of such in PD are found in [3]. They are scatter, hand-off and wishful thinking.

It is relevant to ask if VSM can also be applied to PD processes. However, PD consists of a flow of information rather than a flow of physical products. Information can exist in different versions and at different places simultaneously which makes mapping of a PD process different and more complicated. Furthermore, iterations that in manufacturing are considered as waste are in PD a natural part of the process. VSM therefore needs to be adapted in order to be applicable in a PD context.

Traditionally, different kinds of processes were mapped using methods such as IDEF0 [5], Activity Diagrams of UML [6] or Event Process Chains [7]. The actual mapping was often carried out by a dedicated process mapper who put questions to interviewed individuals working in the process and used a computer tool like e.g. MS Visio or ARIS. The required tools and the required knowledge about the mapping method created a threshold for individuals to use the method and to utilize the outcome of the mapping.

This research aims to lower the threshold for getting started with VSM and adapt the methodology to PD processes. The aim is also to make it an integrated part of everyday life of the PD team.

2 Research Questions

Following the introduction above we pose two research questions:

- How is VSM used in PD in Swedish industrial firms?
- How can VSM be adapted to suit a PD process?

3 Theoretical Framework

3.1 Process Mapping in General

The Oxford Advanced Learner's Dictionary of Current English [8] defines a process as "a series of things that are done in order to achieve a particular result" and modeling as "the work of making a simple description of a system or a process that can be used to explain it". During recent decades, a myriad of different ways to model processes have appeared. Examples are phase/stage-based models, activity networks, IDEF0, UML activity diagrams and Design Structure Matrices (DSM) [9]. To this group belongs also VSM. A number of different reasons drive the need for using process models, for example to provide a base for how to plan, execute and manage projects [10], serve as a support for continuous improvement efforts or for creation of a coherent picture of how the work is done [11–13]. The demand for a process model can also be derived from requirements from standards regulating quality assurance systems or internal company policies [10].

3.2 What is VSM?

VSM has evolved from what Toyota calls the material and information flow diagram, which the company has used to teach the Toyota Production System (TPS) to its suppliers [1]. Rother and Shook [4] were the first to convert Toyota's way of working with VSM to a practical guide, called Learning to see. This focuses on mapping in production but the method has later been adapted to and used also in other disciplines such as administration, office processes, healthcare and supply chain [14–16]. VSM provides support for understanding the flow of material and information required to make the final product and for analyzing possible improvements. By focusing on the customer, value creation and the removal of waste, an effective and efficient process flow can be accomplished [4]. VSM can be used as a tool for communication, planning and/or continuous improvement efforts [4, 17, 18].

VSM is a visual method using predefined notation and symbols, for example to indicate movement and storage of material. Accurate metrics are also collected to assess the value or identify bottlenecks of the process. The method basically consists of four steps: (1) set the scope, (2) identify the current state of the chosen process, (3) draw a future, desired state and finally (4) make a work plan to ensure the implementation of identified improvement areas [4] (Fig. 1).

As previously mentioned, it is difficult to use the VSM method from production in a complex PD environment. Furthermore, PD is generally characterized by long cycle times, which makes it difficult and time-wasting to collect accurate metrics [19]. Extensive cross-functional integration is also often required and consequently

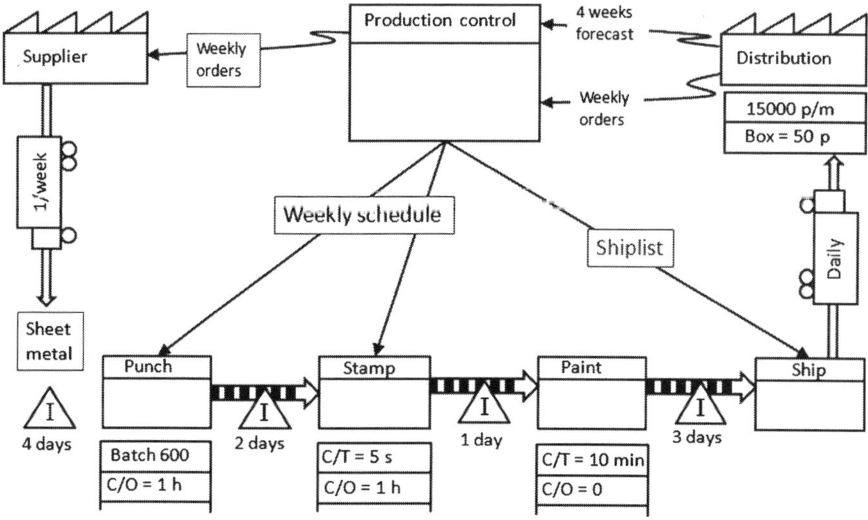

Fig. 1 Example of VSM map, inspired by Rother and Shook, to show the principle of the syntax

puts different demands on how to visualize such a process [17]. However, these should not be reasons for not trying to adapt VSM to PD [20].

One of the first VSM adaptations to PD was made by Morgan [17], who used the production method Learning to see as a basis. Another early effort was made by Millard [18]. His method consists of a Gantt chart or a Ward/LEI map for mapping on a high level combined with a process flow map and a design structure matrix for mapping on a detailed level. McManus [21] developed an extension of Millard's work resulting in a PD VSM manual. Other adaptations of VSM have for example been made by Locher [19] and Mascitelli [20]. Both are strongly influenced by Rother and Shook's original method, but Mascitelli further expands it by mapping on several hierarchical levels.

4 Methodology

4.1 Research Approach

Since the number of practitioners of VSM is still limited, a quantitative study would be difficult to make. A qualitative approach [22] has therefore been selected. It is suitable for developing in-depth understanding, which corresponds to the aim here to understand the application of VSM. The study consists of a literature review, a case study, and a field test of a revised VSM method, see Fig. 2.

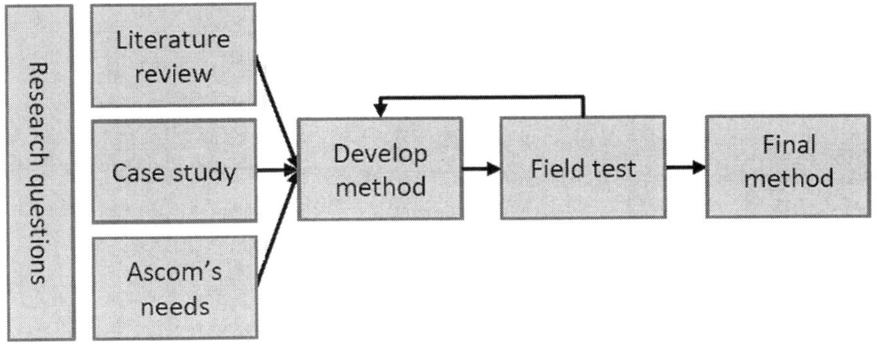

Fig. 2 An overview of applied methods [23]

4.1.1 Case Study

As a complement to the literature review, a multiple-case [24] study of four firms was made to find out how and why they use VSM in PD. The firms are active in respectively radar technology, space technology, information technology and automotive industry. The study consisted of semi-structured interviews with people who had previous experience of using VSM in PD.

4.1.2 Field Test

Action research [25] was used in three separate field tests to gain experience and to develop, test and evaluate an adapted VSM method. The tests were conducted at the telecom firm Ascom Wireless Solutions (Ascom) in Gothenburg, Sweden, which has 1,150 employees.

The field tests followed the procedure below:

1. A VSM method adapted to PD was created through comparison between and evaluations of the findings from the literature review and the case study.
2. Three processes were selected based on suggestions from Ascom.
3. The VSM method was used on one of the processes in the form of a workshop facilitated by two of the authors of this paper.
4. Feedback from the participants after the workshop was used to refine the VSM method.
5. Steps three and four were repeated on the two remaining processes.

The results from the field tests, along with the results from the literature review and the case study, were used to answer the second research question.

5 Results

5.1 General Results

The case study revealed that the firms basically followed the pattern proposed in the literature, but also that they simplified the method to suit the process and the participants. While the literature [17, 18] shows adaptations of VSM to PD by expanding the method, the firms have done the exact opposite and instead use less strict notation and fewer or no metrics at all. Three of them develop software and electronic products, and one of the firms also mechanical ones. The fourth firm develops heavy vehicles. In their VSM operations, three of the firms involve people who work in the actual processes themselves and one also persons working upstream and downstream of the process. The approach of the fourth firm is to make VSM part of the daily work.

The field tests at Ascom indicated that VSM could be simplified even more without losing its main function, i.e. to understand and improve PD processes. Focus was on creating a shared picture of the process with indirect emphasis on value and waste.

An important part of the production oriented method Learning to see is to measure the value adding time for each activity [4]. This is much more difficult to do in PD than in manufacturing. Judging from the literature, there is no good definition of value and waste in PD and it is therefore hard to determine what is value adding or not. Furthermore, the case study indicated that attempts to quantify value could result in empty discussions around it. However, it is of course important to question how separate activities contribute to the end result. As a firm gets more experience from working with VSM, more of the underlying theory can be added to the method. This might even be required since when processes become more efficient it will be more difficult to find new improvement areas.

5.2 New VSM Adapted to PD

We call this new adapted version VSM-PD.

5.2.1 Step 1: Preparation

The relevant key stakeholders agree on the desired scope of the VSM-PD operation, e.g.:

- Create a shared vision of a future state of the process
- Reduce process lead time
- Decide which parts of the process to focus on.

Participants are selected to discuss the above in a workshop. A suitable group size is 6–9 people including:

A facilitator who guides the group through the workshop and acts as a time keeper. The facilitator has knowledge of VSM-PD and shall push and challenge the participants to come up with new ideas.

Participants—the ones mapping the process. The group should consist of:

- A workshop owner responsible for the implementation of improvement suggestions
- People working in the process (not people who only *think* that they know how the work is carried out)
- People working upstream and downstream of the parts of the process focused on (i.e. those who deliver the input to the process and use the output from it)
- People affecting the process without participating in it, for example decision makers
- People necessary in order to achieve a potential future state, i.e. those with mandates to change the analyzed process.

The duration of the workshop depends on the scope and size of the process. Two consecutive half-days are probably sufficient in most cases. An all-day event is exhausting and therefore not recommended, especially since the work that requires particular focus and creativity like future-state mapping and creation of countermeasures is carried out at the end of the workshop. Separating the event into two half-days also gives time for reflection and could hence improve the results, but it is important to use two *consecutive* days in order not to lose any information.

It is recommended that the participants are informed well in advance about the workshop, its purpose and why they should take part in it so that they can familiarize themselves with the VSM-PD method and prepare for the improvement effort.

The following aids are required for the workshop: pens, a roll of wide writing paper, relevant documents and computers to show working procedures etc. and sticky notes (yellow for activities, pink for problems and blue for measures. Other colors can be useful to have in spare if the need arises to visualize for example decisions, iterations etc.).

5.2.2 Step 2: Workshop

The workshop begins with a short introduction by the facilitator including the purpose of doing VSM-PD and a basic description of the method.

This is followed by a discussion regarding the purpose of the process. Both the internal and the external customers are discussed in order to create consensus. The group then defines input to and output from the process. These are written on yellow sticky notes and put on a large sheet of paper placed on a table. The sticky notes here and in the following replace the graphical notation of the more complex

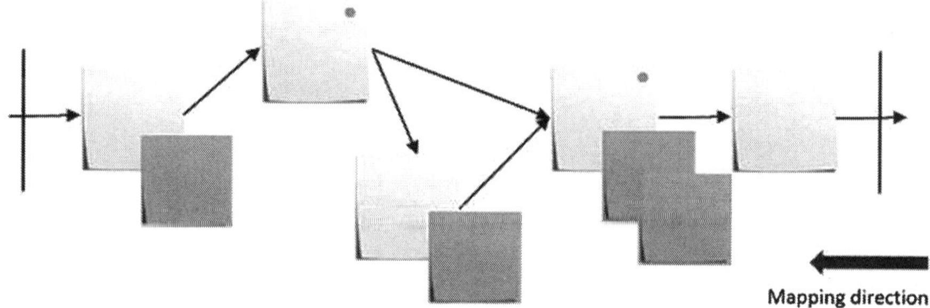

Fig. 3 Mapping is done backwards. Main process steps are marked with *green* dots (*dark grey*). *Yellow* notes (*light grey*) denote activities and *pink* notes (*dark grey*) root cause analysis

VSM methods found in the literature. The borders of the process are marked with black lines and dependencies on other processes are discussed to avoid sub-optimization.

The mapping of the current state is now carried out by starting at the end of the process in focus and working backwards in the flow. By doing so, the participants are forced to consider the customer perspective (Figs. 3, 4).

The current state map shall be created as a teamwork where all participants contribute by placing sticky notes on the paper. This is in contrast to traditional PM where one person often interviews the others and creates the map using a computer tool.

The participants explain their activities on yellow sticky notes. If possible, they show each other the documents and tools that they work with. The level of detail depends on the time available, but it is important that all participants clearly understand each activity and who performs it. The opinions may vary regarding activities and their internal order in the process, and a single current state map might therefore not even exist. However, discovering that employees have different work procedures (and discussing why they have that) can alone be worth the time spent.

The last task of day one is to identify problems by discussing the process activity by activity. Problems are elicited by identifying wasteful and other circumstances which limit the yield of the process. Solutions to problems should not

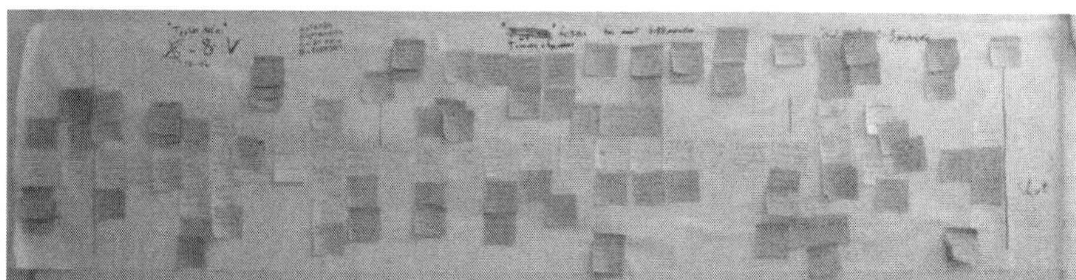

Fig. 4 Current state map from field test at Ascom

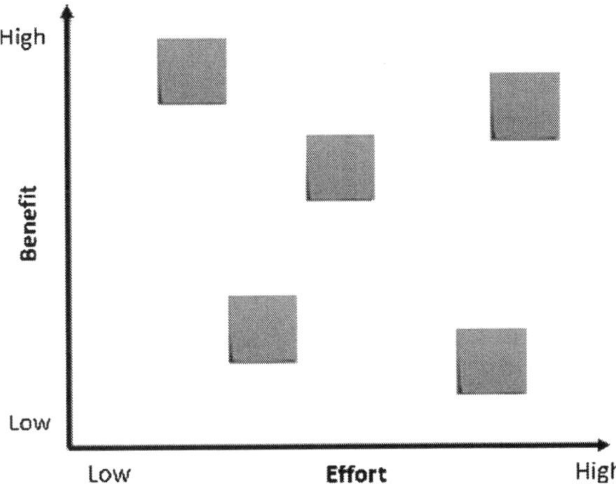

Fig. 5 Improvements, written on blue sticky notes, are placed on the pick chart according to the effort to realize them and their expected benefit. Alternatives in the *top left* corner consequently have the lowest (best) effort/benefit ratio

be discussed at this stage though, before the group has a clear and complete picture of the focused process.

It is important not only to detect problems, but also to find their root causes. The facilitator can for example ask 'why' a couple of times. The identified root causes are written on pink sticky notes which are put at appropriate locations in the flow.

When all root causes are identified, the group decides on the most serious problem areas. Finally, the most important activities in the process are marked with green dots (Fig. 3) and will be used as a basis for the future state map.

The second day begins with the creation of a future state map where the problem areas shall be eliminated. With the sticky notes marked with green dots as a framework, new sticky notes in blue are written and arranged to form an improved flow. A trick to force the group to think differently is to ask "How can you achieve this in half the time?" It is important that the participants can picture themselves working like this within 3–6 months. Otherwise there is a chance that the group ends up with improvement suggestions that are too difficult to implement. Finally, the suggestions are synchronized with the identified problems in the current state map so that no relevant issues remain untreated.

The suggested improvements shall now be ranked by the team using a pick chart (Fig. 5), and people responsible for implementing those chosen shall be appointed. It is the workshop owner's duty to make sure that these tasks are followed up.

5.2.3 Step 3: Implementation and Follow-Up

A follow-up meeting, which the workshop owner is responsible for arranging, shall be held to evaluate the improvement efforts and to resolve problems that may have occurred during implementation. All participants in the workshop, regardless of if they are involved in the implementation efforts or not, shall receive feedback on this in order for them not to lose their interest to participate in similar future activities. The same information should preferably also be communicated at department meetings to create motivation among other employees for further improvement efforts.

5.2.4 Example of Application in One of the Field Tests

Ascom's design and handling procedure of components regulated by law—so called critical components—is one example where VSM-PD was used. The map created contained sticky notes of five different colors representing activities, information, problems, countermeasures, and future state. In total 37 problems were found along with 25 countermeasures, which were ranked using a pick chart. The participants also reached consensus on the process. A future state map was created in a very short time, i.e. in about 15 min.

6 Discussion

Three different sources of information were used in the study, i.e. a literature review, a multiple-case study and action research in the form of a field test. The enquiry in the case study did not contain any questions suggesting modification of VSM, but all of the studied firms have anyway adapted VSM in a similar way. This together with the fact that the field test of VSM-PD produced a considerable number of improvements to the processes in focus strengthens the results of the study and contributes to the trustworthiness of the results.

There is at least one aspect that *may* perhaps have affected the outcome of the study: the VSM methods in the existing literature are developed mainly by American researchers who have investigated American firms, while the case study was carried out by Swedish researchers in Swedish firms. Without further investigation the possibility cannot be ruled out that there are cultural and/or other differences between the two countries that have influenced the results, but that is outside the scope of this paper to find out.

7 Conclusions

From the results of this study we conclude the following regarding the application of VSM to PD:

- In the studied Swedish firms VSM was used in different simplified ways compared to what was found in the literature.
- The further modified VSM-PD method proposed and described in this study was found to be not only suitable for, but also very useful in PD work.

Acknowledgments The authors are very grateful for the help from Ascom Wireless Solutions to conduct this study, and the support from the Swedish Foundation for Strategic Research to publish the results.

References

1. Liker J (2004) The toyota way—14 management principles from the world's greatest manufacturer. McGraw-Hill, New York
2. Womack JP, Jones T, Roos D (1990) The machine that changed the world: the story of lean production. Rawson Associates, New York
3. Ward AC (2009) Lean product and process development. The Lean Enterprise Institute, Hungary
4. Rother M, Shook J (1998) Learning to see—value stream mapping to add value and eliminate muda. The Lean Enterprise Institute, Hungary
5. US Air Force (1981) Integrated computer aided manufacturing (ICAM), Architecture Part II, vol 5—Function modeling manual (IDEF0). Air Force Material Laboratory, Wright-Patterson AFB
6. www.omg.org, access date July 29th, 2012
7. Scheer A-W (1999) ARIS business process modeling. Springer Verlag, Berlin
8. Hornby A (2010) Oxford advanced learner's dictionary of current english. Oxford University Press, Oxford
9. Ulrich KT, Eppinger SD (2012) Product design and development. McGraw-Hill, New York
10. Browning TR, Fricke E, Negele H (2006) Key concepts in modeling product development processes. Syst Eng 9(2):104–128
11. Galloway D (1999) Mapping work processes. ASQC Quality Press, Milwaukee
12. Ulrich KT, Eppinger SD (2008) Product design and development. McGraw-Hill, New York
13. Damelio R (1996) The basics of process mapping. Quality Resources, Florida
14. Tapping D, Shuker T (2003) Value stream management for the lean office. Productivity Press, New York
15. Keyte B, Locher D (2004) The complete lean enterprise: value stream mapping for administrative and office processes. Productivity Press, New York
16. Graban M (2009) Lean hospitals: improving quality, patient safety, and employee satisfaction. CRC Press, Florida
17. Morgan JM (2002) High performance product development: a systems approach to a lean product development process. Univ Michigan
18. Millard RL (2001) Value stream mapping for product development. Massachusetts Inst Technol
19. Locher DA (2008) Value stream mapping for lean development. Productivity Press, New York
20. Mascitelli R (2007) Lean product development guidebook—everything your design team needs to improve efficiency and slash time-to-market. Technol Perspect
21. McManus HL (2005) Product development value stream mapping (PDVSM) manual. Lean Aerosp Initiative
22. Bryman A, Bell E (2011) Business research methods. Oxford University Press, Oxford
23. Fritzell I, Göransson G (2012) Value stream mapping in product development—adapting value stream mapping at Ascom Wireless Solutions. Master's thesis, Chalmers University of Technology
24. Yin RK (2009) Case study research: design and methods. SAGE Publications Inc, USA
25. Wallén G (1996) Vetenskapsteori och forskningsmetodik (in Swedish). Studentlitteratur, Sweden

Lifecycle Challenges in Long Life and Regulated Industry Products

S. A. Srinivasa Moorthy

Abstract With the rapid commoditization of electronics, its content in the products are going up and this has resulted in shorter lifecycle of the systems and components. As the performance per dollar increases, speed of obsolescence of components is quicker. Most of the industries are adjusting to this rapid change to stay in business even though this puts lots of strain on their business. Changes can be replacement for obsolete parts, compaction, aesthetic changes etc. While this may be fine with industries like Consumer, Automotive etc. there are class of products and industries (also known as domains) which have a different set of requirements. Products in the Avionics and Medical domain not only have a very long life but also very tightly regulated by the bodies like Federal Aviation Administration (FAA) and Food and Drug Administration (FDA). This puts additional challenge of getting the product approved and certified whenever changes are made. This paper looks at the different challenges that product with long lifecycle encounter and how some of those issues are addressed. In many cases, impact of a small change can be very profound and unless a well designed process is in place. Otherwise addressing the life cycle challenges can be tricky. Added to this is the unique development processes used in Medical Devices and Avionics Devices development. In both these cases the uniqueness comes into effect due to the safety and reliability requirements of the products as they deal with patients and passengers. This paper deals with two types of challenges: (1) First one are the issues that designers need to address when they are designing the system so that they don't have issues later due to long life (Preventive), (2) Challenges that need to addressed when product has lifecycle issues (Reactive).

S. A. Srinivasa Moorthy (✉)
VP-Design Engineering and Head Sanmina-SCI India Design Center, Sanmina-SCI Technology India Private Limited, A3- Phase II, MEPZ Special Economic Zone, NH 45 Tambaram, Chennai 600004, India
e-mail: srinivasa.moorthy@sanmina-sci.com

A. Chakrabarti and R. V. Prakash (eds.), *ICoRD'13*, Lecture Notes in Mechanical Engineering, DOI: 10.1007/978-81-322-1050-4_66, © Springer India 2013

Keywords NPI (new product introduction) · Outsourcing · WEEE · FDA · FAA · Domains · Avionics · Medical devices · MDSS · EASA · COTS · Obsolescence

1 Introduction

With increased focus on outsourcing, product design process has undergone a paradigm shift. Today about 95 % of electronics product manufacturing is outsourced. Primary driver for the outsourcing is the cost of product, effective utilization of capital invested and above all globalization of markets. With the increased outsourcing product management has undergone drastic change. With the reduced time-to-market and increased competition understanding the Product Life Cycle (PLC) has become very important. PLC was not attracting attention when a product design and manufacturing is done in house under one roof. Most transactions were informal and unstructured. With increased competition and global market pressure companies have started outsourcing the manufacturing to save on cost and reduce the capital expenses. In a scenario like this managing product becomes very critical and especially managing the product till it is withdrawn is even more challenging. This especially became critical when the product had a long service life. Compounding this problem are the issues that arise when the products and its usage are covered by regulatory bodies. Medical Devices and Avionics Products are two such categories that are regulated by Government bodies. In this paper some of the Product Life Cycle issues of long life product are discussed. This paper outlines the strategies for managing the products whose lifecycles are long and controlled by agencies and especially in the context of outsourced manufacturing.

2 Brief Introduction to Product Lifecycle

Human's have a life cycle which is very predictable and defined as shown in the Fig. 1. We can see that human life goes through defined phases and finally fade away.

Likewise products do have very clearly defined phases of life cycle before the product is withdrawn. In fact, it is interesting to note that withdrawing a product in

Fig. 1 Human life cycle

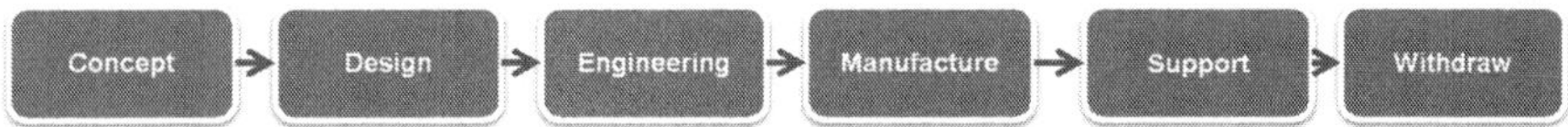

Fig. 2 Product life cycle

a formal way was never a topic of discussion and now this has become the prime driver for lots of Product Life Cycle issues [1]. Primary reasons are the environmental laws and the demand for safe disposal of electronic waste when the product is pulled out of service. Typical electronic product life cycle is represented is the Fig. 2.

Products have very clear defined phases of lifecycle starting from Concept to final withdrawal. Understanding this lifecycle allows us to design a product and manage it well throughout its life. In the above figure Life Cycle looks well defined and discrete with clear interfaces between phases. But when a product is under development most of the activities need close interaction with the participating teams like Industrial Design, HW and SW Design, Manufacturing, New Product Introduction (NPI) and Operations. When the design and manufacturing is under one roof this process was simple and easy to implement. As the outsourcing to external partners caught on, managing the product became a challenge. This inter relationship is explained in the Fig. 3. We can see in each phase corresponding to the design team activity there is a corresponding activity from the manufacturing which is crucial for the product development and managing.

While most of us are familiar with the activities in each phase one crucial thing to note is the explicit product withdrawal phase which is driven by the new environmental standards like Reduction of Hazardous Substances (RoHS) and Waste Electronic and Electrical Equipment (WEEE) directive. These standards

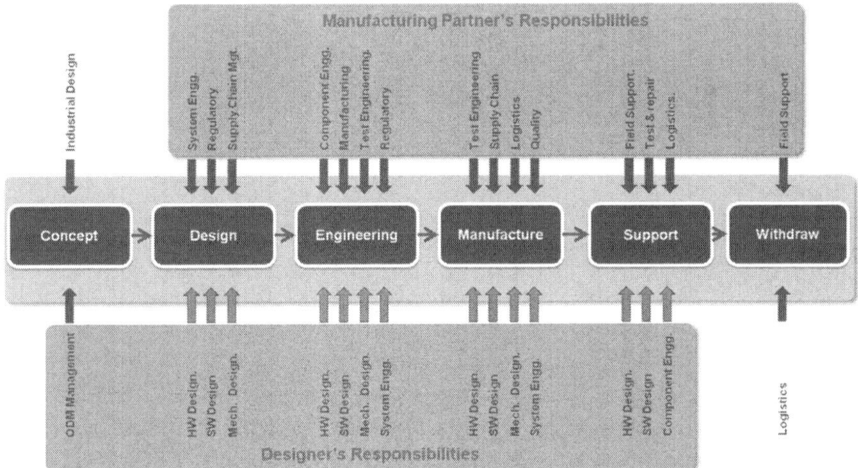

Fig. 3 Lifecycle and activity interactions

explicitly put the onus of product recycling as well as disposal in the hands of the manufacturer and expect written compliance to the same. They also need documentation and procedures to demonstrate the safe disassembly and safe disposability of products. It is this aspect that is becoming a big issue in older products which need a refresh or upgrade or ability to dispose to meet the requirements. In most cases meeting the environmental requirements will have to be addressed from the design phase, while that will be a challenge in old and matured products as there is little flexibility to make them meet the requirements. Assuming older products are subjected to product refresh, outsourced manufacturing complicates the refresh/redesign process.

Looking at the Life Cycle from this perspective it becomes very clear that in the scenario of outsourcing, a completely different approach to the product life cycle management is needed. To understand the intricacies of the in house and outsourced design and manufacturing let us refer to the Table 1 which compares some of the issues between these two. While the list is not exhaustive one nevertheless show the importance of issues.

Looking at the above table we can get an overview of the difference in approaches between in house and outsourced product design and manufacturing. Also we can see the emergence of the simple fact that outsourcing needs extreme discipline and rigorous process for success. In fact it is this aspect that brings the Product Lifecycle Management challenges.

Table 1 Characteristics of in house and outsourced product development and manufacture

S. no	In house development and manufacturing	Outsourced design and manufacturing
1	Most of the discussions and activities happen in house in a very informal way and decisions sometimes based on personal relationship	Discussions need to be structured and documented and signed off. Identification of a clear owner with responsibility is a must
2	Focus of development is on technology and not on the feasibility of manufacturing resulting in the manufacturing team doing what the design team wants leading escalated manufacturing cost.	Focus is on manufacturability, availability, cost and product support. Elaborate checklist and guidelines need to be followed for a product release
3	Multiple iterations are a norm with no or very minimal processes and documentation leading personal time of developers being spent in the assembly line	Extremely defined process with each stage reviewed and signed off. Difficult to implement on the fly changes and needs complete documentation
4	*"Technology First—Rest is Next"* attitude. This results in manufacturing aspects not getting priority	Focus is on manufacturing and all activities like assembly, testing, vendor development, packaging gets equal priority
5	Less structured process and with minimal documentation. Most of the times informal with dependence on tribal knowledge	Rigorous process following and proper documentation to the extent lack of documentation can stall the progress

Products are classified into multiple domains such as Automotive, Consumer, Industrial, Telecommunication, Networking, Avionics and Medical Devices depending on where they are used. It is important to note that each domain has distinct characteristics based on its usage and the product development process. While a long life of the product creates Life Cycle issues, the domain they operate (especially Medical and Avionics domain) have a larger impact on the product life cycle due to process intensity that is mandated. Both Medical and Avionics development mandates strict process and documentation requirements for both old and new products, due to the safety aspect of the products.

3 Regulated Industry Products Development and Manufacture

While the conventional Product development process is linear, development processes of Medical and Avionics products have lots of checks and balances. Also design and manufacturing data have to be documented and submitted to the regulatory agencies like FAA and FDA. While passenger aircrafts avionics are regulated by Federal Aviation Administration (FAA) Medical Devices are regulated by Food and Drug Administration (FDA) in USA. Primary reason for this is the safety of humans involved both in Aircrafts and Medical Devices. In both cases the developers have to prove the process they have followed is sound and meets the safety standards and also all the products are produced in conformance with the standards. The data pertaining to this have to be retained for audit purpose any time. Typically most countries use standards defined by these two bodies as a de facto standard. Europe does have its own standards European Aviation Safety Agency (EASA) and Medical Device Safety Service (MDSS) to be used in European Union.

Both the agencies have a very rigorous process for the approval of product that are either new or products which are in use is and due for enhanced for functionality. Regulated industry products are classified into different groups based on the risk during use. For products to be used in USA, typically there are 3 classes in Medical Devices and 5 classes in Avionics devices as mandated by FDA and FAA respectively. In the case of Medical Devices [2] FDA classifies the products as Class I, Class II and Class III with Class I being highest risk device capable of causing fatality (like Pacemakers etc.) and Class III being the lowest risk device like Thermometer etc. Similarly in Avionics FAA [3] have 5 classes of safety with Class E being the highest risk device and Class A being the low risk device. When a regulated industry product faces Life Cycle issue (e.g. component obsolescence) the development process for alleviating this has to follow the development process mandated by the regulatory agencies which is very important.

In general the processes followed for development are similar. The Fig. 4 shows the generic product development cycle for regulated industry products.

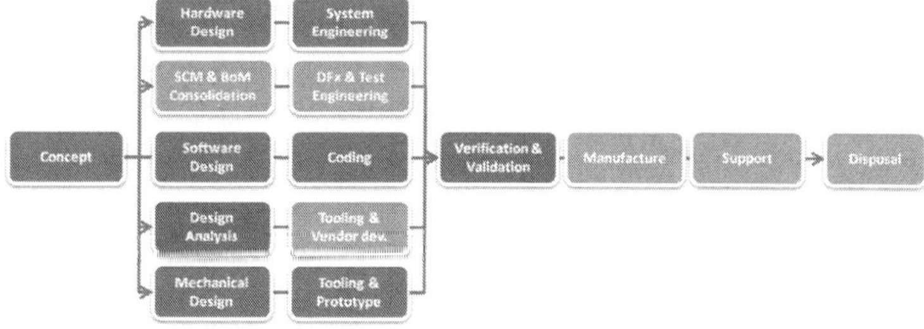

Fig. 4 Generic regulated industry product development process

When Medical and Avionics product are developed, there are two unique processes which are important. First process is ***Design Analysis***. Design Analysis is a process which is done to ensure the product safe and reliable and the second set of processes is ***Verification and Validation*** which ensures product meets all the Safety and Reliability goals. While this is a generic process depending on the medical or avionics domain process varies in a subtle way.

For Medical Devices, Fig. 5 shows the development process for development.

From the above we can see the significant process is the review process at every stage. Additionally this review process is applicable for both the new products and existing products which undergo changes. It is this aspect that creates challenges in managing the product life cycle of Medical Devices when the product is matured and needs a change only.

In the similar way Avionics product do have a rigorous product development process as shown in the Fig. 6. Avionics Product development is much more detailed and rigorous with reviews and validations done by the agencies and experts approved by the FAA.

From the two process flow diagrams we can see that two essential elements that need critical attention. First one is thorough review in every stage and the second is

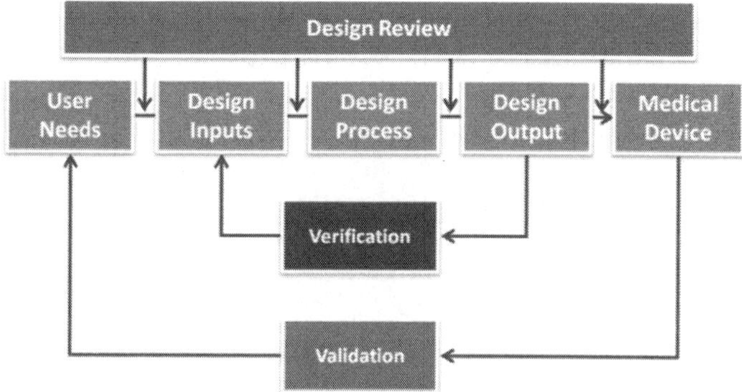

Fig. 5 Product development process for medical devices

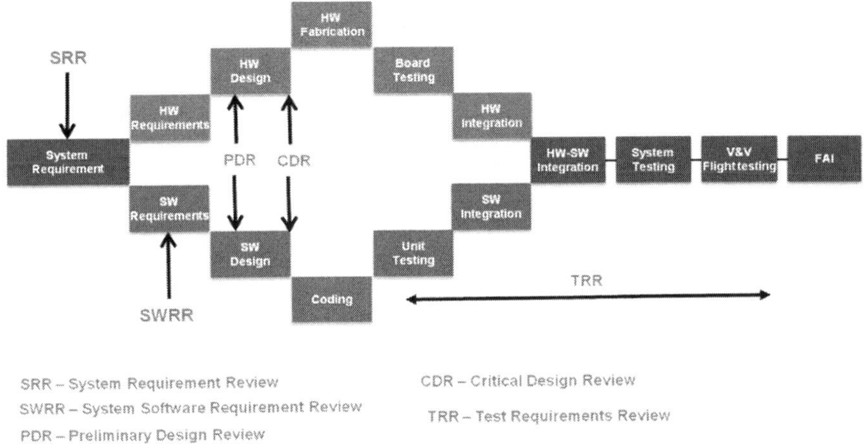

Fig. 6 Product development process for avionics devices

documenting the results in every stage. A generalized process flow is listed as below;

1. Document every stage of the development process
2. Review every stage of development both entry and exit of every stage
3. Verify the results at every stage
4. Validate the output at every stage
5. Maintain the history of development at every stage.

Effectively the above process ensures the history of the product development including the decisions taken and the review records need to logged in and recorded. Finally these logs and reports form the main part of documentation that need to be submitted to the agencies like FAA and FDA for certification purpose. In the case of Medical Devices FDA calls these records as Device History Files (DHF) and Device Master Record (DMR). In the case EU certification these are called as Technical Construction File (TCF).

4 Product Life Cycle Issues

We had looked at the generic product development cycle and regulated industry specific development processes especially for long life products. Let us now look at the key aspects that impact the Product Life cycle. Out of multiple issues which impact the Life Cycle of a product; the following are the key issues that have the maximum impact and needs attention.

1. Component Obsolescence
2. Component Availability

3. Technology Churn
4. International Standards like environmental compliance
5. Skill and knowledge
6. Product Support issues
7. Merger and Acquisition.

4.1 Component Obsolescence

Component Obsolescence is the single biggest contributor for PLC issue. Most of the times we assume the obsolescence of a part take a long to happen. But with the technology development and market conditions component vendor obsolete the parts much earlier. This puts a stress on constant lookout for any obsolescence by the components vendors. While most vendors have standard process of obsolescence based on market conditions and age of the part this can be show stopper. Only Electronic Manufacturing Service (EMS) vendors and some of the large Original Equipment Manufacturer (OEMs) have process to address this successfully. In most cases ability to address Obsolescence is skill that is very unique to companies.

4.2 Component Availability

Most of the designers think that when they select a part they assume the component availability is a non issue. However in reality unless proper procurement and supply chain management strategy is followed even simple parts can become a problem due to market conditions [4, 5]. In fact commodity parts like Memories, Capacitors experience frequent supply chokes (also called as parts on allocation where component manufacturers ration the sale). This can lead to production stoppage or increase in product cost and in some cases both. It is essential that designers select parts and alternatives during the design phase to ensure component is available without any issue.

4.3 Technology Churn

One of the new entrants into the list of issues is the Technology Churn. A decade back the speed of change of semiconductor technology was not as rapid as it is today. The rate of change in the semiconductor technology today doesn't guarantee that a part which is manufactured with the latest technology need not be backward compatible to the same part with an older technology. Devices manufactured with the current technologies have faster rise times leading problems in the older design. In many cases while they functionally work, they fail in compliance standards like Electro Magnetic Compatibility/Electro Magnetic Interference (EMC/EMI).

4.4 International Standards and Environmental Compliance

Environmental compliance brought in the 21st century has been the biggest contributor to lifecycle issues. Global standards like RoHS and WEEE and related country specific standards have created life cycle issue for products which were designed before these legislations were framed. Their continued manufacture and sale now forces them to meet the new environmental standards. In most cases the effort needed to implement the compliance is very high and companies struggle with this. This is becoming a biggest lifecycle support issue for both OEMs and EMS vendors.

4.5 Skill and Knowledge

Next biggest non technical challenge is the product knowledge management. Product knowledge typically gets lost as the time progresses and gets aggravated when the manufacturing is outsourced and the organization loses it's connect with the product. This is especially becoming critical when the product has a long life. In most cases the engineers who were involved in the original product development have either moved to other projects or have left the organization. In most product companies product sustenance strategy is never done consciously and when a lifecycle issue (like component obsolescence) crops up the situation turns panic. Sustenance of matured products needs deep engineering skills for successful revamp of a product.

4.6 Support Issues

This is an interesting problem where clients insist on retaining the product and expect the OEM support as long as they use it. Most hospitals and aviation industry do have this requirement as the equipment cost a lot and they will be profitable only when the product is used for 15–20 years. In most cases the support will be contract bound and OEMs are obliged to support and this becomes a challenge when the product is too old to support and retrofitting needs certification approvals.

4.7 Merger and Acquisitions

Unique products come out of startups but this also comes with a risk of start up getting sold out to a bigger company. In most cases the acquisition happens because the buyer feels threatened by the products from start up so they buy the start up only to kill the product. This leaves the early customers with a product which has no support leading unhappy customers. Another fallout of the M&A is the core engineering team of the startup leaves as soon as the acquisition is over leaving the product without any support for further improvement or follows up products.

5 Strategies for Managing Product Life Cycle Issues

Having seen the lifecycle details and the issues associated with that let us briefly see some of the strategies that we should use to tackle the product Life Cycle issues. For the best management of Product Life Cycle primary requirement is the thorough understanding of the product life cycle of the product and the domain. If a product is designed with Product Life Cycle in mind many problems can be addressed easily up front during design as well as during its active life. In this section we will outline some of the strategies that we can use to address the lifecycle issue.

- When designing electronic products select critical parts like CPUs from a vendor who has a very clear road map especially drop in replaceable enhanced part. Many vendors keep enhancing their offering with little or no care for the existing users leaving the existing users high and dry. Connected to this issue ensure is the tools chain associated with the CPUs need to be well established and supported.
- Put in place a mechanism which proactively tracks the obsolescence well in advance and leaves enough time for redesign if needed. Most EMS vendors have robust mechanism. But they typically use that for ensuring smooth supply chain. Taking help from them proactively will go a long way in addressing the life cycle issues.
- Track the performance of the critical parts from the field carefully. Have a communication mechanism connecting the development team, manufacturing partner and the field support teams. The reason is in most organizations field support team, development team and manufacturing team will not be connected in real-time. So components failures as well as performance issues don't get the due visibility in time. This blinds the teams in addressing issues on time and by the time problem scales up things become unmanageable.
- Connect with your EMS partner's supply chain to understand the component availability trend and up front work with them to design in alternate parts to avoid single vendor situation. Most companies look for an alternate only when

the existing part runs into availability issue and the alternate part invariably needs tuning and this leading to burning midnight oil. Sensible thing will be to make provision for one or two steps higher capacity parts as in most cases as the volume increases higher capacity parts will be cheaper than the currently used ones!

- If programmable devices are used always make provision to program them after assembly. This allows easy assembly process as well as field upgrades. This is especially true for Flash memories and Programmable Logic parts. Also ensure when multiple programmable parts are used in a single system compatible versions are captured in the system and verified by the software so that if any one part has a different program system detects it easily and ensures compatible versions run in the system.
- If the system needs calibration implement the system in such a way it can calibrate itself. Complex calibration process complicates the product support very much and building this as a part of the product will be beneficial. In most cases calibration is the least addressed issue and comes as a post manufacturing issue and not as a design parameter.
- Incorporate a dedicated Life Cycle tracking process as part of product's technical documentation capturing decisions taken to address life cycle related issues.

While the above captures some of the strategies (not exhaustive) there are a few non technical strategies that help in tackling the life cycle issues. Let us see a few of them.

- Document the tribal knowledge existing in the organization in whatever form so that it will help in tackling the product issues later
- Focus on functionality not on technology, most long life products work in the field immaterial of what technology they use. The key is the functioning of the product not what technology it is made of
- Obsolescence is inevitable and can't be avoided, so build your designs anticipating that in a proactive fashion rather than reacting to obsolescence. The same is true to globalization of market for the products
- In a regulated industry product, strike a balance between hardware and software. Hardware changes are easier to verify and validate as compared to verification and validation of software and hence making hardware only changes are much easier to get approved. Best approach is to have the architecture as a platform based one which helps addressing lifecycle issues and product functionality doesn't depend on the underlying hardware.

6 Conclusion

Finally to sum it up, longer the products life more the lifecycle issue and if the products are from regulated industry it gets more complicated. Best solution is to design the product with lifecycle in mind. As the product matures proactively address the issue and make incremental changes which helps in managing lifecycle issue much more effectively.

References

1. Saaksvuori A, Immonen A product life cycle management (ISBN 978-3-540-78173-8)
2. King PH, Fries RC Design of bio medical devices (ISBN 0-8247-0889-X)
3. Hilderman V, Baghai T Avionics certification (ISBN 978-1-885544-25-4
4. http://www.authorizeddirectory.com/pdfs/fake-parts-threaten-electronics-market.pdf
5. http://www.wnd.com/2012/05/fake-electronics-feared-undermining-u-s-defense/

Idea Management: The Importance of Ideas to Design Business Success

Camille Chinneck and Simon Bolton

Abstract Ideas are the life-blood of design. This paper presents a theoretical review of current trends in Idea Management (IM) within the front-end of New Product Development (NPD) literature. It identifies 28 success factors for managing and generating ideas within organizations and five emerging research themes. This is important to design as the Fuzzy Front-End (FFE) is one of the greatest opportunities for improving the overall innovation process [1]. Three idea management trends are discussed: (1) quantity versus quality: a shift from generating as many new ideas as possible to maximizing the number of good ideas (2) internal versus external practices: an increasing importance placed on implementing external ideas, and (3) ad-hoc versus systematic: companies are starting to apply a systematic approach to idea generation aligned with corporate design strategy. It will conclude by discussing future research opportunities in idea management and highlights the implications for design managers.

Keywords Idea management · Front end of innovation · Idea generation · Importance of ideas · New product development

C. Chinneck (✉) · S. Bolton
Center for Competitive Creative Design, Cranfield University, Cranfield, Bedfordshire MK43 0AL, UK
e-mail: c.chinneck@cranfield.ac.uk

S. Bolton
e-mail: s.bolton@cranfield.ac.uk

A. Chakrabarti and R. V. Prakash (eds.), *ICoRD'13*, Lecture Notes in Mechanical Engineering, DOI: 10.1007/978-81-322-1050-4_67, © Springer India 2013

1 Introduction

In this paper, we identify the themes and emerging trends in Idea Management (IM) within the front-end of innovation literature. The knowledge contribution is the identification of five IM themes and their level of establishment in the field. It is accepted that the innovation process is too important to be left to chance. The same rationale applies to ideas, as they need to be effectively managed as an antecedent to innovation. Prior studies have shown that IM and idea generation are in serious need of improved management [2]. Managing ideas is a complex issue [3] but when implemented effectively, idea management has been found to increase sales from new products by 7.2 % [4].

The context of this research is the front-end of innovation, also known as the "fuzzy front-end" [5], which is generally accepted to be the phase where initial product concepts are conceived [1, 6]. Research on the fuzzy front-end (FFE) has been receiving an increasing amount of attention since it was identified to be one of the greatest opportunities for improving the overall innovation process [1]. The FFE activities are important to design as they typically involve high levels of technical and market uncertainty, impacting on idea generation, idea screening and concept development. These front-end activities are critical to design management practices as they lay the foundations of the subsequent NPD process, prior to an idea being fully defined.

Innovation has been defined as the successful implementation of creative ideas within an organization [7]. Idea Management is the process of recognizing the need for ideas, generating and evaluating them [3]. Definitions such as these have been criticized for neglected elements such as idea sourcing [8]. In order to avoid ambiguity, definitions for the key constructs were found from the literature and are presented in Table 1. There is a lack of universally accepted definitions for these terms within the literature [8]. For example, the terms 'fuzzy front-end' and 'front-end of innovation' are one and the same [9].

The rest of the paper is structured as follows; Sect. 2 describes the methodology within a two-element literature review and key topics, Sect. 3 discusses emerging IM themes, Sect. 4 details 28 success factors for IM and idea generation followed by Sect. 5 which outlines three IM trends. The paper concludes with implications for design managers as well as future research opportunities. These findings are of particular interest to academics and practitioners involved in design management and front-end innovation.

2 Literature Review

2.1 Method

The literature review was split into two elements: an exploratory subject review of 86 papers and a systematic review of 134 papers. The literature review and resulting analysis identified best practices, key topics and emerging trends in IM

Table 1 Literature definitions of idea management, fuzzy front-end, idea and concept

Construct	Author	Definition
Idea management	Vandenbosch et al. [3]	"…the concept of idea management, defined as the process of recognizing the need for ideas, and generating and evaluating them" (p. 260)
	Bakker et al. [35]	"The knowledge management system (as a tool for idea management) can only facilitate the capturing, selection and enhancing of ideas among members of the organization" (p. 302)
	Boeddrich [36]	"… (idea management = phase before the project decision)…" (p. 275)
Fuzzy Front-End	Koen et al. [1]	"…those activities that take place prior to the formal, well-structured New Product and Process Development or "Stage-Gate" process…" (p. 46)
	Smith and Reinertsen [5]	"It is the fuzzy zone between when the opportunity is known and when we mount a serious effort on the development project" (p. 49)
	Reid and Brentani [37]	"…the time and activity prior to an organisation's first screen of a new product idea…" (p. 170)
	Murphy and Kumar [29]	"…from the generation of an idea to its approval for development…" (p. 5)
	Kim and Wilemon [38]	"…the period between when an opportunity is first considered and when an idea is judged ready for development." (p. 270)
	Reinertsen [39]	"…the time between when you could have started development and when you actually do, the fuzzy front-end." (p. 4)
Idea	Montoya-Weiss and O'Driscoll [40]	"An idea is defined as the initial, most embryonic form of a new product or service idea-typically a one-line description accompanied by a high-level technical diagram." (p. 154)
	Knudsen et al. [15]	"…ideas are general concepts of what might be technically or economically feasible…" (p. 124)
	Brem and Voigt [9]	"…an idea is a proposal for an action, which either reacts to recent developments or proactively utilizes them." (p. 360)
Concept	Ulrich and Eppinger [41]	"A product concept is an approximate description of the technology, working principles, and form of the product." (p. 108)
	Montoya-Weiss and O'Driscoll [40]	"A concept…is defined as a form, technology, plus a clear statement of customer benefit" (p. 145)
	Backman et al. [42]	"…'concept' has the meaning of a 'development concept', that is, a set of proposed solutions complying with a set of fixed constraints." (p. 20)

Source Provided in table

and idea generation. The exploratory review consisted of a broad search via subject area covering all available journals in four databases; Scopus, ABI, EBSCO, and Web of Knowledge. The purpose for this two-element approach was so that the knowledge from internationally recognized journals as well as highly relevant subject knowledge in other journals was utilized.

The key topics and emerging trends were identified from the systematic review in three key innovation journals; *Journal of Product Innovation Management*, *R and D Management*, and *Technovation*. These journals were selected for two reasons; (1) all of them are graded by the ABS Journal Quality Guide [10] as grade three or four star world elite innovation journals, and (2) these journals revealed the highest number of relevant papers from the performed keyword search strings.

Of the 134 papers analyzed from three top ranked journals, 39 % were theoretical, 33 % research, 27 % practice and 1 % policy. The literature type was determined according to the characterization established by Wallace and Wray [11]. The literature type is fairly evenly split between theoretical, research and practice, however, there is a very low percentage of policy literature. This is most likely due to the academic audience the papers are targeting. A total of 107 (80 %) papers provided empirical evidence to support their arguments.

2.2 Key Topics

Figure 1 below illustrates these emerging topics in representative circles, with their relative size indicating the number of relevant papers reviewed in that topic. The analysis of the reviewed papers reveals three dominant topics, with NPD being the most common with 22 papers, followed jointly by the fuzzy front-end and open innovation with 18 papers each. It should be noted that the shown number of papers addressing these broad topics is not necessarily an indicator of their importance.

The results suggest that these topics are more established within the literature. Ideation (or idea generation), IM and Knowledge Management (KM) had equal numbers of relevant papers. In order to identify and gain a deeper understanding of the important areas within the research, sub-themes were grouped into overarching themes, further detailed in Sect. 3.

3 Emerging Idea Management Themes

The results of the thematic coding analysis, as described by Robson [12], are illustrated in Table 2. These particular sub-themes were filtered from a list of over 320 topics. These topics were recorded in an Excel database. In order for a topic to survive the filtering process it had to satisfy two decision-making criteria; (1) the topic had to be mentioned in three or more papers within a top journal, and (2) the sum total for each topic across all three journals had to amount to 10 or more separate paper mentions. This filtering process ensured that no theme was included that was not identified as important by at least 10 authors in top journals. The resulting 15 sub-themes were then grouped together into five overarching themes of *process*, *ideas*, *knowledge*, *innovation type*, and *people*. These extracted themes and sub-themes provide insight into important areas discussed within the literature in relation to Idea Management and idea generation.

The sub-themes marked with an asterisk were acknowledged as important across all three journals. This suggests they are accepted and relevant across a broader range of disciplines. It could also mean that these marked sub-themes are well established due to their clear definitions within the literature. It also demonstrates the growing importance of *process* and *ideas* as key factors in Idea Management and idea generation. Figure 2 displays the five research themes and number of relevant papers published in each year addressing that theme.

All of the identified themes relate to the role of Idea Management and can be grouped under organizational culture. Within the papers reviewed, the theme of *ideas* was the most well established theme covering the widest range of years. In comparison, *innovation type*, *people*, and *knowledge* are less well established covering 14 years overall. The median years for each theme are similar with 2006 being the median year for three themes. The most published years for the themes are all fairly recent. These results support the statement that this research addresses areas receiving greater attention and importance in the literature in recent years.

The majority of papers discussing *process*, *ideas* and *innovation type*, did so in a high level of detail as opposed to mentioning the topic or discussing it briefly. This trend does not apply to the themes of *knowledge* and *people*. The number of papers that mention *knowledge* and the number discussing it in detail were equal, whilst papers discussing the theme of *people* were equally spread over a brief discussion of the topic and a detailed discussion. This suggests that *knowledge* and *people* are less established as influencing factors on the front-end of innovation when compared to the themes of *process*, *ideas* and *innovation type*. This result could be influenced to differing levels of specificity in labeling the sub-themes and the keyword strategy used to collect the papers.

4 Idea Management and Idea Generation Success Factors

A total of 28 success factors have been found for Idea Management and idea generation, detailed in Table 3. As previously stated, these particular factors were collected from a systematic review of three top innovation journals and an exploratory subject search. These factors were explicitly stated to be important to IM and idea generation practices by authors in the front-end literature. It should be noted that this is not a complete set of success factors for IM and idea generation. We acknowledge that other relevant success factors may not be covered. We suspect that the reason for this is due to the product development context of this

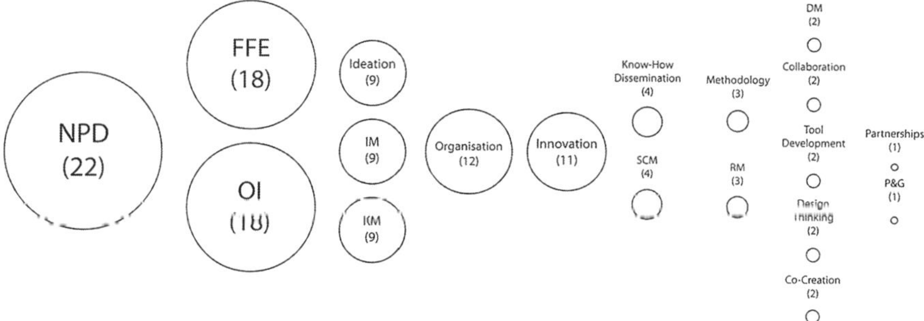

Fig. 1 Key topics from reviewed literature. *Source* Authors

research and the lack of a common language and understanding about how ideas are managed during the uncertain fuzzy front-end phase [1].

The resulting success factors were grouped into two broad categories of process and people. These factors include having a clear business purpose for the ideation event, understanding the window of opportunity, tapping into a diverse pool of idea contributors, and developing an idea through collaboration [13]. Additional success factors include having creative employees, a systematic idea generation process, available time for generating ideas, and a selection and idea evaluation method [4].

Two common mistakes are made in the innovation process; not understanding the difference between incremental versus radical ideas, and not recognizing the value of idea fragments [14]. Successful innovations are often not the result of a single idea, but of a 'bundle or ensemble of ideas and knowledge' [3]. Users are more prone to providing incremental ideas born from deep understanding of their needs rather than suppliers who tend to suggest more technical based solutions aligned to their capabilities [15]. Idea generation techniques that generate the highest numbers of actionable ideas allow for the 'natural role playing of personality types' [16]. This is important if companies want to generate as many actionable ideas as possible. The main role of IM is in ensuring these ideas are captured and managed effectively.

5 Emerging Idea Management Trends

5.1 *Idea Quantity Versus Quality*

Despite the fact that all innovation starts with an idea, little attention has been paid in the literature to understanding the phase of idea creation [17]. Rather than managing the ideation process, the most common approach has been to generate a large number of ideas [18]. This method was originally known as Osborn's principle that quantity yields quality [19]. There is a shifting focus from generating

Table 2 Emerging themes and sub-themes breakdown

Overarching themes	Sub-themes				
Process (99)	NPD (31)	Formal versus informal process[a] (20)	FFE (22)	Network effects (13)	Creativity (11)
Ideas (70)	External sources of ideas[a] (24)	Internal sources of ideas[a] (17)	Idea generation (15)	Evaluating ideas (14)	
Knowledge (31)	Tacit knowledge (18)	Knowledge dissemination[a] (13)			
Innovation type (29)	Radical innovation[a] (15)	Incremental innovation[a] (14)			
People (27)	Communication[a] (14)	Intrinsic versus extrinsic motivations[a] (13)			

Source Authors
(x) number of papers addressing topic in 3 journals
[a] topic addressed across all 3 journals

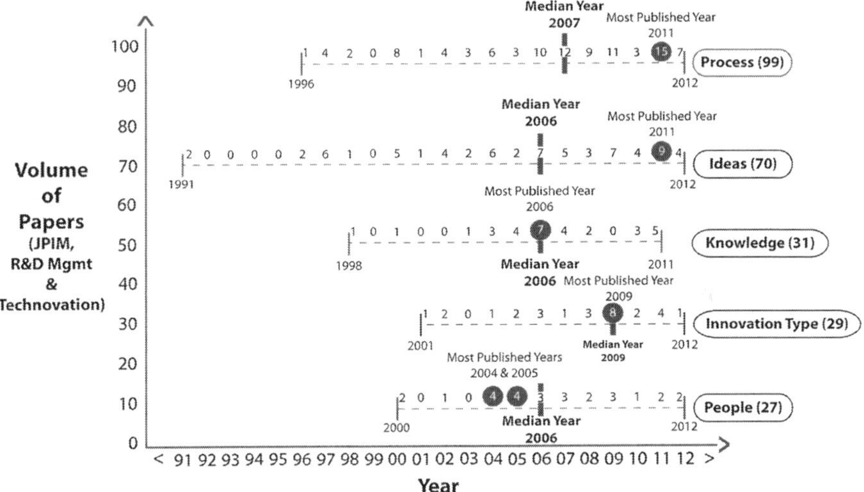

Fig. 2 Timeline reveals the establishment of idea management themes by publication volume. *Source* Authors

as many new ideas as possible to maximizing the number of good ideas that are fed into the NPD pipeline.

Idea quality has been identified to be an important, yet neglected, factor to consider with innovation ideas [20]. There is debate in the literature as to how many raw ideas are required to generate one commercially successful innovation. A vast range of answers have been proposed from 3,000 [21], 100 [22] to 6.6 [1]. This suggests that it is more important that good ideas are effectively managed and implemented over and above generating a large quantity of ideas [1].

5.2 Internal Versus External Practices

Increasing importance is being placed on the sourcing and implementation of external ideas, particularly from collaboration with customers, suppliers and partners. An organization's ability to identify, acquire, and utilize external ideas can be seen as a critical factor in regards to market success [23]. External impacting factors have been identified as an organization's network capabilities, industry structure, competition, and rate of technological change [14].

A key emerging trend is the move from a closed innovation paradigm to an open innovation paradigm in which external ideas are exploited for competitive advantage [24]. Seeking ideas from outside the firm applies best to consumer goods industries [25]. This trend is particularly important to IM as ideas from external sources will need to be managed differently from internal ideas due to barriers such as the 'not invented here' (NIH) syndrome [26]. The absorptive capacity is a critical part of innovation capability [27] in order to internally implement external ideas.

5.3 Ad-Hoc Versus Systematic

Innovation in organizations is often left as an informal ad-hoc process and has been stated to occur by serendipity rather than deliberate management [28]. This view that innovation can only be achieved through ad-hoc processes appears to be changing. This emerging trend reflects how companies are recognizing the need to apply a more systematic approach to idea generation and evaluation, aligned with corporate design strategy. Ideas that are more closely aligned to strategy, which tend to be formally formed, are more likely to lead to a commercially successful product for the marketplace [2].

Actual go/no go decisions are affected by highly subjective criteria such as top management's 'gut feel' [29]. It was found that a large multinational company spent between 5,500 and 11,000 h of total manager assessment time to evaluate 20,000 ideas [30]. This means that organizations have the potential to save a significant amount of time by implementing an effective IM system. All in all, this trend has to do with approaching the problem of finding the right balance between intent and spontaneity with an elusive phenomenon [31].

Table 3 28 idea management and idea generation success factors

Theme	Idea management	Source	Idea generation	Source
Process	Use a systematic process	Little [4]	Generate by association	Wagner and Hayashi [43]
	Give available time for idea generation	Little [4]	Build idea chains	Wagner and Hayashi 43
	Employ an idea selection and evaluation method	Little [4]	Use theme-based generation	Wagner and Hayashi 43
	Establish a clear business purpose	Gamlin et al. [13]	Ensure strategy alignment	Cooper and Edgett [44]
	Understand window of opportunity	Gamlin et al. [13]	Use external stimuli	Anderson [45]
	Reframe challenges	Gamlin et al. [13]	Use a loosely structured technique	Callaghan [16]
	Understand the differences between incremental and radical ideas	Davila et al. [14]	Utilize highly networked individuals	Bond III et al. [46]
	Recognize the value of idea fragments	Davila et al. [14]	Include prompting questions in tools	Wagner and Hayashi 43
People	Have creative employees	Little [4]	Use tools which break conceptual and cognitive sets	Brennan and Dooley [47]
	Tap into diverse pool of idea contributors	Gamlin et al. [13]	Allow role playing of personality types	Callaghan [16]
	Facilitate collaboration	Björk and Magnusson [20]	Make analogies	Ulrich and Eppinger [41]
	Continual idea commitment	Griffiths-Hemans and Grover [51]	Involve lead users	von Hippel [48]
	Collective ownership	Vandenbosch et al. [3]	Group communication	Van Dijk and Van Den Ende [49]
	High idea submission participation	Vandenbosch et al. [3]	Encourage knowledge sharing	Calantone et al. [50]

Source Provided in table–full references available upon request

6 Conclusion and Implications

This paper has presented a set of emerging Idea Management trends and themes from reviewing the front-end innovation literature. A timeline was presented which demonstrated the growing importance of the five identified emerging IM themes. The trends are important to design because of their close relation to Idea Management and front-end design activities. In addition, 28 best practice success factors were identified from the literature: 14 for Idea Management and 14 for idea generation. We argue that IM is an important emerging research area in need of better tools, methods and processes.

Ideas are important to business success because they are the starting point to all innovations [32]. The value of idea fragments and the differences between incremental and radical ideas were found to be important among other factors. An integrated and effective IM system has benefits such as helping to foster an innovation culture in which employee ideas are valued.

A surprising finding was that 'gut feel' of senior executives emerged as a common influencing factor during critical go/no go decisions. We feel that highly subjective criteria should be avoided when evaluating important innovation opportunities. The decision-making process should be made explicit in order to disseminate knowledge to employees, in particular, the criteria being used to judge good ideas from the bad ones. This in turn, should facilitate the generation of more good quality ideas that can meet these criteria.

6.1 Implications for Design Managers

There are several implications for design managers planning to integrate an effective IM system. It is recommended that loosely structured techniques should be used for idea generation, which allows for participants to be creative whilst facilitating suggestions in-line with an aligned innovation strategy. Strategic alignment is especially important when sourcing ideas externally [22]. It is recommended that design managers recognize that the idea source impacts the types of ideas generated [15, 34].

The political processes taking place during the transfer of an idea through from conceptualization to NPD is also of importance. Design managers may consider providing employees with idea selling training to ensure that ideas are given an equal chance to be evaluated. Another benefit of an effective IM system is that it will save managerial time spent on evaluating ideas. The final success of IM strongly depends on the right process structure for the different kinds of ideas and the corresponding organizational implementation [9].

6.2 Future Research Opportunities

There is a general lack of clarity, definition and understanding within the FFE regarding the nature of ideas. There is an opportunity to develop effective idea evaluation and ranking methods to save a substantial amount of managerial time. It has been found that traditional financial analysis techniques do not work for early, embryonic ideas [22].

Organizations generate a higher quantity of ideas [26] but their dynamics hinder the number of novel ideas coming to light. Although research exists on organizational barriers to creativity, there is less on how to effectively minimize these barriers to allow more novel ideas to survive the evaluation process. A poor idea evaluation scheme will kill excellent ideas and favor the survival of less innovative ideas. This involves balancing the need for more innovative ideas and an organization taking on greater risk.

The impact of multiple strategic orientations on market information and its implementation is not currently known [33]. The literature suggests that people are more creative and dedicated when they are motivated by intrinsic factors such as enjoying the idea generation process rather than monetary rewards. An interesting research question is how to encourage employee's intrinsic motivations to increase their rate of idea submission [15].

A users' knowledge of underlying technology has an effect on their ability to contribute with incremental or radical ideas [34]. There is an opportunity to better manage this trade-off in novelty versus feasibility in order to incorporate the best of both. These gaps can be addressed through implementing appropriate IM and idea generation tools and templates. The next step of this research will be to undertake a series of controlled experimental case studies [12], focusing on empirically evaluating the impact of the success factors in Fast Moving Consumer Goods (FMCG) scenarios.

Acknowledgments The authors would like to extend their thanks to Procter & Gamble and the Engineering and Physical Sciences Research Council for sponsoring this research.

References

1. Koen O, Ajamian G, Burkart R, Clamen A, Davidson J, D'Amore R, Elkins C, Herald K, Incorvia M, Johnson A, Karol R, Seibert R, Slavejkov A, Wagner K (2001) Providing clarity and a common language to the "fuzzy front end". Res Technol Manag 44:46–55
2. Barczak G, Griffin A, Kahn KB (2009) Perspective: trends and drivers of success in NPD practices: results of the 2003 PDMA best practices study. J Prod Innov Manage 26:3–23
3. Vandenbosch B, Saatcioglu A, Fay S (2006) Idea management: a systemic view. J Manage Stud 43:260
4. Arthur DL (2005) How companies use innovation to improve profitability and growth. Innov Excellence Study. Boston. Available at: [http://www.adlittle.uk.com/reports_uk.html?&no_cache=1&extsearch=innovation&view=53]

5. Smith PG, Reinertsen DG (1991) Developing products in half the time. Wiley, New York
6. Khurana A, Rosenthal SR (1998) Towards holistic "front ends" in new product development. J Prod Innov Manage 15:57–74
7. Amabile TM, Conti R, Coon H, Lazenby J, Herron M (1996) Assessing the work environment for creativity. Acad Manag J 39:1154–1184
8. Soukhoroukova A, Spann M, Skiera B (2012) Sourcing, filtering, and evaluating new product ideas: an empirical exploration of the performance of idea markets. J Prod Innov Manage 29:100–112
9. Brem A, Voigt K-I (2009) Integration of market pull and technology push in the corporate front end and innovation management—insights from the German software industry. Technovation 29:351–367
10. Harvey C, Kelly A, Morris H, Rowlinson M (2010) Academic journal quality guide. The association of business schools, Version 4, London
11. Wallace M, Wray A (2006) Critical reading and writing for postgraduates. Sage, London
12. Robson C (2011) Real world research: a resource for users of social research methods in applied settings. Wiley, London
13. Gamlin JN, Yourd R, Patrick V (2007) Unlock creativity with "active" idea management. Res Technol Manag 50:13–16
14. Davila T, Epstein MJ, Shelton R (2006) Making innovation work: how to manage it, measure it, and profit from it. Wharton School Publishing, New Jersey
15. Knudsen MP (2007) The relative importance of interfirm relationships andknowledge transfer for new product development success. J Prod Innov Manage 24:117–138
16. Callaghan E (2009) Personalities of design thinking. Des Manag J 4:20–32
17. Dahl DW, Moreau P (2002) The influence and value of analogical thinking during new product ideation. J Mark Res 39:47–60
18. Goldenberg J, Mazursky D, Solomon S (1999) Toward identifying the inventive templates of new products: a channeled ideation approach 36:200–210
19. Osborn AF (1963) Applied imagination: principles and procedures for creative problem-solving. Scribner, New York
20. Björk J, Magnusson M (2009) Where do good innovation ideas come from? Exploring the influence of network connectivity on innovation idea quality. J Prod Innov Manage 26:662–670
21. Stevens G, Burley J (1997) 3,000 raw ideas equal 1 commercial success! Res Technol Manag 40:16–27
22. Cooper RG, Edgett SJ (2007) Generating new product ideas: feeding the innovation funnel. Product Development Institute, Ancaster
23. Van Aken JE, Weggeman MP (2000) Managing learning in informal innovation networks 30:139–149
24. Chesborough HW (2003) Open innovation: the new imperative for creating and profiting from technology. Harvard Business Press, Boston
25. Huston L, Sakkab N (2006) Connect and develop: inside Procter & Gamble's new model for innovation. Harvard Bus Rev 84:58–66
26. Majaro S (1992) Managing ideas for profit: the creative gap. McGraw-Hill, New York
27. Tang HK (1998) An integrative model of innovation in organizations. Technovation 18:297–309
28. Desouza KC, Dombrowski C, Awazu Y, Baloh P, Papagari S, Jha S, Kim JY (2009) Crafting organizational innovation processes. Innov Manag Policy Pract 11:6–33
29. Murphy SA, Kumar V (1997) The front end of new product development: a Canadian survey. R and D Manag 27:5–15
30. Reitzig M (2011) Is your company choosing the best innovation ideas? MIT Sloan Manag Rev 52:47–52
31. Conway S (1995) Informal boundary-spanning communication in the innovation process: an empirical study. Technol Anal Strateg Manag 7:327–342

32. Koc T, Ceylan C (2007) Factors impacting the innovative capacity in large-scale companies. Technovation 27:105–114
33. Magnusson PR (2009) Exploring the contributions of involving ordinary users in ideation of technology-based services. J Prod Innov Manage 26:578–593
34. Spanjol J, Qualls WJ, Rosa JA (2011) How many and what kind? The role of strategic orientation in new product ideation. J Prod Innov Manage 28:236–250
35. Bakker H, Boersma K, Oreel S (2006) Creativity (ideas) management in industrial R&D organizations: a crea-political process model and an empirical illustration of corus RD&T. Creativity Innov Manage 15:296–309
36. Boeddrich HJ (2004) Ideas in the workplace: a new approach towards organising the fuzzy front end of the innovation process. Creativity Innov Manage 13:274–285
37. Reid SE, De Brentani U (2004) The fuzzy front end of new product development for discontinuous innovations: a theoretical model. J Prod Innov Manage 21:170–184
38. Kim J, Wilemon D (2002) Sources and assessment of complexity in NPD projects. R and D Manage 33:16–30
39. Reinertsen D (1994) Streamlining the fuzzy front-end. World Class Des Manufacture 1:4–8
40. Montoya-Weiss MM, O'Driscoll TM (2000) From experience: applying performance support technology in the fuzzy front end. J Prod Innov Manage 17:143–161
41. Ulrich KT, Eppinger SD (2000) Product design and development. Irwin McGraw-Hill, USA
42. Backman M, Börjesson S, Setterberg S (2007) Working with concepts in the fuzzy front end: exploring the context for innovation for different types of concepts at volvo cars. R and D Manag 37:17–28
43. Wagner C, Hayashi A (1994) A new way to create winning product ideas. J Prod Innov Manag 11:146–155
44. Cooper RG, Edgett S (2008) Ideation for product innovation: what are the best methods? PDMA Visions Magazine 32:12–17
45. Anderson BF (1975) Cognitive Psychology. Academic Press, New York
46. Bond EU, Walker BA, Hutt MD, Reingen PH (2004) Reputational effectiveness in cross-functional working relationships. J Prod Innov Manage 21:44–60
47. Brennan A, Dooley L (2005) Networked creativity: a structured management framework for stimulating innovation. Technovation 25:1388–1399
48. Von Hippel E (1988) The Sources of innovation. Oxford University Press, New York
49. Van Dijk C, Van Den Ende J (2002) Suggestion systems: transferring employee creativity into practicable ideas. R and D Manag 32:387–395
50. Calantone RJ, Cavusgil ST, Zhao Y (2002) Learning orientation, firm innovation capability, and firm performance. Ind Mark Manage 31:515–524
51. Griffiths-Hemans J, Grover R (2006) Setting the stage for creative new products: investigating the idea fruition process. J Acad Mark Sci 34:27–39

The Role of Experimental Design Approach in Decision Gates During New Product Development

Gajanan P. Kulkarni, Mary Mathew and S. Saleem Ahmed

Abstract Experimental Design Technique (EDT) in combination with stage gate strategy is a powerful tool for providing critical information to designers during New Product Development (NPD) as well as product redesign activities. It systematically evaluates new product design strategy and facilitates redesign of existing products. The benefit of applying this technique to NPD is to speed up the development process by allowing product design team to make more informed decisions based on the generated experimental design data. The risk that designers face every time when they take decisions at every stage of NPD is high. Going with right decisions and refusing the wrong ones should be the driving force throughout product development process. To understand the decision making challenges of a designer, this paper illustrates go-no-go decisions in a sample of graduate students who attempted to design a "green stool" as part of their class assignment. Factors common to and factors exclusive to the various design stages are analyzed and described using EDT. Although this integrated "experimental design go-no-go approach" was not used initially by the design students, our analysis on the process is done post factor.

Keywords New product development · Experimental design technique · Go-no-go decision gates

G. P. Kulkarni (✉) · M. Mathew · S. S. Ahmed
Centre for Product Design and Manufacturing, Indian Institute of Science, Bangalore, India
e-mail: gajanankulkarni87@gmail.com

M. Mathew
e-mail: mmathew@mgmt.iisc.ernet.in

S. S. Ahmed
e-mail: saleem@cpdm.iisc.ernet.in

A. Chakrabarti and R. V. Prakash (eds.), *ICoRD'13*, Lecture Notes in Mechanical Engineering, DOI: 10.1007/978-81-322-1050-4_68, © Springer India 2013

1 Introduction

New product development is one of the most fundamental activities bringing long term success to business firms. Developing new products and redesigning existing ones are difficult procedures as these are extensive, expensive and laborious ventures [1, 2]. According to Leavitt et al. [3] an estimated one third of the average organization's sales is derived from new products. Companies are always in search of new products which can potentially capture markets, provide competitive edge over others, and grab customer's attention with less time to market [3].

Designers start from generating new product ideas and converting potential ideas step by step into successful products by performing trial and error during New Product Development (NPD). Experimental design is an efficient technique for conducting statistically designed experiments which are important components of product and process design and development [4, 5]. Thus, experimentation is the true guide for designers to show check posts in terms of analyzing performance parameter, critical factors and evaluation criteria in order to move in the right direction of a product development process map. Blake et al. states that, "Experimental Design Technique (EDT) is a powerful tool that can be used in designing robust products, reducing time to market, improving product quality and reliability and reducing life-cycle cost" [5, 6].

Designers have to make decisions in order to move to the next product development stage and this is a mandatory part of any NPD process from idea-to-launch; which decides the product's success or failure. To avoid the decision making errors which lead to product failure, designers and product developers navigate through go and no-go decisions throughout the NPD process. This evaluation of every product development stage is better understood as 'gates' or 'convergent points' [7–9]. Over past thirty years the Stage Gate strategy proposed by Robert Cooper, has continually emerged as a useful tool for business firms in launching new products to market.

1.1 Background

EDT has application in planning during various NPD stages. Albeit that this method is around for decades, it is argued that the operational implementation of EDT has been a challenge for designers. The post facto study of a green stool design project is done using EDT partially; to identify performance parameters (factors common to) of all stool development stages and critical factors (factors exclusive to) for individual stages.

From the case of green stool design project it is apparent that; the designers did not rigorously follow the stage-gate system [10] while carrying their individual projects but they did logically experience the stool development stages and critical decision making points at those stages. The decisions taken at every product development stage worked as a turning point and guided them towards successful completion of the green stool design project. The "go" decision is like a green signal of confirmation from an appropriate expert reviewer for a specific stage gate which allows designers to keep moving to the next stage gate until the final stage is reached [2].

1.2 Objective and Content of Paper

Considering the green stool design project from experimental design and stage-gate system perspective; we realized that design students would have used these techniques integratedly during their actual project runs. From designers and product developers viewpoint; the mentioned techniques go hand in hand and excellently support each other towards overall thoughtful understanding of NPD process. Taking this as a motive, we have developed post facto analysis for four sample stool design projects in this paper. Using these case study analyses; we have written this paper has the following objectives

1. Understanding the literature on EDT in a designer's context and illustrating the use of this technique in the context of product design and development stages.
2. The post facto analysis of a stool project is done for three reasons
    a. To identify performance parameter and critical factors of various green stool development stages by partially following experimental design stages
    b. To determine the quality assessment criteria for evaluation of four stool cases using Garvin [11] eight critical dimensions of product quality
    c. Recognizing decision gates during the green stool development stages taking into consideration the critical factors identified using EDT.

In the subsequent sections; the authors would take readers to go through the journey of case studies on green stool design project. Section 2 demonstrates a brief description of the stool project covering an actual project task given to the design students, tracking of the activities performed by them during this project and an in-depth narration of selected four sample stools considered for this paper. In Sect. 3 literatures on EDT is studied from designers point of view. Section 4 aims at the post facto analysis of four stool cases where factors and evaluation criteria are found out as an outcome of partial application of the experimental design technique. Development stages and go-no-go gates in stool development process are identified in Sect. 5. Conclusions are made in later section.

2 Description of Green Stool Design Project

The green stool design project was a major part of a graduate level three credit course titled Product Design, conducted at Centre for Product Design and Manufacturing at Indian Institute of Science, Bangalore. Eleven graduate design students enrolled for this course and individually completed the green stool design project over the period of two months as a part of this course. The instructor's objective behind offering this project was teaching students to experience the process of form development by journeying through the NPD process. To be precise designers selected a particular letter form as metaphor for the stool. Using this metaphor as a point of inspiration they explored and evolved various interesting product forms and finally shortlisted one potential form and developed a meaningful product which was the green stool. The involvement of higher levels of technology and mechanisms was not the aim of this assignment. Design students generated new ideas, created various creative forms, used different materials, manufacturing processes and developed their final product. The role of the expert reviewer throughout this project was played by the instructor having industrial exposure and experience in the product development field. Designers consulted with the instructor during various stool development phases for making an appropriate decision and proceed further. A market survey was not done for this project, but designers designed their green stools taking into account a suitable customer and a market profile. The final evaluation of the green stool was done by the instructor, as an end customer would do.

2.1 Project Task

The task given to design students for this project was, "Design a 'Green Stool'. It need not be green in colour, but be made with eco-friendly material, cost effective manufacturing process, and must be made with minimum possible energy. The new design of the stool shall be for anyone who will feel proud to possess one due to its inherent values and style statement. It can be for any user in any market of designer's choice, but should definitely be suitable for mass manufacturing. Designer can make suitable assumptions and state them in his/her product brief before proceeding with the design process. It should adopt a theme in form of a character which should communicate environmental consciousness in its form and meaning".

2.2 Green Stool Development Activities

The brainstorming session was done by designers to come up with several themes as a metaphor. Students thought about diversified themes such as—*Symbols, Stones, Colors, Trees, Musical Notes, Birds, and Animals*. After considering all the theme options; the instructor finally selected '*Symbol*' as a metaphor for the development of stool design. Designers independently formulated their own stool design briefs. Eleven designers divided into three groups for carrying out ergonomic, technical and cultural studies of stool. The data collected from this study was shared amongst the eleven designers. Table 1 shows various stool development phases and the activities conducted by the designers during these phases.

Table 1 Development phase wise description of stool design project

Stool development phases	Description of activities during phases
Preliminary phase	**Green stool design brief declaration** • Identifying primary and secondary stool design requirements • Defining customer/end user profile and market profile • Selecting a specific symbol as metaphor **Data collection** *Ergonomic aspects* • Seating ergonomics • Anthropometric aspect of seat design—seat height, seat depth, seat width, stool weight, stool stability etc *Technical aspects* • Materials for stool—wood, metal, plastic, other materials and physical and mechanical properties of materials • Manufacturing processes for wooden, metal and plastic stools *Cultural aspects* • Indian activities connected to stool—listed activities done by men, women, children/teenagers and elder people • Indian culture and beliefs • Visual demands—from environment, market, customer and cost perspective *Design trends* Designers looked at various stools types like four and three legged stools, bar stools, foldable stools, Eco-friendly stools etc.
Ideation and concept development	Concept sketching, concept selection, exploration of concept selected, building mock up/soft models for concepts, concept evaluation
Prototype development phase	• Appropriate material and manufacturing process selection • Manufacturer selection • Green stool manufacturing and its testing
Assessment phase	Final green stool evaluation by expert reviewer /evaluator

2.3 Description of Green Stool Design Cases

For this case study we are considering four sample stool design cases developed by the graduate design students for the following activities described in Table 1. Explanation of four green stools developed by designers as the project outcome is shown in detail here in Table 2.

3 EDT in Designers Context

The credit of pioneering experimental design approach goes to Sir Ronald Fisher who firstly used this technique for performing experiments in agricultural field during 1920s. Presently, the revolutionary impact of EDT in various industries has

Table 2 Detail description of four green stool sample cases

Stool name	The Raga	Curlicue	Eco stool	On stool
Symbol selected as a metaphor	Letter '**R/r**'	Letter '**S/s**'	**Recycle symbol**	**Power button symbol**
Meaning of metaphor and inspiration	Designer interested in Symbolizing '**3Rs**' of environment—**Reduce, Reuse, Recycle** in stool	The inspiration behind letter '**S**' which stands for **sustainability** in stool design	**Recyclability** expression in stool remind people responsibilities towards environment	Used power button as metaphor which can evoke a thought of **energy usage or saving**
Customer and market profile	Adults, unisex Global market	Adults, unisex Global market	Adults, unisex Global market	Adults, unisex Global market
No of concepts generated	30	40	25	20
Mock up models made	Paper model	Cardboard model CAD model	Paper model CAD model	Clay model CAD model
Materials and manufacturing processes	Cane strings Stainless steel pipe Bending, Welding, Knitting	Rubber wood Stainless steel pipe Bending, Welding, Cutting, Drilling	Polypropylene sheet Stainless steel pipe Bending, Welding, Cutting, Drilling	Wooden plank Stainless steel Bending, Welding, Cutting, Drilling
Final green stool prototype	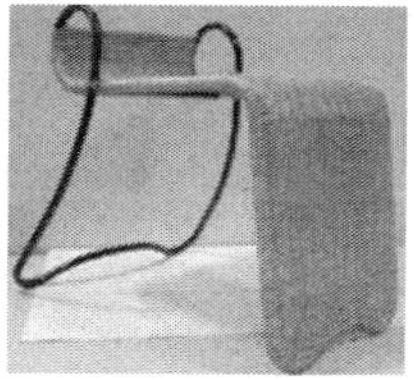			

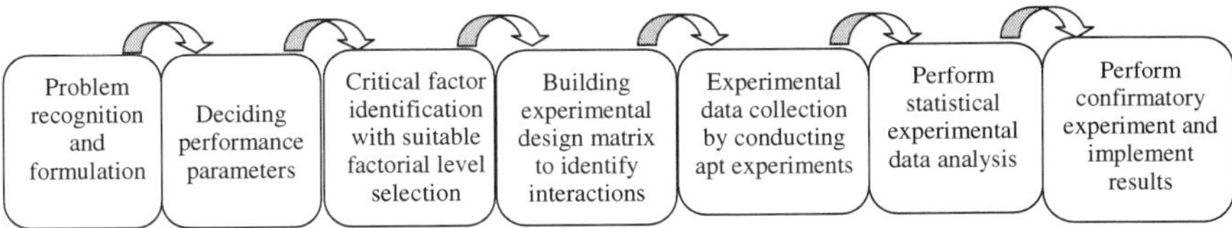

Fig. 1 Main stages of experimental design during NPD

broadened its application for NPD, redesigning existing product/process and product quality improvement. Experimental design is one of the most significant techniques, denotes plan for performing experiments during various stages of NPD process [12]. It helps in deciding 'performance parameters'—factors remain common throughout all stages of NPD and 'critical factors'—factors specific to individual stages. After setting the factorial levels for critical factors, 'experimental matrix' can be generated. This matrix can create enormous valuable information about possibilities in designing experiments for new innovative products development. By running appropriate limited experiments efficiently or checking possibilities from experimental design matrix; designers can save large portion of cycle time and resources in physical product development [13, 17]. Referring Antony et al. the seven stages involved in experimental design process during NPD is shown in Fig. 1 [5, 15, 16].

Ellekjær and Bisgaard [17] elaborated the objective of experimental design during four product development stages namely, conceptualization, prototype building, manufacturability/pilot production, and final production stage as shown in Table 3.

Though many authors have vastly described the application of experimental design, in reality there is a gap between ED and its practical application which is restricting designers and product developers to use this technique on real time product development problems. As per Carlsson [14], even the best industries fail to use experimental design into practice due to poor awareness, knowledge and use of experimental design [5, 14].

Table 3 Objectives of EDT during four product development stages [17]

Ideation and conceptualization	Prototype development	Manufacturing and pilot production stage	Final production
New idea generation	Product performance	Product manufacturability	Process simplicity and process yield
Test existing and establish new theories and concepts	Robustness and reliability	Sensitivity to component variation	Quality of product and process
Validating proof of concept	Sensitivity to component variation	Tolerancing (parameters allowed limit specification)	Reliability of product and process
Handling if-else situation	Simplicity	Simplicity, reliability and cost	Cost

4 Post Facto Analysis

Our post facto analysis for this case study considers four green stool design cases reported in Sect. 2.3. This analysis was conducted on the basis of investigation on stool project task, the design brief made by the four design students, understanding the development stages of green stool project, final functional stool prototypes, interviews of designers and discussion with the instructor. As a part of post facto analysis EDT is partially (first three stages shown in Fig. 1 excluding factorial level selection) used to identify major performance parameters and most prioritized critical factors. The outcomes of EDT and stage gate strategy are integrated to propose the role of experimental design based go-no-go approach. This approach has a scope for understanding inputs and outputs to decision gates during NPD and product/process improvement during redesign.

'Performance parameters' were the major factors/features which designers strictly tried to imbibe in their final green stools prototypes and carried throughout all development stages. These were strict constraints which closely guided the final presentation of the green stool. 'Critical factors' were those prioritized factors/ features which were very specific for each stool development stage that may influence performance parameters.

The critical factors in evaluation are nothing but the core features, that expert evaluator considered for the evaluation of four stool designs. For seeking the market potential of green stool prototypes, the expert evaluated these stools from customer's point of view understanding the customer profiles chosen by the designers in their project design brief. Though form development was the major objective for this project; evaluation of the final green stools was based on—sensory, functional and emotional properties of the stool which customers usually seek during such product purchase. Table 4 shows performance parameters and critical factors identified and common for all four green stool design cases as per their priority.

For quality based evaluation of a product; Garvin [11] has proposed eight critical dimensions of the product quality related to the product development and these serve as the pillars for strategic analysis namely,—performance, features, reliability, conformance, durability, serviceability, aesthetics and perceived quality [18]. The common quality based evaluation criteria for these functional green stool prototypes are identified after a rigorous investigation of the stool prototypes and discussed with the evaluator to understand his evaluation perspective. Finally, with common evaluation thinking; identified quality evaluation criteria are categorized according to Garvin's eight critical dimensional quality aspects as described in Table 5. Actual evaluation of the four sample stool prototypes was done based on most of the critical features mentioned below.

Table 4 Performance parameters and critical factors for various stool development stages

Performance parameters—factors/features common to all stages of green stool design

- *Sustainability*—Usage of eco-friendly materials and manufacturing processes, minimum use of resources (materials, manufacturing processes, energy etc.)
- *Durability*—Robust stool design for long service life and it should be strong enough to take weight up to 100 Kgs in sitting and standing position
- *Portability*—Easy to move from one place to other place or easy transportation
- *Mass manufacturability*—easily bulk produced
- *Emotional and religious sentiment*—finally developed stool should not hurt emotional and religious feeling of user using it
- *Critical factors*—factors/features completely exclusive to individual stages to green stool design

Critical factors in ideation and conceptualization stage

Critical factors in theme (as a metaphor) selection

- *Method of ideation*—precise selection of ideation method for generating metaphors
- *Expressive Style of theme*—fashion, manner, mode, trend of theme selected etc.
- *Feasibility of theme*—extensive possibility of theme exploration during ideation
- *Form variety*—to understand diversity range of shape, figure, outline, structure etc.
- *Theme flexibility*—capability to undergo with change as per designers willingness

Critical factors in concept development and selection

- *Method used for Conceptualization*—thumbnail sketches, doodling, paper/clay models, playing with scrap
- *Supporting symbol form*—retaining the form of stool concept as per symbol selected
- *Concept novelty value*—stool concept newness and avoiding duplication of existing ones
- *Concept transforming viability*—practicability of concept transforming into actual stool
- *Concept Catchiness*—potential of stool concept to attract viewers/customers
- *Concept selection criteria*—evaluation criteria set by designer and expert reviewer

Critical factors in green stool development

- *Stool performance analysis*—refined concept analyzed for performance prediction using various s/w tools
- *Mock up modeling*—Stool concept feasibility checking by making clay, soft models, CAD models, paper models etc.
- *Materials and manufacturing process selection*—selecting materials and processes which lead towards green /eco-friendly stool design
- *Resources/design variation sensitivity*—thinking about alternatives materials and manufacturing processes or alternative modified design in case of non availability of manufacturing facility
- *Simplicity*—predicting ease of use, manufacture and maintenance

Critical factors in manufacturing and production

- *Manufacturer selection*—checking availability of manufacturing process facilities and feasibility to manufacturer to develop stool as per finalized technical specifications
- *Method of testing*—to test stool strength and performance—force application while sitting and standing, joint checking, weight balance, fittings etc.

Critical factors in evaluation of sensory properties

- *Visible aesthetical stool characteristics*—customers attention spots in stool prototype
- *Stool features*—features that customers can grab just by using the stool
- *Impression of eco-friendliness of stool*—specific characteristics showing sense of green stool design

(continued)

Table 4 (continued)

• *Simplicity in design*—ease of use, green stool depicts obvious metaphor behind its design
Critical factors in evaluation of functional properties
• *Basic performance features*—most fundamental functions of stool
• *Durability, Reliability, Conformance and serviceability*
Critical factors in evaluation of emotional properties
• *Emotional attachment towards product*
• *Hidden insights about product*
• *Perception about product quality and features*

Table 5 Identification of stool quality evaluation criteria

Critical features related to product concept quality	Quality evaluation criteria related to green stool
Performance	Sitting comfort, Weight balance (in standing and sitting position), Portability
Features	Colour, Shape, Texture, Form, Rhythm, Sleek, Eco-friendly design, Contemporary/Modern design trend, Skillfulness in design,Customer sensitive design, Harmony, Sustainable resource usage, Elegance, Light weight and sturdy, Easy to clean, Cost
Reliability	Mass production and storage, User friendliness, Seat design, 3R's (reduce, reuse, recycle), Stool life
Conformance	Adherence to specifications and symbol, Conformance on shape /configuration /contour, Sense of robustness, Sense of comfort, Healthy sitting, Product completeness, Damage proofness, Sense of green stool
Durability	Stool strength, Stability and support
Serviceability	Manufacturing supportability, Stool maintenance and repair, Stool component replacements, Competence
Aesthetics	Stool beauty, Catchiness, Appearance, Style, Artistic sense, Ratio proportion, Symmetry
Perceived quality	Affordability, Sense of pride, Inherent cultural values, Customer adaptability, Deep product thinking, Intelligence in product, Market stool trends, Visual unity of stool design

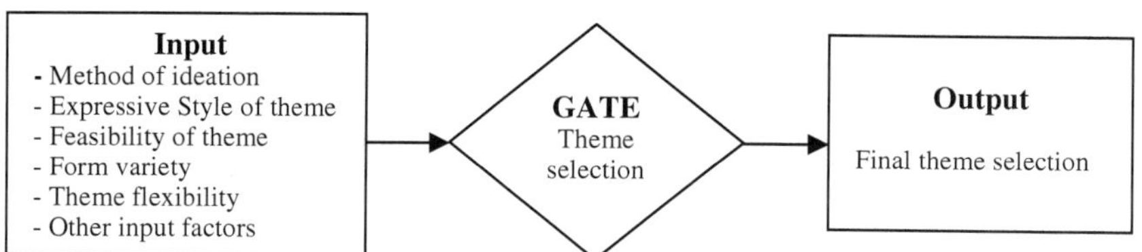

Fig. 2 Input–output diagram for decision gate

5 Experimental Design Based Go-no-go Approach in Green Stool Project

The Go-no-go decision gates are product development stage driven decision points where customer–market requirements, evaluation criteria and product lifecycle activities are closely reviewed by expert reviewers from product idea-to-launch [2, 7, 19]. During stool design project; design students navigated through stool development stages which were interpolated by number of go-no-go decision points. Each previous stool development stage had not proceeded to the next stage without mutual 'go' decisions of student and instructor (expert reviewer) while 'no-go' was a point of disagreement for the designer from reviewer to reaffirm the stage-gate again. In this product form development project; from theme selection to final green stool launch process was a funnel with seven development stages and a number of go-no-go decision points.

Using Table 4 where experimental design is used for stool development process; we can clearly identify the stool development stages and the important decision gates for designers. The identified critical factors have given more clarification towards the inputs and the outputs of green stool project decision gates. For example—'selection of theme as a metaphor' was one of the important decision gates (decision making point) for the designers during ideation stage, where inputs and outputs to decision gates can be highly predicted by considering critical factors (refer Table 4) in theme selection. The general input output diagram for 'theme selection' gate is shown in Fig. 2. Similarly the input–output diagrams for other decision gates can be made for better decision making.

The development stages and the decision gates were confirmed with designers and the reviewer. EDT based stage gate strategy is used as a representation tool for showing go-no-go gates as shown (Fig. 3), the right side shows stool development stages with go-no-go decision gates in between depicted by a diamond outline [7, 9, 10, 19].

6 Conclusions

Experimental design technique in combination with stage gate strategy is used for a post facto study of the green stool design project of graduate students, who used a generic NPD process. We tried to explore the EDT approach and described it during various NPD stages from designer's perspective. Using integrated experimental design based go-no-go approach, designers can understand development stages, decision gates in NPD process and can take decisions in better sense at particular gates to improve their decision making abilities. This can be done by understanding inputs and outputs to particular decision gate. Inputs to decision gate can be highly predicted by looking at the most prioritized critical factors identified using EDT at that particular stage. This 'input-gate-output' insight from

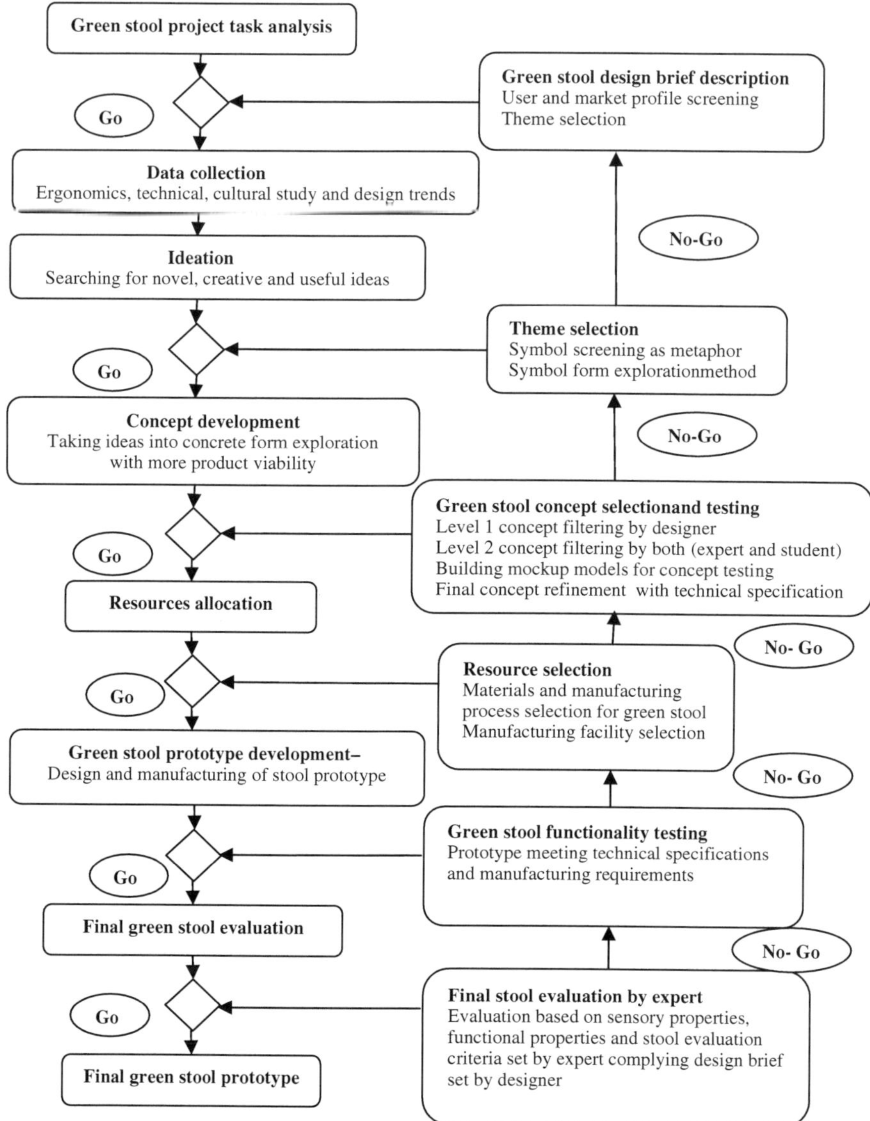

Fig. 3 Experimental design based go-no-go approach for green stool project [9] (inputs and outputs to decision gates are not shown in this figure)

experimental design can help to decide 'go' and 'no-go' decisions at a particular stage-gate in the NPD process. More analysis of this project with all stake-holders as decision-makers will throw insights regarding the decisions made at various stages. This combined approach can be used for any NPD and product redesign activity.

Acknowledgments We appreciate the work of designers Anand, Ankit, Pragati and Subin at CPDM, IISc.

References

1. Clark KB, Wheelwright SC (1995) The product development challenge: competing through speed, quality, and creativity. In: A Harvard Business Review Book. Harvard Business School Press, Boston
2. Cooper RG (1993) Winning at new products—accelerating the process from idea to launch, 2nd edn. Perseus Books, Cambridge
3. Leavitt P, Brown M, Wright S (2004) New product development: a guide for your journey to best-practice processes. American productivity and quality center, Huston
4. Montgomery DC (1999) Experimental design for product and process design and development. R Stat Soc Stat 48(2): 159–177
5. Zhang Z (1998) Application of experimental design in new product development. TQM Mag 10(6):432–437
6. Blake S, Launsby RG, Weese DL (1994) Experimental design meets the realities of the 1990s. In: Quality progress, pp 99–101
7. Cooper RG (1990) Stage-gate systems: a new tool for managing new products. In: Business Horizons, pp 44–54
8. Hart S, Baker M (1994) Learning from success: multiple convergent processing in new product development. Int Mark Rev 11(1):77–92
9. Tzokas N, Hultink EJ, Hart S (2004) Navigating the new product development process. Ind Mark Manag 33: 619–626
10. Cooper RG (2009) How companies are reinventing their idea-to-launch methodologies. Res Technol Manag 52(2): 47–57
11. Garvin DA (1987) Competing on the eight dimensions of quality. Harvard Bus Rev 65(6): 101–109
12. Almquist E, Wyner G (2001) Boost your marketing ROI with experimental design. Harvard Bus Rev 79(9): 135–141
13. Page AL (1993) Assessing new product development practices and performances: establishing crucial norms. J Prod Innov Manag 10:273–290
14. Carlsson M (1996) Conceptual and empirical aspects of TQM implementation in engineering organizations. In: The R&D management conference: quality and R&D. Twente Quality Center, Enschede, pp 84–99
15. Antony J, Perry D, Wang C, Kumar M (2006) An application of Taguchi method of experimental design for new product design and development process. Assembly Autom 26(1):18–24
16. Antony J et al. (2001) 10 steps to optimal production. Quality 40(9): 45–49
17. Ellekjær MR, Bisgaard S (1998) The use of experimental design in the development of new products. Int J Qual Sci 3(3): 254–274
18. Naes T, Nyvold TE (2004) Creative design: an efficient tool for product development. Food Qual Prefer 15: 97–104
19. Cooper RG (2005) Stage-Gate is a registered trademark (www.stage-gate.com) of the Product Development Institute Inc (www.prod-dev.com)

Design Professionals Involved in Design Management: Roles and Interactions in Different Scenarios: A Systematic Review

Cláudia Souza Libânio and Fernando Gonçalves Amaral

Abstract: Design management has gained relevance among practitioners and researchers worldwide, and many have been the attempts to conceptualize it formally over last years. However, the role of design professionals acting in design management, as well as their interactions with other team members are seldom explored. This article examines the characteristics of design professionals in design management at both Brazilian and international levels. A literature systematic review was carried out in order to map these scenarios, highlighting different professional profiles according to their working environment and the organizational structure. Through this study, the various roles of design professionals involved in design management were identified, their personal profiles were examined, as well as the main differences arising from two different scenarios.

Keyword Design management · Human factors · Design activity · Designers

1 Introduction

Over the past decades design management has been discussed and conceived as a multidisciplinary activity that brings together working partners and integrates design within an organizational environment [1]. Aiming at a formal conceptualization and

---

C. S. Libânio (✉)
Business Administration UniRitter, Porto Alegre, Brasil Rua Orfanotrófio, 555 Porto Alegre, RS, Brazil
e-mail: clasl@terra.com.br

C. S. Libânio · F. G. Amaral
PPGEP UFRGS, Porto Alegre, Brasil Av. Osvaldo Aranha, 99, Porto Alegre, RS, Brazil
e-mail: amaral@producao.ufrgs.br

A. Chakrabarti and R. V. Prakash (eds.), *ICoRD'13*, Lecture Notes in Mechanical Engineering, DOI: 10.1007/978-81-322-1050-4_69, © Springer India 2013

attempting to arrive at a meaning that can be effectively used in companies, some design management definitions have been proposed that highlight not only the levels of organizational activities but also their actors along with their respective roles and forms of action. The Design Council [2], for instance, defines design management as the total activity of design, which goes from implementing and organizing the entire development process of new products and services to managing and achieving a company's best performance. The Portuguese Design Center (CPD) [3] defines design management forms of action inside companies in a twofold manner: first, at project level, in which a design manager's role is to manage people, services and products during all phases of a project; and, second, at corporate level as a whole, in which a design manager will foster the development of new products favorably.

Regardless all the attempts at conceptualizing design management, the role of designers within design management, as well as their interactions with other team members and their personal profiles are seldom explored. In such a context, this article examines the characteristics of a design professional at both Brazilian and international levels through a systematic review of research literature available on the subject that specifically refers to the role of designer in design management. First, an analysis of a design professional profile as suggested by a few authors, including their competences, skills, and personal characteristics. The main differences perceived within local and international scenarios are pointed out, not only with respect to a designer's profile but also to his or her professional training. Finally, a few remarks are made and suggestions for further research are presented.

1.1 Design Professionals

In companies, design management consists on managing all aspects of design at two levels: the corporate and the project level [3, 4]. At corporate level, a design manager acts on the company's strategies by fostering a design culture and aligning it with the corporate goals. According to CPD [3], at this level a manager will be responsible, among other tasks, for connecting activities with the company's strategies. At project level, however, the design professional is focused on managing the company's projects operationally.

At corporate level, therefore, design professionals can act either in close collaboration with the company's senior staff or in a design department, being it a department internal to the organization or an outsourced design service. Borja de Mozota [5] argues that a few design tools should be used in corporate decision-making processes, such as design input at senior staff level or within a company's own design department, among others. She also notes that at senior staff level design activities can be the sole responsibility of a design director, a design manager, or a director who belongs to a quality control and design area or a communications and design area. It is inside a design department that the positions of project manager and designers are to be found.

The roles and responsibilities of a design manager are strictly related to the size and structure of a particular company. According to Pereira [6], in large sized companies the design manager's role is that of supervising teams engaged in each project, adapting him/herself to the particularities of each. He or she has to demonstrate a broad view of the whole enterprise and foster integration among the company's operational and strategic sectors. In small sized companies, on the other hand, the author argues that a design manager's task is to identify actions that contribute to integrate projects with a company's pre-established goals, even though he or she is not responsible for supervising each business unit. The design manager must pursue actions leading to the integration of project development with the corporate goals. Regardless the company's structure, Best [4] states that the design manager needs to promote the best design strategy as possible when conceiving an organization as a whole. The author also argues that for this to happen stakeholders should first be convinced of the significance of a particular design strategy for the company. Several are the abilities or skills required for a design manager position. Among them we will find leadership, entrepreneurship, vast technical knowledge, managerial skills, and a proactive profile, to mention a few. A design manager should be able to coordinate, motivate and persuade a team. Bruce et al. [7] highlight the individual competences among the essential features required for good designers, arguing that lacking them is a major cause for failure in design projects. Some views of a designer's competence are described below.

1.2 Designer's Competences

Nedo *apud* Bruce et al. [7] mention the following competences for designers: objective creativity; technical knowledge; color and conceptual design; organizational, planning, problem solving, and commercial skills; commitment; enthusiasm; self-confidence; results orientation; team orientation; strategic thinking; consumer/customer focus; relationship-building skills; presentational skills; and flexibility. Among them, we can find competences related to knowledge, abilities, and attitudes.

Ruas et al. [8] argue that the notion of individual competences bears different lines of thinking based on different approaches, such as the Anglo-Saxon and the French. Anglo-Saxon scholars share a more pragmatic perspective, while members of the "French school" add elements from Sociology and Labour Economics to the notion of individual competences. According to the authors, the concept of competence must not be mistaken for that of performance, for the latter can be seen as a quantification of performance, while the former consists of a tool for achieving a desired performance. With reference to competences and skills, Borja de Mozota [5] suggests five competences in design, such as: compromise, enthusiasm, self-confidence, results and team orientation, high standards, creativity, technical and conceptual ability, organizational, planning, problem solving,

Table 1 Designer competence model, according to Ruas and Mozota

Competences	Description
Knowledge	Technical, scientific, concept, and color
Skills	Creativity, strategic thinking, presentations skills, commercial skills
Attitudes	Commitment, enthusiasm, self-confidence, results orientation, relationship building, problem solving

Source Adapted from Ruas et al. [8] and Borja de Mozota [5]

commercial skills, gathering and using of information, strategic thinking, consumer focus, relationship building, influence, presentation skills and flexibility. Based on the classification of competences proposed by Ruas [8], it is possible to group together the designer skills referred to by Borja de Mozota [5], as shown in Table 1.

According to Ruas' and Mozota's competence model for designers (Table 1) and Moura and Bitencourt's understanding of competences [9], all the technical and scientific competences can be represented as knowledge, or knowing, such as, for instance, color and concept knowledge. Skills, in turn, are those related to knowing how to do, such as creativity, strategic thinking, as well as presentation and commercial skills. Attitudes, conceived as knowing how to act, relate to commitment, enthusiasm, results orientation, self-confidence, relationship building, and problem solving.

In the scenario view briefly described, a systematic review of research literature on design management was carried out aiming, first, to have a better understanding of the profile of design professionals in companies, and, second, to identify different perspectives on design professionals' role in Brazil and abroad. In the following sections, the methodological procedures are described, the main findings are presented, and a few guidelines for future research are suggested, as well as some paths to be followed.

2 Methodological Procedures

A systematic review of applied and exploratory nature was carried out for the purposes of the present study. Both Brazilian and international research literature were reviewed in an attempt to contrast Brazilian and foreign views on the subject of design management. For the Brazilian literature review, theses and dissertations dealing fully or in part with the role of design professionals in design management written within the past 20 years were searched over postgraduate programs offered at local universities. For the international literature review the Science Direct Info website was the main source for searching articles published over the same period.

This search was carried out initially on the WWW using digital databases as a primary search engine. In the Brazilian search the following databases were accessed for finding theses and dissertations: CAPES Website, Domínio Público Website

(Public Domain), and the Brazilian digital library for theses and dissertations available at the Brazilian Institute for Information in Science and Technology (IBICT) Website. The keywords used for searching materials in Brazilian Portuguese were: gestão de design (design management), design estratégico (strategic design), designer, and gestor de design (design manager). A refined search was made among the 278 publications retrieved aiming at excluding the ones that did not fall under our subject of research. As a means of checking whether those research works dealt fully or in part with the professional designer's role in design management, a selection was made based on the reading of paper abstracts, keywords, and introductions whenever they were available on-line. After refining the selection of theses and dissertations, their respective bibliography was analyzed as an attempt at identifying works that had not been found on the initial search. Two dissertations and one thesis were found during the analysis.

The Science Direct Website was the main scientific database used for international research literature retrieval, and the following search options were applied: journals only, titles published within the last 20 years, containing the terms "design management," "strategic design," "design manager," and "designer," and within the subject area of design management and related fields, such as business, management, decision sciences, arts, humanities, engineering, and finance. The search resulted in 913 publications retrieved, of which 880 had to be excluded for not dealing specifically with our theme of interest. The large number of works excluded was due to a wide range of meanings of the term "design". This refinement was made through a careful reading and analysis of the title, abstract, and keywords of each publication.

In brief, the Brazilian search has identified 40 research publications partly or fully addressing the role of the design professional in design management. Figures for the international literature search amounted to 33 articles for the same criteria.

3 Results

Forty Brazilian publications—3 theses and 37 dissertations—and 33 articles published in international journals were found that dealt with the role of designers in design management. All publications retrieved and analyzed were organized as follows: journal of publication; region of publication; keywords; year of publication; and reference to designers in design management in Brazilian and internationals publications.

3.1 Journal of Publication

Design Studies is the main publication journal comprising 17 out of a total of articles. Computers in Industry appears with 3 articles. Technovation, Automation in Construction and Long Range Planning contributed with 2 articles each. Finally, Journal of Materials Processing Technology, International Journal of Production Economics, International Journal of Industrial Ergonomics, International Journal of Human–Computer Studies, Procedia–Social and Behavioral Sciences, Computers and Industrial Engineering and Artificial Intelligence in Engineering appear with 1 article each.

3.2 Region of Publication

In Brazil, most of the research works were written in the South region, amounting to 24 publications. The South–East region appears next with 14 publications and the North–East with 2 publications addressing designers in design management, as shown as Fig. 1.

Figure 2 shows the countries of origin of the international publications. The United Kingdom has contributed with 22 of the authors, followed by China with 9, and the Netherlands with 7 authors. Then, there are 6 authors from France and Spain, 5 from Italy, and 4 from the United States and Taiwan. Thailand has contributed with 3 authors, and Iceland, Denmark, Korea, Australia, Finland, Poland, and Turkey with 1 author each.

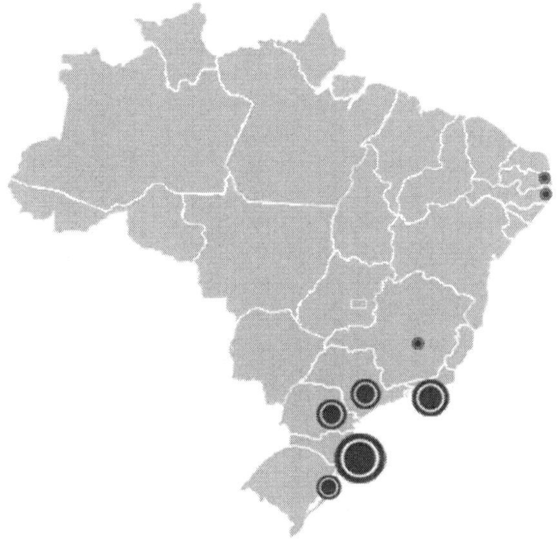

Fig. 1 Number of theses and dissertations per State in Brazil. *Source* The authors

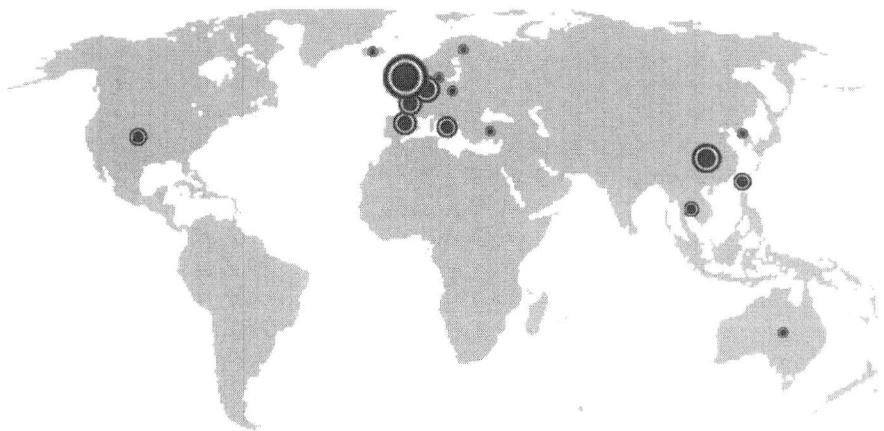

Fig. 2 Authors per country (except Brazil). *Source* The authors

3.3 Keywords in Brazilian Publications

The most frequently used keywords in the Brazilian publications searched were "design", "gestão de design," and "gestão do design" (both are translated as "design management"), with 18, 12, and 8 occurrences, respectively. These keywords plus "estratégia" ("strategy"), "competitividade" ("competitiveness"), "inovação" ("innovation"); and "design estratégico" ("strategic design") correspond to 80 % of total occurrences. The other keywords "gestão" ("management"), "projeto" ("project"), and "MPEs" (an abbreviation standing for Micro and Small Businesses), "Design industrial" ("industrial design"), "metodologia de projeto" ("project methodology"), "processo de design" ("design process"), "moda" ("fashion industries"), and "polo moveleiro" ("furniture industries") appear to be less important in the research publications analyzed.

3.4 Keywords in International Publications

The most frequently used keywords in the international articles can be seen in Fig. 3. The term "design management," the keyword with the greatest number of occurrences, appears in 16 out of 33 international articles, which is indicative that more than 50 % of the publications has focused directly on the subject of design management. The keyword "case study(ies)" appears next, pointing to a research procedure used in some research works. "Collaborative design," and "product design" are used in 4 publications each. "Design process(es), "Design cognition", "Communication", and "Design strategy(ies)" are used in 3 publications each. All of these terms represent 80 % of the total keywords found in the publications, as we can see in the line in Fig. 3.

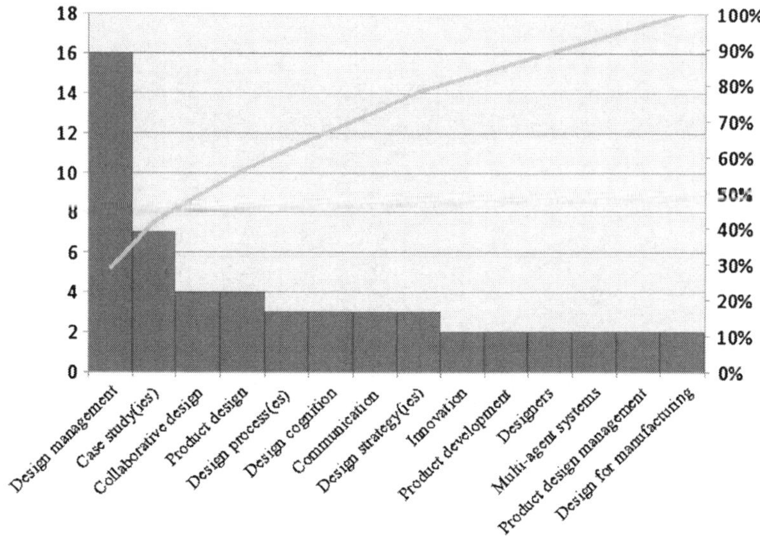

Fig. 3 Most frequently used keywords in the international articles. *Source* The authors

Keywords used in common by both Brazilian and international studies are: "design management", "innovation", and "design process(es)".

3.5 Year of Publication

For the years 1994 and 1996 only Brazilian publications were found according to the search criteria used, while in 1999 the international publications begin to appear. On the whole, the last five years have concentrated more than one half of the publications under research as shown in Table 2.

3.6 Reference to Design Professionals in Brazilian and International Publications

The roles and functions of design professionals in design management were referred to in 16 Brazilian publications. In the 24 remaining ones neither were those professionals clearly identified and acknowledged, nor were their roles and functions clearly defined.

Seventy-one percent of the international publications have mentioned precisely the essential characteristics and requirements for design professionals, such as competences (knowledge, skills, attitudes), leadership, entrepreneurship,

Table 2 Comparison of publications in Brazil and abroad per year

Years	Brazilian publications	International publications	Total number of publications
1994	1	0	1
1996	1	0	1
1999	1	1	2
2000	1	2	3
2001	3	3	6
2002	1	0	1
2003	3	6	9
2004	2	1	3
2005	4	4	8
2006	6	2	8
2007	1	5	6
2008	4	2	6
2009	7	3	10
2010	5	4	9

proactivity, communication, and integration abilities. In the searched articles the designer is constantly mentioned, but the figures of a manager or a design manager are referred to only in 27 % of those studies.

4 Discussion

The Brazilian theses and dissertations published over the 1990 do not address in a clear and well defined manner the design manager's role, and sometimes they deal briefly with the functions of that professional within design management. It can be noted that the studies are focused on the phases of productive processes, giving no emphasis at all to the integration of a design team with other departments inside a company. Nevertheless, the designer seems to be assuming an important role in the scope of product development. There is only one international article published during the 1990s (in 1999) that presents a framework for design management and includes the design manager in that framework. This article also addresses the characteristics of a designer and strongly highlights the designer's competences. The figure of a design manager, or of a designer with a wider access and integration to a company's high staff, has gradually begun to appear in the Brazilian publications since the 2000s. In that period emphasis has been given also to a greater integration of the teams involved in each project.

Among the researched articles published from 2004 to 2010 it is possible to observe an increasing debate around the figure of the design manager. This suggests that sometimes in large scale companies there will be more room for a design professional who acts directly in cooperation with the company's high staff and who is directly linked to another design professional: the design team's project manager. On the other hand, in small-sized companies, the professional in charge of design management can be the project management team leader him/herself, who also has a direct connection and access to the company's high staff. Martins [10] asserts that a design manager's task includes all the creation process of tangible and intangible products. For large companies, the author states that it is a design manager's task the supervision of teams involved in each project and that he or she should adapt him/herself to the particularities of their actions. Pereira [6], in turn, states that a holistic view of the business scope is needed in order to integrate the operational units into the context of a company's global strategy. Martins [10] also suggests that, in small-sized companies, the task of a team supervisor is excluded due to the absence of business units. The author emphasizes, however, that a design manager has to define actions guiding the project process integrated to the organization's goals and strategies.

In the 2004–2010 Brazilian publications [10–12], concepts of leadership and competence are related to the design manager's profile. Those attributes are relevant, therefore, and point to an urge for a broader understanding not only of a designer's position dimensions, but also of their significance to a successful performance in that function. In the international publications, a deeper understanding is shown with regard to the significance of the design team's integration with other participants of a project [13, 14]. Concepts such as shared knowledge [15], organizations' and professionals' competences [7, 16], design team leadership [17] are mentioned, as well as concepts of design as a knowledge agent and knowledge integrator inside organizations [18, 19]. In some study cases [14, 18], it is possible to notice that the researched companies have a sound knowledge of design management and demonstrate to use design policies as well. A large part of the international articles, however, do not mention specifically the figure of a manager or a design manager.

A major issue addressed in some of the Brazilian publications [20–22] refers to local academic training programs in Design. More specifically, these publications deal with professional training for young designers entering the job market and focus on the poor knowledge acquisition of management concepts in the scope of undergraduate programs in Industrial Design. Issues are raised concerning design manager's training, and some authors argue that such a professional could not necessarily be a designer. According to Gallina [21], it is of primordial importance that the professional responsible for a design department inside a company should not only have a clear prospect of a sector potentialities and opportunities, but also take them into account and use them as values to be built into every aspect of a project. In the international articles analyzed, on the other hand, designer's training does not seem to be an explicit issue. Bertola and Teixeira [18] argue that, in Italy, many designers are entrepreneurs, which explains a tradition of applying design competences to manage available resources and guide business strategic decision-making processes. Candi [23] goes further by stating that quite often there is design even without the figure of a designer, or without acknowledging his or her figure as such. Our reading and analysis of those articles reveal that such a little mentioning of the figure of a manager or design manager is due to the design professionals profile abroad. Not only professional training, but also the personal characteristics and skills of international designers help them to have a job profile that fully matches the requirements of a design manager position.

5 Conclusions

The data collected in our research suggest that there has been an increasing and gradual change in the Brazilian and international understanding of the roles, responsibilities, and skills needed for a design professional over time. The figure of the design manager, however, has not been clearly defined, nor has his or her role in design management been clearly understood yet.

Certain professional characteristics have been discussed in specialized journals first at international level, and later at Brazilian level in academic publications. Leadership, autonomy, competences (knowledge, skills, and attitudes), entrepreneurship, pro-activity, communication, integration and teamwork have been highlighted as major characteristics of design professionals, both in the Brazilian and international publications examined. Regardless of their mentioning, those characteristics have not been studied in depth in the searched publications. Furthermore, a great professional development can be noticed abroad when contrasted to the Brazilian context. It is possible to conclude that the difference between the Brazilian and the international contexts is a result of the professional environment in which a designer works, but a great deal of that difference is due to our current professional training, as well as to the individual personal profile.

In face of the relevance of a deeper understanding of design management by designers, a question may be raised whether Brazilian professionals are appropriately prepared to assume leadership and other management roles inside organizations. A further question may also be raised, then, with respect to a more appropriate training for future design managers. This brings to the fact that we lack a wider perspective on the subject, for very few are the Brazilian undergraduate and postgraduate schools that include disciplines in the field of management in their regular course programs.

In conclusion, this research points to an urge to intensify the exchange of knowledge among Brazilian researchers in different regions of the country, thus increasing the number of papers presented in local conferences and research articles published in Brazilian journals, for this type of publication is easier to be accessed than theses and dissertations. By fostering the spread and exchange of knowledge on the understanding of design management from a Brazilian perspective, studies developed here shall not only succeed in keeping up with leading research works abroad and global trends on the field of design management, but also contribute more effectively to the development of the designer profession as a whole at Brazilian level.

References

1. Libânio CS, Amaral FG (2011) Aspectos da Gestão de design abordados em Dissertações e Teses no Brasil: uma Revisão Sistemática. Revista Produção Online 11(2):565–594
2. Design council (2010) Available at: www.designcouncil.org.uk. Accessed Aug 2010
3. CPD—Centro Português design (1997) Manual de Gestão do design, Porto, Portugal
4. Best K (2006) Design management: managing design strategy, process and implementation. Ava, Switzerland
5. Borja de Mozota B (2003) Design management: using design to build brand value and corporate innovation. Allworth, New York
6. Pereira D (2009) Competências da Gestão Estratégica do Design no Pólo Moveleiro do Alto Vale do Rio Negro (SC), Master Dissertation, Universidade Técnica Federal do Paraná, Ponta Grossa
7. Bruce M, Cooper R, Vazquez D (1999) Effective design management for small businesses. Des Stud 20(3):297–315
8. Ruas RL, Antonello CS, Boff LH (2005) Os novos horizontes da gestão: aprendizagem organizacional e competências. Bookman, Porto Alegre
9. Moura MCC, Bitencourt CC (2006) A articulação entre estratégia e o desenvolvimento de competências gerenciais. In: RAE, São Paulo: vol 5(1)
10. Martins RFF (2004) A gestão de design como uma estratégia organizacional: um modelo de integração do design em organizações, Teses, Universidade Federal de Santa Catarina, Florianópolis
11. Brito da Silva S (2009) As forças da gestão do design nos níveis estratégico, tático e operacional: um estudo de caso na Electrolux do Brasil S.A, Dissertação (Mestrado), Universidade Federal do Paraná, Curitiba
12. Dickie IB (2010) Gestão de design aplicada: estratégias de comunicação no contexto do design sustentável, Dissertação (Mestrado), Universidade Federal de Santa Catarina, Florianópolis
13. Reid FJM et al (2000) The management of electronics engineering design teams: linking tactics to changing conditions. Des Stud 21:75–97
14. Lauche K (2005) Job design for good design practice. Des Stud 26:191–213
15. Kleinsmann M, Valkenburg R (2008) Barriers and enablers for creating shared understanding in co-design projects. Des Stud 29:369–386
16. Belkadi F, Bonjour E, Dulmet M (2007) Competency characterisation by means of work situation modeling. Comput Ind 58(2):164–178
17. Lee KCK, Cassidy T (2007) Principles of design leadership for industrial design teams in Taiwan. Des Stud 28:437–462
18. Bertola P, Teixeira JC (2003) Design as a knowledge agent: how design as a knowledge process is embedded into organizations to foster innovation. Des Stud 24(2):181–194
19. Girard P, Robin V (2006) Analysis of collaboration for project design management. Comput Ind 57:817–826
20. Avendaño LEC (2002) Resgate do Protagonismo do Desenhista Industrial Através da Gestão do Design. In: 5° CIPD, Anais..., Rio de Janeiro
21. Gallina MC (2006) Ações do líder na gestão design como auxílio na formação do branding: um estudo de caso em uma empresa do setor do mobiliário, Dissertação (Mestrado), Universidade Federal do Paraná, Curitiba
22. Andrade MB (2009) Análise da Gestão de Projetos de Design nos Escritórios e Prestadores de Serviços em Design de Porto Alegre: Proposta baseada em estudos de caso, Master Dissertation, Universidade Federal do Rio Grande do Sul, Porto Alegre
23. Candi M (2010) The sound of silence: re-visiting silent design in the internet age. Des Stud 31(2):187–202

Design Professional Activity Analysis in Design Management: A Case Study in the Brazilian Metallurgical Market

Cláudia Souza Libânio, Giana Carli Lorenzini, Camila Rucks and Fernando Gonçalves Amaral

Abstract This study aims to highlight the role of Design Management in two metallurgical firms in Brazil. Considering the importance of combining theoretical concepts to the business reality, the methodological procedures were divided into two exploratory steps. In the first one, a literature review was performed by bibliometric analysis about the theme 'Design Management', mapping the international scientific production developed over the last 20 years. In the second step, two case studies were developed. The companies chosen belong to the metallurgical industry in Brazil, considered a traditional segment in the country that needs to understand design as a strategic tool. The bibliometric analysis results showed the evolution on the subject, as well as highlighted the multidisciplinary nature of design management, encouraging a pluralistic view. In the research inside companies it was possible to realize the importance of design teams' integration with other areas that participate in the Design Management process.

Keywords Brazilian metallurgical firms · Designer · Design management

C. S. Libânio (✉)
Business Administration UniRitter, Porto Alegre, Brazil
e-mail: clasl@terra.com.br

G. C. Lorenzini · C. Rucks · F. G. Amaral
Production Engineering Post-Graduation Program, School of Engineering, Federal University of Rio Grande do Sul, Porto Alegre, Brazil
e-mail: giana@producao.ufrgs.br

C. Rucks
e-mail: camila.rucks@ufrgs.br

F. G. Amaral
e-mail: amaral@producao.ufrgs.br

A. Chakrabarti and R. V. Prakash (eds.), *ICoRD'13*, Lecture Notes in Mechanical Engineering, DOI: 10.1007/978-81-322-1050-4_70, © Springer India 2013

1 Introduction

The debate about Design Management (DM) is evolving and has been spread mainly by studies carried out in institutions such as the Design Management Institute (DMI) [1], the International Council of Societies of Industrial Design (ICSID) [2], the Portuguese Design Center [3], and the European Institute of Design (IED) [4]. The theoretical debate promoted academically can also be reflected in corporate practices, promoting several interpretations about the role of design inside enterprises [5–7]. The design can be understood as a multidisciplinary activity that combines creativity and product development processes within the organization. More than just associate design as an aesthetic feature of finished products, companies are investing in design as a strategic tool that helps to produce new concepts and innovative ways of using products [8–12]. Thus, definitions of the Design Management are arising and trying to find an explanation specifically applied to business.

ICSID adopted Tomas Maldonado's [13] concept, dated in 1977, noting that the act of designing something includes the coordination and integration of all factors involved in the process of giving form and meaning for a product. In 1990, Gorb [14] considered the relationship between the design features that a company has and the strategic objectives of this company, suggesting a definition of Design Management that puts together design and business.

Nowadays, the Design Management is understood as a broad and global action, able to position the design on a higher level of responsibility inside a corporation. In this sense, DM establishes the appropriate insertion of the design in the company through the management of human and material resources—from the conception phase of a project until the launch of a new product, integrating specialized areas to senior management [15]. In the context of this research, the Borja de Mozota's definition [11] of Design Management was adopted.

Visualizing DM as a "bridge between design and business", McBride [16] defended the leadership focused on design and strategy, transforming experiences, companies and opportunities. Thus the design becomes an agent of organizational knowledge, which enables companies to sustain strategic partnerships and better understanding of users, supporting innovation processes [17]. Therefore, the collaborative work of design professionals that encourages the design management is essential to create special conditions to innovate in an organizational arrangement [18].

In Brazil, some government programs have been implemented—such as the Brazilian Program Design (PBD), Program Design Brazil, Brazil Innovation and Technological Development Support Program (PRODETEC)—in an attempt to align practices with design process and innovation management, but there is still a certain lack of design professionals prepared to work in strategic positions within companies, as identified in national studies [19–21].

This research aims to highlight and analyze the role of Design Management in two metallurgical industries in Brazil. The choice for the metallurgy was done because of its national importance as a traditional export industry linked with another important industry in the country—the furniture industry which deals in a significant way with the main aspects of design—and because there was identified the need to implement best practices in product process development and design. This is also an important sector, which applies a lot of complex process and technologies, influencing suppliers, industries, and the segment as a whole. To enable the study, there was done a conceptual review through a bibliometric analysis on the subject of Design Management, followed by two case studies according to the Brazilian reality.

2 Methodological Procedures

This paper begins presenting the description of the methodological procedures performed, emphasizing the steps of the bibliometric analysis. The results are presented in conjunction with the discussion of case studies developed in two metallurgical firms.

2.1 Step 1: Bibliometric Analysis

As a first step, it was conducted a literature review, guided by an exploratory bibliometric analysis. The systematic evaluation on a particular branch of knowledge can show how some knowledge areas or topics have evolved and how they have contributed to the science and the development of others studies. Thus, this type of analysis procedure helps to identify trends and research centers; studies the dispersion and obsolescence of scientific literature; provides productivity and quality of studies and analyzes processes of citation and co-citation; measures the growth of certain areas and the emergence of new subjects [22, 23].

The bibliometric analysis in this paper was structured by mapping international scientific production about Design Management published in English. The period of time used in the present research was set from 1992 to 2010, considering the challenges in the design expertise, as pointed by Borja de Mozota et al. [24]. The analysis was started from a research question, which guided the sequence of subsequent proceedings. Each step has been defined, measured and held by two independent researchers in order to ensure the accuracy of the data obtained. Some other systematic reviews and bibliometric analysis were applied as sources of reference for the proceedings [25]. These studies investigated topics related to processes of innovation and new product development [26], innovation in manufacturing [27], the manufacturing strategy [28]. The Table 1 shows the pre-established criteria, in accordance with the objective proposed.

Table 1 Steps of the bibliometric analysis

Standard steps for the review of literature with bibliometric analysis	Steps performed in the review of literature with bibliometric analysis about design management
(1) Defining the scientific question	What is the current stage of intellectual production on the theme design management?
(2) Identifying the databases to be searched; keywords	The database researched was the Web of Science with the keyword 'design management'
(3) Establishing criteria for inclusion/ exclusion of items	The papers included must follow the criteria: – Article or review article; published from 1992 to 2010 – Article should present: design Management as a business strategy and area of knowledge in business, empirical studies on the strategic role of design in business and its relationship with organizational systems, specialized professional and skills development in the area – Articles related to the areas, selectable in search engines: management, industrial engineering, multidisciplinary engineering, manufacturing engineering, operations research management science, business, multidisciplinary sciences, development planning, social issues, communication, education scientific disciplines
(4) Conducting search in the databases and on the strategy set	From October to November (2011) searches were made on the papers, conducted by two independent researchers
(5) Comparing and setting the selection of articles	The search of the three researchers was compared, with identical results, with a total of 95 articles found
(6) Applying criteria for selection and exclusions	58 articles were eliminated, leaving 35 articles for full reading
(7) Reviewing the studies included and preparing critical summary	Critical summary containing the identification number of each article; database; English title, author(s), journal, year of publication, abstract, business sector, aim, method, major results/conclusion
(8) Quantifying and qualifying the data	Data were quantified from the support of analysis capabilities available at the database (Web of Science) and through the utilization of three different softwares (Sitkis, Histcite, and Ucinet)
(9) Presentating the results	Results organized from the main data extracted of the studies, their main objectives, countries of publication, and conclusions

Source Adapted of Brereton et al. [40]

2.2 Step 2: Analysis Performed—Case Studies

Analysis were carried out in two companies of the metallurgical market defined by convenience. Companies participating in this study were chosen because they are the two largest metallurgical companies in Southern Brazil, developers of products for the furniture industry and fashion, with award-winning design products nationally and also regulated by quality certification. Both companies are leaders in their market segment and retain professionals able to supporting a culture of strategic design. Although these companies do not represent the entire Brazilian metallurgical market, they are representative in this segment because of their performance over the years; their relation with other national industries; their influence in regional market.

The in-depth interviews were conducted [29] with design professionals holding senior positions in the Department of Product Development. Each interview was conducted individually, and it lasted approximately 90 min. For conducting the interviews, it was used a semi-structured and qualitative script. The set of questions in the script was based on information extracted from the main topics of the bibliometric review and current data of the industrial sector. The subjects investigated were: the perception of design by senior managers of the company, formalization and design processes, conflicts in design teams, requirements to integrate the design team. Data analysis was performed using the methodology of content analysis as proposed by Bardin [30].

3 Results and Discussion

The results presented here show the emphasis drawn from the literature regarding the proposed approach. Subsequently, the empirical results collected in two metallurgical enterprises were contrasted with the literature review to identify the current position of Design Management in these two Brazilian companies.

3.1 Design Management: Concepts and Propositions

Through the bibliometric analysis of literature it was possible to find ninety-five potentially relevant studies. After the reading of these studies, a new filtering was applied considering the research objectives initially proposed. It resulted in thirty-five international studies selected to final evaluation. The data extracted from these articles were supported by specific softwares (Sitkis, Ucinet, and Histcite).

According to the data collected, it was perceived a certain conservatism in both the methodological procedures and in the definition of the objectives proposed in literature. Of the 35 articles, 18 of them proposed principles, guidelines, methodologies or program implementation of the Design Management; 13 analyzed the Design Management from the concepts and processes for competitive advantage, and 11 studied the professional training of designers and teams in the process of Design Management. Regarding the methodological choices, 26 were case studies or experimental studies, and 15 were the propositions of models or modeling concepts from theoretical frameworks.

The results obtained in the first stage showed a positive development and a growing body of research in the field of Design Management. It was also noted that the implementation process of the design is interesting not only because of the current situation of a company, but also because of the development of management related to creative design skills, based on the real value of design as an alternative source of income and competitive advantage.

According to the studies, the design is part of a dynamic industrial perspective, driven by social, environmental, economic enterprises that move and change their processes towards innovation. Long processes of development in competitive markets may require coordination of multiple design activities in product development. Each activity requires analysis and recognition of problems and generation of relevant information [31], which implies Design Management to contribute to the design strategy.

The ways that the company chooses to organize its design relations with other industries, suppliers and their position in the chain proved to be decisive for the competitiveness achieved [32]. In larger companies, with processes formalized and design professionals integrated and well trained, the literature showed positive results for design within firms [33]. The measurement of performance under the guidance of design efficiency and effectiveness was also highlighted as an important point [34] when considering the different design activities in companies.

The volume of cases analyzed showed the positive impacts of design that has its activities managed in support of corporative and strategic objectives. Sectors such as furniture and clothing were recurring sources of research, however, there was noticed the lack of studies in other economically important sectors—such as the metallurgical sector—which encourages further research in this area.

3.2 Case Studies

Both companies evaluated are from the same market segment—metallurgy—but with different characteristics and histories. In order to maintain the identity of these companies preserved, these were addressed in this article as 'Company A' and 'Company B'.

Company A was founded 34 years ago and, nowadays, it represents a joint venture between a Brazilian and an Italian companies, producing metal accessories for the sectors of decoration and fashion. This metallurgical company is an innovative medium size enterprise, which focuses on quality standards and follows labor and environmental legislations. The design team of Company A is composed by: a professional with a degree and a Postgraduate course in Design, a professional with a degree in Design, three Undergraduate students in Design and eight professionals with technical course in Mechanics. The respondent has served the Company A for 13 years as Design Manager, and he was responsible for all product development.

Company B was founded 27 years ago and has been developing innovative accessories in metal for footwear, fashion clothes, and decoration. This enterprise is a traditional Brazilian metallurgical company, with medium size structure, which follows labor and environmental legislations. Two owners who work in the corporative management compose the executive board structure. The design team in this company has a Technologist in Fashion and Style with a Postgraduate degree in Design and a professional with a Postgraduate degree and a Master degree in Design. The respondent served as a Technical Analyst and was responsible for the development of products for the decoration's line, for the main product's collections development and for some specific projects.

3.3 *Product Development Processes in Two Metallurgical Companies*

It was observed that the two companies have been dealing with innovation in an incremental way, with focus on material innovation and improvement in quality, however these companies do not cover the entire production chain. It was pointed out by the companies to be an important factor the region where they are located, with other similar companies around using the same universe of materials, suppliers, references (in relation to international trends) and even the same manpower. It was also noted that innovation and new product releases still occur more often to follow national and international trends instead of an attitude to search for innovative solutions to supply the market and amaze their customers.

In Company A the request of a product development starts with the customer, who makes the order to the Commercial Area. The Commercial and Design Departments, then internalize the market demands within the company. Thus the Design Department develops the product project, preparing it for production. After the production phase, the Commercial Area participates once again, making contact with the customer, collaborating to launch and to promote the product. It is important to note that this company used to imitate competitor's products, once it still had no support to develop original products lines. Over time, nevertheless, the company has started to invest in its own creation and including design on this, looking for references in international events and catalogs. Nowadays, the partnership with external designers to create some product lines is a common practice.

In Company B, references are similar to those of Company A, constituted basically by materials collected from international fairs. Comparing to Company A, in this company external designers are not usually contracted, focusing internally throughout the process of product development, from the initial conception to the final product. The new product collections follow customer demands with deadlines set by the company. The process of new product projects begins with an information research from the commercial area and the Design Department. So the designers generate alternatives and then they selected alternatives to build the prototypes. The product is then developed and launched to the market through the commercial staff.

3.4 Design Management Within Companies

The design is present in the product concept, and is seen as a relevant resource for the generation of competitive advantage, in a positive relationship between investment and performance [35], although it was observed that in both companies there is still a lack of formalization in the new products development process. Companies A and B tend to use the international quality standards, without having a clear structure, which would include the processes of creative steps of generating and selecting ideas. The presence of trained professionals in the area and the investments made to have these professionals in the company have shown interest of the companies about design. So it is possible to say that the Design Management concept is not completely applied as advocated by Borja de Mozota [11] and Neumeier [36] in these companies.

It is also noticed that the design can not be managed appropriately in these companies because of not having clearly defined boundaries with areas like Marketing, Research and Development (R&D) and Engineering Departments. Furthermore, design is not part of the corporate strategic planning, which ultimately compromises its real importance in achieving goals at the organizational level. Another point that is directly reflected in the Design Management is the lack of structured functions and competencies regarding to the concern of the professional designer and design manager roles, hindering the understanding of who are professionals involved in the Design Management as well as what are the roles they have in the processes of both companies.

3.5 The Role of the Design Professional

Concerning to sharing information and knowledge among professionals and sectors involved in a certain project, this does not happen so often in the companies surveyed. This is due to organizational structures and excessive staff division and activities with rigid and specific functions. As a result there are teams working with little exchange of information with sectors participating in a project.

An intervening factor listed by respondents was the lack of autonomy and freedom to do their jobs and make decisions. Respondents were allocated in the areas of project management or product development and the projects were approved or disapproved by the senior managers. As stated by Libânio [37], design may be understood as a strategic element, and it must be connected directly to the senior managers, at a strategic level, and taking an active part in the products decisions. As reviewed in literature, Borja de Mozota [11, 12] considered that the design activity must be planned and formalized, with the coordination of the design features in all activity levels of the organization.

Such relationships were highlighted by Gorb [14], arguing that the Design Management aims to employ effectively the design according to the company's strategic objectives. This is corroborated by Borja de Mozota [11, 12], who has emphasized that the Design Management searches for achieve corporate goals.

Thus, it is clear that the integration of the teams involved in a project is progressive, but should be encouraged, both by professionals involved in projects as the company's top management in an ongoing dialogue between design professionals, teams and network of suppliers, also very important in the design process, as identified by Twigg [38], including tools and systems that can facilitate such processes [39].

4 Conclusion

Based on the set of information collected it was concluded that the Design Management is increasingly understood in a multidisciplinary way, encouraging an exchange of knowledge and strengthening a pluralistic view. It is still possible to observe an increasing volume of publications as well as a gradual evolution in the understanding about the theme internationally. It was found out that Design Management is an important research area, with multidisciplinary research centers that study the subject. However, there is discussion of certain aspects of research first abroad and later in Brazil.

The case studies allowed to think about the importance of integration of design teams with other areas and coworkers who participate in the Design Management process. As presented in this paper, this integration is hindered by some intervening factors such as lack of incentives for the development of Design Management in these studied companies, as well as the lack of design professionals and strategic partnerships. Considering the field of design research, the bibliometric analysis provides an overview of theoretical concepts related to everyday life experienced by traditional markets. Therefore, it is hoped that designers or design professionals with competencies in design can work together with managers, top managers and other partners in the projects, encouraging collaboration among teams. The research has the limitation to be applied in only two cases studies, so it is not possible to generalize the results and conclusions obtained in the metal chain as a whole. Consequently, it is suggested further research studies alike that may show industrial patterns in the metallurgical market in Brazil and abroad.

References

1. Design Management Institute (DMI) (2011) Available at <http://www.dmi.org>. Accessed on Aug 2011
2. International Council of Societies of Industrial Design (ICSID) (2011) Available at <www.icsid.org>. Accessed on 05 Sept
3. Centro Português de Design (CPD) Manual de Gestão do Design. Porto, Portugal

4. Instituto Europeu de Design (IED) (2011) Available at <http://www.ied.edu/>. Accessed on Aug 2011
5. Walsh V (1996) Design, innovation and the boundaries of the firm. Res Policy 25:509–529
6. Libânio CS, Amaral FG (2011) Aspectos da gestão de design abordados em dissertações e teses no Brasil: uma revisão sistemática. Revista Produção Online, Florianópolis 11(2):565–594
7. Lorenzini GC, Libânio CS, Wolff F, Amaral FG (2011) Design management with innovation: a case study in the furniture chain of Brazil. In: Proceedings of Tsinghua-DMI international design management symposium, Hong Kong, SAR, China
8. Kotler P, Rath GA (1984) Design: a powerful but neglected strategic tool. J Bus Strat 5(2):16–21
9. Roy R, Riedel JCKH (1997) Design and innovation in successful product competition. Technovation 17(10):537–594
10. Lojacono G, Zacai G (2005) The evolution of the design-inspired enterprise. Rotman Mag, Winter, pp 10–15
11. Borja de Mozota B (2003) Design management: using design to build brand value and corporate innovation. Allworth Press and D.M.I., New York
12. Borja de Mozota B (2006) The four powers of design: a value model in design management. Des Manage Rev 17(2):44–53
13. Maldonado T (1977) El diseño industrial reconsiderado. Gustavo Gili, Barcelona
14. Gorb P (1990) Design Management. Van Nostrand Reinhold, New York
15. Rodrigues RB (2005) A atividade de gestão do design nas organizações: um estudo no polo moveleiro de Santa Catarina", Master dissertation, Universidade do Vale do Itajaí
16. McBride M (2007) Design management: future forward. Design Manage Rev 18(3):18–22
17. Bertola P, Teixeira JC (2003) Design as a knowledge agent: how design as a knowledge process is embedded into organizations to foster innovation. Des Stud 24:181–194
18. Dell'Era C, Verganti R (2011) Diffusion processes of product meanings in design-intensive industries: determinants and dynamics. J Prod Innov Manage 28(6):881–895
19. Libânio CS (2011) O papel do profissional de design e suas interfaces na gestão de design: Um estudo de caso. Master dissertation, Porto Alegre: UFRGS
20. Fascioni L (2008) Considerações sobre a formação dos gestores de design no Brasil. Anais do P&D Design, São Paulo
21. Avendaño LEC (2002) Resgate do Protagonismo do Desenhista Industrial Através da Gestão do Design. In: 5° Congresso Internacional de Pesquisa em Design, Anais…, Rio de Janeiro
22. Spinak E (1996) Diccionario enciclopédico de bibliometría, cienciometría e informetría. Montevideo
23. Vanti NAP (2002) Da bibliometria à webometria: uma exploração conceitual dos mecanismos utilizados para medir o registro da informação e a difusão do conhecimento. Ciência da Informação 31(2):152–162
24. Borja de Mozota B, Klöpsch C, Costa FC, Da X (2011) Gestão de design: usando o design para construir valor de marca e inovação corporativa. Bookman, Porto Alegre
25. Chai K, Xiao X (2012) Understanding design research: a bibliometric analysis of design studies (1996–2010). Des Stud 33(1):24–43
26. Garcia R, Cantalone R (2002) A critical look at technological innovation typology and innovativeness terminology: a literature review. J Product Innov Manage 19:110–132
27. Becheikh N, Landry R, Amara N (2006) Lessons from innovation empirical studies in the manufacturing sector: a systematic review of literature from 1993 to 2003. Technovation 26:644–664
28. Dangayach GS, Deshmukh SG (2001) Manufacturing strategy: literature review and some issues. Int J Oper Prod Manage 21(7):884–932
29. Malhotra N (2001) Pesquisa de marketing: uma orientação aplicada, 3rd edn. Bookman, Porto Alegre
30. Bardin L (2005) Análise de conteúdo, ed. rev. e atual, Lisboa: Edições 70

31. Ahmadi R, Roemer TA, Wang RH (2001) Structuring product development processes. Eur J Oper Res 130(3):539–558
32. Roy R, Potter S (1996) Managing engineering design in complex supply chains. Int J Technol Manage 12(4):403–419
33. Lauche K (2005) Job design for good design practice. Design Stud 26(2):191–213
34. O'donnell FJ, Duffy AHB (2002) Modeling design development performance. Int J Oper Prod Manage 22(11):1198–1221
35. Chiva R, Alegre J (2009) Investment in design and firm performance: the mediating role of design management. J Prod Innov Manage 26(4):424–440
36. Neumeier M (2008) The designful company. Des Manage Rev 19(2):10–15
37. Libânio CS (2009) "Design como Elemento Estratégico para a Melhoria da Competitividade das Empresas da Cadeia Moveleira de Bento Gonçalves", Final Postgraduate Work. UniRitter, Porto Alegre
38. Twigg D (1998) Managing product development within a design chain. Int J Oper Prod Manage 18(5–6):508–524
39. Whitfield RI, Duffy AHB, Coates G, Hills W (2002) Distributed design coordination. Res Eng Design 13(4):243–252
40. Brereton P, Kitchenham BA, Budgen D, Turner M, Khalil M (2007) Lessons from applying the systematic literature review process within the software engineering domain. J Sys Soft 80(4):571–583

Analysis of Management and Employee Involvement During the Introduction of Lean Development

Katharina Helten and Udo Lindemann

Abstract The introduction of Lean Development (LD) requires change efforts within the organisation. People need to be convinced and motivated, development processes are questioned and adapted, eventually structures are changed. This paper compares the LD introduction processes within three SME with respect to three aspects: the top management involvement, the Lean leadership, and the employee involvement. Within those dimensions, main patterns are identified and interpreted to allow the derivation of implications for a successful introduction of LD.

Keywords Lean development · Organisational change · Management involvement · Employee involvement

1 Introduction

In order to improve product development processes and to manage them more efficiently, the idea of Lean Thinking [1] is applied to product development. The Lean philosophy focuses on the customer value, and therefore the elimination of any form of waste. Companies which introduce the Lean philosophy and Lean Development (LD) undergo a significant organisational change. This change process needs to be well managed to allow for a long-lasting implementation of LD and an improvement of engineering processes.

K. Helten (✉) · U. Lindemann
Institute of Product Development, Technische Universität München,
Boltzmannstraße 15 D-85748 Garching, Germany
e-mail: helten@pe.mw.tum.de

U. Lindemann
e-mail: lindemann@pe.mw.tum.de

A. Chakrabarti and R. V. Prakash (eds.), *ICoRD'13*, Lecture Notes in Mechanical Engineering, DOI: 10.1007/978-81-322-1050-4_71, © Springer India 2013

This paper therefore investigates the change processes during the LD introduction in detail. The aim is to identify relevant patterns and mechanisms to derive appropriate management models in the future. The findings base on empirical data gained during a project in three small and medium-sized entreprises (SME). The projects within each company ran between 12 and 18 months.

The authors analysed three main dimensions of management and employee involvement—top management involvement, Lean leadership and employee involvement. The identified patterns are compared based on an evaluation of the success of the introductory process in each company. Thus, relevant aspects and their impact are derived which are crucial for the management of organisational change during the LD introduction.

2 Background

2.1 Lean Development

The idea of Lean Thinking as defined by Womack and Jones [2] has attracted many engineering companies. Lean in general strives for the creation of customer value while eliminating wasteful tasks. After Womack et al. [1] had described the strengths of the Toyota Production System, and thus defined main ideas of Lean production, Womack and Jones derived five main principles of Lean: value, value stream, flow, pull and perfection [2].

Applying findings from Lean production to product development has proven to be difficult. Whereas the manufacturing process creates a tangible product or artifact, Lean development (LD) is mainly characterised by the transformation of information [3]. Siyamet al. claim that a deeper investigation on information value and waste will lead to more successes in LD, and define a guideline that relates information waste types to appropriate value methods [4]. However, several authors propose different frameworks to differentiate types of waste in LD. Oehmen and Rebentisch come up with a condensed overview of several authors' findings and derive eight types of waste: over production of information, over processing of information, miscommunication of information, stockpiling of information, generating defective information, correcting information, waiting of people, and unnecessary movement of people [3].

2.2 Change Management and Change Models

Change management in general can be described as "the process of continually renewing an organization's direction, structure, and capabilities to serve the ever-changing needs of external and internal customers" [5]. Change in an organisation can be based on three different aspects. According to Todnem By who refers to previous work of Senior [6] the main categories to characterise change are the rate of occurrence (e.g., (dis) continuous or incremental), how it comes about (e.g., planned, emergent) and the scale. Numerous authors have tried to describe or enhance change processes [7]. Often literature refers to the change model of Kotter to plan the main phases of change processes. Kotter's model consists of eight steps [(1) Establishing a sense of urgency, (2) Forming a powerful guiding coalition, (3) Creating a vision, (4) Communicating the vision, (5) Empowering others to act on the vision, (6) Planning for and creating short-term wins, (7) Consolidating improvements and producing still more change, (8) Institutionalizing new approaches] [8].

2.3 Lean Product Development as a Change Process Through Management and Employee Involvement

Moran and Bright man emphasise that people in general wish to have a dialogue about the coming change, and that management avoids such a dialogue. It means for management to commit publicly and to lead the change. Furthermore, the change within the organisation needs to be pushed from both directions, top-down and bottom-up. The authors depict the "Leadership paradox". Anticipating tomorrow's stage, leaders must promote the change and ensure stability at the same time. Effective change leaders act as role models and interact as much as possible with people to communicate the why, how, when etc. [5]. Emiliani and Stec list eleven errors of senior management during the Lean transformation. Among others, leaders do not show sufficient personal engagement. Moreover, senior management behaves inconsistently, exhibiting wasteful behaviour and asking the elimination of waste at the same time [9].

Nightingale and Srinivasan define seven principles of enterprise transformation. One principle specifically addresses leadership commitment. The authors emphasise the need to drive the transformation from the highest levels, i.e., senior management. Other principles address for example a holistic approach and organizational learning. The Enterprise Transformation Roadmap addresses among others the following aspects which are of relevance for this paper. The element "Engage Leadership in Transformation" (Strategic Cycle) includes a step to obtain executive buy-in, and to establish an executive transformation council. Later, change agents need to be empowered (Planning Cycle), as well as the transformation plan needs to be communicated (Execution Cycle). While implementing and coordinating the transformation plan, employees are provided with education and training. Important learnings are then diffused within the enterprise [10].

In a survey from 2010, participants were asked for the importance of different aspects regarding the expectations from leaders during change processes. Among the top answers was the ability of leaders to drive employees, to communicate actively the change was, and to initialise and shape the change process. The three most relevant success factors for successful change initiatives are to ensure mobilisation and commitment, to analyse and understand the environment, and to support leadership [11].

Taking the model of Kotter, in this regard especially the creation of a guiding coalition, the communication of the vision as well as the empowerment of people to react according to the change are of utmost importance [8].

Graebsch et al. have shown in their study that even though the idea of LD attracts lots of people, its introduction is perceived to be rather difficult. Participants of their study therefore agreed on the positive impact of pilot projects, followed by the use of a bottom-up approach [12]. As stated by Helten and Lindemann, literature doesn't provide a framework for conducting such pilot projects. Such a framework needs to indicate the necessary involvement of people as well as the definition of responsibilities. In their paper, they present a pilot scheme that includes four important phases (analysis, synthesis, realisation, and implementation), as well as important levers to lead the introduction. One of their relevant levers is "persons", thus the involvement of people. The modelling approach presented in the paper underlines the importance of the involvement of people by integrating a domain "People/Person" and two sub-domains (Management level, Department). Also lots of the proposed relations indicate the interaction of people, such as "is responsible for", "takes part in", or "knows about". Literature mentions the importance of leadership and employee involvement, but does not show specific patterns, their comparison and implications for the LD introduction [13].

3 Research Approach

The structures that are depicted and analysed in this paper are based on experiences during a research project with industrial participation. The aim of the project was the first-time introduction of LD into SME. Three companies are supported and monitored over the run of twelve to 18 months. Researchers and industry partners were in contact by means of meetings (approx. monthly), and calls and emails (approx. weekly). Some companies had experiences with Lean production. They differ among others in their product type (consumer vs. capital goods), the number of engineers in PD (about 5 to about 80) and management structure (owner vs. managing directors). It needs to be emphasised that due to the specific setting of SME and the first-time introduction of LD, the change process was characterised by the engagement of single persons or small groups, and the implementation of specifically defined actions, e.g., the creation of a knowledge database or new forms for the requirements management. In some cases more extensive actions followed after gaining some experience in LD.

Each company defined a core team, consisting of three or four persons of different hierarchical levels. The research team supports the introduction by delivering academic and literature input, participating and moderating relevant meetings, and supporting important steps such as the waste analysis or the action definition. Therefore, the research method follows the action research approach. The underlying thesis is that complex social processes can be investigated best, if changes are introduced into the system and their effects are observed [14]. The researcher is therefore both observing and participating. Susman and Evered add an aspect that is relevant in the context of the overall goal of this research to support the self-help approach during LD introduction. The authors state that action research aims "to develop the self-help competencies of people facing problems". They propose a cyclic approach that consists of five phases (Diagnosing, Action planning, Action taking, Evaluating, Specifying learning) [15].

Over the run of the project, the researchers monitored all activities in a monitoring sheet to allow for a detailed analysis. Therefore, in addition to the activities, the date and the involved persons are noted. The sheet differentiates activities that were project logical and those that solved problems. Finally, any feedback from industry during the introduction was documented. Of importance were e.g., activities between project meetings, whether the procedure is clear, or which suggestions the companies had to continue and to adapt the process to their specific internal conditions. In addition, the core teams of all participating companies were asked for their overall assessment in an interview. Based on this feedback and the evaluation by authors the success of the LD introduction can be identified. Success refers to the introduction process itself, not to the (financial) success of the LD actions.

4 Results

Based on a literature review, the analysis of the monitoring sheet with respect to management and employee involvement leads to the identification of three main dimensions and several sub-dimensions, as shown in Table 1. As described above, the involvement of top management as well as employees is essential for the success of the change. Furthermore, one category concentrates specifically on the Lean leadership.

Due to the limitations of the qualitative research, the differences are assessed more in a relative manner. In case of companies A and C the pilot project focused on product development in one business unit, in case of company B on all business units. The successes can be described as follows:

Table 1 Relevant dimensions for management and employee involvement during LD change processes

Dimension and sub-dimensions		
Top management involvement	Lean leadership	Employee involvement
• Initialisation	• Composition of core team	• Waste analysis
• Awareness		• Action definition
• Communication	• Responsibility for Lean pilot project	• Information/ Communication
	• Action taking	
	• Internalisation	

Company A and *Company B* : Based on a comprehensive waste analysis, actions were defined, detailed, and to large parts already implemented. Several employees were directly involved, most/all development engineers are at least informed. In the following of the pilot phase, the company develops further actions and follows continually the realisation of previously defined actions. The evaluating interview showed that the term LD is used within the company, in one case is also part of the company goals. The state after the pilot phase with respect to LD is to be qualified assuccessful and promising.

Company C: Based on a comprehensive waste analysis, actions were defined, detailed, and some are implemented. For those ones future steps were defined. After being involved in a questionnaire during the waste analysis, employees besides the core team are involved in the specific implemented actions, not in other initially defined actions. The core team members feel to be much more sensitised towards processes. The LD pilot project enabled several continuous improvement efforts. The term LD is not specifically used as a term throughout the entire company. Nevertheless, the company is striving for success.

Table 2 shows the patterns of relevant dimensions. The interpretation refers to the overall assessment of the LD introduction as described above. Those elements and relations are chosen that represent the situation the best according to the authors. The challenge is to find the right level of abstraction that allows a comparison of the three companies. Therefore, e.g., the descriptions of the positions of each person are slightly adapted to allow a comparison. By means of a qualitative comparison of the companies and their results, the relevance of a single sub-dimension as well as the implications for a general pilot scheme for introduction are derived.

Table 2 Models of LD change situations, their interpretation and implications for a pilot scheme

Top management (TM) involvement		
Initialisation		
Company A	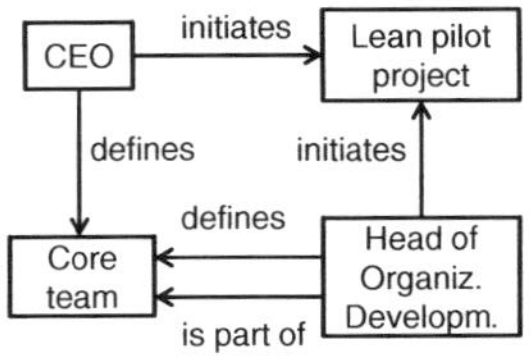 CEO; initiates; Lean pilot project; defines; initiates; defines; Core team; is part of; Head of Organiz. Developm.	**Interpretation**: In all companies the Lean pilot project is initialised by the TM. Since the CEO is part of the core teams in companies B and C, which differ slightly in their success of introduction, the CEO's participation seems not to be crucial. The Head of Organizational Development (HOD) participates in the core team of company A. Since the projects run over a period of a year or longer after the initialisation and showed the implementation of actions, any form of TM's initialisation seems to have a positive impact. **Implications**: The initialisation by TM ensures at minimum management attention, and thus a positive start into the Lean journey.
Company B and company C	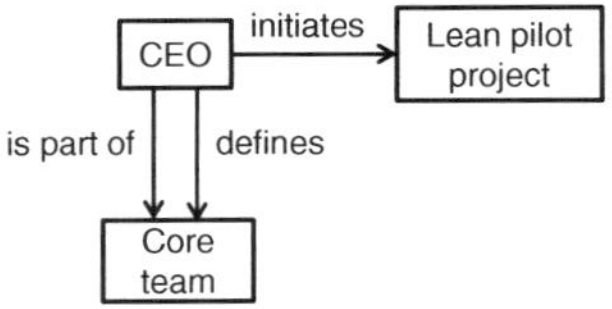 CEO; initiates; Lean pilot project; is part of; defines; Core team	
Awareness		
Company A	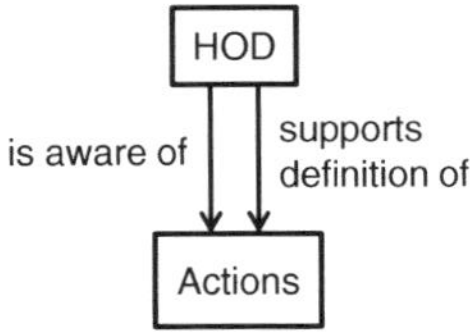 HOD; is aware of; supports definition of; Actions	**Interpretation**: Even though TM has initialised the project and takes part in the core team, its awareness of ongoing actions to eliminate waste differs. The awareness scale varies from a continuous participation in as well as definition of actions (B) to awareness of the current state and support of the actions' definition (A), to an awareness and support if required (C). In TM's awareness the continuity seems to be important. Company A and B are rated the same in the success of the introduction. Even though they show slightly different pattern, they ensure the necessary continuity. Similar patterns can be found with focus on the participation. **Implications**: A continuous awareness and support of TM is required, the depth can vary.
Company B	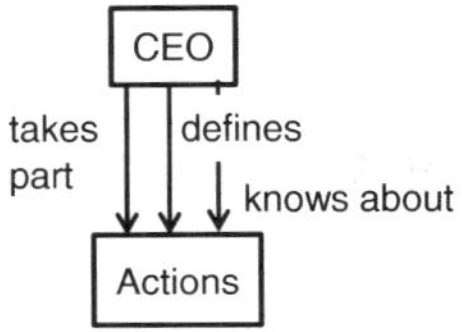 CEO; takes part; defines; knows about; Actions	

Table 2 (continued)

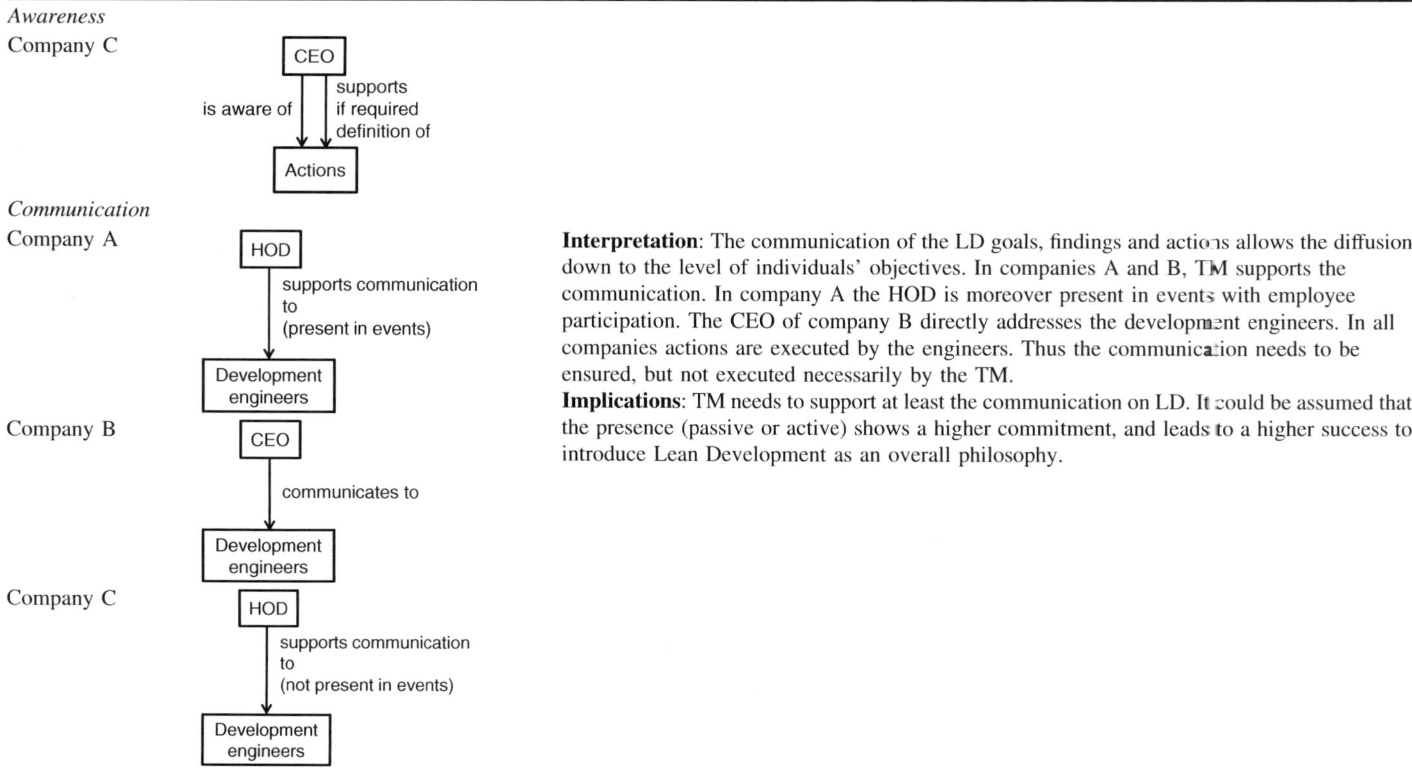

Top management (TM) involvement		
Awareness		
Company C	CEO –is aware of→ Actions; CEO –supports if required definition of→ Actions	
Communication		
Company A	HOD –supports communication to (present in events)→ Development engineers	**Interpretation**: The communication of the LD goals, findings and actions allows the diffusion down to the level of individuals' objectives. In companies A and B, TM supports the communication. In company A the HOD is moreover present in events with employee participation. The CEO of company B directly addresses the development engineers. In all companies actions are executed by the engineers. Thus the communication needs to be ensured, but not executed necessarily by the TM. **Implications**: TM needs to support at least the communication on LD. It could be assumed that the presence (passive or active) shows a higher commitment, and leads to a higher success to introduce Lean Development as an overall philosophy.
Company B	CEO –communicates to→ Development engineers	
Company C	HOD –supports communication to (not present in events)→ Development engineers	

Table 2 (continued)

Lean leadership

Composition of core team

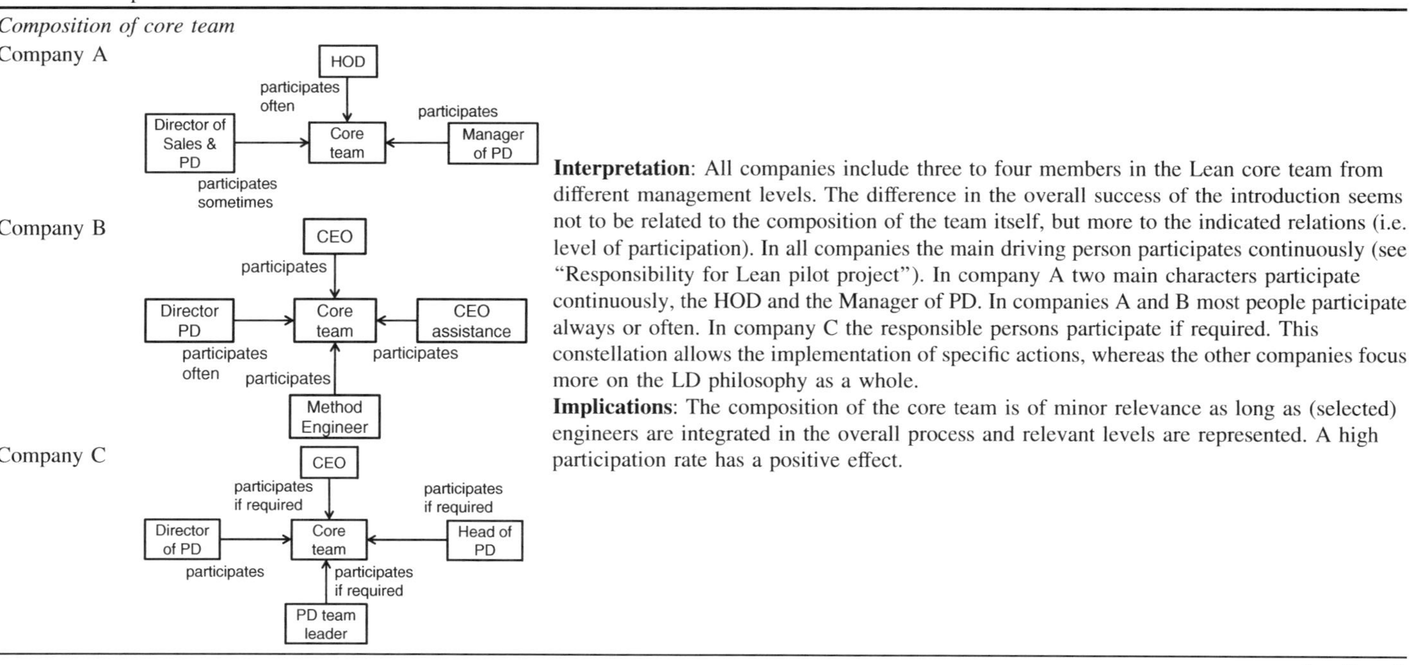

Interpretation: All companies include three to four members in the Lean core team from different management levels. The difference in the overall success of the introduction seems not to be related to the composition of the team itself, but more to the indicated relations (i.e. level of participation). In all companies the main driving person participates continuously (see "Responsibility for Lean pilot project"). In company A two main characters participate continuously, the HOD and the Manager of PD. In companies A and B most people participate always or often. In company C the responsible persons participate if required. This constellation allows the implementation of specific actions, whereas the other companies focus more on the LD philosophy as a whole.

Implications: The composition of the core team is of minor relevance as long as (selected) engineers are integrated in the overall process and relevant levels are represented. A high participation rate has a positive effect.

(continued)

Table 2 (continued)

Lean leadership

Responsibility for Lean pilot project

Company A

Company B

Company C

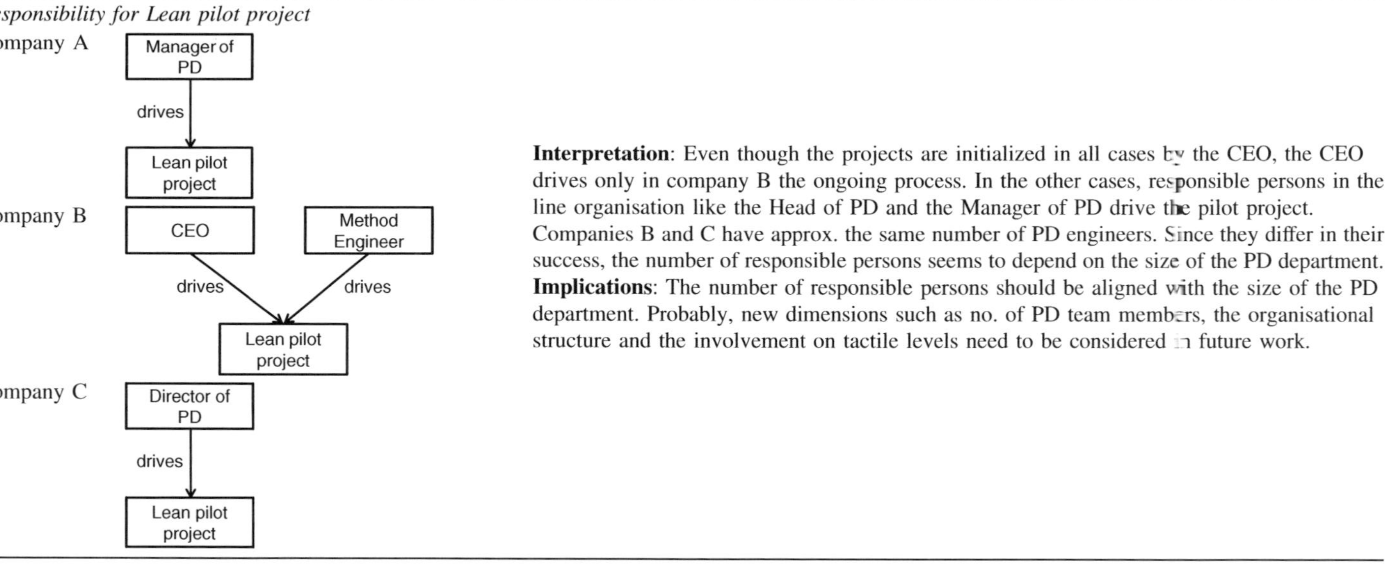

Interpretation: Even though the projects are initialized in all cases by the CEO, the CEO drives only in company B the ongoing process. In the other cases, responsible persons in the line organisation like the Head of PD and the Manager of PD drive the pilot project. Companies B and C have approx. the same number of PD engineers. Since they differ in their success, the number of responsible persons seems to depend on the size of the PD department. **Implications**: The number of responsible persons should be aligned with the size of the PD department. Probably, new dimensions such as no. of PD team members, the organisational structure and the involvement on tactile levels need to be considered in future work.

(continued)

Table 2 (continued)

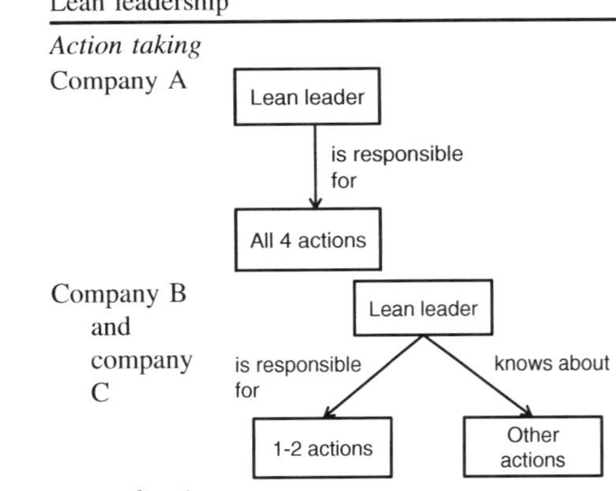

Lean leadership	
Action taking	
Company A	**Interpretation**: The Lean leaders of company A is responsible for all four actions, and thus automatically informed. In company B and C the Lean leader(s) is(are) responsible for 1–2 actions and know about the other actions. The general structure doesn't seem to be relevant for the success. Companies B and C show the same structure, but vary in their overall success. The difference is addressed to some extent in the previous factor, since two persons are leading the pilot project in company B. Again the size of the PD department in company C seems to allow for one strong Lean leader who is responsible for all actions. **Implications**: The size of the PD department seems to necessitate a certain number of responsible persons. Since all companies show a certain success after the pilot project, it could be assumed that this is due to the fact that the Lean leaders themselves are in every company responsible for at least one action.
Company B and company C	
Internalisation	
Company A and company B	**Interpretation**: This factor describes to what extent the internal team embraces the Lean initiative and drives it without external push (i.e. the research team). Companies A and B show the same frequency, and the same success. Company C meets internally if required during the pilot project. The frequency of meetings seems to correlate with the level of outcome—from the implementation of specific actions to the LD philosophy. The companies focussing more on LD need to meet often internally. **Implications**: The more often core team members meet internally, the higher the probability that the company introduces Lean as a philosophy instead of a part of the continous improvement process. The frequency of meetings seem to be an indicator for the depth of embracing the LD approach.
Company C	

Table 2 (continued)

Employee involvement

Waste analysis

Company A

Company B and company C

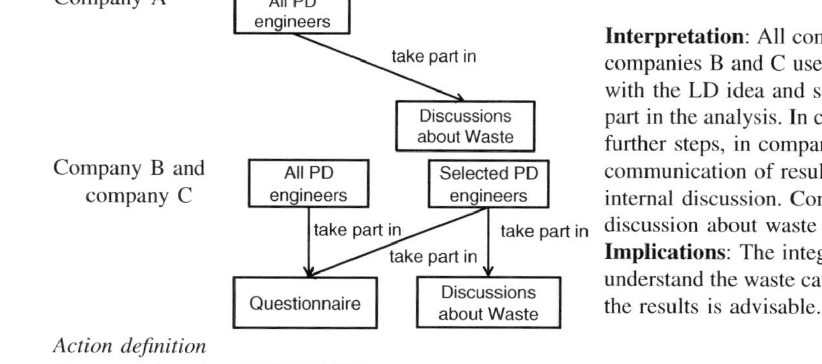

Interpretation: All companies integrate several engineers during the analysis phase. The larger companies B and C use questionnaires in addition to discussions. To allow for an identification with the LD idea and success of the introduction, engineers need to be integrated after taking part in the analysis. In company B e.g. employees are informed in an event about the results and further steps, in company C there is not such an event. Company C focuses more on the communication of results whereas company B presents also preliminary results to enhance the internal discussion. Company A seems to be successful by having a continuously intensive discussion about waste within the whole team.

Implications: The integration during the waste analysis has a positive impact. It is essential to understand the waste causes to participate in the following Lean change. Thus information about the results is advisable.

Action definition

Company A and company B

Company C

Interpretation: Whereas companies A and B involve selected engineers in the definition of all actions, specific actions are defined with the support of selected engineers in company C. The PD engineers give a feedback on a tactile level and allow for a higher applicability and more tangible improvements.

Implications: The definition (in detail) of all actions should involve PD engineers to allow feedback and suggestions from all hierarchical levels. It is assumed that a higher rate of involvement leads to a more visible commitment of the company to LD and higher success in terms of LD during the introduction.

(continued)

Table 2 (continued)

Employee involvement		
Information/Communication		
Company A and company B	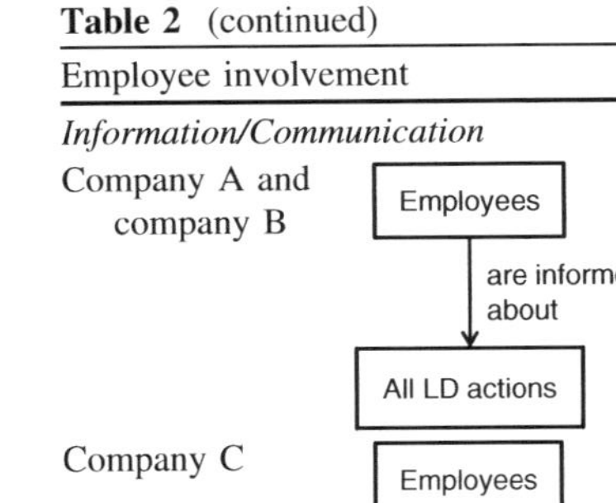	**Interpretation**: All companies inform the employees about the actions. In contrast to companies A and B, company C focuses more on the information about specific actions. In companies A and B all defined actions are presented. Employees have the chance to understand the current project state and to ask questions. The companies allow thereby the participation in the Lean journey for everybody, ensure transparency and show commitment to the Lean effort. **Implications**: Employees are to be informed to ensure an open and constructive attitude towards LD. Such a public commitment drives the internal expectations to succeed.
Company C		

5 Conclusion and Outlook

The analysis of the LD introduction in three SME identifies three dimensions of utmost importance to understand management and employee involvement during the change process. These categories are top management (TM) involvement, Lean leadership, and employee involvement. The in-depth analysis of these qualitative cases allows the identification of several main factors and implications. Regarding the TM involvement, continuity seems to be the most important aspect, e.g., of awareness. In addition, it is assumed that the communication must be at least supported by the TM. Thereby the management commits publicly the need for change and paves the way for employee engagement.

With respect to the Lean leadership, it seems to be important that the core team and its members take over the Lean initiative as early as possible to embrace the idea, e.g., by driving themselves some actions. A specific composition of the core team is less important than the representation of different levels and a high rate of participation. It is assumed that the size of the PD department has some influence, e.g., on the number of responsible persons for LD. But a further investigation is needed.

If it comes to employee engagement, it seems to be of utmost relevance for the success that employees get the chance of involvement. Once involved, people need to get feedback during the whole process. Furthermore the constructive feedback of the PD engineers allow for a high understanding and applicability of the Lean actions.

The further research will investigate on further factors such as the complexity of actions and the extension of LD within the company to allow for an overall view on the LD introduction. Thus also the characterization of the companies will be considered. Moreover, the success needs a more detailed discussion. Aspects such as the fact that different numbers of business units were involved require further inquiry.

The identified alternative patterns need a deeper discussion with industrial partners. Their feedback on the patterns allows an evaluated overview and the derivation of a descriptive model. Other companies could be interviewed to gain insights into the relevance and impact of the presented dimensions.

References

1. Womack JP, Jones D, Roos D (1991) The machine that changed the world—the story of lean production. Harper Perennial, New York
2. Womack JP, Jones DT (2003) Lean Thinking. Banish waste and create wealth in your corporation. Simon & Schuster, New York
3. Oehmen J, Rebentisch E (2010) Waste in product development. In: LAI Paper Series Lean Product Development for Practitioners, Massachusetts Institute of Technology, Cambridge

4. Siyam GI, Kirner K, Wynn DC, Lindemann U, Clarkson PJ (2012) Relating value methods to waste types in lean product development. In: Proceedings of International Design Conference—Design 2012, Dubrovnik, 21–24 May 2012
5. Moran JW, Brightman BK (2001) Leading organizational change. Career Dev Int 6(2):111–118
6. Senior B (2002) Organizational change, 2nd edn. Prentice Hall, London
7. Todnem R (2005) Organisational change management: a critical review. J Change Manage 4(5):369–380
8. Kotter JP (1995) Leading change: Why transformation efforts fail. In: Harvard Business Review, Vol 73(2)
9. Emiliani ML, Stec DJ (2005) Leaders lost in transformation. Leadersh Organ Dev J 26(5)
10. Nightingale DJ, Srinivasan J (2011) Beyond the lean revolution: achieving successful and sustainable enterprise transformation. AMACOM, New York
11. Capgemini Consulting (2010) Change Management Studie 2010: Business Transformation—Veränderungen erfolgreich gestalten, Munich
12. Graebsch M, Lindemann U, Weiß S (2007) Lean Development in Deutschland—Eine Studie über Begriffe, Verschwendung und Wirkung. Dr. Hut, Munich
13. Helten K, Lindemann U (2012) Structural modelling and analysis of organizational change during lean development implementation. In: Proceedings of International Design Conference—Design 2012, Dubrovnik, 21–24 May 2012
14. Baskerville RL (1997) Distinguishing action research from participative case studies. J Syst Inf Technol 1(1):24–43
15. Susman GI, Evered RD (1978) An assessment of the scientific merits of action research. Adm Sci Q 23:582–603

ICT for Design and Manufacturing: A Strategic Vision for Technology Maturity Assessment

Mourad Messaadia, Hadrien Szigeti, Magali Bosch-Mauchand, Matthieu Bricogne, Benoît Eynard and Anirban Majumdar

Abstract Based on the EU-FP7, ActionPlanT project aims at assessing and ranking Information and Communication Technology (ICT) that will have the most impact on European competitiveness. One of the outcomes is a classification of ICT for design and manufacturing. ICT, used inside companies, can be classified in different levels according to company's organization. The paper describes layered software architecture for design management and manufacturing execution of company which intensively uses ICT. The ICT classification can link two perspectives which are developed industrial strategy and used design and manufacturing technology. Results of this work were used in the FP7 project for integrating strategy with design and manufacturing levels.

Keywords ICT · Manufacturing systems · Product development process · Information systems · Industrial strategy

---

M. Messaadia · M. Bosch-Mauchand · M. Bricogne · B. Eynard (✉)
Université de Technologie de Compiègne, Compiègne, France
e-mail: benoit.eynard@utc.fr

M. Messaadia
e-mail: mourad.messaadia@utc.fr

M. Bosch-Mauchand
e-mail: magali.bosch@utc.fr

M. Bricogne
e-mail: matieu.bricogne@utc.fr

H. Szigeti
Dassault Systèmes, Delmia, Vélizy-Villacoublay, France
e-mail: Hadrien.SZIGETI@3ds.com

A. Majumdar
SAP, Dresden, Germany
e-mail: anirban.majumdar@sap.com

A. Chakrabarti and R. V. Prakash (eds.), *ICoRD'13*, Lecture Notes in Mechanical Engineering, DOI: 10.1007/978-81-322-1050-4_72, © Springer India 2013

1 Introduction

The contribution of Information and Communication Technology (ICT) to the manufacturing industry has become unquestionable over the 20 years. ICT will be increasingly merged with future manufacturing processes and support the development of efficient business processes. To ensure the sustainable competitiveness of European manufacturing industry, a continuous improvement of the use of R&D expertise as well as of manufacturing and technology resources is required. The proposed analysis will consider technology and business trends as well as politics, environmental, and society needs [1].

The Lisbon Strategy adopted in March 2000 set the European Union the goal of becoming the knowledge economy the most competitive and most economy in the world. Consequently, the EU has adopted plans for Use of ICT in European industrial companies (first "e-Europe 2002" in June 2000, second "e-Europe 2005" in June 2002, and finally the strategy "i2010") [2]. The ActionPlanT project is co-funded by the European Commission under the Private–Public Partnership (PPP) "Factories of the Future" within the Seventh Framework Programme (FP7) [3]. ActionPlanT aims to develop a vision on the short, medium, and long term role of Information and Communication Technology (ICT) in the European manufacturing industry [4].

Advanced systems such as Product Lifecycle Management (PLM) Enterprise Resource Planning (ERP) or Customer Resource Management (CRM) systems are nowadays unquestionable in the strategic, tactical and operation management of the company. Based on a single database, ERP allows managing the business processes of a company and sharing information among its teams for sales and purchases, and also with the departments of "finance", "planning", "marketing", etc. CRM allows managing all customer relationships in a same process: collecting and distributing information to other services and customer analysis for marketing (pricing, promotional sales organization, choice of distribution channels).

The following section presents a literature survey concerning the growing of ICT part in the enterprise activities which is summarized on Fig. 1. Section 3 introduces the ICT classification based on outcomes proposed on Sect. 2. Section 4 is the deployment of the classification on ActionPlanT project. The conclusion discusses results of the project.

2 ICT Evolution: From the Plant to Enterprise

For over three decades, it has been seen the unquestionable increasing of ERP implementation in manufacturing industry. The use of ICT started long before, production management systems have evolved from Material Requirements Planning (MRP) systems in the 1970s and Manufacturing Resource Planning (MRP II) systems in the 1980s, to ERP systems in the 1990s, which is based on a

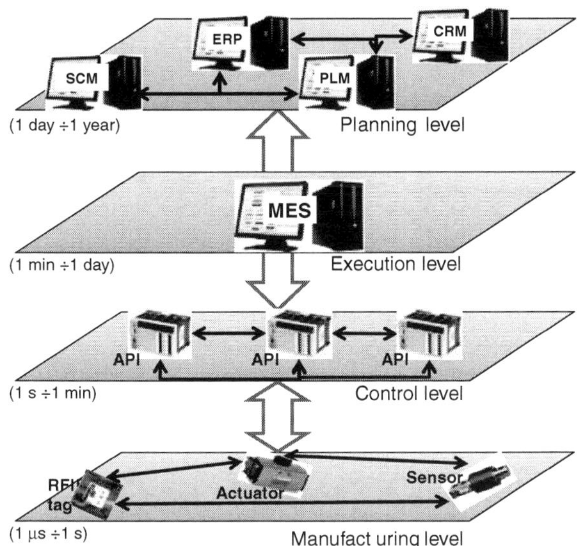

Fig. 1 Levels ICT areas source

single source of data for internal processing of financial assets, production planning and control, human resource management, raw materials management as well as traditional manufacturing process planning. In the 1990s, other functions began to be integrated to ERP, such as and customer relationship management (CRM) [5] and supply chain management (SCM) [6]. These "extended enterprise systems" expanded the scope of enterprise information system beyond the boundary of the company covering relationships with suppliers, partners and customers [7].

For another example, although Computer-Aided Design (CAD) systems have made it easy to create and modify drawings, these powerful systems also often create new problems with uncontrolled proliferation of technical documents and their versions. Then, PDM systems have been used by manufacturing companies to manage the data and documents accumulated in the design of their products [8]. Those systems are intended to support product data structuring and management throughout the product development process. They manage information through document management and especially product data evolution using predefined workflows [9, 10]. As an extension, Product Lifecycle Management (PLM) systems introduce new functionalities such as project management or requirements engineering [11]. Current PLM systems integrate Internet-based technologies and offer groupware-like functionalities) for proactive collaboration between project team members. One of the first definitions of engineering data management was in 1980s by Konstantinov, his work was situated in the design management area, and the vision of PLM has hugely changed since then [12]. Currently, PLM system manages all contributions and information exchange of global team members, like to business partners, suppliers and customers [13].

The management system of company usually consists of an Enterprise Resource Planning (ERP) and a Computer Aided Management program (CAM); the monitoring system providing real-time control of production. The Manufacturing Execution System (MES) is born of a lack of communication between the computer and functional layer. It therefore contributes to the data integration for Computer Integrated Manufacturing (CIM). The MES system bridges the gap between the production planning system and the control and supervision system using on-line data to manage the current application of manufacturing resources: people, equipment and inventory (Fig. 1).

The MES is an information system specified and implemented for the control of manufacturing systems. Nowadays, most manufacturing enterprises use production planning based on ERP/ERP II [14, 15] to define what and when products have to be manufactured but they have no close control and supervision of executed operations.

For many industrial companies, information systems encountered, consist of multiple levels architecture as presented on Fig. 1. They are able to carry out the products manufacture and assembly with several variations in performing operations on multiple workstations, automated or not. The physical flow are controlled, monitored and plotted based on the multiple levels and the integration of ICT.

Manufacturing level is classically split up into: (1) an operating sub-system, which ensures material processing (space, time or shape) for producing the desired added value; (2) a control sub-system, able to supervise a planed processes e and to control the operating sub-system according desired objectives [16].

Input data of the information system can be broken down into instructions sent by the human operators and messages sent from other components of the ICT architecture (MES, etc.).

The shop floor is connected with MES, which has a number of functions (supervision, launch and track work orders, performance analysis, traceability, etc.) and allows the integration to the upper ICT levels (including order processing and scheduling of manufacturing and assembly orders). The MES is implemented based on web services, allowing a remote access to all functionalities and in particular the control of the production lines and operations. In this way, it is possible to access archived production data for use as current operating data, for example, validate models. Gradual extension towards the upper level (ERP, PLM, etc.) is ensured, as well as the integration or interoperability with specific systems [17].

3 ICT Classification

According to ICT survey and their application areas, it has been identified a four levels architecture (Manufacturing, Control, Executive and Planning) see Fig. 1.

Based on these levels and the survey, ICT's for manufacturing have been classified according to Fig. 2. The proposed classification aims to link the low level of the company, which is composed of machine-tools, conveyors, sensors, actuators and controllers with the highest level which is a suppliers and customers

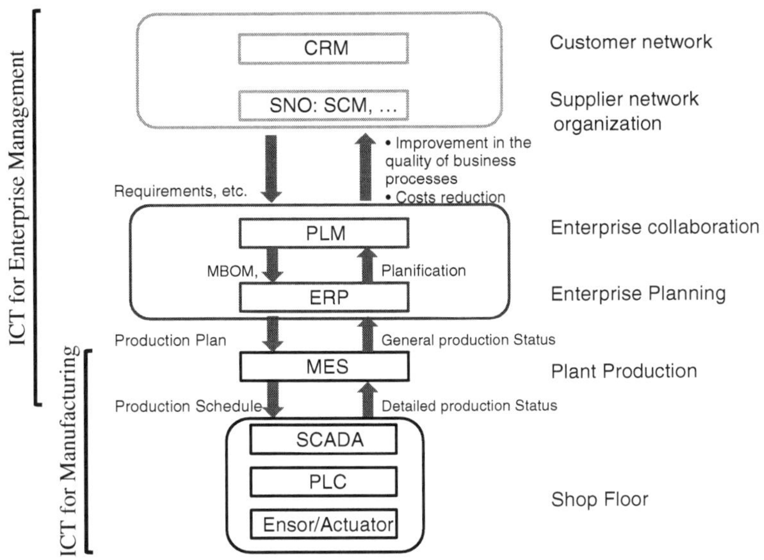

Fig. 2 ICT classifications according to enterprise architecture

oriented management strategy. Figure 3 shows a detailed view of ICT layer in an automated production system.

ICTs are classified according to their scope of application. The classification starts from the small part of enterprise, shop floor, the first level of monitoring system is the Supervisor Control and Data Acquisition (SCADA) that supervises the working productive area [18]. MAS (Machines, Automation, and SCADA) includes sensors/actuator, PLC and SCADA. MES is the bridge between the production plant—shop floor and the whole Enterprise. It coordinates and synchronizes manufacturing processes, the information flow between ERP/PLM and production plant/shop floor (PLC, Robots...). It also distributes operations and tasks to other sub-level of information system. MES receives information from ERP/PLM (process planning, operations sequence) and sends process scheduling as machine commands for manufacturing sequences. MES gives the visibility on what happens in the production and operation area. ERP system manages information on values and costs for manufacture and assembly of products [19].

ID	Optimisation	On Demand	Innovation	Green	Better Plant	Improvement Goal
RT10	2	0	0	0	1	3
RT11	0	1	0	0	0	1
RT12	0	0	0	1	0	1
RT13	2	0	0	0	0	2
RT14	0	0	0	0	0	0

Fig. 3 Research topics (RT) weighting

PLM an ERP are at the highest enterprise level (Strategic planning). ERP manages and controls information of the whole enterprise. It includes several modules supporting a large range of business functions such as finance, administration, logistics, manufacturing, marketing and so on [13]. ERP exchanges data and information with other systems of CAx family (CAD, CAE and CAM) or Supply Chain Management (SCM) to Customer Relationship Management (CRM) via Product Lifecycle Management (PLM).

PLM manages all documents and information exchange from all project team members, like to suppliers (via SCM) and customers (via CRM). PLM increases communication, coordination and collaboration with partners of an "extended enterprise" [20]. PLM makes the link with the highest level of management, from enterprise to extended enterprise, by providing collaborative and interoperable share space where suppliers/clients/OEM can access and share, in real time, product lifecycle information.

CRM enables us to collect many types of data. For example, it can be known about the customer's interaction history with us, when the customer lodged a complaint, when we reply, etc. Overall, CRM leads to better insights about customers, which help to see how it can be improved and planed for services that are of added value to customers [21].

4 Case Study

ActionPlanT project aims at defining an agenda for leveraging "ICT in Manufacturing" in the next ten to 15 years by creation of a vision and roadmap [1]. The ActionPlanT vision will serve as a precursor to the ActionPlanT roadmap. The former will focus on highlighting the key manufacturing challenges and their corresponding ICT enabling production technologies. This is reflected by the establishment of effective link between the strategic approach (Ambitions) and technologies deployment in the manufactory (Technology Readiness Level: TRL[1]).

Westkämper identified several "inefficiencies" in the use of ICT in the "practice of manufacturing". The risk of disrupting a functional production line by adopting a disruptive technology is the foremost. Second is the issue of obsolescence whereby systems and middleware of varying maturity deployed at the levels of shop floor (plant connectivity), plant/site (MES, PLM), and enterprise (ERP) do not seamlessly interact with one another thereby forcing system landscapes to be inflexible. The third is the issue of prohibitive cost of developing in-house systems or purchasing off-the-shelf systems. Modern day enterprises, especially SMEs,

[1] TRLs are a set of management metrics that enable the assessment of the maturity of a particular technology and the consistent comparison of maturity between different types of technology [22].

require low-cost solutions [23], open systems with standard interfaces, and those that are easily configurable.

4.1 Ambitions

The European Environment State and Outlook Report 2010 (SOER 2010) outlined a cross-cutting assessment of how key global megatrends are going to impact European denizens in the near to mid-term future. The premise for the ActionPlanT Vision is based on these multidimensional changes which are taking place socially, technologically, economically, ecologically and politically. In the project we cluster the eleven global megatrends in SOER 2010 into three prominent European Megatrends relevant for European industries: (1) Impact of demographics (people) on consumption and employment; (2) Global competition and push for innovation; (3) Increasing importance of sustainability issues.

Based on the list of clusters of R&D topics provided by state-of-the-art analysis of over 35 roadmaps, we first identified which/how different factors impacted the R&D topics in a positive or a negative way. These identified factors were further streamlined and grouped into a set of 25 STEEP Factors, each taxonomized into distinct Factor Clusters (Social, Technological, Environmental, Economical, and Political/Legal). The final step of the analysis consisted of mapping these factors to ICT topics (distinguished by their topic codes) which were presented to the ActionPlanT experts in the form of scenarios for evaluation and ranking. Based on STEEP analysis approach [4], the first axis identifies five ambitions: Optimization, On Demand, Innovation, Green and Better Plant. These ambitions also relate to the strategies outlined implicitly or explicitly in all roadmaps analyzed under the inventory analysis work of ActionPlanT.

- ON-DEMAND: to deliver on-demand custom products faster than competitors through a network of manufacturing partners
- OPTIMIZED: to provide optimal product quality, safety, durability and additional services for a given cost
- INNOVATIVE: to industrialize new product/process technologies faster than competition
- GREEN: to track and reduce the environmental footprint for products and processes
- BETTER PLANTS: to provide workers a better place to work, adapted to them, safer and compliant with all regulations.

Second axis based on TRL approach, identifies the maturity of technologies in different topics of the roadmap. This work aims to give a maturity level to each RT in order to evaluate the technologies maturity used in the RT. Results are evaluated by experts (Industrial and scientists) during workshops.

4.2 Research Topics

The international experts involved in the workshops organized by the ActionPlanT Consortium suggested more than 130 specific Research Topics (RTs); all capable of significantly impacting the European manufacturing industry through ICT focused research.

In order to evaluate the impact in a more structured way, each RT has been evaluated against the five ambitions (On-demand, Optimized, Innovative, Green, Better Plants), using a mapping approach with an index of a scale of the number of "sub-ambitions" impacting the RT.

Improvement goal is the result of RTs weighting. Improvement goal is an indicator that will allow us to evaluate the RT and compare it to other RT. This will allow us to assess its maturity and develop a strategy (future) of improvement.

4.3 ICT Classification

The ICT classification by levels will be used for linking the two project axis i.e. "ambitions" and "technologies maturity". The deployment of the approach started with mapping topics according to the ICT levels (column) and different other axis (line). As an example it can be mentioned case of 5 M axis (lean manufacturing) line, which is specific to the manufacturing workshop environment [4]. As result it is obtained an ICT distribution according to their field of use in the workshop.

Once ambitions are defined, each one was detailed on different "sub-ambitions". Initially we assign weights to sub-goals which corresponded to the average weight of ambition. Thus, we could have a qualitative assessment. A second work was to give weight to the ambitions for the number of sub-goals. Finally we compared these two approaches and found that they converged. However, deployment should be done in order to have more accurate weight (Table 1).

Table 1 Ambitions and their "sub-ambitions"

Optimization	On-demand	Innovation	Green	Better plants
Yield	Reusability	Supplier integration	Waste	Work environment
Quality	Quick development	Time information	Recycle	Stress
Reliability	Scalability	User innovation	Resource usage	Human intervention
Cost	Adaptive		Safety	
	Small size batches customization		Energy	

	Measure	Machine	Method	Worksapce	Material	Man
CRM	0	0	4	0	1	2
SNO	1	2	6	0	1	2
PLM	4	7	14	5	3	3
ERP	3	7	12	4	1	4
MES	4	10	16	4	1	4
M.A.S	4	10	15	4	1	4

Fig. 4 ICT concentrations in the production plant

During the mapping, we have chosen the weight of the ambition equal to the sum of "sub ambition" impacting the RT. Once the weight of the RT defined, we made a mapping of RTs according to their application field in manufacturing.

All RTs were regrouped and evaluated according to five ambitions (Fig. 4). The aim is to link these den RT and manufacturing, especially the application in

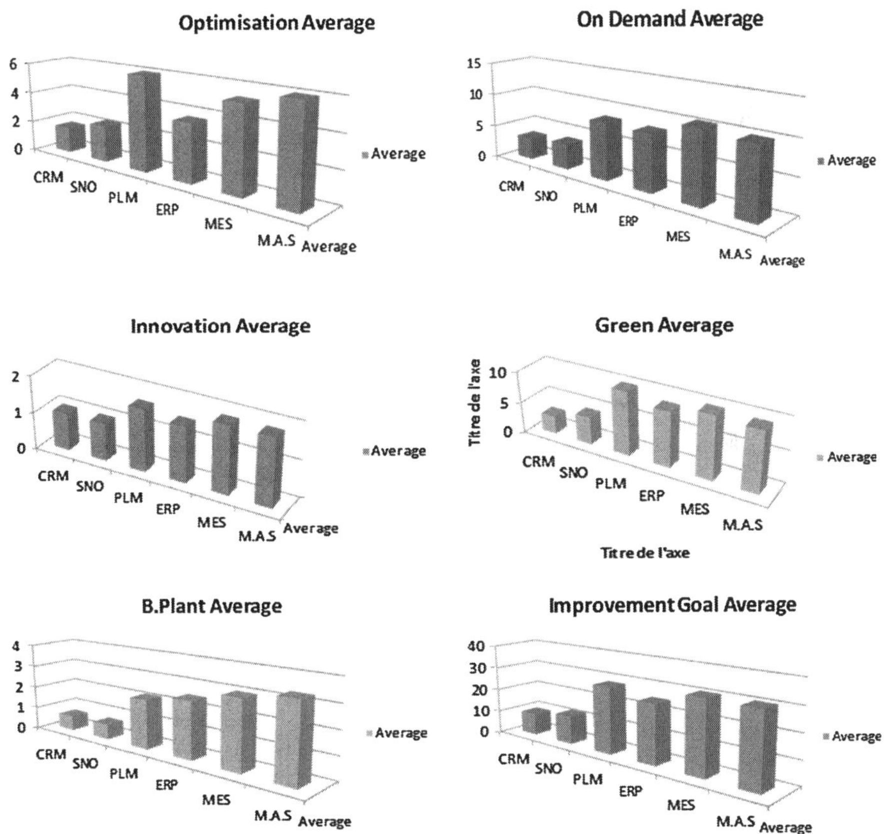

Fig. 5 ICT levels applied to ICT projects roadmap according to the five ambitions

	Measure	Machine	Method	Worksapce	Material	Man
CRM	0	0	5	0	1	4
SNO	0	2	7	0	1	4
PLM	2	7	15	5	2	5
ERP	2	5	8	2	0	5
MES	3	8	14	2	0	5
M.A.S	3	8	16	4	0	5

	Average
CRM	1,66666667
SNO	2,33333333
PLM	6
ERP	3,66666667
MES	5,33333333
M.A.S	6

Fig. 6 ICTs averaging approach

manufacturing. For this, we identified different approach that we present the 5 M (standard in Lean Manufacturing).

On Fig. 4, dark red areas underline field where the density is huge, those lighter red areas are less of density, etc.: CRM, SNO (supplier network organization), are less important in the Research Topics (RT). The first conclusion is that Organization and Human Services (CRM, SNO) are less treated in Research Topic (RTs). Considering Fig. 4, it can be found that the highest density is that Methods in the interval [MAS, PLM]. This outcome is explained by the limitations of ICT (or rather their current maturity). Indeed the "Challenge histogram Technologies Standards" according to ambitions is presented on Fig. 5.

It is worth noticing that the above described components, the Ambition axis, Average Factor, and TRL, are not dependent on each other, but combined they give a significant overall. For example, a low TRL level does not necessarily mean that the RT should be researched at later stages of the Horizon 2020 timeframe while on the other hand a high TRL level does not indicate that the RT should be researched immediately in the future (Fig. 6).

Summarized it can be said that the TRL measures the current technological maturity of the RT, the ambitions, especially their "sub-ambitions", are the potential impact of a successfully researched RT, and the average factor can be used as a base for an indication of the proposed time for carrying out research activities.

5 Conclusion

In a global and competitive context it is important to have a link between strategic vision and technologies used in the manufacturing industry. The work presented in this paper aims to link the two stages of the ActionplanT project. Hence ICT are classified according to their area of use for data and information management, from the manufacturing cell in the production plant until the whole extended and collaborative enterprise based on PLM. It provides a synthesis view of limitations of these information systems and a clarification of the link between technology maturity and strategic targets of development (Ambitions).

Future work will focus on the development of new prospects, and more complete deployment of the proposed approach by integrating ICT standards survey and not only systems and applications.

References

1. ActionPlanT (2012) Roadmap and vision for manufacturing 2.0. http://www.actionplant-project.eu/index.php?option=com_content&view=article&id=51&Itemid=56. Accessed 30 Apr 2012
2. Eurostat (2012) http://epp.eurostat.ec.europa.eu. Accessed 30 Apr 2012
3. ActionPlanT (2012) Project. http://www.actionplant-project.eu. Accessed 30 Apr 2012
4. Szigeti H, Messaadia M, Majumdar A, Eynard B (2011) STEEP analysis as a tool for building technology roadmaps. In: Internationale challenges e-2011 conference, Florence, 26–28 Oct 2011
5. OECD (2012) http://www.oecd-lirary.org. Accessed 30 Apr 2012
6. Choy KL, Lee WB, Lo V (2003) Design of an intelligent supplier relationship management system: a hybrid case based neural network approach. Expert Syst Appl 24:225–237
7. Møller C (2005) ERP II: a conceptual framework for next-generation enterprise systems? J Enterp Inf Manag 18(4):483–497
8. Kovacs Z, Le Goti JM, McClatchey M (1998) Support for product data from design to production. Comput Integr Manufact Syst 11(4):285–290
9. Liu DT, Xu XW (2001) A review of web-based product data management systems. Comput Ind 44:251–262
10. Eynard B, Gallet T, Nowak P, Roucoules L (2004) UML based specifications of PDM product structure and workflow. Comput Ind 55(3):301–316
11. Terzi S, Bouras A, Dutta D, Garetti M, Kiritsis D (2010) Product lifecycle management: from its history to its new role. Int J Prod Lifecycle Manag 4(4):360–389
12. Guerra-Zubiaga DA, Ramon-Raygoza ED, Rios-Solter EF, Tomovic M, Molina A (2009) A PLM tools taxonomy to support product realisation process: a solar racing car case study. In: Tomovic M, Wang S (eds) Product realization a comprehensive approach. Springer, London
13. Consoli D (2011) A layered software architecture for the management of a manufacturing company. Informatica Economică 15(2):5–15
14. Botta-Genoulaz V, Millet PA, Grabot B (2005) A survey on the recent research literature on ERP systems. Comput Ind 56(6):510–522
15. Bond B, Genovese Y, Miklovic D, Wood N, Zrimsek B, Rayner N (2000) ERP is dead—long live ERP II. Gartner Group, New York
16. Khedher AB, Henry S, Bouras A (2010) An analysis of the interaction among design, industrialization and production. In: 7th IFIP international conference on product lifecycle management, Bremen, 12–14 July 2010
17. Tursi A, Panetto H, Morel G, Dassisti M (2009) Ontological approach for products-centric information system interoperability in networked manufacturing enterprises. Ann Rev Control 33:238–245
18. Kilpatrick T, Gonzalez J, Chandia R, Papa M, Shenoi S (2008) Forensic analysis of SCADA systems and networks. Int J Secure Netw 3(2):95–102
19. Lalsingh D, Raut N (2012) Data management among PLM-ERP application: an integrated PLM–ERP technique. In: AIMS international conference on technology and business management. Dubai, 26–28 Mar 2012
20. Eynard B, Troussier N, Carratt B (2010) PLM based certification process in aeronautics extended enterprise. Int J Manuf Technol Manag 19(3–4):312–329

21. Thompson SHT, Devadoss P, Pan SL (2006) Towards a holistic perspective of customer relationship management (CRM) implementation: a case study of the housing and development board, Singapore. Decis Support Syst 42:1613–1627
22. TEC-SHS (2008) Technology readiness levels handbook for space applications
23. Bosscha P, Coetzee R, Terblanche P, Gazendam A, Isaac S (2006) SmartFactory: the challenges of open and low-cost ICT in the small manufacturing industry. S Afr J Sci 102(7–8):335–338

Part VIII
Enabling Technologies and Tools (Computer Aided Conceptual Design, Virtual Reality, Haptics, etc.)

Approaches in Conceptual Shape Generation: Clay and CAD Modeling Compared

Tjamme Wiegers and Joris S. M. Vergeest

Abstract We compared the methods of modelers who modeled in clay and those who modeled in CAD. We gave special attention to the size and shape characteristics of the model, and to the differences in approach between individual modelers. Four modelers made three different objects in clay and four other modelers made the same three objects in CAD. As a measure of success of the modeling method, we used the quality of the generated model, based on a set of criteria. Generally, the overall appearance of the clay models was better than that of the CAD models. Individual modelers applied different approaches for the same shape, not only when using clay, but also during CAD modeling. The quality of the models varied greatly. We conclude that the most appropriate modeling method depends on the size and shape character of the model, and also of the preferences and skills of the subjects.

Keywords Shape ideation · Conceptual design · CAD · Clay modeling

1 Introduction

For designers in the phase of idea generation, different tools and methods are available for shape exploration. New shapes can be sketched, modeled in clay, foam or card board, or generated with Computer-Aided Design (CAD) systems.

T. Wiegers (✉) · J. S. M. Vergeest
Department of Design Engineering, Delft University of Techno.ogy, Delft, The Netherlands
e-mail: t.wiegers@tudelft.nl

J. S. M. Vergeest
e-mail: j.s.m.vergeest@tudelft.nl

A. Chakrabarti and R. V. Prakash (eds.), *ICoRD'13*, Lecture Notes in Mechanical Engineering, DOI: 10.1007/978-81-322-1050-4_73, © Springer India 2013

For many shapes, one can deliberately choose one of these methods. However, it can be expected that particular shapes can best be made by one method and other shapes by other methods. Conceptual modeling can be considered as early prototyping, in the sense of Lim's definition [1]. Lim defines prototyping as creating a manifestation that, in its simplest form, filters the qualities in which designers are interested, without distorting the understanding of the whole. As purposes, Lim mentions evaluation and testing; understanding of user experience, needs, and values; idea generation; and communication among designers. Her Economic principle of prototyping tells that the best prototype is one that, in the simplest and the most efficient way, makes the possibilities and limitations of a design idea visible and measurable. Additionally, Grady [2] shows that quick and inexpensive prototypes make designers less defensive and users freer to criticize. Therefore, the economic principle of prototyping can function as a criterion to decide between different shape modeling methods in the ideation phase of design. For one modeling method may be more appropriate for a particular strategy than another one. And the same shape may be modeled more satisfactory with one method than with another one. Robertson and Radcliffe [3], for example, found that CAD is often not applicable for the creative phase of design. The complexity of CAD systems is a challenge for the design of interaction. Issues in this area are, for example, visibility, feedback, size of workspace, direct manipulation and context recognition [4]. Many of these are not big issues in physical modeling. Direct manipulation, for example, is obvious when modeling in clay or foam, and visibility is just a matter of turning the object in the hands, or moving around it. Elkaer [5] advocates the use of physical models, because their ambiguity makes them well suited for early discussions, whereas the digital models often close down the creative process if used from day one. The focus of this study is on idea generation, the creative phase of shape generation, in which many alternative shapes are explored in a short time. In this phase, I have often seen design students making small clay models within a few minutes, to make their ideas tacit. Such models are good examples of the economic principle of prototypes. Clay has as advantages that modelers can feel the model they are working with and can see it from different viewpoints. But there are drawbacks, too: for large models this craft is time consuming, and tiny parts may collapse. We compared the modeling processes of subjects that modeled in clay and subjects that modeled in CAD. We gave special attention to the size and the shape character of the model, and to the differences in approach between individual modelers. In this paper, we will indicate clay modeling and CAD modeling as modeling *methods*. When applying one of these modeling methods, different sequences of activities can be performed to achieve a particular shape modification. We will mention such an activity sequence an *approach*. So, subjects using the same modeling method (e.g. clay modeling), can still use different approaches (e.g. hitting and cutting are possible approaches for making a flat surface).

About the sizes and shape character of the models, it is believed that clay modeling will be more appropriate if the model is small, or has double curved faces, and CAD modeling will be more appropriate for large models and models that have geometrically well-defined shapes (e.g. blocks and cylinders). To verify this, in our tests we used multiple objects that differ in size and shape characteristics.

About the modeling approaches, students often believe that, for generating a particular shape in CAD, there is one best approach they should apply. On the other hand, when clay modeling, they often just start to knead the clay, without thinking a particular strategy is required. Because of this, we want to verify whether CAD modeling subjects indeed apply the same strategy if they model the same shape. And if a variety of approaches can be observed among clay modelers who generate the same shape. To investigate these issues, we formulate the following research questions:

- Do different modeling methods differ in appropriateness if they are used to model the same shape?
- Do subjects use different approaches if they use different modeling methods for the same shape?

Which modeling approaches can be expected? For CAD modelers, important factors are the available tools and functions offered by the CAD system. In general, CAD systems offer a multitude of tools, much more than a modeler will use for generating the rather simple objects that are used in this test. Therefore, we will not list all possible functions of the CAD system used, but only mention a subset. Functions that are frequently used by design student are: sketch, extrude, transfer, scale, rotate, cut, trim, and also functions that support viewing and evaluating the model-under-construction, such as zooming, panning and rotating the view.

For clay modeling, the situation is different. Which activities are applied, depends mainly on the preferences and skills of the user. van Dijk and van Veldhoven [6] observed clay modelers and identified ten clay modeling activities (Fig. 1). We used this categorization for our analysis. Apart from this, the type of clay plays a role. Water based clay behaves different from oil based clay. From the latter, the softness at room temperature is important. Styling clay is hard at room temperature, so modelers can apply sand paper and files. For molding styling clay, it should first be heated. We will use plasticine that is soft enough at room temperature to deform it by hand, without tools. However, the subjects are allowed to use tools, because they can be useful to make sharp cuts and edges and to flatten surfaces and make transitions smoother.

Fig. 1 Some of the identified clay modeling activities

2 Method

To answer the research questions, we designed an experiment. In this experiment, subjects will model shapes using different methods. As methods we choose clay modeling and CAD modeling. During the experiment, subjects have to model an object. Some subject will use clay for the modeling, others will use CAD. Different objects will be used, to be able to test the influences of size and shape characteristics. We choose objects from which we think they can be modeled in clay and in CAD. However, we also want to verify this. For this reason, students are asked which modeling method they would prefer to model the object. However, if this question is asked to the subjects that actually have to model the object in clay or in CAD, the results can be biased. Therefore, this question is asked to students that do not participate in the test as modelers. The modeling subjects had to finish their task in 15 min. They filled pre and post experimental questionnaires.

2.1 Subjects

Subjects were 24 students Industrial Design Engineering (IDE). Twelve subjects modeled objects in clay. These subjects had at least 1 year experience in clay modeling. Twelve other subjects modeled in CAD. The applied CAD software was Solid Works. The subjects had at least 2 year experience with Solid Works. Three objects were modeled. Each object was modeled by eight subjects. Four of them modeled the object in clay; the other four modeled the object in CAD. Eight more IDE students participated in the test. Their task was to evaluate the quality of the modeled shapes. They were also asked which modeling method they would prefer if they had to model the particular object, as described above. The assessment was based on a list of criteria. These criteria concerned the recognizability of the form of the model, the proper volume impression, the perceived quality of the model, the recognizability of characteristic details and in how far the model could serve to estimate if the product can easy be used (user friendliness). The latter is important for designers that generate a quick model to get a first feedback on their generated ideas. Figure 2 shows the question list for the evaluators.

2.2 Objects

Because the appropriateness of the modeling method may depend on whether a shape is large or small, organic or geometric, and tiny or solid, we selected objects that differ in these qualities, see Fig. 3. The mobile phone is small and the stool is large. The hand set of an old telephone has an organic shape. In contrast to this, the stool has geometric shapes. And the stool has tiny parts, but the hand set is rather

Which method do you prefer for Modeling this object in 15 minutes?	CAD / Clay / Hard foam / Drawing / Other
In what level do you recognize the form of the phone/horn/stool?	Bad 1 2 3 4 5 Good
In what level do you get an impression of the volume of the object?	Bad 1 2 3 4 5 Good
What do you think about the quality of the object?	Bad 1 2 3 4 5 Good
In what level do you recognize the details of the phone/horn/stool?	Bad 1 2 3 4 5 Good
In what level can you test the user friendliness of the model?	Bad 1 2 3 4 5 Good

Fig. 2 Questions for the evaluators

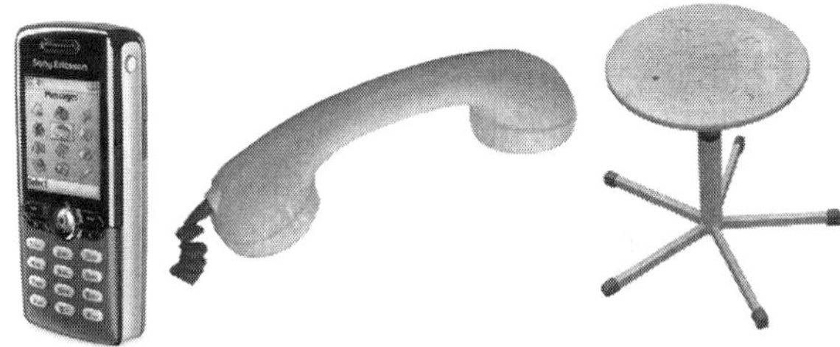

Fig. 3 The three objects for the experiment

compact. We expect the hand set will best be modeled with clay, but the stool with CAD. The small size of the mobile phone is an advantage for clay modeling; however, its buttons can easy be copied and pasted in CAD. Therefore, we expect that the approaches in clay and in CAD will be different, but the quality of the results will differ less than those of the other objects.

We choose a method in which multiple subjects model the same shape, because we want to be able to compare individual results. This method, however, has also a disadvantage. In fact, we ask subjects to copy a shape, not to design it. In this aspect, the test differs from the real design situation. Therefore, we should be aware that there can be other factors that play a role if a designer is creating new shape. For example, the skills of the designer and the ease of use of the applied CAD software may influence the ease of working, and, implicitly, the creativity of the designer.

2.3 *Interviews*

The subjects were asked questions before and after the experiment. The pre experimental questions were about their CAD experience and about their expectations of the modeling task. They are shown in Fig. 4. The post experimental questions are about the result of the modeling task (Fig. 5).

Which method do you prefer for modeling this object?	Clay CAD
How much time do you expect you need to model this object in:	Clay? ...... (min)
	CAD? ...... (min)
Which difficulties do you expect when you model this object in clay?	
Which difficulties do you expect when you model this object in CAD?	
How experienced are you in working with SolidWorks?	Bad/Average/Good/Excellent

Fig. 4 Pre experimental interview

Name:	
Start year at IDE:	
What were the difficulties in the modeling of the object?	
Afterwards, would you prefer modeling this object in clay or in CAD?	Clay CAD
What do you think about the quality of your own model?	(Bad) 1 2 3 4 5 (Good)
What do you think about the details of your object?	(Bad) 1 2 3 4 5 (Good)

Fig. 5 Post experimental interview

3 Results

The mobile phone was completed very fast by all subjects. Most time was spent on the detailing. In general, the evaluators found the overall impression of the clay models better than that of the CAD models (Fig. 6). According to the averages of the grades given by the evaluators, the clay models were better on Volume Impression and Testability of user friendliness, however, the CAD models are better on Recognizability and Quality of the shape, see Table 1. The averages for Recognizablity of Details hardly differs between clay and CAD models.

For the telephone hand set, some clay modeling subjects first created both ends and the midsection separately, and then assembled them into a hand set. Others molded the hand set as one whole. Some CAD modelers used extrudes, sweeps and fillets, others worked with lofts and assemblies. The differences in approaches are obvious in Fig. 7. Within the time limit of 15 min, the CAD subjects were not able to blend and bend the different part sufficiently to resemble the hand set properly. For all aspects that were assessed by the evaluators, the average grades for the clay models were higher than those of the CAD models.

The stool was quite a challenge for the clay modelers. They all made a small scale model and exaggerated the diameters of the thin profiles. Yet their models were not strong enough for testing. The CAD modelers had fewer problems. They had enough time left to make a render with realistic materials. Figure 8 shows the

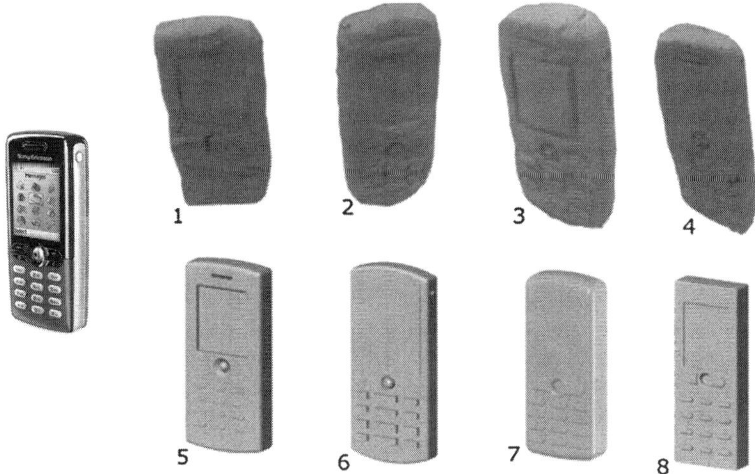

Fig. 6 The results of the mobile phone

results. For all aspects that were assessed by the evaluators, the average grades for the CAD models were higher than those of the clay models.

We also compared the scores of all clay models to those of all CAD models. On average, clay models score higher on Volume Impression and Testability of user friendliness; CAD models score better on Recognizability and Quality of the global shape. The Recognizability of details does not differ very much between clay and CAD models. This is exactly what we found for the results of the mobile phone. The results of the hand set and of the stool, however, slightly differed from the overall results.

From the questions asked to the evaluators, we found the following frequencies of preferred modeling methods:

- 10 times CAD;
- 5 times clay;
- 5 times foam;
- 3 times drawing;
- 1 time no preference was expressed.

Apparently, CAD is the preferred modeling method in most cases. Nevertheless, the subjects estimated the required modeling time for CAD 5 times higher than the modeling time for clay modeling. Remarkably, the subjects who modeled in clay expected long times for CAD modeling, while the subjects who used CAD estimate long times for clay modeling. Subjects with more than average experience with Solid Works expected longer modeling times. Those with 2 years experience actually needed those times, however, those with 3 years experience were quicker than they expected. The subjects who have more experience with SW scored better on Recognizability of details and on Testability of user friendliness.

Table 1 Results of the questionnaires

Subject	1	2	3	4	5	6	7	8	9	10	11	12	13	14	15	16	17	18	19	20	21	22	23	24
	Cell phone								Hand		Set						Stool							
Applied method[a]	1	1	1	1	2	2	2	2	1	1	1	1	2	2	2	2	1	1	1	1	2	2	2	2
Observers																								
Pref. method[a]	3	3	2	2	2	3	.	3	1	1	4	2	1	1	3	1	2	4	2	2	2	4	2	2
Recognizability	5	4	2	3	5	5	5	5	5	5	5	4	4	2	3	3	2	4	1	3	4	5	5	5
Volume impr.	5	4	1	5	1	1	2	2	2	4	4	4	2	1	1	1	1	2	1	1	1	2	3	3
Quality	4	4	1	2	3	4	4	3	2	4	4	2	3	3	2	2	3	3	2	3	3	4	4	5
Details recogn.	5	3	3	3	2	4	5	2	3	4	3	3	1	3	1	1	1	3	2	1	1	5	4	4
Testability	3	4	2	4	1	2	1	1	2	2	4	2	1	1	1	1	1	1	1	1	1	2	1	1
Average grade	4	3, 8	1, 8	3, 4	2, 4	3, 2	3, 4	2, 6	2, 8	3, 8	4	3	2, 2	2	1, 6	1, 6	1, 6	2, 6	1, 4	1, 8	2	3, 6	3, 4	3, 6
Modelers																								
Cohort (+2000)	2	1	2	1	1	1	1	1	4	3	4	1	1	2	1	1	1	1	3	4	1	1	1	1
Average grade	4	2, 5	2, 5	2, 5	4	3, 5	4	2	4	1, 5	3	2	3	3	1, 5	1	1	3	2, 5	3	1, 5	3	3, 5	2, 5
SW experience	2	3	3	3	2	2	3	1	2	1	3	2	3	3	2	2	1	3	3	1	2	2	3	3
Exp. clay time ′	15	8	10	10	10	10	10	15	10	10	10	10	45	20	20	30	10	.	2	10	15	10	20	30
Exp.CAD time ′	240	15	100	3	30	20	20	30	300	60	90	90	75	240	30	60	60	30	45	60	10	10	10	90
Actual time ′	12	10	10	6	15	15	15	15	15	15	13	12	15	15	15	15	6	5	4	4	15	15	15	15

[a] 1 = clay, 2 = CAD, 3 = foam, 4 = drawing

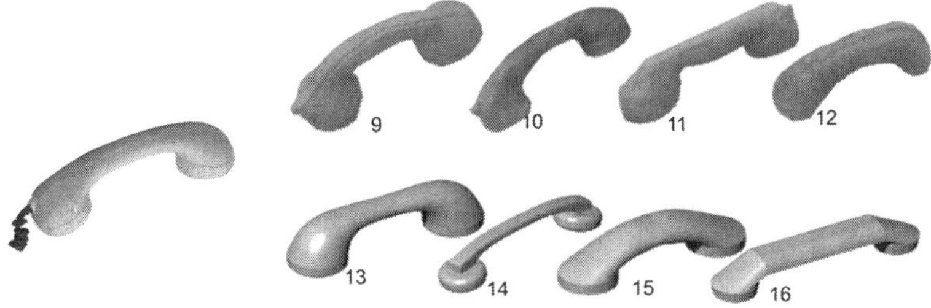

Fig. 7 The results of the hand set

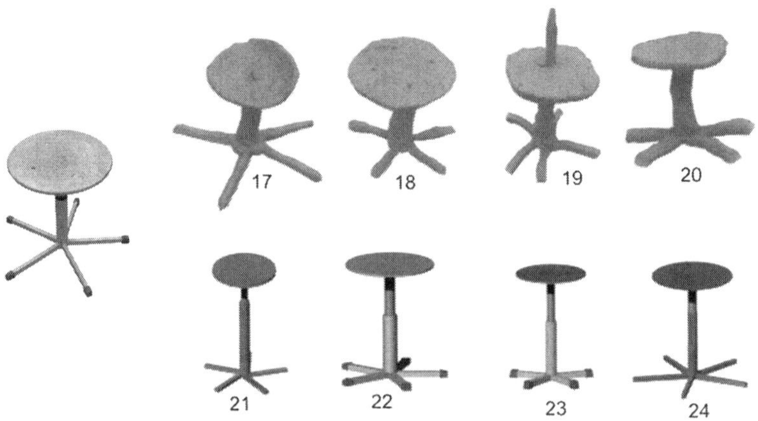

Fig. 8 The results of the stool

Subjects with scores below 2 had short modeling times; they scored in particular low on Volume impression and Recognizability of details. Many of them had no realistic expectations of the required modeling time. They estimated the modeling time either too high or too low.

Which activities were performed? Typically, a clay modeling process started with pressing a volume of clay for making the global shape of the object. Next, the activities differed among subjects. Some started detailing the global shape, others first check if the volume and its ratios are correct. The following step is the modeling of the smaller shape elements. Pinching is often used for that, and tools are applied to make sharp details. Individual subjects performed different activities, such as hitting, pounding, pressing and cutting. Several subjects rolled clay into small cylinders. After the modeling of the smaller shape elements, the global shape was optimized. Also fir this purpose, individual subjects differed in the activities they performed for that. We observed smoothening, hitting, adding clay, detailing with the help of a tool, etc. Summarizing, subjects did not much in the modeling of the global shape. However, for the finishing touch, we observed different strategies, in particular for the modeling of flat surfaces and for the fine tuning of the global shape.

For CAD modelers, extrusion of a sketch was the typical activity for generating the global shape. Some subjects followed with making cut extrudes for particular local shape elements. Next, fillets were made. Some subjects applied other functions, such as chamfering, line patterns and the dome function. Also the mirror function was used by several subjects.

From the above, we see clear differences in approach between clay modeling and CAD modeling subjects. We mention a few. Generally, clay modelers start from a volume and press it into the right shape. Solid Works modelers, however, sketch a cross section and extrude it. They also make fillets, because otherwise the edges will be too sharp. In contrast, clay modelers sometimes use tools to sharpen the edges, because the shapes they generate are often too smooth. However, we did not only see different approaches between clay modelers and CAD modelers, also subjects who used the same modeling method differed in their strategies.

Comparing the results of the clay modeling subjects to those of the CAD modeling subjects, we can summarize that the clay modeling subjects delivered better handsets, and the CAD modeling subjects delivered better stools. For the mobile phones, the outcome is less simple. Some aspects of the mobile phones were better in the clay models, while other aspects were better in the CAD model.

From the hand sets, the clay models had better global shape and gave a better impression of the volume distribution. Also, the separate parts were more realistic, and they were more smoothly connected to each other. Furthermore, the clay models contained more details. The assessors found the clay models of the hand sets, compared to the CAD models, were more appropriate to test the user friendliness of the product.

The stools were much better when modeled in CAD. Although the global shapes of the clay models were appropriate to indicate a stool, the details were far less convincing than those of the CAD models. None of the clay models, for example, contained the telescopic shape of the pole, nor the caps at the ends of the legs. The legs did not have clear profiles, where the CAD models had oblong cross sections, just as the stool that served as the example for the modeling task. On the other hand, the seat of most clay models was slightly hollow, while the CAD models neglected this shape aspect and just contained a flat disk.

These results give us qualitative insights into the appropriateness of different modeling methods. However, the numbers of subjects and objects in the test are too low to do quantitate claims. Another restriction is caused by the applied modeling software. Solid Works was used, because nearly every design student is familiar with it. However, a wide variety of modeling software exists, from which several are more appropriate for modeling organic shapes, e.g. because they contain better functions for surface modeling. In addition, for designers who are more experienced in computer modeling will have fewer problems with modeling organic shapes.

4 Conclusions

Revisiting the research questions, we can conclude that different modeling methods do differ in appropriateness for modeling the same shape. Secondly, the approach a modeling subject applies depends on the modeling method the subject is using. Different methods have their own pros and cons. With clay modeling, for example, it is possible to give a good overall impression of an organic shape in short time, in particular if the detailing does not need to be very precise. In such cases, clay modeling is a good choice according to Lim's Economic principle of prototyping [1]. If CAD is used for the same modeling task, the modeler might either spend more time or end up with a model that has some aspects worked out in a very detailed way, while other shape aspects are missing. In the latter case, the model does not tell a consistent story and may confuse the observer. Other types of shapes, however, may be easier to model in CAD, for example if they consist of thin elements with simple geometric cross sections. It is difficult to work with such elements in clay, because they easily bend or break, sometimes already because of their own weight. CAD models don't have such problems, because they have no mass at all.

Summarizing, we conclude that the best modeling approach to be applied does indeed depend on the size and shape characteristics of the model. Furthermore, individual modelers apply different approaches for modeling the same shape, not only when using clay, but also during CAD modeling.

Acknowledgments The authors like to thank Stephan Maaskant and Koen Vorst for performing the experiments and gathering the data.

References

1. Lim Y-K, Stolterman E, Tenenberg J (2008) The anatomy of prototypes: prototypes as filters, prototypes as manifestations of design ideas, ACM transactions on computer-human interaction, Vol. 15(2), pp 1–27
2. Grady HM (2000) Web site design: a case study in usability testing using paper prototypes. In: Proceedings of IEEE professional communication society international professional communication conference and proceedings of the 18th annual acm international conference on computer documentation: Technology & Teamwork, IEEE, Cambridge pp 39–45
3. Robertson BF, Radcliffe DF (2009) Impact of CAD tools on creative problem solving in engineering design. Comput Aided Des 41:136–146
4. Ghang Lee G, Eastman CM, Taunk T, Ho C-H (2010) Usability principles and best practices for the user interface design of complex 3D architectural design and engineering tools. Int J Hum Comput Studies 68:90–104
5. Elkær TN (2009) Using computers to aid creativity in the early stages of design—or not! http://www.salle.url.edu/sdr/info/ ©ARC, Enginyeria i Arquitectura La Salle, Universitat Ramon Llull. In: Computation the new realm of architectural design 27th eCAADe conference Proceedings, 1st edn. Cenkler Printing, Istanbul, pp 761–768
6. van Dijk O, van Veldhoven E (2003) Research on clay modeling, advanced modification analysis. Dynash research report, Delft University of Tecnology, Faculty of Industrial Design Engineering

Optimization of the Force Feedback of a Dishwasher Door Putting the Human in the Design Loop

Guilherme Phillips Furtado, Francesco Ferrise, Serena Graziosi and Monica Bordegoni

Abstract The aim of the research work described in the paper is to enable designers to optimize the force feedback of a dishwasher door, in order to improve the user experience with the product at the moment of purchase. This is obtained by allowing the user to test the product since the beginning of the design process through the use of interactive Virtual Prototypes based on haptic technologies. A commercially available dishwasher is used as case study. The mechanical system producing the force feedback is modeled in a multi-domain simulation environment, and in parallel a parameterized simplified simulation is made available to the user through a force feedback haptic device. That feedback can be easily modified on user's requests and the desired behavior can be sent back to the multi-domain simulation, which optimizes the system to behave in the desired way. How to correctly involve humans into the proposed design framework is also discussed, highlighting their key role in determining product characteristics.

Keywords Virtual prototyping · Product virtualization · Human in the loop · Product design

1 Introduction

It is known that what makes consumers like and buy a product has not a simple answer. Anyway it can be assumed that the most successful product appeals on both the rational and the emotional levels [1, 2].

G. P. Furtado (✉) · F. Ferrise · S. Graziosi · M. Bordegoni
Dipartimento di Meccanica, Politecnico di Milano, Milan, Italy
e-mail: francesco.ferrise@polimi.it

A. Chakrabarti and R. V. Prakash (eds.), *ICoRD'13*, Lecture Notes in Mechanical Engineering, DOI: 10.1007/978-81-322-1050-4_74, © Springer India 2013

Most of the emotions, which are part of the user experience, rise during interaction with the products. All the human senses concur to create the final estimate, even if the buyer is not aware of this at the moment of purchase [3]. Consequently from a design perspective, there is a big issue in designing appropriately those product features that consumers will interact with at the point of sale: good design can make pleasant and efficient the consumers' interaction with the product as well as the buying experience [4].

Therefore, it is now strategic for a company to understand which are the multisensory characteristics of the products that mainly influence customers' delight. This is mainly the job of marketing experts who have to collect information from or about customers' likes/dislikes. That job is not a trivial task to perform, for: (1) on one hand, it is mandatory to create a valid context for collecting information so as to avoid "unduly discouraging" and useless innovation [5], conversely, (2) it is fundamental to find an effective way to transform such "fuzzy" product attributes into engineering technical specifications. In order to be useful these specifications should be quantitatively measurable criteria that the product design is expected to satisfy [6]; unclear targets can determine longer time to market and increase the probability of not correctly meeting the customers' expectations.

In a previous work [7], authors have put efforts in addressing the first challenge, proposing a methodology, based on the use of mixed prototypes and multisensory interaction models, to build a valid context for collecting product information related to users' experience with the product. That approach enables the creation of high fidelity and flexible representations of the product without the need of developing costly and time-consuming physical prototypes, and allowing us to evaluate several multisensory product variants (e.g. different sounds, touch and visual effects) [3]. In that work it has been used a fridge freezer as case study, which has been used to analyse the perception that users have of the frontal doors and subsequently of the internal elements of the products [8]. Using such a multisensory product representation it is then possible to rapidly change and tune the combined stimuli (e.g. the sound produced when opening a door and the force required to perform the opening action) until the user perceives an optimum condition. The final outcome is a virtual representation of the "ideal" product characterized by the desired behavior and the preferred appearance.

In this paper, the second challenge is discussed, which is how to transform such virtual representation of the ideal product, or better its "meta" characteristics, into technical specifications necessary for starting the design phase. To reach such purpose the methodology has been additionally developed by integrating an initial phase of advanced modeling of the physics at the basis of the haptic interaction modality by using the LMS-AMES im simulation environment [9]. Then the resulting outcome consists of a double but totally integrated view of the product: one view is technical and expressed in terms of physical laws, and the other one is experiential and expressed as a perceptual model. A change in the perceptual model should correspond to a change in its physics model. This new methodology has been tested on a case study provided by the Indesit Company (www.indesitcompany.com), a company operating in the field of domestic appliances. The research work consists in the optimization and re-design of the haptic feedback of a dishwasher door with a particular attention to the user experience.

2 Related Work

Especially for market-driven innovation projects, the role of marketing experts is to capture and interpret customers' expectations (i.e. what the product should do and how), but also to transfer that knowledge to the company, mainly to designers, who will have to develop the new product. Performing such interpretation is not a trivial task since customers' requirements are often expressed through imprecise, non-technical terms [10]. Even if up to now a great effort has been put to define mechanisms to enhance communication between marketing and engineering teams [11, 12] difficulties are still present since their working domains, control variables and product representation models are completely different [13].

Thus, the complex work of designers is to make use of their experience and intuition to translate such qualitative requirements into product features: the final result is however a product capable of satisfying their interpretation of the customer needs, rather than the real ones. Furthermore, several levels of costly and time-consuming refinements are necessary: during design review sessions, specifications have to be verified usually by means of testing sessions on physical prototypes. These ones work as the experimental set-up of end users' response evaluation tests, enabling them to physically see, touch and interact with the product. However, mainly due to costs and time saving targets, these artefacts are not always made with the same material that will be selected for the final product, and neither they usually implement the final construction solutions. For this reason it is not guaranteed that visual and sounds effects will be the same of the final product.

Due to the cost of the physical prototypes, one of the recent trends in the Product Development Process (PDP) is to substitute these ones with their virtual replica (Virtual Prototypes—VP), which should at best look like, work like and behave like as the final product or at least as the physical prototypes; VPs are flexible to changes, especially during design review sessions, and less expensive than the physical ones [14, 15]: intermediate versions of the new product concept can be verified.

Interaction with Virtual Prototypes happens through Virtual Reality (VR) technologies. Visualization VR technologies have now reached a high rendering quality and in some cases cost reduction [16]. Thanks to the interest they attracted within the research community for decades they have been studied, developed and applied more than haptics [17]. However, the evaluation of the physical interaction by means of "non-physical" components is definitely poor and does not provide useful results: visual prototypes, i.e. prototypes based on pure visualization technology, are mainly aesthetic in the sense that they enable people to see in a realistic manner how the final product will look like.

Simple haptic control devices, such as knobs and buttons, have been developed for testing user interaction with the interface components of consumer products [18, 19]. The fidelity of the representation of the haptic behavior of the knob is high, but the test of the user interaction with the product is limited only to these elements. Other custom 1-Degrees of Freedom (DOF) haptic devices have been developed to help engineers to design and evaluate the haptic feedback of doors of refrigerators [20] and car doors [21] during the design activity.

However, a method is lacking on how to correctly translate the haptic interaction results into technical specifications useful to optimize the complex initial mechanical system (that is the one responsible for the interaction). Effort has been put to overcome this lack for what concerns the sound stimuli since studies [22, 23] have concentrated on correlating them with specific product features, and then with engineering specifications.

In [24] an attempt has been made to merge haptic, sound and visualization stimuli: a multimodal VR system has been designed to test and reproduce the human multisensory experience through interactive Virtual Prototypes (iVPs, i.e. prototypes made to test human-product interaction features [7]). The haptic feedback is returned through a 6DOF haptic device equipped with a generic handle, and a single sound for each colliding component is recorded and played. This pilot study has driven the research activity described in [7] where the analysis performed on the real product is more accurate.

In this paper authors add a detail to the design of the multisensory experience, i.e. an approach to help in translating marketing qualitative requirements into engineering specifications. The final output should bea multi-domain representation of the system under analysis, equipped with appropriate optimization algorithms. In particular, even if the paper is focused on describing the optimization procedure for the force feedback stimulus (i.e. the haptic stimulus), it takes into account that this feedback is always part of a multisensory experience, which the user will test through an interactive Virtual Prototype (iVP) as described in [7]. The intent of focusing the work only on the force feedback is justified by the fact that, as previously said, while for vision and sounds methods are available on how to correctly translate the interaction results into technical specifications, for the haptic feedbacks these methods are not present in literature.

3 Designing the Multisensory User Experience: A Design Framework

The term *User Experience*s defined by the ISO 9241-210 [25] as the user perception of the use of a system. It represents what the user feels interacting with the system and how that interaction affects his/her personal perception of it. In order to make that perception as much pleasant as possible objective methods are necessary to enable the quantification of user's preferences and then the translation of these into clear design specifications.

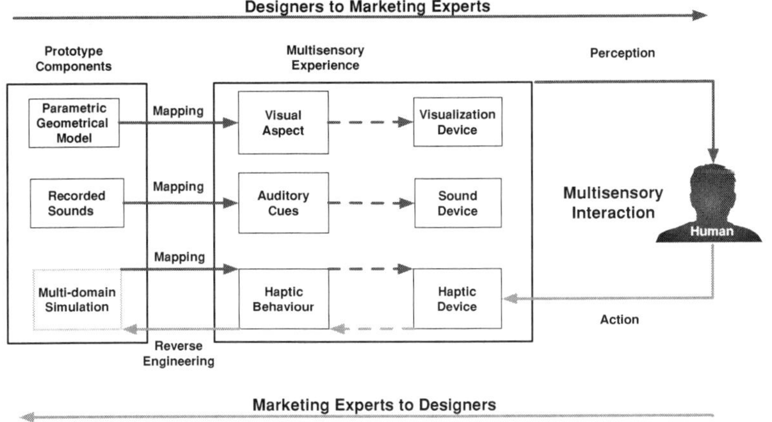

Fig. 1 The proposed design framework: how translating user experiences into measurable parameters to use in the product design

To address this need a design methodology has been defined. The idea behind the methodology is to make use of iVP which, even if the product does not exist physically, can however model its effects and functions. As shown in Fig. 1 the multisensory experience of the product is realized mixing three different sensory stimuli: (1) *vision*, using a rendering device which reproduces the appearance of the product; (2) *audio*, involving a sound device able to generate auditory cues; (3) *haptic*, by means of a device which reproduces the haptic behavior of the product. That mix generates a perception on the user that will react interacting with the virtual product itself: the final result is then an action-perception loop.

A new loop will be generated once changes are implemented on the virtual prototypes thus creating a new multisensory experience to explore. However, since the research purpose is to quantitatively monitor the feeling that users can get by interacting with and observing the virtual replica of the product, the multisensory interaction loop alone is not sufficient. To completely address the research purpose, another loop (i.e. mapping-reverse) has been inserted in the proposed design framework. This loop is based on linking each sensorial stimulus of the iVP to the specific functional representation or technical source that generates it. The intent is to let users interact with the virtual prototype, express their preferences and ask for changes to be applied to the interactive Virtual Prototype. Knowing the physical laws and the dimensional characteristics that are behind the iVP, is then possible to perform a reverse engineering of these preferences in order to update the technical models in accordance with users' insights. The final resulting framework is itself a loop between two different perspectives, the technical and the marketing one.

The proposed methodology is then able to support the experience creation and modification: the interaction with the virtual prototype is parameterized so that needs and preferences can be translated into something measurable. Furthermore, in case of technology push projects, where the implementation into the product of a new technology is the driving innovation force, the methodology enables designers to virtualized their technical solutions or their product variants, and then ask market experts to choose the most appealing.

In this paper the discussion will be focused on demonstrating the potentials of correlating the haptic behaviour of the virtual product with its functional model (i.e. the multi-domain simulation indicated in Fig. 1) since the other elements of the framework have been already tested and discussed in previous papers (see [7] for the multisensory interaction loop, and [22, 23] for the sound feedback).

4 Case Study: Force Feedback Optimization of a Dishwasher Door

The physics of the dishwasher door opening highly influences the user's first perception of the product since it is the first element the user interacts with, at the point of sale.

Depending on how the door and its opening mechanism have been designed, the force required to open it can vary. To open the door, typically, the user has to pull hard enough to un-lock a spring-loaded mechanism. A traditional qualitative rule states that the door should close and lock without heavy pressure, to prevent the user has to force the dishwasher to shut. The critical point is how to correctly translate the qualitative indication of "avoiding heavy pressure" into a quantitative force/pressure value to consider as input when designing the door lock mechanism. The dishwasher door force feedback optimization has been considered as an interesting case study to test the proposed design framework and in particular, the haptic loop (see level 3 of Fig. 1).

4.1 Interaction Problem Analysis

The primary aspect that a designer takes into account when designing the opening system of a dishwasher door is to be compliant with the technical requirements defined by reliability issues and normative validation analysis.

In a dishwasher the door is attached to the front side of the cabinet by means of a hinge placed at the bottom part of the door (Fig. 2). The hinge provides an appropriate balancing force, generated by the cumulative effects of the spring and of the frictions, in order to guarantee the stability of the doors during its movement from the vertical to the horizontal position. The latch mechanism used to lock the door, and specifically the component that clips into the locking mechanisms, can be represented as a leaf spring. From the user interaction point of view, the haptic feedback of the door opening is the combination of the initial leaf spring (i.e. the

Fig. 2 Mechanical components that contribute to the creation of force feedback on user's hand

click effect perceived at the beginning of the opening) plus the inertia of the door together with frictions plus a compression spring.

To merge technical requirements with customer's needs the first step is to build the dynamical model of the door opening system. The model is necessary to simulate how the different parameters that compose the mechanisms affect the overall behaviour of the door opening system. It can be described by a set of differential equations, whose constants are the parameters that mostly affect users' perception: these ones can then be used to tune the mechanical behaviour of the system.

The model has been designed by using the LMS-AMESim suite. Figure 3 illustrates the sketch of the system with highlighted the main parts that contribute to the interaction (i.e. frictions, spring and door collisions). The dishwasher door has been modeled as a bar mechanism, while the friction devices present in the system were considered as producing dry friction. The latch mechanism was modeled as a linear spring able to work only on a narrow range, when opening and closing the door. The effects of air compression and circulation (when opening and closing the dishwasher) were considered as not influent. The human force (i.e. the input force, Fig. 3) exerted to open the door, has been also inserted in order to simulate the human behaviour. That behaviour is based on the results described in the work of Jain et al. [26], and on the experimental acquisitions performed on the door of the fridge described in [7]. The effect can be effectively treated as a ramp function.

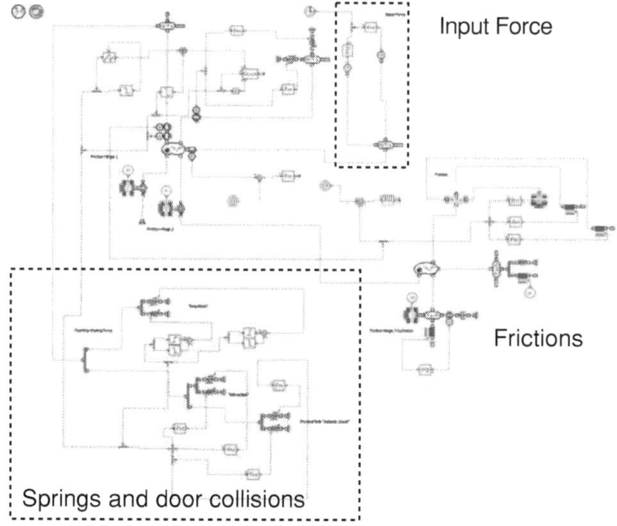

Fig. 3 The main contributions to the force exerted on user's hand highlighted in the sketch of the door mechanism

4.2 Making the Simulation Interactive: The Use of the Haptics

The next step of the proposed design framework (Fig. 3) consists of mapping the model built with the multi-domain simulation into functions that control the haptic device in order to make the user able to test the model.

In reality, the behaviour reproduced by the haptic device might not be necessarily described using the same mathematical model selected to represent the system in the multi-domain simulation environment. This might occur for several reasons, among them undesired vibrations on the device caused by a high stiffness, or by the fact that it might be simpler to program the device using a different approach, like creating a magnetic surface to constrain the movement of the end-effector instead of reproducing mathematically the effect of a reversed pendulum.

Before developing the application, a typical scenario of use has been figured out (Fig. 4) in order to guarantee that the test is performed in a realistic condition: the dimensions of the visual and of the haptic models should be the same of the dimensions of the real product.

Specifically, to control the haptic manipulation and then propose to the user different opening behaviours, the scripting language Python integrated with H3DAPI (www.h3dapi.org) has been used for developing the interaction model and for retrieving the values necessary to quantitatively describe the interaction phase in terms of displacement, acceleration and velocity. The dishwasher door trajectory has been reproduced faithfully in the haptic device by creating a cylindrical magnetic surface with high attraction force. The device is then only able to follow a semi-

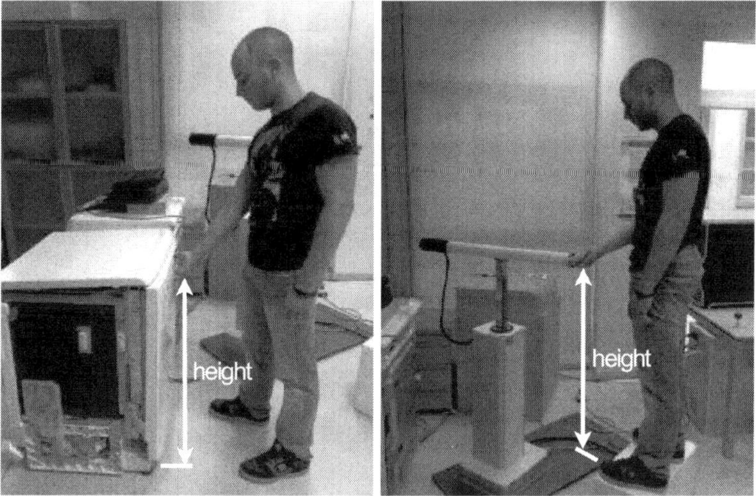

Fig. 4 User interacting with the real door and the simulated one

circumference trajectory. The tuneable opposing force F_h, generated by the haptic device, is composed by a dry friction component and a viscous damping, both acting on the direction of the trajectory. The coordinate θ describes the angular position of the end-effector with respect to the horizontal plane, while R is the length of the door. The dry friction component can be modeled as $D * \text{sgn}(\dot{\theta})$, D being the magnitude of the force, and $\text{sgn}(\dot{\theta})$ the sign function, a discontinuous function that can be approximated by a continuous hyperbolic tangent, $\tanh\left(v\dot{\theta}\right)$, when $v \gg 1$. The viscous damping is considered proportional to the velocity of the door. Therefore, the equation describing the magnitude of the force acting tangent to the trajectory, opposing the force F_u applied by the user, is:

$$F_h = D\tanh\left(v\dot{\theta}\right) + c\dot{\theta} \tag{1}$$

The latch mechanism that acts when opening and closing the door can be modeled as a spring, where both the stiffness and the direction of the force depends whether the user is closing or opening the door. The equation that computes the magnitude of the force from the action of opening the door is:

$$F_o = k_o(\theta - \alpha) + F_h \tag{2}$$

where α and k_o are heuristically chosen to tune the resistant force when trying to open the door.

The magnitude of the force from the action of closing the door is computed by a similar equation:

$$F_c = k_c(\theta - \beta) + F_h + c_c\dot{\theta} \tag{3}$$

The main difference between (2) and (3) is the addition of a damping effect. The use of the damping force is to avoid the end-effector from bouncing back when closing the door. Additional terms could also be added to the force equations to express additional effects.

All the parameters can be changed in real time: they can be increased or decreased while the user is testing, accordingly to his/her preferences. Once the parameters are chosen, the force applied by the user together with the position of the haptic device end-effector is extracted (in function of time). The force F_u applied by the user is used as the input for the AMESim model (Fig. 3). Then, the position $\theta(t)$ obtained from the haptic device is compared with the position $\theta^*(t)$ obtained from the simulation. The comparison is described as the maximum square difference between the two values:

$$\varepsilon = \max(\theta^*(t) - \theta(t))^2, t \in \{0, T\} \tag{4}$$

The objective is to find the parameters on the differential equation that minimizes ε. LMS-AMESim offers tools that can be used to solve this optimization problem. If the result of the optimization is adequate, a physical prototype can then be constructed for validation.

5 Conclusions

The design of the emotional response is one of the key factors of a company's success because it influences consumers' purchasing behaviour. Starting from this consciousness, the work described in this paper aims at proposing a new design framework able to objectify the interaction aspects that determine consumers' choices, and to make easier the translation of qualitative customers' expectation into quantifiable design specifications.

To reach such purpose the haptic feedback, coming from the interaction of the user with the Virtual Product (by means of an haptic device), has been correlated with the physics and then the equations of those specific mechanical systems or sub-systems of the product involved in the interaction. The LMS-AMESim environment has been chosen to define the dynamical/mathematical model used to represent the system behaviour. The algorithms used to control the haptic device have been shaped according to the previously defined model. Correlations between these two mathematical representations have been identified in order to come out with a mapping-reverse design loop: a change in the algorithms that control the haptic device, and then in the interaction of users with the Virtual Prototype, can be translated into a change of the mechanical system responsible for the interaction. An optimization activity of the emotional response of a product can then be planned so as to make easier and more effective the collaboration of designers with product final users (or company marketing experts).

The proposed framework has been tested on a case study provided by the Indesit Company.

Acknowledgments The authors would like to thank the Indesit Company, and in particular Eng. Dino Bongini, for providing the case study and the feedback during the development of the case study

References

1. Norman DA (2003) Emotional design: why we love (or hate) everyday things. Basic Books
2. Neff J (2000) Product scents hide absence of true innovation. Advertising Age 7:22
3. Spence C, Gallace A (2001) Multisensory design: reaching out to touch the consumer. Psychol Mark 28(3):267–308
4. Meyer C, Schwager A (2007) Understanding customer experience. Harvard Bus Rev 117–126
5. Veryzer RW (2005) The roles of marketing and industrial design in discontinuous new product development. J Prod Innov Manage 22(1):22–41
6. Otto KN, Wood KL (2001) Product design: techniques in reverse engineering and new product development. Prentice-Hall, Upper Saddle River
7. Ferrise F, Bordegoni M, Graziosi S, (2012) A method for designing users' experience with industrial products based on a multimodal environment and mixed prototypes. Comput-Aided Des Appl (in press)
8. Bordegoni M, Ferrise F, Cugini U (2012) Development of virtual prototypes based on visuo/tactile interaction for the preliminary evaluation of consumer products usage. CIRP Design Conference, Bangalore
9. Marquis-Favre W, Bideaux E, Scavarda S (2006) A planar mechanical library in the AMESim simulation software. Part II: library composition and illustrative example. Simul Model Pract Theor 14(2):95–111
10. Harding JA, Popplewell K, Fung YK, Omar AR (2001) An intelligent information framework relating customer requirements and product characteristics. Comput Ind 44(1):51–65
11. Griffin A, Hauser JR (1996) Integrating R&D and marketing: a review and analysis of the literature. J Prod Innov Manage 13(3):191–215
12. Leenders M, Wierenga B (2002) The effectiveness of different mechanisms for integrating marketing and R&D. J Prod Innov Manage 19(4):305–317
13. Michalek JJ, Feinberg FM, Papalambros PY (2005) Linking marketing and engineering product design decisions via analytical target cascading. J Prod Innov Manage 22(1):42–62
14. Buchenau M, Fulton J (2000) Experience prototyping symposium on designing interactive systems 2000, ACM Press, Brooklyn, pp 424–433
15. Zorriassatine F et al (2003) A survey of virtual prototyping techniques for mechanical product development. Proc Instit Mech Eng, Part B: J Eng Manuf 217(4):513–530
16. Hainich RR, Bimber O (2011) Displays: fundamentals and applications. A K Peters/CRC Press
17. Hayward V, Astley OR, Cruz-Hernandez M, Grant D, Robles-De-La-Torre G (2004) Haptic interfaces and devices. Sens Rev 24(1):16–29
18. Bordegoni M, Colombo G, Formentini L (2006) Haptic technologies for the conceptual and validation phases of product design. Comput Graph 30(3):377–390
19. Kim L et al (2008) A haptic dial system for multimodal prototyping. In: 18th international conference on artificial reality and telexistence (ICAT 2008)
20. Sunghwan S et al (2012) Haptic simulation of refrigerator door, Haptics Symposium (HAPTICS), 2012 IEEE, pp 147–154

21. Strolz M, Groten R, Peer A, Buss M (2011) Development and evaluation of a device for the haptic rendering of rotatory car doors. Ind Electron IEEE Trans Impact Factor 58(8): 3133–3140
22. Parizet E, Guyader E, Nosulenko V (2008) Analysis of car door closing sound quality. Appl Acoust 69(1):12–22
23. Van der Auweraer H, Wyckaert K, Hendricx W (1997) From sound quality to the engineering of solutions for NVH problems: case studies. Acta Acustica Unit Acustica 83(5):796–804
24. Ferrise F, Bordegoni M, Lizaranzu J. (2010) Product design review application based on a vision-sound- haptic interface. In: Haptic and audio interaction design, vol 6306. Nordahl R, Serafin S, Fontana F, Brewster S (eds) Lecture notes in computer science. Springer, Berlin/ Heidelberg, pp 169–178
25. ISO DIS 9241-210:2008. Ergonomics of human system interaction—part 210: human-centred design for interactive systems. International Organization for Standardization (ISO), Switzerland
26. Jain A et al (2010) The complex structure of simple devices: a survey of trajectories and forces that open doors and drawers. In: 2010 3rd IEEE RAS and EMBS international conference on biomedical robotics and biomechatronics, pp 184–190

Cellular Building Envelopes

Yasha Jacob Grobman

Abstract The paper argues that the digital revolution in architectural design and manufacturing, particularly the new possibilities offered for the design and manufacture of complex geometry, calls for a re-examination of the traditional concept of the layer-based building envelope which serves only as a barrier. The paper presents a framework for developing building envelopes based on a complex cellular or sponge-like geometry and preliminary design experiments that examine various tectonic approaches to cellular envelopes. The new envelope types, inspired by both cellular/spongy envelopes in nature and monocoque structures in the aviation, automotive and naval industries, are based on simple materials that can be manipulated to generate a complex geometry. The complex geometry of the cellular grid and the cells is developed using parametric digital modeling.

Keywords Cellular envelope · Parametric design · Freeform · Biomimetics

1 Introduction

The building envelope has changed significantly from ancient times to the modern era. It has shifted from being made of massive elements, which were used both for climate control and for structural purposes, into thin elements occasionally made of state-of-the-art materials that do not necessarily have a structural role.

Y. J. Grobman (✉)
Faculty of Architecture and Town Planning,
Technion Israeli Institute of Technology, Haifa, Israel
e-mail: yasha@technion.ac.il

A. Chakrabarti and R. V. Prakash (eds.), *ICoRD'13*, Lecture Notes in Mechanical Engineering, DOI: 10.1007/978-81-322-1050-4_75, © Springer India 2013

However, during the entire history of construction, the basic structure of the building envelope, a laminated entity made of different layers that are used as a barrier, has remained unchanged.

Today, the building envelope must cope with increasing demands for performance. The common solution is changing the dimension (mainly thickness) and/or the material of one or more layers that constitute the envelope. This often involves adding advanced high-tech—and thus, usually costly—materials, which pushes up the cost of the entire building. Moreover, the envelopes of contemporary buildings are treated mainly as a threshold that must dispose of rainwater as quickly as possible and avoid vegetation growth (green wall or roof) within the envelope itself. When a green wall or roof is designed, it is added as yet another external layer to the envelope, further increasing the envelope's cost.

The paper argues that the digital revolution in architectural design and manufacturing, and particularly the new possibilities offered for the design and manufacture of complex geometry, calls for a re-examination of the traditional concept of layer-based building envelopes that are used only as a barrier. The paper presents the preliminary results of a research study that develops a building envelope based on complex cellular or sponge geometry. The suggested cellular envelope type, inspired by envelopes in nature, is made from state-of-the-practice simple materials (such as concrete), which can easily manipulated to construct complex form. The final aim of the research is both to develop prototypes for cellular building envelopes and to show that a high-performance façade can be produced by the joint effect of the envelope's material properties and the micro-climate that is being created close to the envelope's surface due to the complex form.

2 Free-Form Design and Manufacturing in Architecture

There is a strong connection between the ability to design a form and the ability to fabricate it. In fact, according to William J. Mitchell, "[a]rchitects draw what they can build and build what they can draw" [1]. Free-form design has been widely used by architects since the end of the 1990s with the introduction of commercial design tools that allow design and manipulation of surfaces based on Non Uniform Rational B-splines (Nurbs). The current decade is witnessing the assimilation of parametric design and Building Information Modeling (BIM) tools and concepts that expand ever further the designer's ability not only to manipulate complex form but also to fabricate it. One can clearly argue that architects today have very few (if any) limitations in formal or geometrical design and manipulation.

One of the most salient advances of the use of parametric design and BIM is the direct connection to fabrication. This allows direct information exchange between architectural design and manufacturing without the need for mediators (construction drawings made by consultants or contractors) [2]. The use of CNC milling machines and other computer-controlled manufacturing machines is being increasingly assimilated into the building industry's standard manufacturing process. Moreover, even 3D additive manufacturing machines have reached the size and material capacities of building scale elements and end products.

Indeed, the cost of fabricating a complex form in general and a complex geometry façade in particular is still far more expensive than a traditional orthogonal-layer-based façade. Moreover, the building industry is still oriented toward mass production of standardized elements, and the shift to mass customization, not to mention customized construction, will clearly take some time.

However, it seems safe to argue that the shift toward computer manufacturing and especially large-scale CNC milling and 3D printing will continually reduce this gap. Therefore, given the understanding that the cost of computer-based manufacturing will drop in the near future and the difference between orthogonal form and freeform computer-based manufacturing will diminish if not totally disappear, there is both an opportunity and a need to examine the performance potential of freeform envelopes in architectural buildings.

3 Inspiration from Nature

Building envelopes have numerous distinct functions. Hutcheon [3] organized these functions into two groups with a total of 11 functional requirements.

The first group consists of the items that relate to the facade as a barrier for the control of heat flow; air flow; water vapor flow; rain penetration; light, solar and other radiation; noise; and fire. The second group consists of overall requirements, such as providing strength and rigidity; being durable; being esthetically pleasing; and being economical.

Similar functional requirements exist in the natural world. During evolution, living organisms developed various approaches and strategies to fulfill these requirements. Architecture has a long history of looking at nature for inspiration. Some approaches concentrated on the rather formal aspects of nature or natural form. These approaches include, among others, art nouveau architecture [4], organic architecture [5] and zoomorphic architecture [6]. The focus of this research is a different approach, generally called biomimicry, which examines the performative aspects of natural form and tries to extract insights for creation of architectural form and processes [7, 8]. More specifically, this research examines skins and envelopes in flora and fauna as a possible inspiration for the performance of a building façade.

A recent review by Gruber and Gosztonyi [9] presented a summary of the sparse existing academic research and studies related to biometric façade and compared the functions of skins of organisms and their analogy in architecture. A more specific study by Badarnah et al. [10] examined various strategies for thermoregulation based on insights from nature and shading strategy based on organizational feature in leaves [11]. Laver et al. suggested a cellular structure for a high-performance masonry wall system based on insight from termites and barrel cacti [12]. None of the above described research suggested an overall framework or an argument for a shift to a cellular approach in building envelopes.

4 Why Cellular or Spongy Envelopes?

Ever since the modernist separation between the structure and the building envelope, the development of building envelopes has concentrated mainly on finding new materials, combining materials, or shear optimization of the performance of the building envelope's various layers and their combined performance.

Knippers and Speck [13] argue that traditional architecture and civil engineering define construction in two separate categories: material and structure. They claim that this separation is impossible in natural world structures, which could be divided into five to twelve interconnected hierarchical levels in different scales/levels (biochemical level, microscopic level and up to the ultra-structural level). They define an important characteristic of natural system as being multi-layered and having a "finely tuned and differentiated combination of basic components which lead to structures that feature multiple networked functions."

Comparing building envelopes and natural envelopes or skins, one can clearly see that one of the main differences between the two has to do with the cellular-based structure. Natural skins or envelopes—and in fact, a large percentage of natural tissues—are based on cellular units [14]. These cells are characterized by complex 3D freeform (as opposed to the flat envelopes of buildings), which is based on geometric and material logic; multi-functionality; structural and formal heterogeneity; and multilevel hierarchical structure that consists of both isotropic and anisotropic structure according to local needs (the characteristics are based on Knippers and Speck's design principles of natural systems).

As opposed to the complex cellular structure of natural skins, traditional building envelopes are typically based on flat (extruded 2D) orthogonal geometry, repetition, limited functions (usually as a barrier) and structural homogeneity (frequently the envelope does not have a structural role). Developing cellular building envelopes that are based on a number of natural cellular skins principles and cellular/sponge-like geometry [15] could facilitate a multifunctional envelope system that could offer the following advantages:

- A single spatial structure—This could function as a barrier, water collector, shading mechanism and green wall. This represents a shift into more efficient building structure based on ideas implemented in monocoque structures, which are currently used in the naval, aviation and automotive industries.
- More than a threshold—The suggested envelope changes the narrow perception of the building envelope, which is currently regarded almost exclusively as a threshold. It challenges the perception that rainwater must be avoided and/or

disposed of rapidly in building envelopes by allowing a certain amount of water to be collected inside the cavities, where it will be used for cultivation of plants. Thus, the envelope itself also turns into a green wall (as opposed to the current need to construct a special layer for plants). Previous research has shown that green walls offer considerable benefits by reducing heat islands, helping to conserve animal habitats and saving on infrastructure costs (by retaining some of the water and reducing demands, especially in extremely rainy conditions) [16].

- Microclimate—There is a possibility of using the air flow close to the envelope's surface to create a microclimate. As opposed to the traditional layering approach, a parametric complex geometry approach to the building envelope is fundamentally based on a cellular or perforated surface in which the spatial relationship between the filled spaces and the hollow spaces is controlled parametrically and is used to create a microclimate. The microclimate can be optimized for insulation, ventilation, light, draft, water conservation and the cultivation of vegetation (green wall) according to the demands [17, 18].
- Form heterogeneity—There is a possibility of creating a variation of envelope cells that would be customized to deal with changing local conditions within the building envelope.
- Simple materials—In terms of materials, the suggested approach suggests a shift to building envelopes based on a small number of simple, widely used materials, such as concrete. This could have significant ramifications, since the creation of high-performance, low-cost envelopes could considerably decrease the buildings' energy consumption.
- Decrease of the environmental impact/footprint of building—This would occur due to the increase in performance and the possibility of embedding green walls and using storm water collection [19].

A shift to building envelopes based on freeform cellular geometry and logic also entails some challenges or disadvantages. One of the main challenges has to do with the programmatic flexibility of customized complex forms. As it is suggested that the envelope would be tailored to fit both the external and internal needs of a specific program, one can assume that during the buildings' lifetime the internal program is liable to change. This might change the demands, for example, for natural illumination. In traditional buildings where all openings are similar, this would not be a problem, but in customized buildings, the opening demands of one programmatic function might not well serve other functions.

Another disadvantage is cost. Although it is expected that the cost of fabricating complex geometry will be reduced substantially when computer-based fabrication becomes widespread, it is logical to expect that there would be a cost difference as compared with manufacturing an envelope based on repetitive elements.

Other possible challenges have to do with the fact the living envelope has to be carefully maintained and that complex form might not be well accepted by the client that who is accustomed to traditional orthogonal buildings.

5 Precedents for Cellular or Spongy Building Envelopes

Although freeform architectural design in general and freeform cellular or spongy form in particular demands computer-based manufacturing for its realization, the notion of cellular buildings and building envelopes is not a new one. The following sections will briefly describe precedents for cellular or freeform envelopes.

While freeform architecture was not common in the post-industrial revolution period, architects such as Antoni Gaudi, Eladio Dieste and others were able to design and build highly articulate building forms and building envelopes. However, even though the entire form of some of their buildings was complex, the envelopes of these buildings were still based on traditional building methods and did not try to postulate better performance as a result of the form.

At first glance, one might consider Gaudi's well-known Casa Mila project (La Pedrera in Barcelona, Spain, 1910) as an example of a complex cellular façade due to the formal complexity of the envelopes. Nonetheless, a deeper examination reveals that the façade design is driven by solely formal aspects and that no argument was suggested by the designer for the performative aspects of this type of envelope.

Erwin Hauer's work and research on complex 3D wall systems (mainly for interiors) can be considered one of the early examples of cellular complex 3D logic in building walls [20]. He developed and implemented complex 3D repetitive units, mainly from concrete, back in the 1950s (see Fig. 1). His walls are principally orthogonal, but the units or cells that populate the grid he creates within the wall are formally complex. His work has been an inspiration to later research that tried to use parametric design tools to examine possibilities of creating both complex wall systems (as oppose to Hauer's orthogonal walls) and replacing the repetitive grid and tile with parametrically modified ones [21].

A different perspective on cellular approach to building envelopes is derived from Leatherbarrow and Mostafavi's idea [22] of the "denial of the frontality of the façade" in relation to Le Corbusier's introduction of the *brise-soleil*. The façade's frontality and flatness is replaced in this case with a space that acts as a light control mechanism but also challenges the notion of the flat building envelope. A more general perspective regarding this notion could refer to kinetic building

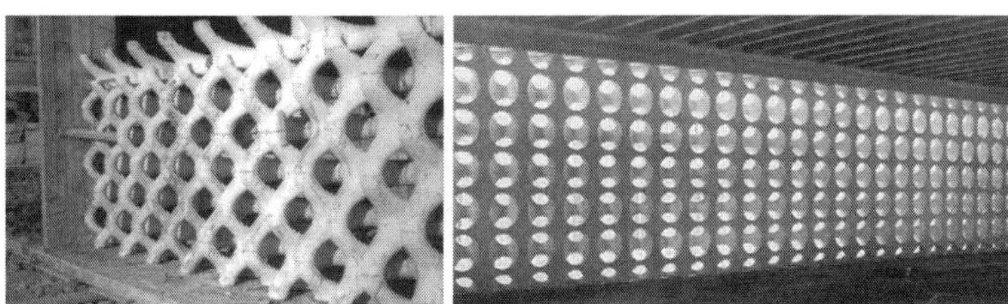

Fig. 1 Erwin Hauer—church in Liesing, Vienna, Austria, 1952 (*left*). Church in Erdberg, Vienna, Austria, 1954 (*right*). *Source* www.erwinhauer.com

envelopes. A well-known example in this realm is Jean Nouvel's Arab World Institute (Paris, France, 1987). An earlier but comprehensive discussion on kinetic building and envelopes can be found Zuk's book on kinetic architecture [23].

The introduction of computers to architectural design and particularly the use of parametric design "offers a high degree of geometric control combined with ability to rapidly generate variations" [24]. According to some researchers, the assimilation of parametric design methods and tools in architectural design and manufacturing has introduced a new "style" called Parametricism to architectural design and stimulated experiments in both urban and building (mainly envelope) scales [25, 26]. Parametric tools such as ParaCloud generative modeler (GEM) and Grasshopper (generative modeling tool for Rhino) have made it possible to generate complex geometry and to connect the architectural form to simulation software [27]. Parallel to research that concentrated on the geometric aspects of building envelopes, a considerable amount of research has been dedicated to the idea of performance in architectural [28, 29] and computational material [30, 31]. The new direct data exchange between these ideas and tools and computer-aided manufacturing tools, such as CNC milling machines and laser cutters, has fostered a flurry of parametrically designed and computer-manufactured structures, mainly in pavilion or installation scale, over the last 5 years [32, 33].

At the outset of the computer's assimilation to architectural design and manufacturing in the late 1980s, design experiments initially concentrated on creating new types of building layers, such as inflated materials (for example, the Beijing Olympic swimming pool by PTW Architects (Fig. 2) [34], Allianz Arena by Herzog and De Meuron architects [35] and the Eden Project by Grimshaw-Architects [36]. Subsequent experiments with complex geometry façades concentrated on the new potential for manipulating complex forms that required almost no attention to the envelope's performative aspects [37]. See also, for example, Migrating Formations wall by Contemporary Architecture Practice (Fig. 3) [28], KOL/MAC Architecture's INVERSAbrane building envelope [28], Greg Lynn's Blobwall [28] and Gramazio Kohler's The Dissolved Wall/Screens projects [37].

Fig. 2 PTW architects—watercube—Beijing National Aquatics Center, Beijing, China, 2003. *Sources* http://www.ptw.com.au, http://www.terrywier.com/, http://www.flickr.com/photos/xiaming/484446352/lightbox/

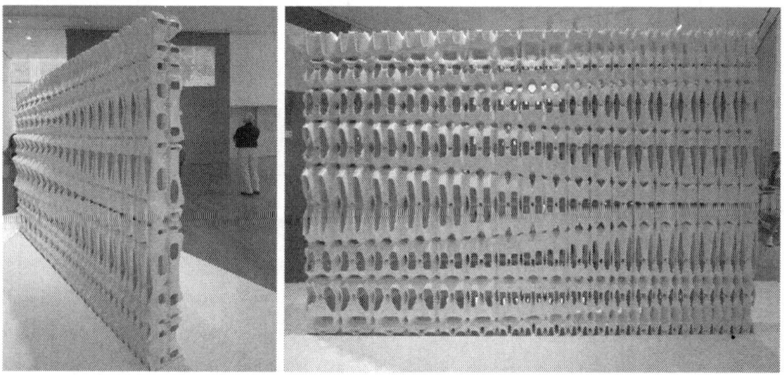

Fig. 3 Contemporary architecture practice (CAP)—migrating formations, New York, USA, 2008. *Source* Grobman and Neuman [28], p. 97

6 Cellular Envelope Design Experiments

The following section presents preliminary design experiments that examine the potential and trajectories in the design of cellular building envelopes. The method used in the design process of these projects combines digital form-finding methods with more traditional formal design methods. It thus combines ideas from research by design approach [38, 39] and digital and non-digital form-finding [27]. The design method used for these experiments is based on "populating" cellular elements on the cells of a grid that was generated for each of the experiment's envelopes. The rather complex grid that is used in each of the experiment is developed from initial regular grid that was modified according to performance criteria such as orientation, program (of the spaces behind the façade) and function of the specific areas of the facade. For example, an area which is intended for utility equipment does not usually needs a similar amount illumination as areas which are used for offices. Each of the final cells in the grid was populated by a different cellular element according to its location and function (type of space served by the specific cell of the envelope).

Each one of the three different experiments examines a different approach to cellular envelopes. In the first approach, the grid is used as a structural element and the cellular elements are inserted in the spaces created by the grid. As opposed to the duality characterizing structure and cells suggested in the first experiments, in the second and third experiments the grid serves as both a structure and a barrier.

6.1 Experiment No. 1

A cellular unit is populated inside a Voronoi based geometry grid (Fig. 4). Each unit is unique and made to fit a specific position in terms of size and performance.

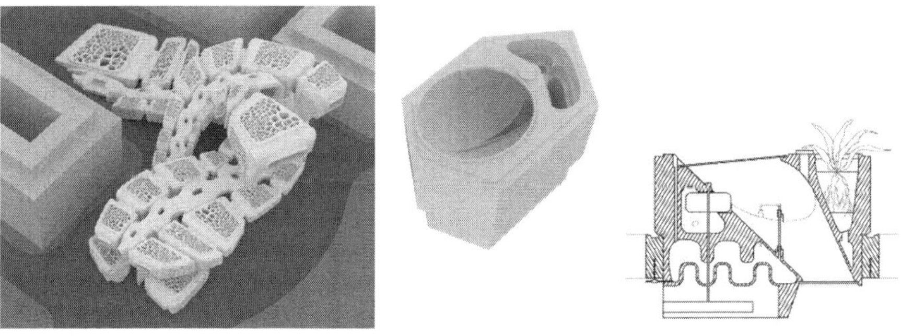

Fig. 4 Experiment no. 1 envelope with a structural cellular grid based on Voronoi algorithm geometry (*left*). Isometric view of a cell unit (*middle*). Section of a cell unit (*right*)

The unit presented in Fig. 4 contains the following elements: a place for a plant; a solar radiation system based on heating water by means of focusing the solar radiation using a circular surface; and a ventilation heat-exchange system based on a turbine. A cell unit can contain these entire features or any a combination of them, based on local need.

6.2 Experiment No. 2

The envelope is created from a family of cellular units, which are used both as a structure and as an infrastructure for functions such as shading, growing plants and isolation. The envelope's front view in Fig. 5 shows an example of the parametric approach to populating the cells in which the designer can choose a specific member from a unit family for every position in the envelope. The units' function can vary; it can serve as a passage, a room/space or a balcony.

Fig. 5 Experiment no.2—Isometric view of an envelope unit (*left*) and a front view of an envelope (*right*)

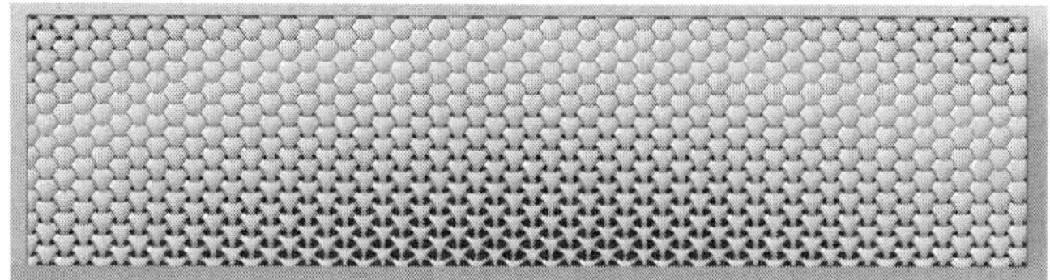

Fig. 6 Experiment no.3—envelope, front view

6.3 Experiment No. 3

This experiment presents a similar system to the one developed for the previous experiment in terms of the parametric population of the cells within the grid and the multifunctionality of the cellular unit. The main difference between the two experiments is that the current cell system is based on a singular unit that allows a gradual change in its dimensions. This allows the creation of a continuous variation in the envelope units, which in turn gives the envelope a more organic formal expression (see Fig. 6). The system is built from a structure of fiberglass and metal, which create the structure for Ethylene Tetrafluoroethylene (ETFE) air cells that are used both for thermal isolation and for transferring natural light (see Fig. 7).

Fig. 7 Experiment no.3—section (*left*) and isometric rendered section in the envelope (*right*)

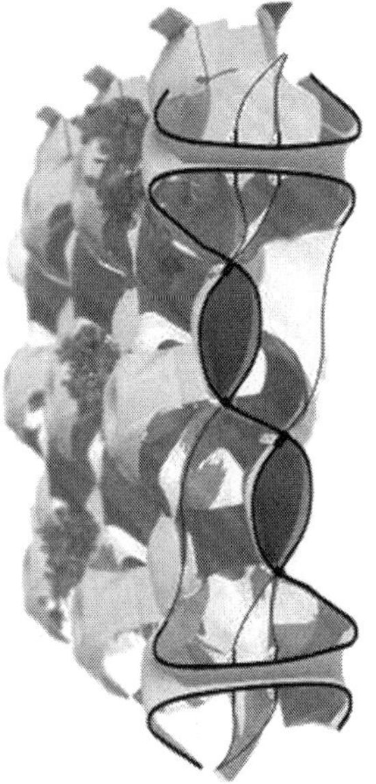

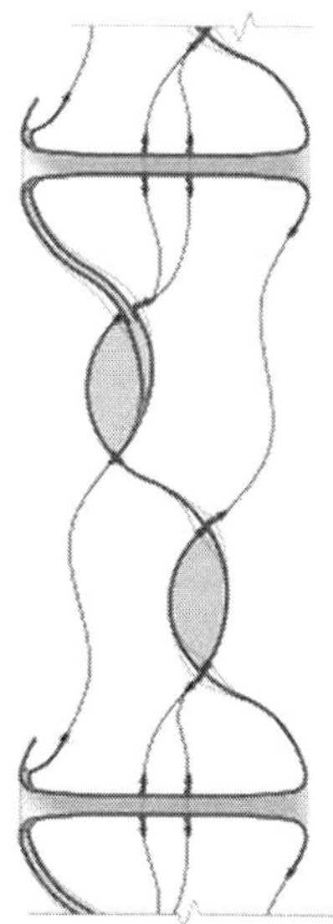

7 Conclusion and Future Research

The approach and the design experiments described above present the initial framework and possible trajectories for developing cellular building envelopes. Although several design directions have been developed and the concept seems plausible from the design and manufacturing viewpoints, the next stages of the research has yet to prove the possibility to reach similar performance in various environmental criteria as in traditional envelopes.

The significance of the proposed approach lies in the centrality of the building envelope to the design, manufacturing and performance of buildings. The resulting shift in the traditional concept of building envelopes could potentially improve the building's overall energetic performance, decrease urban heat islands by allowing vegetation to grow over the envelope and reduce the infrastructure needed for handling rainwater. Moreover, the new possibility of creating low-cost complex geometry envelopes that embed vegetation as an integral part of the envelope itself could trigger a dramatic change in the way our built environment looks and behaves. From the current strict division between built and green areas, our built environment would become—to a certain extent, at least—all green.

Acknowledgments The design experiments were developed during a design studio by the students Itay Blaistain (experiment no. 1), Asaf Nevo (experiment no. 2), and Michael Weizmann (experiment no. 3). Their contribution is hereby acknowledged.

References

1. Mitchell WJ (2001) Roll over Euclid: how Frank Gehry designs and builds. In: Gehry FO, Colomina B, Friedman M, Ragheb JF, Cohen J-L (eds) Frank Gehry architect, hardcover. Guggenheim Museum Publications, New York, pp 352–363
2. Eastman C, Teicholz P, Sacks R, Liston K (2011) BIM handbook: a guide to building information modeling for owners, managers, designers, engineers and contractors, 2nd edn. Wiley, New Jersey
3. Hutcheon N (1963) Requirements for exterior walls-IRC-NRC-CNRC. Canadian Building Digest, CBD-48
4. Russell F (1979) Art nouveau architecture. Academy Editions, London
5. Pearson D (2001) New organic architecture: the breaking wave, 1st edn. University of California Press, California
6. Aldersey-Williams H, Victoria AM (2003) Zoomorphic: new animal architecture. Laurence King Publishing in Association with Harper Design International, London
7. Benyus JM (2002) Biomimicry: innovation inspired by nature. William Morrow Paperbacks, New York
8. Gruber P (2010) Biomimetics in architecture: architecture of life and buildings, 1st edn. Springer, New York
9. Gruber P, Gosztonyi S (2010) Skin in architecture: towards bioinspired facades, pp 503–513
10. Badarnah L, Nachman Farchi Y, Knaack U (2010) Solutions from nature for building envelope thermoregulation. pp 251–262

11. Badarnah L, Knaack U (2008) Organizational features in leaves for application in shading systems for building envelopes, vol I. pp 87–96
12. Laver J, Clifford D, Vollen J (2008) High performance masonry wall systems: principles derived from natural analogues, vol I. pp 243–252
13. Knippers J, Speck T (2012) Design and construction principles in nature and architecture. Bioinspiration Biomimetics 7(1):015002
14. Gibson LJ, Ashby MF, Harley BA (2010) Cellular materials in nature and medicine, 1st edn. Cambridge University Press, Cambridge
15. Burt M (2011) The periodic table of the polyhedral universe. Int J Space Struct 26(2):75–94
16. Oberndorfer E, Lundholm J, Bass B, Coffman RR, Dunnett N, Gaffin S, Kohler M, Liu KKY, Rowe B (2007) Green roofs as urban ecosystems: ecological structures, functions, and services. Bioscience 57(10):823–833
17. Laver J, Clifford D, Vollen J (2008) High performance masonry wall systems: principles derived from natural analogues, vol I. pp 243–252
18. Yannas S (2004) Adaptive skins and microclimates, in built environments and environmental buildings. pp 615–621
19. Kats G (2009) Greening our built World: costs, benefits, and strategies. Island Press, Island
20. Hauer E (2007) Continua-architectural screen and walls. Princeton Architectural Press, Princeton
21. Hensel M, Menges A, Weinstock M (2010) Emergent technologies and design: towards a biological paradigm for architecture. Routledge, London
22. Leatherbarrow D, Mostafavi M (2005) Surface architecture. The MIT Press, Cambridge
23. Zuk W (1970) Kinetic Architecture. Van Nostrand Reinhold, New York
24. Schumacher PS (2011) The autopoiesis of architecture: a new framework for architecture, 1st edn. Wiley, London
25. Kolarevic B, Klinger K (2008) Manufacturing material effects: rethinking design and making in architecture, 1st edn. Routledge, London
26. Meredith M (2008) From control to design: parametric/algorithmic architecture. Actar, Barcelona
27. Grobman YJ, Ron R (2011) Digital form finding—generative use of simulation processes by architects in the early stages of the design process. In: Proceedings of eCAADe 2011 respecting Fragile spaces, Ljubljana, pp 107 115
28. Grobman YJ, Neuman E (2011) Performalism: form and performance in digital architecture, 1st edn. Routledge, London
29. Kolarevic B, Malkawi A (2005) Performative architecture: beyond instrumentality. Routledge, London
30. Sass L, Oxman R (2006) Materializing design: the implications of rapid prototyping in digital design. Des Stud 27(3):325–355
31. Menges A (2010) Instrumental Geometry. In: Corser R (ed) Fabricating architecture: selected readings in digital design and manufacturing. Princeton Architectural Press, New York, pp 22–41
32. Iwamoto L (2009) Digital fabrications: architectural and material techniques. Princeton Architectural Press, New York, p 144
33. Krauel J (2010) Contemporary digital architecture: design and techniques. Links International, Ceg
34. PTW [Online] (2011) Available http://www.ptw.com.au/ptw.php. Accessed 17 Oct 2011
35. Grimshaw-Architects [Online] (2011) Available http://www.grimshaw-architects.com/base.php?in_projectid=. Accessed 17 Oct 2011
36. Herzog and de Meuron Architects [Online] (2012) Available http://www.herzogdemeuron.com/index/projects/complete-works/201-225/205-allianz-arena.html. Accessed 4 Oct 2012
37. Gramazio F, Kohler M (2008) Towards a digital materiality. In: Kolarevic B, Klinger K (eds) Manufacturing material effects: rethinking design and making in architecture, 1st edn. Routledge, Londo, pp 103–118

38. Schumacher P (2011) Architecture schools as design research laboratories. In: Hadid ZM, Schumacher PS (eds) Total fluidity: studio Zaha Hadid projects 2000–2010 University of Applied Arts Vienna, 1st edn. Springer, New York
39. Sevaldson B (2010) Discussions and movements in design research. FORMakademisk 3(1):8–35

Development and Characterization of Foam Filled Tubular Sections for Automotive Applications

Raghu V. Prakash and K. Ram Babu

Abstract Crash safety requirement without much penalty in structural weight of automotive structures has provided scope to fill hollow sections with foams. Different classes of foams are used for this purpose—polymer foams and metal foams. Metal foams are prepared out of light metals such as aluminum, magnesium, though occasionally steel foams are also suggested in the literature. This paper presents the results of crushing, bending and damping characteristics of steel extrusions with and without foam filling. Polymeric foam and aluminum foam are considered for this study. Based on the experimental study, the following observations are made: (a) The force verses displacement characteristics of aluminum foam filled tubes show large resistance (at higher loading rates) during axial crush; polymeric foam filling did not show such a marked improvement in energy absorption characteristics, (b) The bending resistance of aluminum foam filled sections shows an improvement in bending by 60–200 % during 3-point bend testing, and (c) Vibration levels are found to be reduced in lateral direction for foam filled sections. This foam filled section was tried on typical section of a two-wheeler component.

Keywords Foam filled sections · Vibration · Crushing capability · Aluminum foam · Polymer foam

R. V. Prakash (✉)
Department of Mechanical Engineering, Indian Institute of Technology Madras,
Chennai 600 036, India
e-mail: raghuprakash@iitm.ac.in

K. R. Babu
TVS-Motor Company, Hosur, India

A. Chakrabarti and R. V. Prakash (eds.), *ICoRD'13*, Lecture Notes in Mechanical Engineering, DOI: 10.1007/978-81-322-1050-4_76, © Springer India 2013

1 Introduction

Occupant safety is a prime concern in case of modern automobiles, and assessment of crashworthiness is an important stage in structural design. Crashworthiness represents a measure of the vehicle's structural ability to plastically deform and yet maintain a sufficient survival space for its occupants in crashes involving reasonable deceleration loads. In this context, use of appropriate materials, design of safe crush zones and other methods such as passenger restraint systems, occupant packaging provide additional protection to reduce severe injuries and fatalities.

To meet the requirements for improved safety, sometimes, thicker steel sheets or additional reinforcements are usually provided, which leads to a heavier body-in-white. Therefore, it is necessary to improve crash safety while at the same time reducing the weight of vehicles for better performance. In order to achieve a safe automobile body in the event of a collision, deformation of the cabin structure should be minimized to protect the occupants, and the collision energy should be absorbed in a short deformation length within the crushable zones. However, the reaction force generally exceeds a certain value when a material with higher strength is used to build a car; new structures and materials are required for building the ideal car body that can absorb the collision energy in a short span and with a constant reaction force. Towards this, in the recent times, foam filled structures are considered to design programed crush zones in automotive structures. Polymeric foams and metallic foams are considered as candidate materials for improving the energy absorption characteristics. This paper presents the results of mechanical property evaluation of metallic as well as polymeric foam filled steel sections subjected to compression, bending and crushing experiments.

2 Literature Review

Foams and other highly porous materials with a cellular structure are known to have interesting combinations of physical and mechanical properties, such as high stiffness in conjunction with very low specific weight or high gas permeability combined with high thermal conductivity. A typical stress–strain curve of metal foams in several stages is shown in Fig. 1; the graph consists of an initial, almost linear deformation, plastic collapse and final densification. It can be seen from the comparison between the stress–strain curve of an aluminum foam and the corresponding curve of an plastic foam, that the two loading curves are similar except that an approximate thirty times higher stress amplitude was found for the aluminum foam as compared with the plastic foam.

In the case of bending, tensile stresses produce fracture in the tensile zone. Consequently, the behavior of the foam alone (with or without skin) is in agreement with the results of tensile tests, which leads to very low energy absorption. The combination of the foam with the tube allows having a composite structure

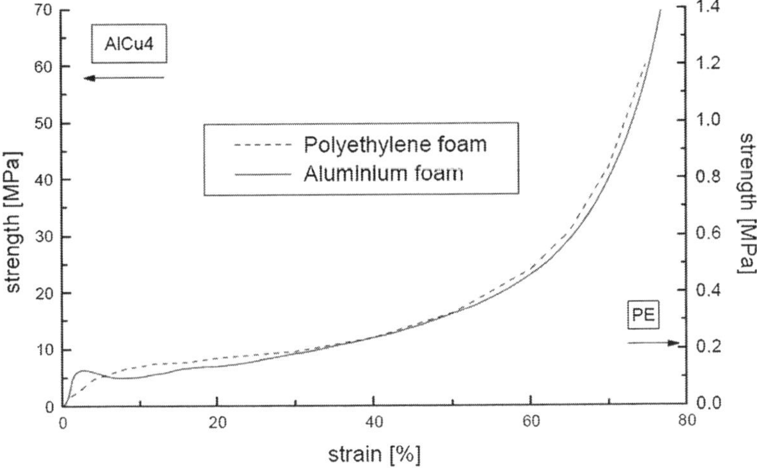

Fig. 1 Stress-strain response of polymeric and metallic foam. *Source* [1]

with high energetic absorption and collapse load, if compared with the basic elements. The foam presence produces a substantial change in the tube collapse mechanism. Even if the foam fractures, the effect of internal constraint of the tube gives some benefits for the energetic absorption thus, avoiding the formation of the well known bending collapse mechanism (with consequent decrease of the resistant load). However, the new collapse mechanism increases the tensile stresses in the tube walls: with higher rotations. This causes the fracture of the tube, with consequent sudden decrease of the load and of the capacity of further energy absorption. This situation is made worse by the reduced fracture deformations of the aluminum alloys for extrusion. Figure 2 shows the bending behavior of different tubular sections.

The aim of this work is to evaluate the crushing characteristics of steel extruded sections with and without foam filling, as well as bend and vibration characterization of foam filled sections. Aluminum foam and polymer foam filled structures are considered for this study.

3 Manufacturing of Metallic Foams

Aluminum foam was developed in-house at the Metallurgy and Materials Engineering Department of IIT Madras as per schematic shown in Fig. 3. Pure aluminum ingots of required quantity are placed in graphite crucible, which was placed inside a resistance melting furnace. About 1.5 wt.% calcium (Ca) (in metal form) was added to the aluminum melt at 680 °C. The melt is stirred for 6–10 min during which the melt viscosity continuously increases by a factor of up to 5 owing to the formation of calcium oxide (CaO), calcium aluminum oxide ($CaAl_2O_4$) and

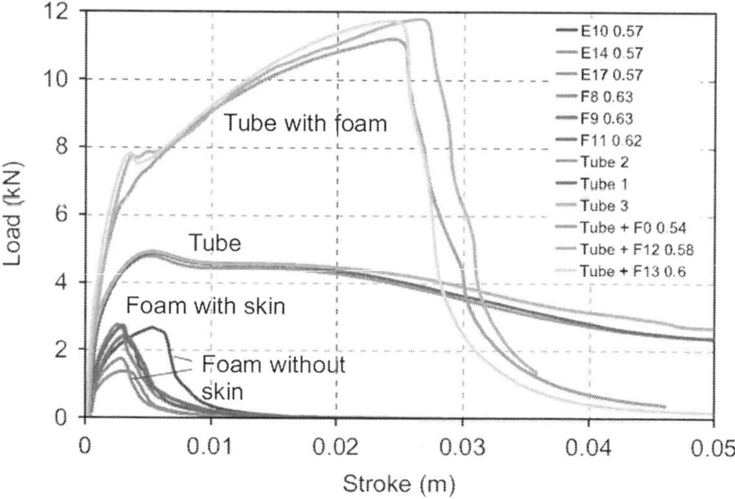

Fig. 2 Bending characteristics of extruded sections with and without foam. *Source* [2]

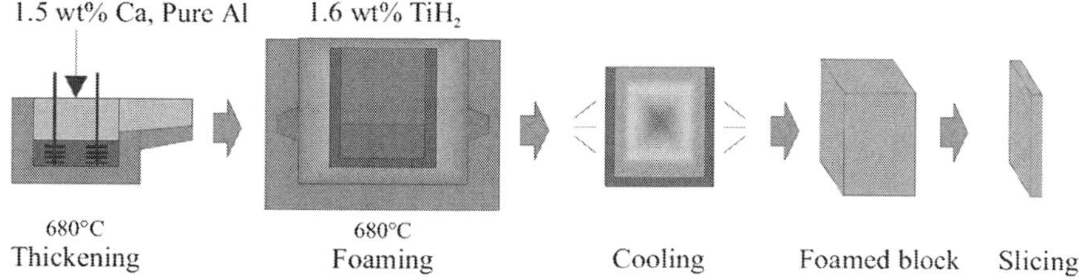

Fig. 3 Schematic of aluminum foam manufacturing in laboratory

in some cases Al_4Ca inter-metallics which thicken the liquid metal, according to the following reactions:

$$2Ca + O_2 \rightarrow 2CaO,\ 2Al + Ca + 2O_2 \rightarrow CaAl_2O_4$$
$$4Al + Ca \rightarrow Al_4Ca$$

Figure 4 shows the effect of stirring time on the viscosity of aluminum melts with various quantities of calcium addition. While the pure molten aluminum does not increase in viscosity on stirring, stirring with the addition of Ca increases the viscosity remarkably. After the viscosity has reached the desired value, a blowing agent is added to the melt. The blowing agent decomposes under the influence of heat and releases gas which then propels the foaming process. In this study titanium hydride (TiH_2) is added (typically 1.6 wt.%) which serves as a blowing agent by releasing hydrogen gas in the hot viscous liquid, according to the following reactions:

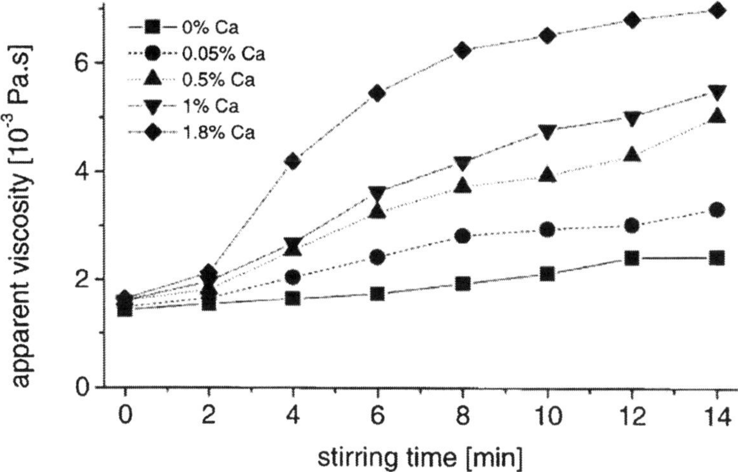

Fig. 4 Effect of stirring time and percentage Ca addition on viscosity of molten aluminum. *Source* [3]

$$TiH_2 \rightarrow Ti + H_2$$

The melt starts to expand slowly and gradually fills the foaming vessel. The foaming takes place at constant atmospheric pressure. After cooling the vessel below the melting point of the alloy, the liquid foam turns into solid aluminum foam and can be taken out of the mold for further processing. The entire foaming process takes along about 15 min for a typical batch of 0.6 m^3. The typical charge chosen for each melt is about 500 g. It has been found through experimental trials that a careful adjustment of process parameters leads to very homogeneous foams. Typical densities after cutting off the sides of the cast foam blocks are estimated to be between 0.18 and 0.24 g/cm^3. Table 1 shows the aluminum foam material composition by percentage weight.

It may be noted that viscosity enhancement of molten aluminum can also be obtained by bubbling oxygen, air or other gas mixtures through the melt, thus causing the formation of alumina, or, by adding powdered alumina, aluminum dross or scrap foamed aluminum or by using metallic viscosity enhancing additives. However, stabilization of process parameters by this method seems to be quite difficult and requires complicated temperature cycles and mechanical agitation.

Table 1 Aluminum foam material composition (by percentage weight)

S.No.	Raw material	Parts by weight %
1	Aluminum ingot	0.9699
2	Calcium metal (Ca)	0.0145
3	Titanium hydride (TiH_2)	0.0155

4 Processing of Polymer Foam

The preparation of polymeric foam involves the formation of gas bubbles in a liquid system, followed by the growth and stabilization of these bubbles; as the viscosity of the liquid polymer increases, solidification of the cellular resin matrix is formed. Polyurethane foam falls under the thermosetting group foams, having no thermoplastic properties.

Castor oil (polyol) and toluene di-isocyante (isocyanate) are the two basic raw materials required for the synthesis of polyurethane foam. When these raw materials are mixed in the appropriate ratio, polymerization reaction takes place resulting in polyurethane foam. Other raw materials used are DABCO® catalyst, poly (di-methylsiloxane), di-chloromethane (solvent) and water. Vigorous stirring was carried out till the creamy stage is reached. Foaming occurs by evaporation of solvent and evolution of carbon dioxide gas due to chemical reaction between excess iso-cyanate and water. Poly (di-methylsiloxane) was used as a surfactant to reduce the surface energy of generated gas bubbles and thereby foam structure was retained. No foaming reaction occurred without addition of catalyst as catalyst acts as a polymerization initiator. Polyurethane foam formation is an exothermic process and the temperature rise due to foaming was measured to be approx. 60 °C; di-chloromethane evaporated during this temperature rise. Optimal amount of solvent and water were added to control the extent of foaming reaction. Excess solvent resulted in boiling rather than foaming. Foaming reaction is given by:

$$R^1\text{-}N{=}C{=}O + R^2\text{-}O\text{-}H \longrightarrow R^1\text{-}\overset{H}{\overset{|}{N}}\text{-}\overset{O}{\overset{\|}{C}}\text{-}O\text{-}R^2 \tag{1}$$

Polyol is a macro molecule of polyhydric alcohols, when reacted with isocyanate, which contains a NCO radical reacts with the OH part of the polyol in the presence of suitable catalysts to form a urethane linkages (–NC). The catalysts accelerate the reaction to the required level. The blowing agents blow the cells, increases its volume to form the light weight polyurethane foam. The surfactants promote and stabilize the polyurethane cells and helps to retain the shape into which it has been blown to. Density of foam obtained 0.35–0.45 g/cm^3. Table 2 presents the polyurethane foaming material composition by percentage weight.

Table 2 Polyurethane foaming material composition (by percentage weight)

S.No.	Raw material	Parts by weight %
1	Castor oil (polyol)	0.5305
2	Toluene di-isocynate (isocynate)	0.2599
3	Poly-dimethylsiloxane (surfactant)	0.0053
4	DABCO catalyst	0.0212
5	Water (blowing agent)	0.0239
6	Dichloromethane (solvent)	0.1592

5 Specimen Preparation

Extruded steel tubes of rectangular cross-section with dimensions of 50 (b) × 25 (d) × 100 (h) mm, and wall thickness of 1 mm is used for axial crushing characterization, specimens with dimensions of 50 (b) × 25 (d) × 160 (h) mm, having a wall thickness of 1 mm were used for 3-point bending characterization in this study. Polymer and metal foam as described in earlier section was filled inside this extruded tube section. Typical polymer foam density was estimated to be approximately 0.4–0.5 g/cm^3 and that of metal foam density was estimated to be approximately from 0.9 to 1.1 g/cm^3.

6 Material Characterization

Foams Axial crushing and 3-point bending testing of specimens have been performed on a 100-kN MTS servo-hydraulic system. Figure 5 presents the view of experimental setup for axial crushing of extruded sections. There are two rigid platens that can be gripped into the hydraulic jaws of the MTS machine and the test section is loaded in between the lower and upper platens. Crush tests were conducted under displacement control mode at a displacement rate of 5 and 100 mm/min and the specimens were compressed by 30 mm from the original height of 100 mm. For each crushing velocity, force–displacement characteristics were recorded during the test and the energy absorbed during the impact was calculated from the area underneath the curve.

Figure 6 shows the photograph of three-point bending test setup, for characterizing the bend performance of extruded sections. The fixture includes two adjustable lower support anvils and one upper center loading anvil. The support beam has graduated lengthways in metric and imperial units for accurate positioning of the anvils. The specimen is supported on machined (free to rotate or stationary) anvils of a defined radius and the bend force is applied centrally. Bending tests were conducted under displacement control mode at a displacement rate of 5 and 100 mm/min and the specimens were deformed for a vertical distance of 25 mm.

7 Results and Discussion

7.1 Axial Crush Tests

Figure 7a presents the deformed shape of empty extruded section subjected to crush test. Shown along with in Fig. 7b is the load–displacement record during crush test. One complete symmetric plastic lobe and one partial lobe are formed

Fig. 5 Photograph of crush test of foam filled extruded section

for the crushing distance of 30 mm. An excellent agreement is found in the overall shape of the deformed tube for two different loading rates (5 and 100 mm/min) and each specimen displays one complete lobe and one partial lobe. Imperfections present in the tubular sections leads to slight disturbance in lobe formation at high loading rates and the formation of first lobe can be anywhere along the length. The area under the load–displacement record is used to estimate the cumulative energy absorption during crushing. The energy dissipation is observed to be higher at higher crushing rates of 100 mm/min. The mean crush force was found to increase with the increase in the rate of loading up to certain displacements, and thereafter instability conditions cause it to be reduced.

Figure 8 presents the deformed shapes of foam filled extruded sections subjected to compression loading. Table 3 presents the energy absorption characteristics of empty, aluminum foam filled and polymer foam filled extruded sections subjected to compression loading. The number of lobes in the side walls of extruded section has increased in case of metal foam filled section compared to empty and polyurethane foam filled tubes; the two complete symmetric plastic lobes are formed with the initiation of third lobe for the crushing distance of 30 mm. At high loading rates, it is observed that aluminum foam reduces the plastic lobe formation length, thereby generating large number of plastic hinges

Fig. 6 Three point bend testing of foam filled extruded sections

compared to other conditions, which results in an increase in energy absorption capacity of structure.

Specific energy absorption rates were estimated based on the final weight of sections, and it was found that polymer foam filling results in maximum energy absorption for a given weight, which is followed by Al-metal foam filled section.

7.2 Three-Point Bend Tests

Figure 9 shows the results of 3-point bend tests of extruded sections. Each specimen was loaded at its mid span and deformed up to 25 mm (vertically), each specimen displays inward curl, which is generated by the round nose of the testing fixture and the plastic hinge is concentrated and the inward curl is slightly deeper at high loading rates. The energy absorption during bending was estimated by considering the area underneath the load–displacement curve for each specimen. Table 4 presents the results of three-point bend testing of empty, foam filled extruded sections. Specific energy absorption (energy absorbed per unit weight of extruded section) for aluminum foam filled tubular sections was found to be more than empty and polyurethane foam filled sections.

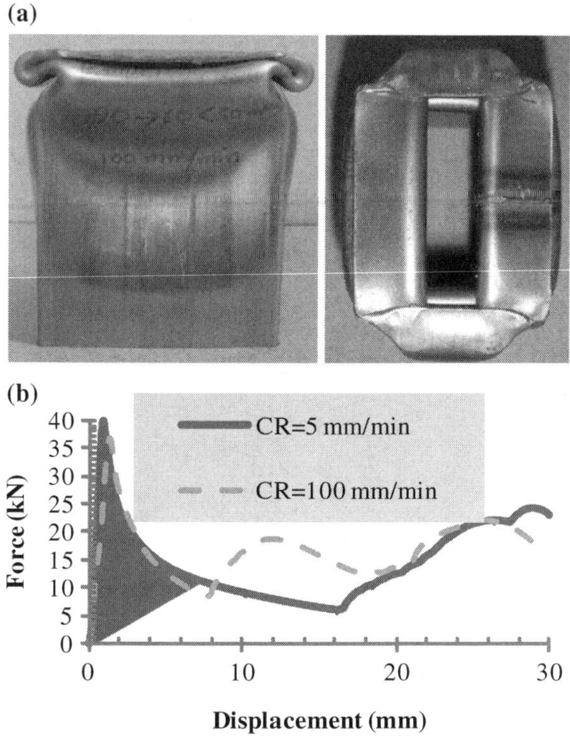

Fig. 7 **a** Deformed shapes of empty extruded sections at 100 mm/min rate of compression. **b** Load–displacement record during axial compression testing of extruded sections

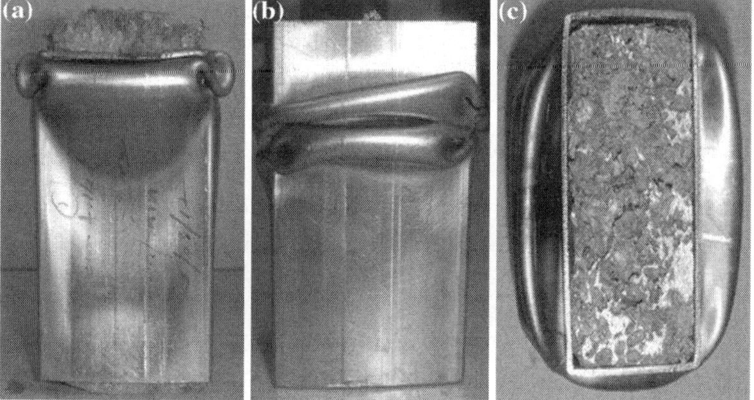

Fig. 8 **a** Crush shapes of polymer foam filled section at 5 mm/min. **b** Al-metal foam filled section at 100 mm/min. **c** Al- foam filled section at 5 mm/min rate of compression

Table 3 Crush characteristics of extruded sections at two different loading rates

S. No.	Loading rate (mm/min)	Peak force (kN)			Energy absorbed (J)		
		Empty	PU foam filled	Al foam filled	Empty	PU foam filled	Al foam filled
1	5	39.84	38.17	44.7	688.24	711.5	919.7
2	100	36.76	35.76	58.45	484.24	668.5	1691

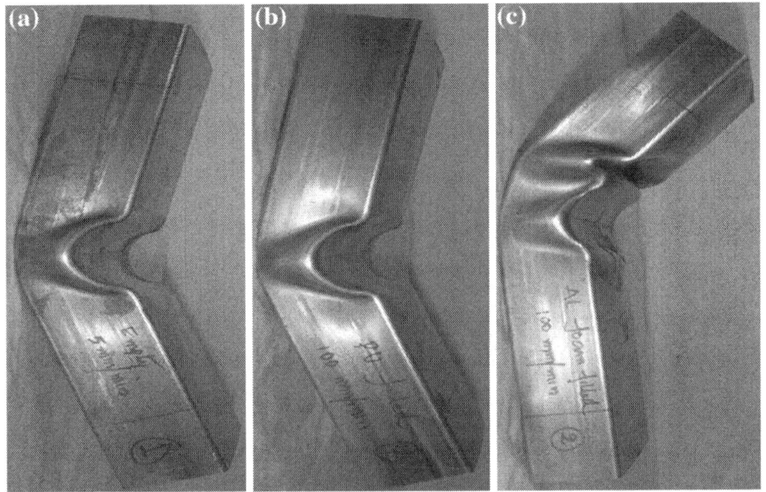

Fig. 9 Deformed shapes of 3-point bend specimens. **a** Empty extrusion, **b** PU foam filled extrusion, **c** Al-metal foam extrusion

The force–displacement response shows that in addition to initial peak force, a secondary peak force (relatively higher than the initial peak force) after 15 mm deflection was noticed, whose magnitude increases with the loading rate. The results show that the rate of loading has a significant influence on the mean crush force, initial peak force as well as energy absorbed by the extruded member.

7.3 Application of Concepts to a Motor Cycle Frame

Both poly-urethane and aluminum metal foams were filled in swing arm of a typical motor cycle and the performance of these sections was evaluated for lateral stiffness, vibration response. It was observed that there is an improvement in load bearing capacity of swing arm as yielding starts at higher loads compared to unfilled swing arm section; it was also noticed that the swing arm stiffness increased approx. 10 % in case foam filled sections. Vibration levels on handlebar measured directly on the vehicle and obtained through shake table testing suggests

Table 4 Bending characteristics of extruded sections

S. No.	Loading rate (mm/min)	Peak force (kN)			Energy absorbed (J)		
		Empty	PU foam filled	Al foam filled	Empty	PU foam filled	Al foam filled
1	5	8.75	9.55	18.82	172	203.5	440
2	100	9.2	9.35	21	177.5	206.4	321

a significant reduction in vibration levels in case of lateral direction when tested on vehicle directly. This offers the advantage of improving the structural dynamic performance.

8 Summary

This paper presented the details of axial crushing and 3-point bending behavior of the empty, polyurethane foam filled and aluminum foam filled tubular sections, and application of foam filling in motorcycle chassis structures, like swing arm. The force verses displacement characteristics of aluminum foam filled tubes shows large resistance at high loading rates during axial crushing. The bend tests of aluminum foam filled tubes shows greater resistance to deformation i.e. about 60 and 200 % improvement at two peak forces, and marginal improvement in polyurethane foam filled tubes over the empty tubes. The energy absorption characteristics improved with foam filling. Thus foam filling can be considered positively in design of automobile structural elements.

References

1. Yu C-J, Banhart J (1998) Mechanical properties of metal foams. In: Proceedings of metalfoam.net, conference paper, pp 37–48
2. Peroni L, Avalee M, Peroni M (2008) The mechanical behavior of aluminum foam structures in different loading conditions. Int J Impact Eng 35:644–658
3. Banhart J (2001) Manufacture characterization and application of cellular metals and metal foams. Int J Prog Mater Sci 46:561–608

The Current State of Open Source Hardware: The Need for an Open Source Development Platform

André Hansen and Thomas J. Howard

Abstract Open Source Hardware (OSHW) is a new paradigm attempting to emulate the Open Source Software movement. While there are several flagship OSHW projects, this product development paradigm has yet to live up to its full potential. This paper reviews the current state of OSHW and reveals the lack of a robust and simple development platform as being a major barrier to the uptake of OSHW. The authors argue that an Open Source, Cloud-based platform would be the most viable direction.

Keywords Open source hardware · Open source hardware collaboration platform · Open source software

1 Introduction

While free software is becoming as commonplace as your very own personal computer, free physical products are more of a rare phenomenon. However, it is our strong believe that this will change in the not too distant future, following the emerging paradigm of Open Source Hardware (OSHW). Free products may only be the minor benefits of OSHW as it might revolutionise the way new technologies are created and the way we interact with physical products in our everyday life.

A. Hansen (✉) · T. J. Howard
Section of Engineering Design and Product Development,
Department of Mechanical Engineering, Technical University, Copenhagen, Denmark
e-mail: mail@andrehansen.me

T. J. Howard
e-mail: thow@dtu.dk

A. Chakrabarti and R. V. Prakash (eds.), *ICoRD'13*, Lecture Notes in Mechanical Engineering, DOI: 10.1007/978-81-322-1050-4_77, © Springer India 2013

The Open Source movement has in recent years managed to challenge even the biggest international software companies, threatening the closed systems [1]. Open Source Software development has been able to capitalize from the development in communication technologies, enabling dispersed individuals and communities to efficiently share information. While Open Source Software (OSS) is motoring ahead and changing the realm of software development, one wonders why engineering design and the development of physical products still seems to be a mere bystander yet to embrace the full potential of the Open Source methodology.

Some obvious barriers arise when trying to transpose the paradigm of Open Source Software into the realm of physical products and engineering design, e.g. problems regarding test and validation. However, engineering design has become more and more digitalised through the use of CAE. This realization should allow a better utilization of communication technologies in engineering design and foster hope for a further adoption of the Open Source methodology in the development of physical products.

A key aspect of the future success of OSHW is the development of a robust collaboration platform. Using the words of Koch and Tumer "Why are robust collaboration tools openly available to the programmer, but not to the designer?" [1]. It seems evident that this must change if OSHW development is to counter OSS in efficiency and success. Another important realization is that, for now, OSHW communities do not work on or share partial designs [2]. If OSHW is to embrace the efficient development of more complex and meaningful products, this must also change. This paper sets out to explore the characteristics of collaboration platforms supporting OSHW development.

2 What is Open Source Hardware?

So what is Open source hardware? Before investigating the questions regarding OSHW collaboration platform, this chapter sets out to give a general overview of the OSHW Paradigm.

2.1 A Framework and Definition

Some conceptual ambiguity seems to be surrounding open source development of physical products. At the moment two terms appear online and in literature: open (source) design and open source hardware. The use of the term open design could result in confusion as it could be both a noun and a verb, which is why this paper will be using the term OSHW as convention. We propose the following Open Source Design Taxonomy (Fig. 1) emphasizing the activity of designing open source elements of all kinds. Note that mechanics also encompasses structures and architecture—these are included in the term for simplicity.

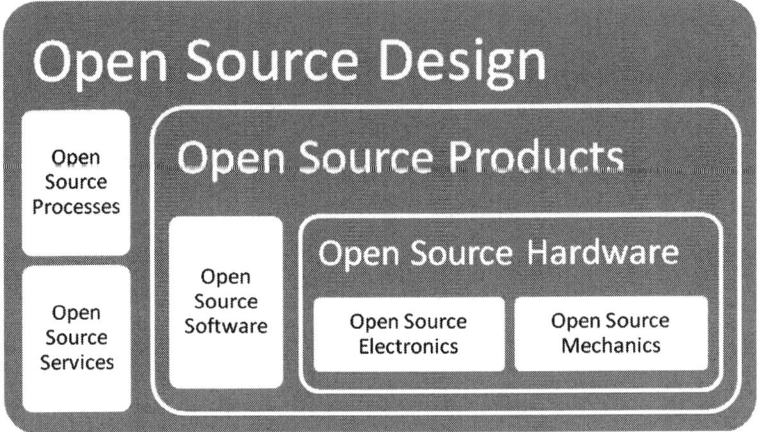

Fig. 1 The taxonomy of open source design

Although OSS has reached a stable definition, work is still being carried out to finalize an OSHW definition. The OSHW definition hosted at freedomdefined.org (definition in 2012) gives hope to this work being concluded in the near future:

> Open source hardware (OSHW) is a term for tangible artefacts—machines, devices, or other physical things—whose design has been released to the public in such a way that anyone can make, modify, distribute, and use those things.

2.2 Protecting and Profiting from Open Source Design

Reaching a stable definition is most important in respect to creating suitable licenses protecting OSHW products from unintentional exploitation. The most used licenses at the moment are variations of the Creative Commons licenses, http://creativecommons.org/. Licenses are a key element of creating sustainable OSHW business models.

According to Fjeldsted et al. [3] there are some fundamental characteristics of open design business models that apply to both OSS and OSHW. The nine building blocks of business models [4], used to describe an archetypal business model, illustrate these characteristics. There are a number of important elements that are needed for a successful implementation and operation of open design, such as attracting and retaining participants, creating a value proposition that appeals to both end-users and participants but most notably, a platform to build the open design activities upon and a strong community [3]. The platform enables, facilitates and empowers interaction and development between participants through a symbiotic relationship. The community involves the company's key resources as well as participants that co-operate through mostly a self-serviced platform, which in addition works as a channel for sales and services. The fundamental business part of the model and one which holds the biggest potentials for improvements is the identification and exploitation of revenue streams.

In this relation Fjeldsted et al. support Fitzgerald's [5] claim of trademarks and brands becoming the next IP mechanism, taking Oracle as an example, using the "unbreakable Linux" slogan. The same approach is being used by the OSHW developer Arduino that licenses their brand and logo to manufacturers for a 10 % share of their sales.

On top of licensing there are other revenue streams to be found such as incorporating manufacturing and direct sales, providing consultancy, writing books, selling merchandise, enabling subscriptions, membership fees, etc. These revenue streams affect other parts of the business model, resulting in an iterative process which may produce different versions of business models for a company to choose from and adopt.

2.3 Commons-Based Peer Production and Open Source

Following the advances in communication technologies, the phenomenon of large-scale collaborations between individuals in non-hierarchical communities is becoming commonplace. Most Open Designs seem to be taking advantage of this new mode of economic production. Commons-based peer production (*CBPP*) [6], is the collaboration among large groups of individuals who effectively provide information, knowledge or cultural goods without relying on market pricing or managerial hierarchies to coordinate their common enterprise.

Common-based peer production as a mode of socio-economic production has proven to be feasible to such an extent that it rivals other modes of economic production such as *market-based* and *managerial-firm based production. CBPP* differentiates itself in two core characteristics from other modes of economic production. The first being decentralization of authority. The authority to act resides with the individual and not a hierarchical based organiser. The second is the use of social cues and motivations to motivate and coordinate the actions of participating agents, rather than prices or commands [6]. Although open design projects do not necessarily need to make use of *CBPP* as a mode of economic production, there is an obvious synergy between the two paradigms. Open source elements seem to be a prerequisite for an efficient exploitation of *CBPP* and vice versa. Most Open Design collaborations take advantage of *CBPP* while you often find it implemented alongside more traditional *firm-based production* exemplified by a somewhat hierarchical structured core foundation.

2.4 OSHW Development Process

Most OSHW projects seem to be initiated by core teams which publish a more or less finished design for a community to start working on. Not having a design to start from might leave new developers too high an entry barrier to start developing. This seems to be the case with the collaborative space travel and research team, cstart.org, a promising project whose activities have now ceased. With the core team facilitating the development, a larger group of perimeter participators contributes with varying commitment. This resembles OSS set-ups where established companies, such as IBM, also undertake open designs development. In this case, the community of developers acts as a virtual development team [1]. To the knowledge of this author no big established company has yet to implement OSHW development in their product development strategy, had their initial product not started out as an open design.

Traditional product development is often guided by a somewhat well defined development process creating a framework for development and collaboration. In many cases having such a well defined process allows for a more aligned development thus heightening efficiency [7]. Many Open Design developments seem to take on a more autonomous approach where formality and standards emerge from the ongoing collaboration and where different processes can coexist within a project. In common practice there would usually be a difference between how core team members and projects perimeter participators would work. This autonomous approach might allow for more people to participate while possible misalignment could affect the projects negatively, e.g. designs made available in different formats thus harming the communication between developers.

In such non-hierarchical distributed processes the trust networks between participants becomes one of the important elements in the development. When dealing with complex products there is a particular need for groups and individuals to be able to trust the input of others to a higher degree. For individuals and groups with better reputations it will be easier for them to get their ideas implemented, while new developers might find it hard to be heard. One of the most interesting things about Open Design development is when the developer also becomes the user (see Fig. 2). The traditional lines between user and developer are being blurred in open design development, which in the end can result in very close feedback loops between usage and development [2] as it is the case in OSS projects. Being both the developer and user is also one of the reasons people contribute to OSS projects. Developers give their time and knowledge and will in return get to use the product for free as they can simply download it. Free use of the product is not always the case in OSHW as someone has to pay for the product to be produced before it can be used. Furthermore, developers using the product are in OSS projects are a key source for improvements as they will report mistakes in the code, "bug reports". Due to the ease by with the developers can test new ideas and solutions, mistakes in the code are corrected with incredible speed while the data management systems automatically keep track of these changes. Most OSHW projects will never match OSS in this aspect as the tangible product often needs to be manufactured before it can be tested and validated. However, this is being somewhat eased through the emergence of DIY manufacturing and electronics projects such as the RepRap (3D printer) and Arduino (microcontroller).

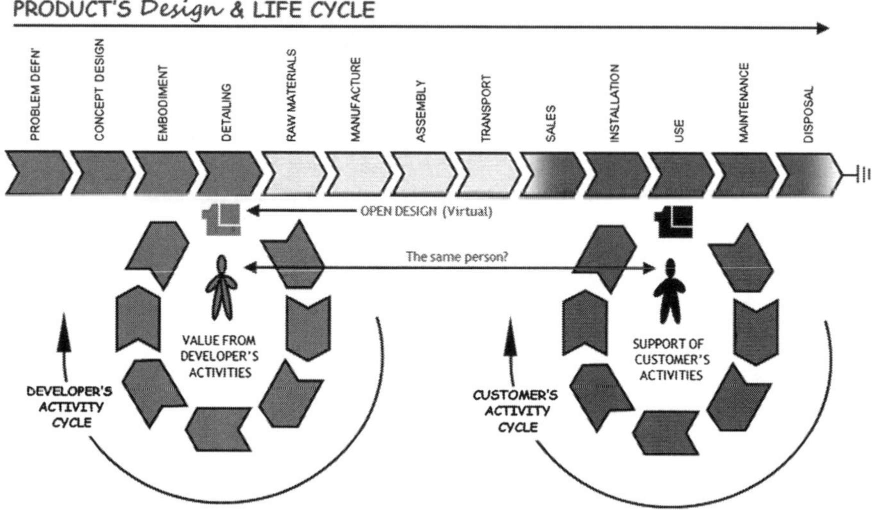

Fig. 2 The life and activity cycles for open source design [2]

2.5 Challenges Facing Open Source Hardware

Three main challenges facing OSHW are presented by Howard et al. [2] that partly explains why OSHW is yet to repeat the success of OSS. The challenges are interdependent so one needs to take a holistic view on these challenges to overcome them.

The first, and most obvious challenge, comes from the fact that OSHW entails a physical product needing allocation of physical resources to be produced. So there is a challenge of manufacturing. OSS products can be copied and distributed at a negligible cost, facilitated by the same system used for its development. OSHW products need a more complex and investment-heavy supply chain set-up to reach its end user. On one hand this can result in some OSHW products never reaching a final form, benefiting society. On the other hand, the manufacturing cost can result in a more sustainable business model for the project originators as consumers will choose to purchase the final product instead of producing it themselves [2].

The second challenge facing OSHW emerges from the need to validate designs. Again there is often a need to produce prototypes to test and validate designs properties to allow close loop iterations. Considering the complexity of common products today, this could be a major obstacle for the realization of OSHW products, not only in respect to the cost but also the know-how needed. Digital simulations are the first step to overcoming this obstacle but creating systems that support and lower the cost of testing/validating design properties is crucial if OSHW is to entail the development of complex products and efficient product development. Such a system could mirror the peer-review systems in OSS projects. Having partially open systems, which allows some parties to capitalize on closed source test documentation, as proposed by the 40 fires foundation (http://www.40fires.org/), a OSHW project focusing on developing energy efficient technology. This strategy could also be part of the answer to overcoming the challenge of validating designs. The concept of "Manufacturing as a Services" (MaaS), community based manufacturing shops which allow individuals to make use of flexible manufacturing equipment, might in time create a base for easier design validation as it allows communities to bypass large operations and manufacturing facilities [1, 8].

Finally, there lies a challenge in allowing OSHW communities to efficiently collaborate on high-tech complex products [2]. This is closely related to the validation of designs because complex products will entail more sub designs in need of validation. Benkler and Nissenbaum [6] present three structural attributes of the CBPP relations which are highly relevant with regards to overcoming the challenge of complexity in OSHW projects. These attributes are: *modularity, granularity* and *low-cost integration.*

For now OSHW communities do not work and share partial designs [2] mainly due to a lack of a suitable collaboration platform. A higher focus on attributing *modularity* to OSHW objects could allow the OSHW system to better handle partial design development and allow communities to work on more complex products. Modular design will also allow subsystem to be tested and validated independently, one of the module drivers of Erixon [9]. Focusing on modularisation will however, also result in higher requirements for the discipline in the development process as modularity will lay additional constraints on the object of development in the form of structural standards.

The *granularity* of the modules is important, as appropriate finely-grained modules allow contributors to contribute in respect to their motivation and skill [1], lowering the barriers of entry to complex projects. Defining modules with the "right" granularity can be a difficult task but several methods guiding the modularisation process are available such as the "design structure matrix" [10].

The *cost of integration* refers to the cost of integrating the modules into a whole end product. OSS benefits from low-cost integration due to efficient integration software, which allows somewhat automated integration (merge features), and established iterative peer-review practises. The latter also plays a big role in design validation and is in some cases formally integrated in the collaboration platform, e.g. sourceforge.com. Although many OSS peer-review practises could easily be adopted in OSHW development, automated integration cannot, partly due to the wide range of digital formats used in many OSHW projects. One way of dealing with a higher cost of integration in OSHW projects could be having a more formal development process, in which stricter requirements are set on the various contributions. The level of formality on OSHW projects seems to vary a lot, not only between projects but also between different participating groups within each project, e.g. as mentioned between core team members and perimeter participators. Creating an OSHW collaboration platform that supports development of designs that contain the three structural attributes is by these authors seen as an essential part of letting OSHW communities deal with the challenge of product complexity. The next chapter will go further into detail with the requirements for an OSHW collaboration platform.

3 OSHW Collaboration Platforms

OSS and OSHW platforms alike seem to have two fundamental functions; to function as a social platform and to provide tools to enable efficient collaboration. Social platform referring to a platform that creates awareness of the OSHW projects and allows projects initiators, participators and stakeholders to meet creating a base for the community. Although no studies have been done on how the size of an OSHW community affect the success of a related project, most successful OSHW projects today seem to have large and active communities.

The second fundamental function is to provide the community with the right tools, collaboration tools such as design repositories and wikis. As mentioned earlier, there is a lot of variation in how OSHW projects are structured so there is a need for the platform to be flexible. OSS platforms, such as sourceforge.com and github.com, provide a lot of flexibility. When projects are created on these platforms it is easy to activate the tools the project creators see a need for. There is no current collaboration platform specifically designed to support OSHW development. The Open design engine.net looks to be one of the only promising OSHW platform projects judging by their guiding values and potential features. The project has not yet reached a state for evaluation and unfortunately the project progress seems to have halted.

Koch and Tumer present 6 main areas that must be integrated into a collaboration platform; Project overview, Documentation and Design Repository, Communication, User Identification Standard, Funding and Licensing [1]. The following section will discuss the first 3 areas as they are the most influential in regard to creating efficient collaboration.

3.1 Project Overview

Providing a good project overview allows for several things. First of all it gives new potential users a good first impression which might result in them participating in the project [1, 11]. Project overview in this sense is very important in relation to allowing the platform to function as a social platform. The project overview also allows for an overall guiding of the project by providing information to the participators, e.g. displaying news and urgent matters. There is a question of how big a role such a collaboration platform should play in regard to the management of the projects? There are open source project management tools out there, why building on this work, implementing a suitable product data management system, seems to be a logic way forward.

More commercialized current OSHW projects e.g. makerbot.com and arduino.com, implement a product website focusing on marketing, providing product introduction and selling their products while maintaining links to the development site, which give a good current view of the project. Other OSHW products such as RepRap.com and Openeeg.sourceforge.net use a wiki as their main page, focusing more on displaying the development of the product. In general it can be said that a good project overview is something that is lacking in OSHW projects. This is especially the case with regards to documenting the development process. One comment on the open source ecology forum gives one reason for this.

> The OSE development process is intended to be completely wide open and accessible to anyone who wants to watch. However, knowing HOW to watch can be challenging. OSE activities take place in a wide variety of web sites, software applications, email, conference calls, and on-site meetings at the Factor e Farm development facility.

This case is repeated in many OSHW projects and results in a high entry barrier for new developers. Especially for widely distributed projects such as RepRap, it is hard to keep track of new derivatives and development tracks.

3.2 Documentation and Design Repository

Most, if not all, current OSHW projects collect most of the development documentation on wikis. The dynamic nature of wikis makes them the perfect tools for collaboration, making knowledge sharing in large groups easy. However, the wiki content should be standardized using page templates to make communication more efficient. Many developers might share the view that documenting your work is tedious work although important, so tools with allow for an easy documentation will most likely heighten the quality of the documentation. By utilizing the script based CAD software, Open SCAD, the RepRap communities are working on implementing automated documentation by embedding additional product documentation, such as assembly instructions into the CAD model as comment threads, making it much easier to keep up-to-date documentation. Open SCAD functionalities make it appealing to the 3D printer communities as it is script based, but will probably not be the dominant CAD software any time soon. The idea of linking CAD and additional documentation would relieve OHWS communities of a tedious and resource demanding task and is therefore very interesting.

The design documentation, e.g. schematic, CAD files etc. in larger OSHW projects, are for now often made available via design repositories on sites such as Github.com and Sourceforge.com while providing some product data management. The Design repositories on these sites are very focused on OSS development, just understanding the syntax can be an obstacle for people without knowledge about software, although the sites have easy to use functionalities.

An important aspect of design repositories is which product data management (PDM) model, or version control system (VCS) as it is called in software development, they make use of as it can have a huge impact on the development workflow. The most widely used model is the centralized. Such a system has a single master repository from where all developers check-out the documentation. However, not all developers will have write-access [12]. The centralized model therefore focuses the project control to the initiated. A decentralized model lets every check-out be a full-fledge repository with complete commit history. With no evident master repository a repository is identified by convention within a community [12]. Clarifying which model is most suitable for OSHW projects is beyond the scope of this paper as it also depends on how the projects wish to structure their product development. There exist several high quality PDM/PLM solutions on the market today although they are probably too expensive for most OSHW projects. Aras corp. provides an open source PDM solution which might be useful as it is currently being used by international companies such as Motorola. A very interesting notion is cloud PDM. Such a thing could be exactly what is missing for OSHW to undertake more complex product development. It could be the Github.com of OSHW. Solidworks n! fuze is leading the way in commercialized cloud PDM but for now no open source counterpart exist.

Thingiverse.com is a site which makes design documentation available in a more accessible form than the OSS focused repositories, and is widely used in the desktop 3D printer communities. Although it has none of the features embedded in more advanced design repositories, such as an actual version control systems, it provides an easy and visual access to the design documentation. As a design repository Thingiverse is not the best choice when dealing with complex product development as it functions more like a showroom. A combination between thingiverse's ease of access and sites such as Github's functionality would be an interesting possibility.

3.3 Communication

Kock and Tumer speak of three main forms of communication; group (forums), one-on-one (mail) and real time (Skype, chat). As communication is essential for an efficient collaboration these three forms should definitely be implemented. Furthermore, automated indirect communication on a community level would be helpful in respect to project overview. For example having voting systems that lets participators vote on promising development tracks could allow other developers anticipate which direction the project is taking so they could streamline their efforts. This is especially useful in the implementation of standards. Following a thread on the RepRap forum, a group had decided on a standard. The next step was to inform the community. Quoting one of the posts, "now it's time to scream, shout and make noise". How do you make a community aware of the "good" ideas? Voting and bidding systems seem to be obvious solutions, alongside competitions. These elements are used to a wide extent at crowd sourcing sites.

3.4 A Platform that Sources the Crowd

Looking at the collaboration system at OSHW projects today three main tools facilitates the development (not looking at the inter-collaboration between the core team members). These three tools are the wikis, forums and design repositories. Between them, these tools make a rather complex system of information which results in a development process easily lacking tractability. Although this system might, to some extent, prove sufficient for the dedicated developers it leaves a high entry barrier for new developers, as for example in the open source ecology example earlier. This results in OSHW projects not fully utilizing the possibilities of CBPP. Even for successful projects such as RepRap this is the case. Looking at the RepRap model "Prusa Mendel" which is focusing on low-cost and ease of sourcing, making it the ideal choice for new developers, only four "pull request" have been made the last year judging by the activities at github.com and only three commits a available under the Prusa improvements overview at RepRap.org. It looks as if the new improvements are not integrated into an "optimal" design. Instead, new derivatives a created which are hard to find or poorly documented. One reason for this could be the high cost of integration as the models are not modular. Another reason could be that many people actually do not share their designs. In this aspect there might be lessons learned from crowd sourcing. A crowd sourcing site like 99designs.com has 113 designs per project (contest). It is very easy to contribute to projects on such sites and it is easy to get an overview of the different contributions due to the use of "design contests". Such a concept could easily be implemented in a platform to attract more perimeter participators who just wants to give small contributions, while more complex collaboration tools should be implemented to suit the need of the more dedicated participators.

4 Conclusion

OSHW projects have definitely gained more awareness following projects such as Arduino and RepRap, but OSHW development does for now not mirror the efficiency in OSS development. This is partly due to the challengers that come from dealing with physical products and the fact that there is no collaboration platform specifically design to support OSHW development. A part of the solution to overcoming these challenges is giving OSHW products the structural attributes of CBPP, creating rightly grained modular products which will lower the cost of integration and allow for independent test and validation of designs.

Because many OSHW projects entail some OSS elements, an OSHW collaboration platform needs to provide the same service as OSS collaboration platforms today, and more, for the complete project to be run on the same platform. A platform should not only play in tact with the development process and work flows, but also support the business models and revenue streams all of which are in need of further research. Two fundamental functions of an OSHW collaboration platform are found to be: to function as a social platform and to provide the community with collaboration tools. For this to happen, better product data management systems need to be made available to the OSHW communities and workflows should be defined which will lower the entry barriers for new developers. Although, many agree that realizing a true OSHW collaboration platform is of utmost importance if OSHW is to emulate OSS in success, there is no apparent solution out there which will truly fill the void any time soon.

References

1. Koch MD, Tumer IY (2009) Towards open design: the emergent face of engineering—a position paper. In: Proceedings of ICED'09, Vol. 3(11), pp 97–108
2. Howard TJ, Achiche S, Özkil A, McAloone TC (2012) Open design and crowd sourcing: maturity, methodology and business models. In: Proceedings of the 12th international design conference
3. Fjeldsted AS, Aðalsteinsdóttir G, Howard TJ, McAloone TC (2012) Open source development of tangible products. In: Nord design 2012, Denmark, pp 22–24
4. Osterwalder A, Pigneur Y (2010) Business model generation: a handbook for visionaries. Game Changers and Challengers. Wiley
5. Fitzgerald B (2006) The transformation of open source software. MIS Q 30(3):587–598
6. Benkler Y, Nissenbaum H (2006) Commons-based peer production and virtue. J Polit Philos 14(4):394–419
7. Ulrich K, Eppinger SD (2007) Product design and development, 4th edn. McGraw Hill, New York
8. Manufacturing as a service (MaaS). Postfully yours
9. Erixon G (1998) Modular function deployment—a method for product modularization. KTH Royal Institute of Technology, Sweden
10. Steward DV (1981) The design structure system: a method for managing the design of complex systems. IEEE Trans Eng Manage (3):71–74
11. Saade RG, Otrakji CA (2007) First impression last a lifetime: effects of interface type on disorientation and cognitive load. Comput Hum Behav 23:525–535
12. De Alwis B, Sillito J (2009) Why are software project moving from centralized to decentralized version control systems? CHASE'09. In: Proceedings of the 2009 ICSE workshop on cooperative and human aspects on software engineering, IEEE, pp 36–39
13. Balka K, Raasch C, Herstatt C (2009) Open source enters the world of atoms: a statistical analysis of open design. First Monday 14(11):248–256

Part IX
Applications in Practice (Automotive, Aerospace, Biomedical-Devices, MEMS, etc.)

Drowsiness Detection System for Pilots

Gurpreet Singh and M. Manivannan

Abstract Though Pupil Diameter (PD) of the human eye has been known to be an indicator of sleep-onset, the exact quantification of the PD were not known. In this study variations of PD which are in excess of +5 % of the mean value were monitored and classified as Event A. In addition, the eye closure beyond −30 % of mean diameter were classified as Eye Blinks or Event B. The duration of eye blink was monitored if it is more than 1 s. Both Event A and Event B were used simultaneously to detect sleep-onset. The algorithm using Open CV and MS VC++ was tested with IR videos PD of subjects on a driving simulator. The Alert subject exhibited several Event As but no Event Bs and the Drowsy subject exhibited both Event As and Event Bs and was successfully classified as "Drowsy" 2 times during the test run.

Keywords Drowsiness detection · Pilot snooze · Puppillometry · Sleep-onset

1 Introduction

THE pilots work in a very demanding environment where they are expected to be alert at all times. They are supposed to monitor the health of different sub-systems of the aircraft throughout the flight. Even when the auto-pilot is engaged, pilots cannot afford to lose situational awareness. If any snag develops in a flight, it

G. Singh (✉) · M. Manivannan
Department of Applied Mechanics/Biomedical Engineering Division, Indian Institute of Technology, Chennai, Tamilnadu 60036, India
e-mail: gurpreet261@yahoo.com

M. Manivannan
e-mail: mani@iitm.ac.in

A. Chakrabarti and R. V. Prakash (eds.), *ICoRD'13*, Lecture Notes in Mechanical Engineering, DOI: 10.1007/978-81-322-1050-4_78, © Springer India 2013

requires immediate corrective measures from the pilot. Also if there is even a minor deviation in flight path of the aircraft due to loss of attentiveness, it could lead to disastrous consequences. Many instances have come to light wherein major mishaps have occurred due to pilot having fallen asleep while at the controls of the aircraft like the crash on 22 May 2010 in Mangalore, India, killing 158 people; in June 2008, an Air India aircraft headed to Mumbai flew past its destination with both pilots asleep [1]. In order to enhance flight-safety, it is imperative to ensure that the pilot does not fall asleep while at controls.

While there have been numerous studies on sleep [2–7], we have to deal with the sleep-onset and not with sleep itself. The reason is that the pilots undergo pre-flight medical examination which ensures that they are fit to fly. However, due to various reasons preceding or during the flight, the pilot may involuntarily tend to fall asleep. Once a pilot falls asleep there are very few physiological changes which can be monitored to reliably activate the alarm. But the transition from alert state to sleep state is a condition which is critical both from physiological monitoring and aircraft controlling point of view. It becomes imperative to identify sleep-onset rather than going into stages of sleep.

1.1 Constraints in Sleep-Onset Detection in Cockpit

Several parameters which can be used to detect sleep-onset are EEG, EOG, change in respiration, change in heart rate, etc. but the stringent requirements in case of pilots, limits the options available for making these measurements in cockpit. These constraints are:

1. There can be no use of intrusive electrodes on any part of pilot's body. This rules out the use of EEG, EOG, ECG, EMG, Blood Pressure monitoring, analysis of exhaled gases to predict sleep-onset.
2. The only parameter available to non-intrusive analysis is eye, using an optical measurement tool. Again the option is further narrowed as use of goggles should be avoided. The reason being that goggles would be required to be custom-fitted to each individual's face structure. Thereafter, every time a pilot flies, it would have to be calibrated. Lastly, it would not be a purely non-intrusive system and hence could be disproved of by Director General of Civil Aviation (DGCA).

1.2 Overcoming Constraints Using Optical Methods

The solution is to use a camera which can get the real-time images of the eye without pilot having to do anything. There is a possibility of using several different eye parameters for analysis. But as the purpose of this paper is to design a robust system with the least number of variables, only two variables were used and were found to be fairly accurate in describing the state of the subject.

Over the years different modalities have been used to study sleep-onset like Respiration [2], Cardiovascular function [3], Electro-Oculography [4], etc. but all these modalities present the basic difficulty of being intrusive in nature. In a cockpit environment, we need to have such a device which will not interfere with the working of the pilot in any way but would still be capable of reliably detecting the onset of sleep in pilot. One non-intrusive method of fatigue-detection has been discussed in [8], which brings out the fact that in case of an Alert subject the deviations in pupil diameter are restricted to +5 % of the mean pupil diameter whereas in case of a Drowsy subject, the variations are way beyond +5 % of the mean pupil diameter.

In the present study, effort has been made to clearly define the various characteristics of pupil diameter which need to be monitored and classified so that it results in very unambiguous criteria for declaring a subject as Alert or Drowsy. As the only physiological parameter being monitored is Eye pupil diameter, there is a need to specify the conditions of classification when the whole pupil is visible and also when only partial pupil is visible due to occlusion by eye lashes/lids.

In order to account for the changes in pupil diameter when the whole pupil was visible, a parameter called Event A was used whereas in cases where pupil was only partially visible due to eye blinks, another parameter called Event B was used, which was considered to have occurred if the pupil diameter reduced more than 30 % of mean pupil diameter.

The scope of this paper is to discuss the algorithm which was used to separate out the events which are encountered while monitoring the eye pupil of a subject. After the separation, the events are stored and the algorithm keeps a track of all the relevant data. Until the stored data meets the specified criteria for alarm activation, the algorithm continues to run through the video file without causing any alarm. However, the instant the laid down criteria for alarm activation are met, the alarm is activated and the video clip processing is halted. In order to prove the concept, two different video files in AVI format were used. Since this method uses a camera to capture the image of eye, it is totally non-intrusive and hence would be more acceptable in terms of cockpit requirements (Fig. 1).

2 Methods

The task of acquiring the pupil diameter begins by first acquiring the video of pupil and then it is subjected to various morphological operations which results in the residual image being the contour of pupil alone. Thereafter an ellipse is fitted to the pupil contour which is then used to calculate the area of the pupil. This in turn is used to calculate the pupil diameter. OpenCV ver2.1 is a very versatile tool to perform morphological operations on the image frames. Many operations like Image

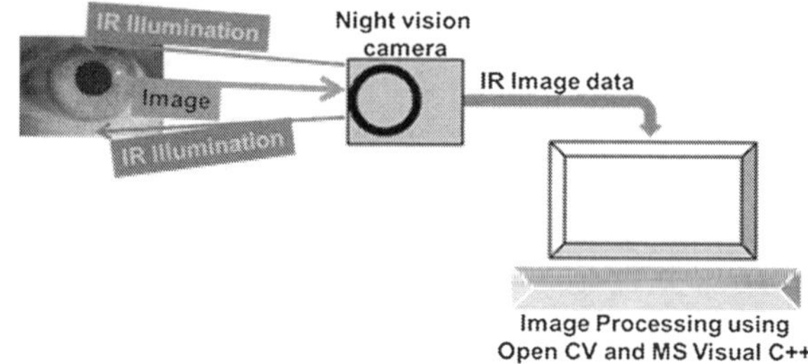

Fig. 1 Block diagram of the set-up for capturing video film for testing of alertness of the subject

Dilation, Image Erosion, Image Opening, Image Closing can be performed with very few lines of code.

Various steps involved in getting the contour of eye pupil are shown in the flowchart below (Fig. 2):

In order to convert video from colour image to grey-scale, Luminosity method has been used. However, if the video is already grey-scale or is in night-vision mode, the algorithm skips this step. Thereafter, Adaptive thresholding is applied in order to cater for illumination gradient. Thereafter, the morphological process of "Closing" is applied to each frame so as to get rid of noise and unwanted images. After all these processes have been sequentially performed, the image consists of only pupil contour and is ready for Pupil detection, as shown below (Fig. 3):

The image available at this juncture consists of only the pupil. So Open CV is used to fit an ellipse to the pupil contour. Actually Open CV provides the option of either fitting a circle to the detected contour or fitting an ellipse to it. But the manner in which Open CV fits a circle leads to introduction of large errors. If a minimum circle is fitted to the contour, then the software just draws a circle with a minimum possible diameter which includes all the points of the contour within itself. Whereas if an ellipse is fitted, Least squares method is used to determine

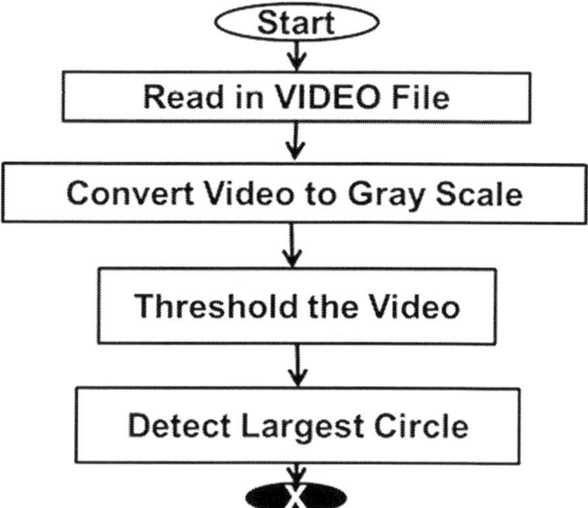

Fig. 2 Flowchart depicting sequence of steps for detecting eye pupil

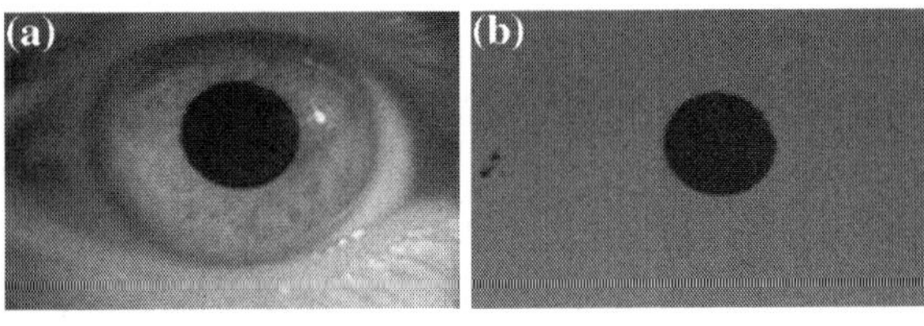

Fig. 3 **a** The input image frame, **b** the residual image after application of morphological operations, consisting of only pupil contour, ready for pupil detection

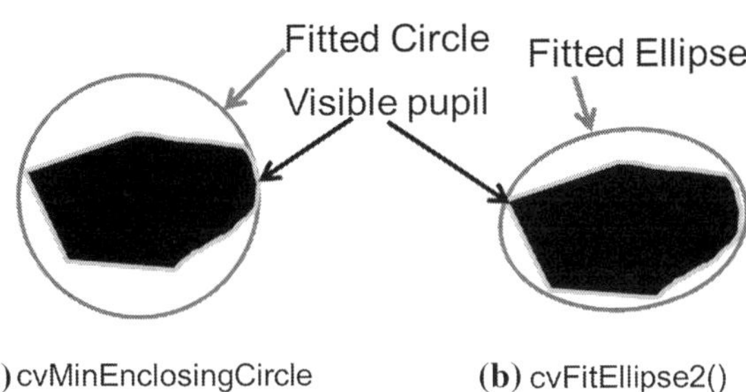

Fig. 4 The comparison of methods of (**a**) fitting a minimum enclosing circle, (**b**) an ellipse to a set of points. *Source* [9]

coordinates of an ellipse such that the ellipse is the best approximation of the detected contour [9]. This results in keeping the errors between the detected contour and the fitted shape to the minimum (Fig. 4).

Another factor of consideration here is that the pupil takes a circular shape whenever the person is looking in front. However, the shape of the pupil tends to seem elliptical only when a person is looking from the corner of his/her eyes. A point to remember here is that when a person is drowsy, the pupil automatically tends to come into the centre of the eyes. Therefore, when an ellipse is fitted to the detected pupil contour, it not only produces lesser errors than the circle but also provides better tracking of the contour and follows the pupil faithfully. Whenever the subject is drowsy, the eyes tend to come to the centre and ellipse can virtually be treated as a circle for all calculations (Fig. 5).

Once an ellipse has been fitted to the pupil contour, there is a requirement to get the diameter of the pupil. But OpenCV does not provide a direct measure of the semi-major and semi-minor axis of the ellipse. Instead, a circuitous method needs to be followed to get the dimensions of the ellipse. This involves using Contour Moments to get the area of the contour [8]. Many different characteristics of the detected contour can be calculated by Contour Moments but as we are interested only in getting the area of the contour which in turn would be used to calculate the diameter of the circle, we need to use only moment (0, 0).

Once the area of the contour has been calculated, the radius of the circle can be calculated as per the following formula:

Diameter $= ((\sqrt{}(\text{Area of Circle})/\sqrt{\pi})*2)$.

This formula faithfully gives the diameter of the pupil till the time the person is looking straight. But if the subject is looking from the corners of the eyes, an error is introduced in diameter value. But as previously discussed, this error in pupil diameter calculation would not affect the activation of alarm as a drowsy person

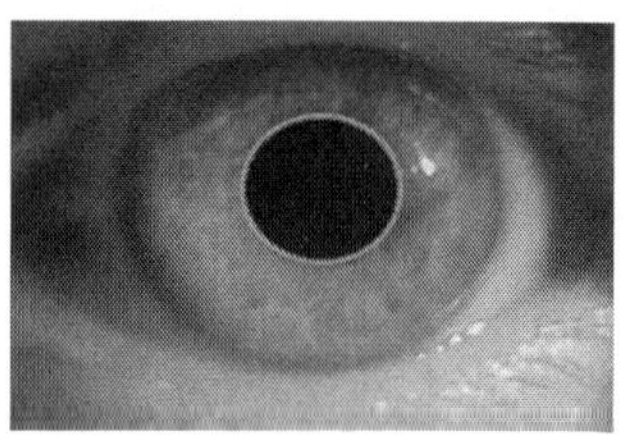

Fig. 5 The ellipse fitted to eye pupil. It is clear that when the subject is looking straight, ellipse takes the form of a circle

has a tendency to have his pupil involuntarily coming into the centre of the eyes and not at the corners. So when a person is drowsy, the pupil would be circular and there would be no error in calculation of the pupil diameter.

Availability of pupil diameter leads to the next step where constraints are set on the input in such a fashion that whenever the set criteria of drowsiness is met, the alarm is activated which alerts the pilot.

It was decided to have two defining events which need to be monitored to reliably predict sleep-onset. These are:

1. Event A: If there is +5 % change in "Pupil Diameter" w.r.t. "Mean diameter" at any instant of sampling.
2. Event B: If there is −30 % reduction in "Pupil Diameter" w.r.t. "Mean diameter".

Event A is the actual variation in the pupil diameter w.r.t. the mean pupil diameter and is calculated every clock pulse to account for different video formats working at different frame rates. However, in view of the computational requirement, only the largest value of pupil diameter detected every 1,000 ms is used for calculation of running average pupil diameter. Event B is Blink of the eye "i.e. closing of the eye-lid and opening of the eye-lid". The OpenCV® function used for fitting the best curve to detected pupil is cvFitEllipse2(). This ensures that even if pupil is partially occluded by the eye-lid, the best possible estimate of the pupil area is still available. The reason for having Event B only −30 % in pupil diameter is that blink can only lead to occlusion of eye-lid thereby leading to reduction in pupil area available for calculation. Even when eye-lid is fully retracted, it cannot have a variation which is higher than maximum pupil diameter.

The diameter of pupil is calculated in every frame. The video clips used in this project are in AVI format with a rate of 29 Frames per Second (fps). This means that each frame is available for about 34.4 ms to derive the information from it. But as OpenCV supports different video formats which may have different fps, so it was decided to sample the video every 7 ms. This ensures that a very high rate video is also processed reliably and at the same time there is no loss of data due to skipping of frames. Also 7 ms time-gap gives a feeling of continuity even though the data input changes at a rate lower than it (Figs. 6, 7).

Once the pupil diameter is available and Event As and Event Bs have been detected, a Criteria needs to be specified which, when satisfied, would lead to

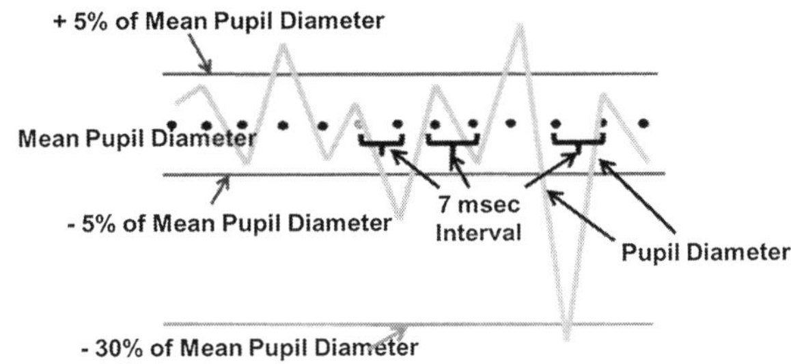

Fig. 6 Different parameters which are calculated from the available eye pupil diameter

activation of alarm. The procedure for detecting Event A and Event B are shown in the flowchart below (Fig. 8):

Once Event A and Event B have been detected, there is a need to have a timer which helps us in keeping a track of the number of Event A and Event B occurring within a specified time frame, which in this case is 5 s. Track is kept of the number of Event As and Event Bs occurring during a period of 5 s and also duration of Event B. It needs to be kept in mind that a window of 5 s is created which is reset every 5 s in case Alarm is not activated. However, it is not reset when the alarm is activated (Fig. 9).

The conditions for activation of alarm can be specified as:

1. If there are more than 3 Event As in 5 s,
2. If there are more than 1 Event B in 5 s AND the duration of any Event B is more than 1 s.

2.1 Reasoning for the Selection of Criteria

Event A can be specified as variations in Pupil Diameter due to drowsiness whereas Event B can be specified as Eye Blink. It has been observed that pupil diameter is very stable in an alert subject rarely going beyond +5 % of the mean diameter whereas for a drowsy person the variations are very frequent [8]. This study utilised two groups, a total of 30 healthy subjects (mean $\pm$SD:28.2 $\pm$ 8.9 years old), for the experiment. One group rated themselves as alert (15 men, a normal night sleep with more than 8 h), and the other group as

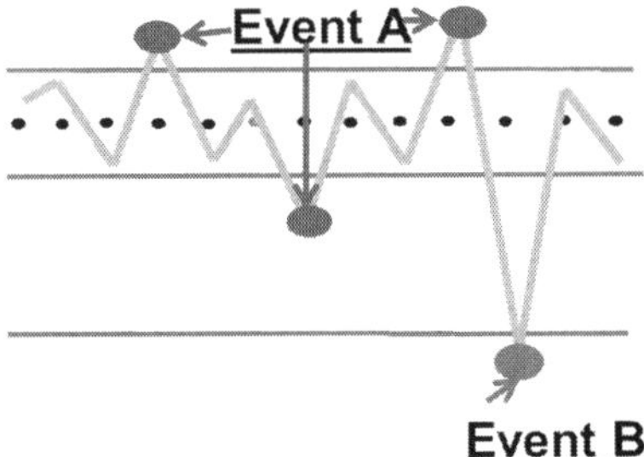

Fig. 7 Occurrence of Event A and Event B w.r.t variations in pupil diameter

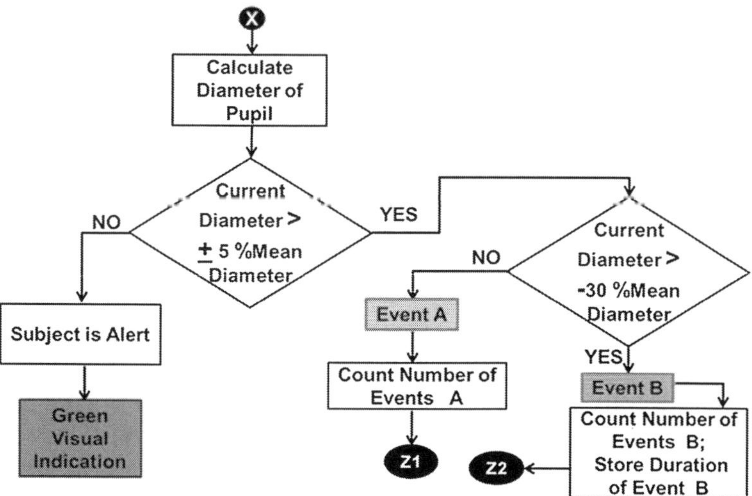

Fig. 8 Flowchart depicting conditions for detecting Event A and Event B

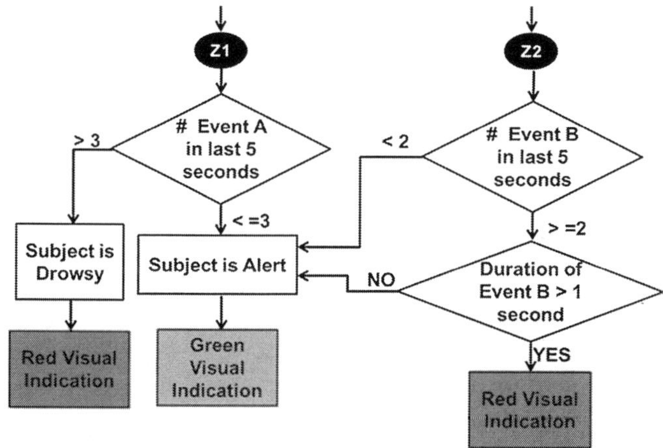

Fig. 9 Flowchart depicting classification of a subject as alert or drowsy based on conditional occurrence of event A & B

drowsy (15 men) [8]. As there is no specified number available as to how many times the pupil diameter might vary for a drowsy person, it was decided to set the threshold at 3 variations every 5 s. If the number of pupil diameter variations (Event A) during 5 s epoch is less than or equal to 3, the subject would be classified as Alert. However, if the number of Event As is higher than 3, the subject would be classified as Drowsy, leading to activation of alarm.

However, if the variation in pupil diameter is beyond the constraints of Event A, it is classified as Event B. The numbers of blinks per minute vary from individual to individual and also as per the task being performed by the individual. Usually a task which requires continuous use of cognitive function leads to reduction in number of blinks per minute to about 5 [10]. But at the same time it needs to be remembered that with increasing drowsiness, the duration of eye closure increases. For alert subject, the duration of eye blink can range from 300 to 450 ms. This duration could increase many folds during drowsiness. It was decided to set a threshold at 1,000 ms so as to avoid false alarm as well to minimise misses. So in case of Event B, the input is subjected to following checks:

1. If the number of Event Bs in 5 s epoch is less than 2, subject is alert,
2. However, if the number of Event Bs in 5 s epoch is more than or equal to 2, the duration of Event B is checked. If the duration of Event B is less than 1,000 ms, the subject is classified as alert. If however, the duration of Event B is more than 1,000 ms, the subject is drowsy which leads to the activation of the alarm. It is pertinent to mention here that Event B, also called as Blink, may also be termed as eye-lid droop which becomes a deciding factor only when the eye-lids take more than 1,000 ms to finish 1 cycle of first covering the pupil and then coming back to the default position. Event B is designed specifically to counter problems posed by blinks.

2.2 *Accounting for Occlusion of Pupil by Eye Lids/Lashes*

Eye lashes occlude the pupil only during eye blinks. So it is important to remember that Event A is not due to eye blink but due to actual variations in Pupil. On the other hand, Event B is the superficial variation in pupil diameter which is caused due to occlusion of pupil due to presence of eye lash over the pupil. We know that eye blinks usually take about 300–450 ms to complete, assuming video rate of 29 fps. This leads us to approximately 10–13 frames of the video clip to complete 1 eye blink. This means that about 5–6 frames are required to capture the closure of the eye and 5–7 frames for capturing opening of the eye. This means that each successive frame would lead to a change of about 15–20 % superficial change in pupil diameter. "Superficial" because it appears to be happening but is actually not. In order to avoid erroneous results due to occlusion, the algorithm is so designed that when the variations in pupil diameter are −30 % of the mean pupil diameter, it is automatically classified as Event B. So now the algorithm already knows that the subject is in the process of blink. Now the algorithm sets up another check for measuring the number of blinks in the 5 s epoch. As is known, if the number of blinks in the epoch is less than 2, it ensures that the subject is alert. However, if the number of blinks (Event Bs) in the epoch is equal to or more than 2, the algorithm checks for another condition i.e. whether the duration of the blink is less than 1 s or not. It has been found that with increased drowsiness the duration of blink increases. So if the duration of blink exceeds 1,000 ms, the subject is classified as drowsy else he is classified as alert.

2.3 Effect of Different Blink Rates

The rate of eye-blinking varies from individual to individual, from 300 to 450 ms. Usually people tend to overlook individuals who have slower blink rates whereas those with higher blink rates are immediately recognised. There are two issues with blink rate:

(a) The blink period is larger than 450 ms.

When a person is drowsy, the blink period increases. The blinks are still there though with reduced frequency and increased period. When a drowsy person falls asleep, blinking stops i.e. blink period becomes ∞. The result is that the blink period keeps on increasing with increased drowsiness. There does not exist any data on what is the exact value or a range of blink-periods which denote stage-I sleep. In order to design the algorithm, it was decided to arbitrarily keep the value at 1,000 ms. This provided a mid-value between a sleeping person with zero blink-rate and a normal person with a nominal blink period of 400 ms.

(b) The blink period is smaller than 300 ms.

There are individuals who are pre-disposed towards smaller blink periods. They tend to blink very frequently and thus present another challenge to the algorithm. The challenge arises due to frequent occlusion of pupil due to eye-lid cover, either partially or fully.

The essential requirement of the algorithm is to get the correct value of pupil diameter once every second. This would be sufficient to satisfy the requirements of both Event A and Event B. However, the algorithm is so designed that it calculates the "running average" of the pupil diameter every clock pulse. Assuming that the video being tested has a rate of about 25 frames per second. This means that the video provides 25 pictures of pupil per second along with its running average. This provides for a lot of redundancy in calculations. So even if the eye-lid occludes the pupil, the results are robust enough to give the correct status of the subject.

3 Results

In order to validate the hypothesis and the set criteria, two AVI format videos, one each of an alert and a drowsy person, were used to validate the algorithm. Both these videos were acquired by Hirata Y [5]. The videos were recorded when subject sat comfortably on a driving simulator equipped with a steering wheel and brake & accelerator pedals (Logicool PRC-11000) in a dark room as elaborated in [5]. The subject wore a goggle (NEWOPTO ET-60-L) with 2 CCD cameras each of which takes infrared images of each eye at the frame rate of 29.97fps (NTSC). The various characteristics of the videos used for validation of the algorithm are tabulated in the following table (Fig. 10):

The video featuring the pupil of Alert subject exhibits very few changes in pupil diameter (few Event As) and no Event Bs whereas the video featuring a drowsy subject exhibits many Event As and Event Bs. The algorithm is so designed that the program execution stops if the Drowsiness alarm gets activated. The summary of the results is given in a tabulated form below (Fig. 11):

The results obtained after subjecting the two videos to the algorithm are shown in pictorial form in Figs. 12, 13. It is clear from both the tables and graphs that Events A and B are far too frequent in a drowsy person as compared to an alert person.

In addition to the above-mentioned videos, one of the authors recorded two videos, each of about 2 min duration, of alert persons using SONY® HANDY-CAM® hand-held camera in night-vision mode. The results were found to be consistent with the hypothesis. However, as no corresponding recordings were carried out for drowsy person by the author, the mention of even the videos of alert person has been kept out of the paper.

The activation of alarm was found to be consistent with the criterion specified. If either of the two criteria as specified in Sect. 2 were satisfied, the alarm was activated.

4 Summary and Future Work

The novelty of this method is that pilot does not have to wear any goggles. The system is totally non-intrusive in nature which could help it get more acceptability in aviation community. The camera is to be fixed on an instrument panel of the aircraft which maintains pilot's face in the view. The challenge lies in ensuring

S. No.	Type of Video	Size (Mb)	Frame Rate (Frames/sec)	Clip Duration (sec)	Frame Size (Width : Height)
1.	Alert Person	1.13	29	20	320:240
2.	Drowsy Person	1.40	29	20	320:240

Fig. 10 The table showing the characteristics of the two video clips used to validate the algorithm

S. No.	Type of Video	No. of Event A detected	No. Of Event B detected	Drowsiness Alarm Activation
1.	Alert Person	5	Nil	Nil
2.	Drowsy Person	8	5	2

Fig. 11 The table showing the summary of the working of the algorithm on the two videos

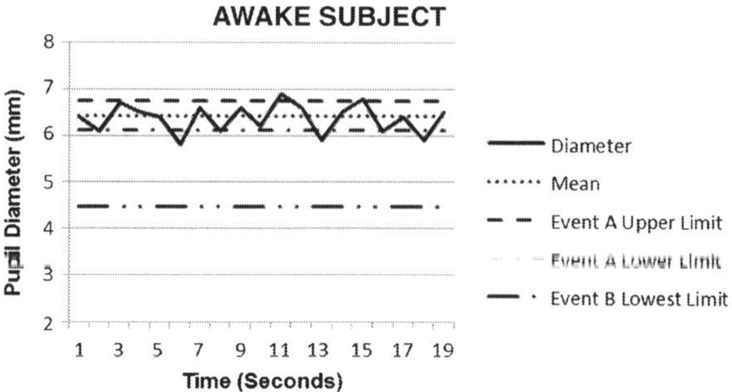

Fig. 12 Graph showing the pupil diameter variations limited to +5 % variations of mean pupil diameter in case of alert subject

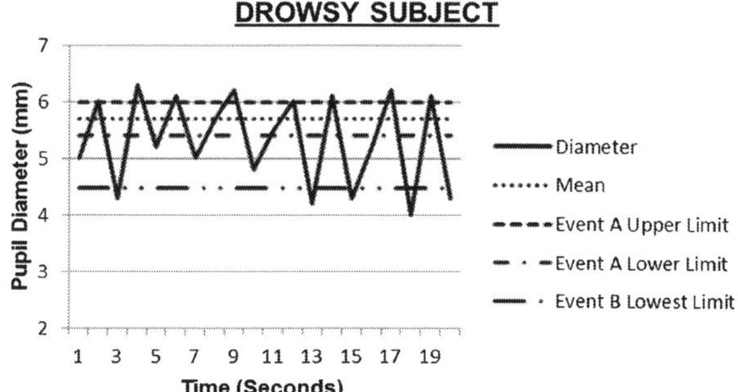

Fig. 13 Graph showing pupil diameter variations in case of drowsy subject. It can be observed that pupil diameter variations are way beyond +5 % variations about mean pupil diameter. Also, there are numerous variations below 30 %

that pilot's face is always in the view of camera but here a few points are pertinent to mention:

1. Any drowsy person's face would be almost motionless so it would not go out of the camera's view. Once it is ensured that the resting position of the face is always captured by the camera, there is a surety of capturing the pupil's image for processing.
2. Whenever a person is drowsy, his/her eye-balls are pulled back to the centre of the eyes by the autonomic nervous system. Thus pupils automatically come to the centre of the eye which ensures the full view of the pupil for the camera. The full view of pupil is always circular. Pupil can assume shape other than circular, like oval, only when a person is looking through the corner of his/her eye. However, a drowsy person cannot look through the corner of the eye.
3. The default position of the eye, when a person is about to fall asleep, is the centre of the eye. This provides an opportunity to test the eyes of the pilot for drowsiness when the pilots head is relatively steady, eye-balls are in the centre and variations of pupil diameter are high with relatively less blinks.

It has been known that the fluctuation of pupil diameter can be a parameter for detection of sleep on-set. However, the exact quantification of these parameters for the detection of sleep-onset reliably is not known. In this work, an algorithm is proposed with +5 % of the mean pupil diameter for alert subjects and in case of a Drowsy subject, the variations more than +5 % of the mean pupil diameter along with the duration and frequency of blinks. The criteria to detect drowsiness were tested with two IR video acquired from subjects on driving simulator. The downside is that the results have been proved just for two video one each of alert and drowsy subjects. The results need to be validated for a much wider group of subjects who are in different states of alertness/drowsiness. The subjects need to be kept in a controlled condition of sleepiness and then the videos need to be acquired and the same need to be tested with the algorithm.

The present study has relied heavily on just 2 videos, one each of alert and drowsy person, which were recorded under controlled conditions. There is a need to increase the sample size to at least 30 to get a much higher confidence in the set criteria. The system has already been designed with clearly defined criteria for alarm activation. It can be considered as a prototype which can be actually fitted on an instrument panel and tested for effectiveness.

Present study was conducted to prove the correctness of the criteria required to set off the alarm. It was conducted on two video clips. The actual deployment of this technique requires that the concept needs to be proved in real-time video streaming and not just on the video clips.

References

1. Levin A (2010) USA Today Updated 18 Dec 1:12 PM URL: http://usatoday30.usatoday.com/news/world/2010-11-18- airindia18_ST_N.htm?csp = 34news&asid = 2f98cfea
2. Ventilation during sleep onset by Colrain IM, Trinder J, Fraser G, Wilson GV (1987) Department of Psychology, University of Tasmania, Hobart, Tasmania 7000, Australia; The American Physiological Society, pp 2067–2074
3. Acute changes in cardiovascular function during the onset period of daytime sleep: comparison to lying awake and standing by Zaregarizi M, Edwards B, George K, Harrison Y, Jones H, Atkinson G (2007) Research institute for sport and exercise sciences. J Appl Physiol 1332–1338, Henry Cotton campus, Liverpool John Moores University Liverpool L3 2ET, UK
4. Electro-oculographic and performance indices of fatigue during simulated flight by Morris TL, Miller JC (1996) Biological psychology, pp 343–360
5. The pupil as a possible premonitor of drowsiness by Nishiyama J, Tanida K, Kusumi M, Hirata Y (2007) Conf Proc IEEE Eng Med Biol Soc 1586–1589
6. Variability of vigilance and ventilation: studies on the control of respiration during sleep Schäfer T (1998) Department of applied physiology, Ruhr-Universität Bochum, D-44780 Bochum, Germany. Respiration Physiology 114 (1998) 37–48
7. Monitoring eye and eyelid movements by infrared reflectance oculography to measure drowsiness in drivers by Johns MW, Tucker A, Chapman R, Crowley K, Michael N (2007) Somnologie, vol. 11, pp 234–242
8. Fatigue detection based on infrared video pupillography by Deng L, Xiong X, Zhou J, Gan P, Deng S (2010) Laboratory of forensic medicine and biomedical information Chongqing medical university Chongqing, China. In: Proceedings of the 4th international conference on bioinformatics and biomedical engineering (iCBBE), pp 1–4
9. Learning OpenCV by Bradski G, Kaehler A (2008) Copyright © 2008, published by O'Reilly Media, Inc., 1005 Gravenstein Highway North, Sebastopol, CA 95472
10. Analysis of blink rate patterns in normal subjects by Bentivoglio AR, Bressman SB, Cassetta E Carretta D, Tonali P, Albanese A (1997) Movement disorders society © 1997, pp. 1028–1034

Discussion About Goal Oriented Requirement Elicitation Process into V Model

Göknur Sirin, Bernard Yannou, Eric Coatanéa and Eric Landel

Abstract The major concern of past and present industry is to find how to make and supply the products for fulfilling customer needs with minimal cost and time. With this motivation, Renault recently decided, in order to improve efficiency and reduce costs, to re-design their distributed and heterogeneous thermal comfort simulation model's activities in order to have a complete simulation environment. The purpose of this work is to represent the preliminary steps in achieving this goal in building a change process model which aid in V model requirement elicitation phase. So, in this work, we propose an extended V model based on Goal Oriented Requirement Engineering (GORE) for complex system's requirement elicitation. Is the existing approach ineffective? What really are the research issues?

Keywords Change process model · Goal oriented requirement engineering · System engineering V model

G. Sirin (✉) · B. Yannou
Ecole Centrale Paris, Laboratoire Genie Industriel,
Grande Voie Des Vignes F-92-290 Chatenay-Malabry, France
e-mail: Goknur.sirin@ecp.fr

B. Yannou
e-mail: Bernard.yannou@ecp.fr

G. Sirin · E. Coatanéa
Department of Engineering Design and Production, School of Engineering,
Aalto University, PO Box 14100,Aalto 00076, Finland
e-mail: eric.coatanea@aalto.fi

G. Sirin · E. Landel
Renault SAS, Techno-Centre, 1 Avenue du Golf, Guyancourt, France
e-mail: eric.landel@renault.com

A. Chakrabarti and R. V. Prakash (eds.), *ICoRD'13*, Lecture Notes in Mechanical Engineering, DOI: 10.1007/978-81-322-1050-4_79, © Springer India 2013

1 Introduction

Engineering changes and lack of requirement elicitation are considered inevitable; especially for complex product producer such as vehicle and plane where customer satisfaction, safety and reliability are key drivers.

According to some research, lack of requirement elicitation and change process model definition could destroy a project's potential value. Furthermore, researches indicated that requirements elicitation or the lack thereof, is often times the chief suspect in the cause of project failures. In a study of complex system engineering projects, McManus and Wood-Harper found that one of the most important factors in the failure of these requirements was "the lack of due diligence at the requirements phase" [1].

Especially, in the 1980s, the V-Model has been used heavily in the automotive industry, and nowadays, V model is always applicable to variant industrial projects which provide a series of steps that make up and simplify the complex system projects. Within Renault Thermal Comfort project, we tend to use V model for project life cycle development. We started Renault Thermic Comfort Project by a standard Requirements analysis phase. In this phase, we collected and grouped all the requirements by analyzing the needs of the user(s). What we propose here is to elicit first the goals (business and operational objectives) before eliciting system requirements. Because good defined objectives, simplify the process for project planning. With a clear understanding of the business and operational objectives, you can then break the work down into the elemental tasks required to meet the system requirements. In addition, the change process model that we prose here by extending V model Requirement analysis phase is provide a well-structured levels in order to identify goals, activities and Information system tools [See Fig. 3]. Thus with this work, we propose an extended V model in where we elicit first of all the system's objectives. The reason to propose an extended V model is that the rigid nature of V model is ineffective in dealing with complex and dynamic situations in where this model is unable to handle change efficiently. So, present work aims at maximizing the combined effect of V model's requirement analysis phase and Enterprise Change Process model where we define first, business and operational goals [2, 3]. In such a way, this paper focuses on explicit goal representations in requirements models. Because goal modeling help to define the inter-dependencies among requirements.

The paper is organized into four sections beyond the introduction, in the second section; we reviewed briefly the literature on change process model integration to System Engineering V model's Requirement Elicitation considerations. In the third section, we introduce our discussion towards Renault thermal comfort simulation model case by introducing its business and operational objectives. Finally, in the last section, a set of conclusions and recommendations for future work are presented.

2 Quick Literature Review on Change Process Model Integration to System Engineering V Model's Requirement Elicitation Considerations

In every organization, regardless of industry or size, there are three organizational elements that both drive change and are affected by change which are Processes, Technology and People. Technology supports the processes designed to respond to changes in market conditions. Ultimately, however, it is the people who must leverage these processes and technology for the benefit of the organization. With the following explanations, we can briefly look at how each of these elements affect by organizational change:

Process: Business processes are defined by process maps, policies and procedures, and business rules that describe how work gets done. This drives the adoption of new technology.

Technology: Technology ensures greater organizational efficiency in implementing the changes. It is a means to process data with greater accuracy, dependability and speed.

People: Generally, organizations excel at designing new or improving existing processes. They also do well at identifying or developing technology to realize the power of new processes. However, most organizations fail to focus sufficient attention on the role people play in the processes and technology used to accomplish the desired organizational changes.

Change management is a cyclic process, as an organization will always encounter the need for change. There are three phases in the Organizational Change Management Life Cycle which are **Identify**, **Engage and Implement**. The elements of change (processes, technology and people) and the phases of the Organizational Change Management Life Cycle are closely linked, and their intersection points must be carefully considered. In this work, we focus only on change's identification phase [4–6].

Identify the Change Phase: In the Identify stage, someone within an organization-typically a senior executive spearheads an initiative to change a current process. This need is then presented to the organization with a general description of the current state of affairs, offset by a high-level vision of the desired future state. While it seems obvious, identifying the change is an absolutely fundamental first step in successful change adoption. It is important that the changed condition be described in a common, consistent language.

The first step to create a change process model is to have a global picture of current engineering system's functionality and structures; second we need to identify the requirements and reasons for change. As illustrated in Fig. 1, *contextual forces* represent external market requirements such as *enter the competition market, increased emphasis on quality, and efficiency of developing product* etc [4, 5]. The idea that we implement here is to integrate change process model to V-model requirement analysis' phase. Because, as we mentioned previously that the biggest disadvantage of V-model is that it's very rigid and the least flexible. If any

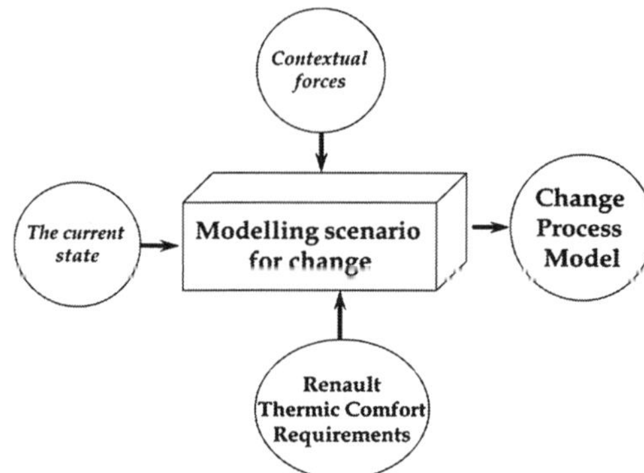

Fig. 1 Change process model

changes happen mid-way, not only the requirements documents but also the test documentation needs to be updated.

As illustrated in Fig. 2, we introduced goal oriented requirement engineering method and change process model to V-model requirement analysis phase. Integration of this change process model to system engineering V model's requirement analysis phase, provide us to share an understanding of how the current project functions. Also it provides to share a vision for whatever change is required, to develop scenario for implementing the change, to develop arguments for and against the various scenario and finally to keep a history of decisions made during the process.

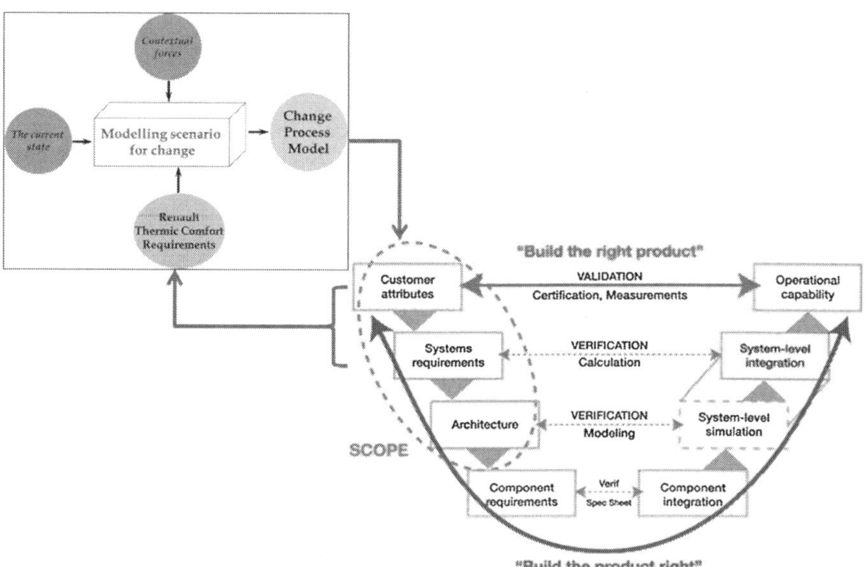

Fig. 2 Change process model implementation into V model

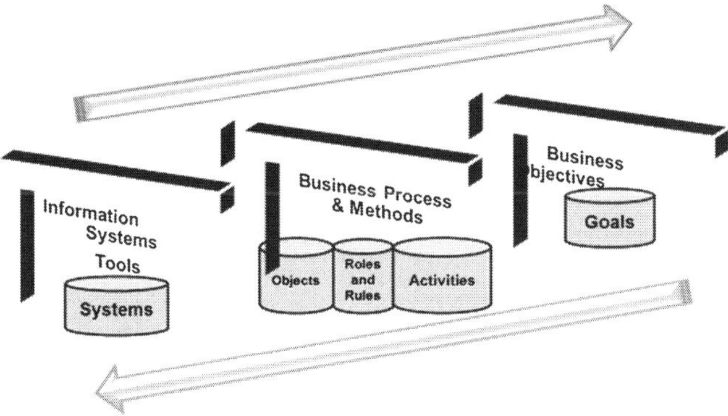

Fig. 3 Change process levels

2.1 Change Process Levels

As illustrated in Fig. 3, enterprises have to form a network which consists of three levels related to change processes. Strategic goals on business objectives level set the direction and purpose of the enterprise. The purpose of the goal model is to describe what the enterprise or department wants to achieve or to avoid. Roles correspond to sets of responsibilities and related activities. The actor/role model aims to describe how actors are related to each other and also to goals. The role/activity model is used to define enterprise processes, the way they consume/produce resources to achieve enterprise objectives. Business objects set the structure of the support systems and their behavior has an identifiable lifecycle. The object model is used to define the enterprise entities, attributes and relationships. Business rules are the way that enterprises function. The business rules model is used to define business rules consistent with the goals model [6]. In this work, we focus on Business Objectives, Goal elicitation phase.

2.2 Goal Identification: Goal Oriented Requirement Engineering

In today view, goals drive the requirements elicitation process and they express a result to reach, an objective to be fulfilled. Goal-oriented requirements engineering has many advantages, such as:

- object models and requirements can be derived systematically from goals;
- goals provide the rationale for requirements;
- a goal graph provides vertical traceability from high-level strategic concerns to low-level technical details;

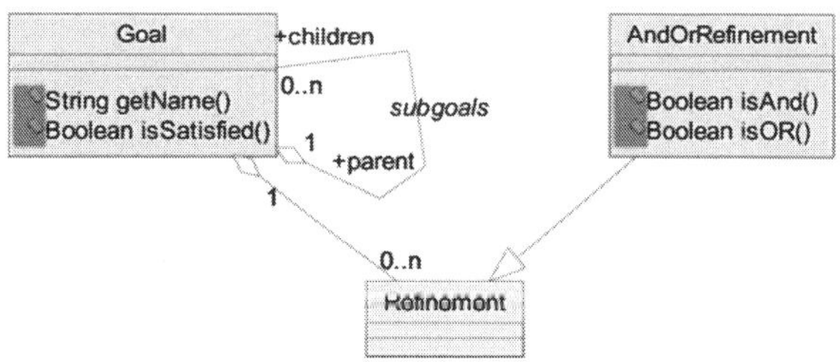

Fig. 4 Goal class diagram

- goal AND/OR graphs provide the right abstraction level at which decision makers can be involved for important decisions etc.... [7, 8].

Thus, the aim of the present work is to reformulate these needs for change by using Goal Oriented Requirement Engineering Method [4–6].

2.2.1 Modeling Goals

The reason of goal(s) modeling is to support heuristic, qualitative or formal reasoning during RE. Understanding the context in which the system will have to function, the rationale for the To-Be system, deriving the system requirements from the explicit representation of the organization's mission & objectives bridging organization objectives to system requirements. The goal model is a goal graph build with AND/OR links (See Fig. 4). Constructing OR links supports identify, evaluate, negotiate and reason on alternatives. AND links mean that the father goal is satisfied only if the children goals are satisfied. It helps in driving the process of goal operationalization [7, 8].

2.2.2 Formulating Goals

Goal could be defined in informal, semi-formal and formal way.

Generally, a goal consists of a verb, a target with a parameter.

Goal: Verb <Target> [<parameter>] [4–6].

Target: verb complement which can be Object or Result

Other parameters: source, beneficiary, destination, means, manner, place, time, quantity, quality, reference

Ex: Improve (Verb), thermal comfort simulation activities (Obj) for vehicle end-user (Ben) by providing a complete simulation model (Man)

In the next section, we implemented our theoretical knowledge to Renault Thermal Comfort case study.

3 Discussion Towards Renault Thermal Comfort Simulation Model Case

We implemented Goal Oriented Requirement Engineering Methods and Change Management Process to Renault's thermic comfort distributed simulation activities.

3.1 Change Reasons

Varying HVAC (Heating, Ventilation and Air-Conditioning) operational modes requires multiple physical testing, which is often not feasible. Thus in the recent past, the only way to properly measure, analyze, and thus manage these thermal issues was to use physical prototypes in climatic wind tunnels—an extremely complex and time consuming process. Renault's distributed thermal comfort simulation model consists of hot loop thermo-hydraulic model, air passenger circuit model, and cold loop model and outside air circuit model (See Fig. 5). Each simulation model has its own set of models, each with its own modeling conventions, nature(0-1D, 2-3D, simulator)and maturity (Draft, validated, obsolete). Various simulation solutions such as Power FLOW, Amesim and Fluent are used in order to reduce the team's usage of physical prototypes by 30–50 %.

On the other hand, heterogeneity and diversity of the simulation model increase the system complexity. The current Renault thermal comfort's problem arises also from multidisciplinary stakeholders, model repetition, silos engineering process,

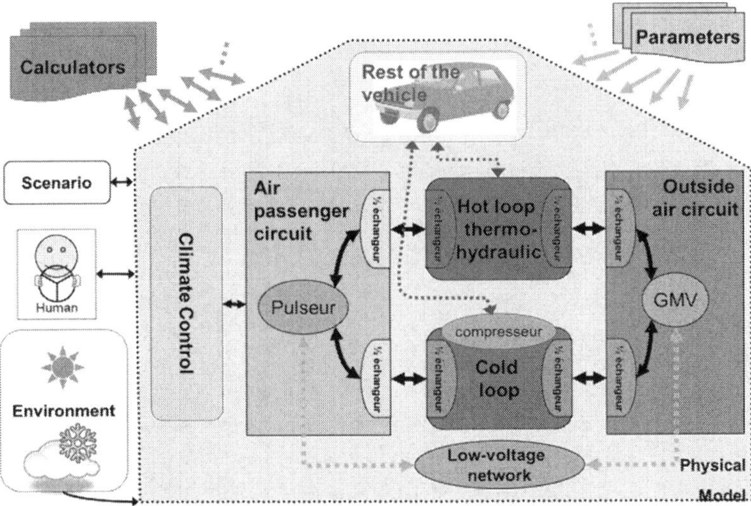

Fig. 5 Renault thermic comfort activities

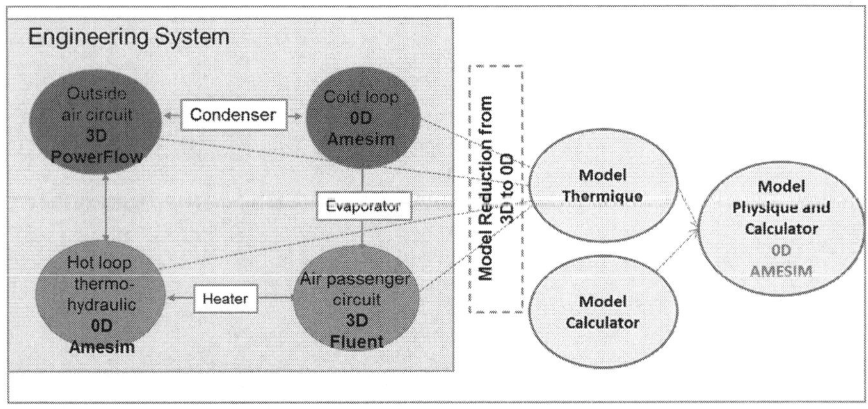

Fig. 6 Coupling of thermal comfort models/entities into a complete model

data and model exchange, etc. The main aim of the change is to reduce the model from 3D to 0D for time saving and to aggregate distributed simulation models in order to have a complete, robust thermic comfort model (See Fig. 6).

The aim of the present work is to reformulate these needs for change by using Goal Oriented Requirement Engineering method. By taking into account the technical process that we have already mentioned in the previous parts, we reformulate the needs for change in the following picture.

Example of requirement elicitation:

- R3: Check Model Consistency
  - R31: Analysis of dependency network
  - R32: Determine decomposition for concurrent engineering
  - R33: Determine workflow
  - R34: …

As illustrated in Fig. 7, goal is divided into two parts which are operational and strategically. AND links mean that the father goal is satisfied only if the children goals are satisfied. It helps in driving the process of goal operationalization. Goal elicitation is on the top level of change process level (See Fig. 3) which called business objectives. This top level determination helps to identify the system's objectives, rules, activities and process because, process is set of activities, each of them be related the role played by an actor, triggered by an event, to produce a result in order to fulfill a goal [4, 5].

We need to manage change, from a current situation As-Is to a future situation To-Be. Change creates a movement from an existing situation captured in an As-Is model to new situation captured in a To-Be model. (See Fig. 8) According to C. Rolland, there are seven types of change goals which are Adapt, Introduce, Cease, Replace, Maintain, Extend and Improve.

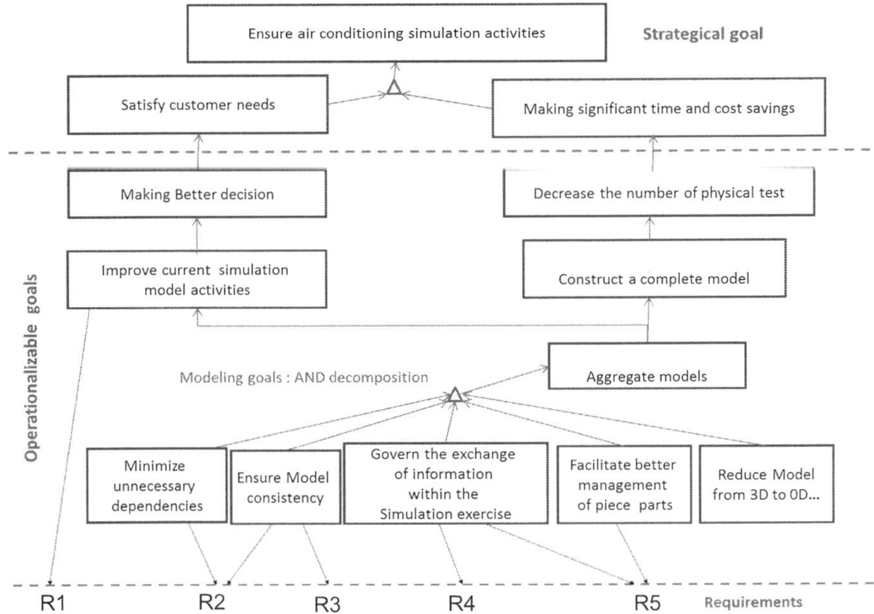

Fig. 7 Goal elicitations for thermal comfort case/as-is

Fig. 8 Goal types to-be

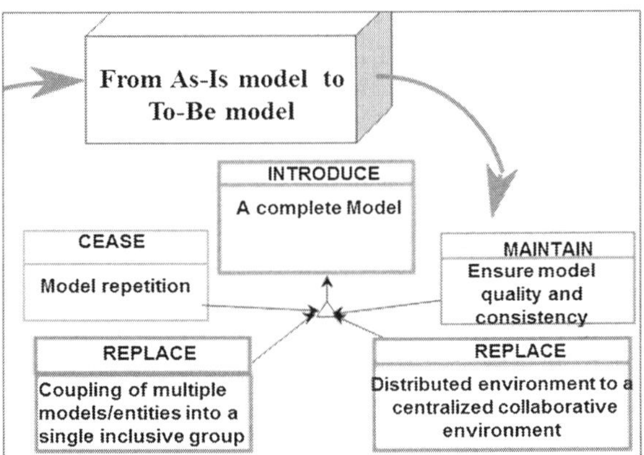

4 Conclusions and Recommendations for Future Work

Thus Renault Thermal Comfort distributed and non-homogenous simulation models were aimed to coupling in order to have a complete developed simulation model for thermal comfort activities. The aim of this work is to re-organize the needs for change by using goal oriented requirement engineering methods. We introduce goal oriented requirement engineering method and change process model to V model requirement analysis phase (See Fig. 2). Integration of this change process model to system engineering V model's requirement analysis phase, help to understand, how the current project functions, to share a vision for whatever change is required, to develop scenario for implementing the change, to develop arguments for and against the various scenario and finally to keep a history of decisions made during the process. Also, the principal aim of this implementation is to minimize inflexibility structure of V model in order to respond to change. Present work does not represent the entire work. This work represents the top level of change process level which is business objectives (goal elicitation).

On the other hand we have some recommendations for future work such as the advantages, needs and obstacles of proposed method for an industrial project.

References

1. McManus J, Wood-Harper T (2010) http://www.bcs.org/content/ConWebDoc/19584
2. Katina PF, Jaradat RM (2011) A three-phase framework for elicitation of infrastructure requirements. Members of IEEE, the 4th annual international conference on next generation infrastructures, Virginia Beach, Virginia, 16–18 Nov 2011
3. van Lamsweerde A () Goal-oriented requirements engineering: A Guided Tour, Département d'Ingénierie Informatique, Université catholique de Louvain, B-1348 Louvain-la-Neuve (Belgium), avl@info.ucl.ac.be
4. Nurcan S, Rolland C (2003) A multi-method for defining the organizational change. Elsevier: J Inf Softw Technol 45(2):61–82
5. Rolland C, Nurcan S, Grosz G (1999) Enterprise knowledge development: the process view. Inf Manage J (Elsevier) 36(3):165–184
6. Nurcan S, Rolland C (1999) Using EKD-CMM electronic guide book for managing change in organisations. In: Proceedings of the 9th European-Japanese conference on information modelling and knowledge bases, Iwate, Japan, 24–28 May 1999, pp 105–123
7. Dardenne A, van Lamsweerde A, Fickas S (1993) Goal directed requirements acquisition. Sci Comput Program 20:3–50
8. Mylopoulos J, Chung L, Yu E (1999) From object-oriented to goal-oriented. Communications of the ACM, vol 42(1), Jan 1999
9. Sirin G et al (2012) Analysis of the simulation system in an automotive development project. Accepted paper to CSDM2012
10. AUTOSAR Seminar WS2008/2009. http://www.advanced-planning.de/advancedplanning-128.htm
11. INCOSE (2006) Systems engineering handbook—a guide for system life cycle processes and activities, version 3. Cecilia Haskins

Prediction of Shock Load due to Stopper Hitting During Steering in an Articulated Earth Moving Equipment

A. Gomathinayagam, B. Raghuvarman, S. Babu and K. Mohamed Rasik Habeeb

Abstract Design of structural members to overcome the failure due to shock loads is a challenging process. In this work, an effort is made to analytically predict the shock load due to hitting of the front chassis stopper on the rear chassis stopper during steering operation in an articulated earth moving equipment. The value of the shock load is obtained by equating the moment due to steering cylinders to the moment due to stopper hitting force. In order to verify the methodology, a prototype plate similar to the stopper was made and strain gauge rosette was pasted to it. Experimentation was done by applying the load to the prototype plate and the strain values are recorded for different load values in a laboratory set up condition. Strain gauging was also carried out by mounting the same prototype plate in the actual field machine. The value of shock load was obtained by calibrating the strain values obtained from the field against that of the laboratory experimental values. It is found that a very good correlation exists between the analytical and experimental values of the stopper load.

Keywords Shock load · Articulation · Earth moving equipment · Steering

A. Gomathinayagam (✉) · B. Raghuvarman · S. Babu · K. Mohamed Rasik Habeeb
Product Development Center, Larsen & Toubro Ltd, Coimbatore, India
e-mail: agnayak@yahoo.com

B. Raghuvarman
e-mail: raghuvarman.b@larsentoubro.com

S. Babu
e-mail: babu.s@larsentoubro.com

K. Mohamed Rasik Habeeb
e-mail: mohamedrasik.habeeb@larsentoubro.com

A. Chakrabarti and R. V. Prakash (eds.), *ICoRD'13*, Lecture Notes in Mechanical Engineering, DOI: 10.1007/978-81-322-1050-4_80, © Springer India 2013

1 Introduction

Earth moving equipment are primarily used for tasks in earthwork, compaction and re-handling. The structures of them are subjected to different kinds of loads like static, dynamic and impact. Designing the structures to overcome failures due to theses loads is a challenging task. The prediction of impact or shock loads is cumbersome than that of static or dynamic loads.

The majority of the earth moving equipment are articulated vehicles. During steering operation the front and rear portions of the structures about the articulation joint come and hit each other causing the shock loads. The stopper plates are generally provided on these equipment to limit the steering angle as well as to avoid these shock loads being transferred to the steering cylinders. Figure 1 shows a few typical failures in the structures due to shock loads during steering stoppers hitting. One of the reasons of these failures is difficulty in prediction of shock loads.

Literatures [1–3] are available mostly for study of operator comfort due to shock loads transferred to the structures in earth moving equipment. A very few literatures like [4] to find the dynamic digging load are available for predicting the shock loads. The individual industry shall use its own methodology based on analytical, simulation or experimental to estimate the shock loads. Many industries use simulation techniques [4–7] like finite element analysis (FEA) and multi-body dynamics (MBD) followed by experimental verification to predict the dynamic and shock loads.

In this paper, an analytical method is developed to predict the shock load due to hitting of the front chassis stopper on the rear chassis stopper during steering operation in an articulated earth moving equipment. The developed methodology is validated experimentally by strain gauging testing at the field.

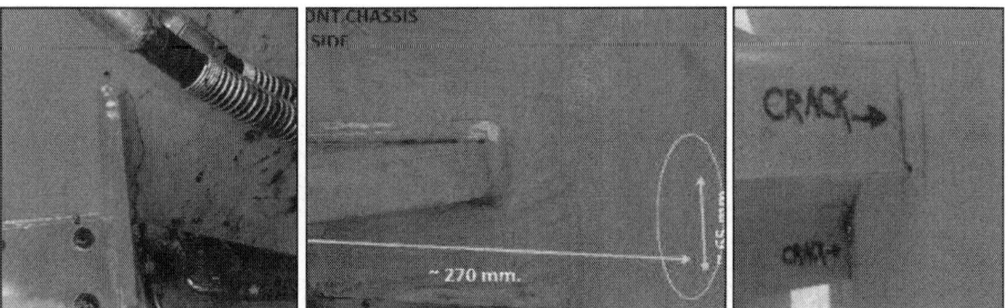

Fig. 1 Failure due to shock loads in the structures

2 Analytical Method

The model of an articulated earth moving vehicle in straight and steered conditions is shown in Fig. 2. It is generally assumed that the front or rear structure with less reaction starts to steer initially. Then both structure together moves and hit each other. The value of the hitting load is obtained by equating the moment due to the steering cylinders to the moment due to stopper hitting force.

During steering one cylinder extends and another cylinder retracts. The force acts on the bore side and the rod side in the extension and retraction cylinders respectively. The operating parameters are shown in Fig. 3 and Table 1.

The hitting force is derived using the equations given below:

$$\text{Area on the bore side of the cylinder,} \quad A_b = \pi * d_b^2/4 \tag{1}$$

$$\text{Area on the rod side of the cylinder,} \quad A_r = \pi * \left(d_b^2 - d_b^2\right)/4 \tag{2}$$

$$\text{Force acting on the extension cylinder,} \quad F_1 = p * A_b \tag{3}$$

$$\text{Force acting on the retraction cylinder,} \quad F_2 = p * A_r \tag{4}$$

$$\text{Moment due to cylinder forces,} \quad M_c = F_1 * d_1 + F_2 * d_2 \tag{5}$$

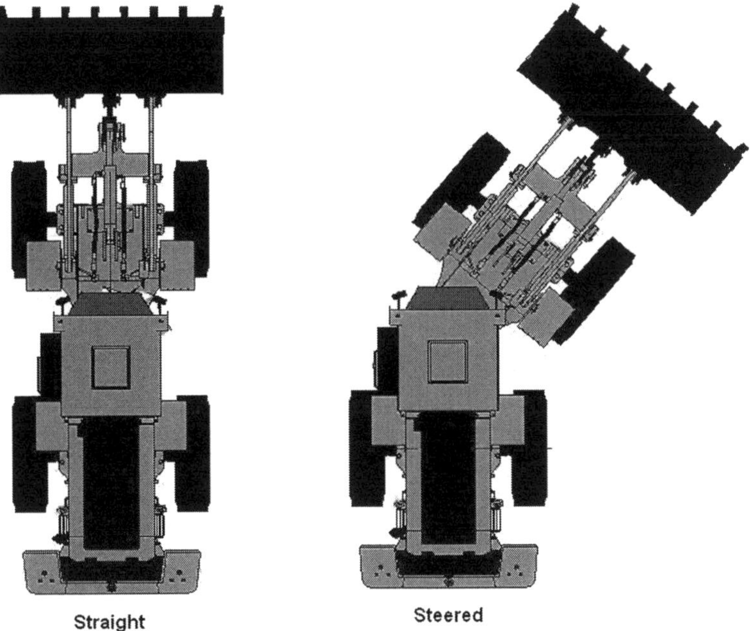

Fig. 2 Articulated earth moving equipment in straight and steered conditions

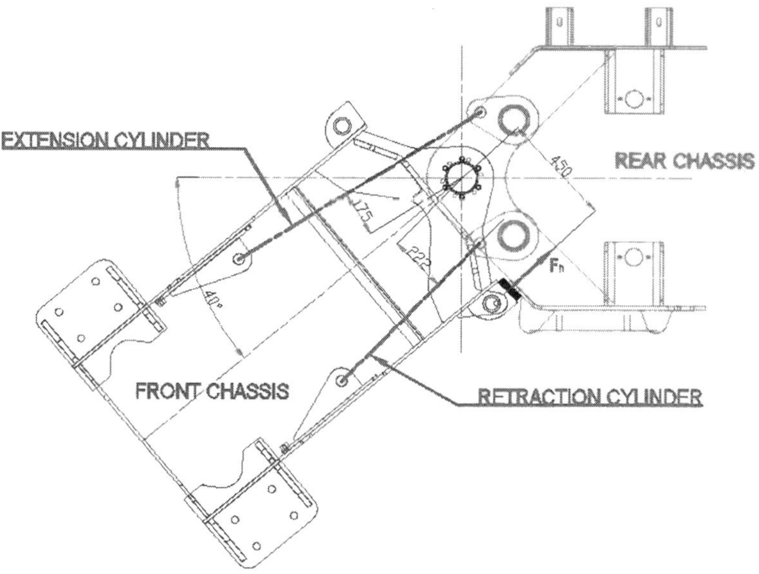

Fig. 3 Details @ steered condition

Table 1 Operating parameters

Parameter	Value	Unit
Articulation angle, α	±40	deg
Bore diameter of steering cylinder, d_b	75	mm
Rod diameter steering cylinder, d_r	45	mm
Perpendicular distance between extension cylinder and articulation joint, d_1	175	mm
Perpendicular distance between retraction cylinder and articulation joint, d_2	222	mm
Perpendicular distance between stopper pad and articulation joint, d	450	mm
Maximum pressure in the steering cylinder at the time of hitting, p	165	bar

$$\text{Moment due to stopper hitting force} \quad M_s = F_s * d \tag{6}$$

$$\text{By equating } M_c \,\&\, M_s\text{, the Stopper hitting force,} \quad F_s = M_c / d \tag{7}$$

Hence the stopper hitting force developed a steering cylinder pressure of 165 bar using the above equations is 5,240 kgf.

3 Experimental Work

In order to verify the analytical approach, experimentation was carried out. A prototype plate similar to the stopper was made and strain gauge rosette was pasted to it as shown in Fig. 4. The trials were conducted by applying the load to the prototype plate in an UTM. The strain values were recorded for different load values in a laboratory set up condition. The load vs strain values for the loading and unloading is shown in Fig. 5.

The prototype plate was fixed to the rear chassis stopper plate in an actual machine working in the field. Strain gauging was carried out in the field machine and the parameters measured at the time of stopper hitting are given in Table 2. It was found that the pressure developed in the steering cylinders was 150 bar which was less than the maximum pressure of 165 bar. The following are the reasons for the less pressure in the cylinders:

- Since the prototype plate was mounted in addition to the rear chassis stopper plate, the full articulation angle of 40° was not achieved. The maximum pressure of 165 bar was achievable only at 40° articulation angle.
- The theoretical maximum pressure of 165 bar was derived based on certain assumed value of coefficient of friction between the tire and the ground. But the coefficient of friction at the tested field might be different than that the assumed value.

The value of shock load was obtained by calibrating the measured strain value against that of the laboratory experimental values. Also the hitting load from analytical approach was found out for the steering cylinder pressure of 150 bar (4,761 kgf). It is found that a very good correlation exists between the analytical and experimental values (Fig. 6). Hence the developed analytical approach shall be used to predict the hitting load in a similar type of applications so that the structures can overcome failures due to shock loads.

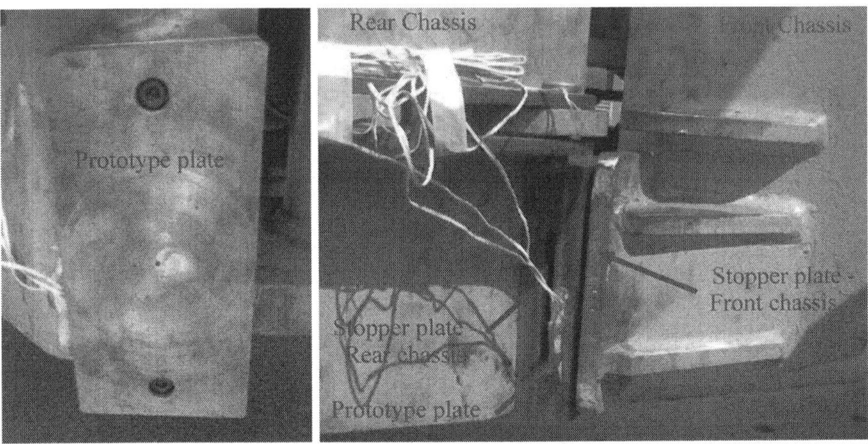

Fig. 4 Experimental setup

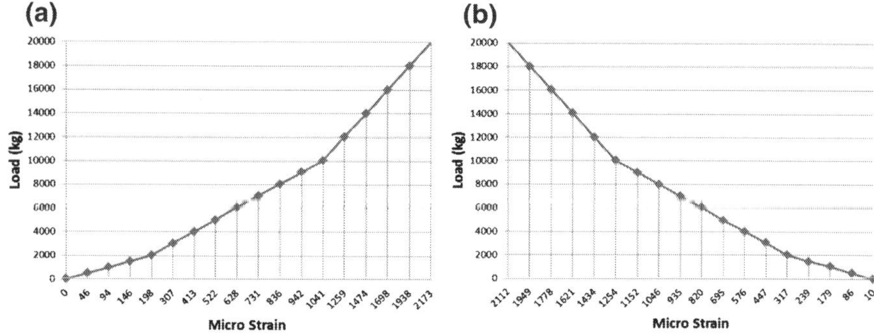

Fig. 5 Load vs strain during loading and unloading

Table 2 Parameters at the time of stopper hitting

Parameter	Value	Unit
Pressure in the steering cylinders	150	bar
Value of micro strain pasted in the plate	477	–
Load value corresponding to Micro strain from the graph (Fig. 5a)	4,591	kgf

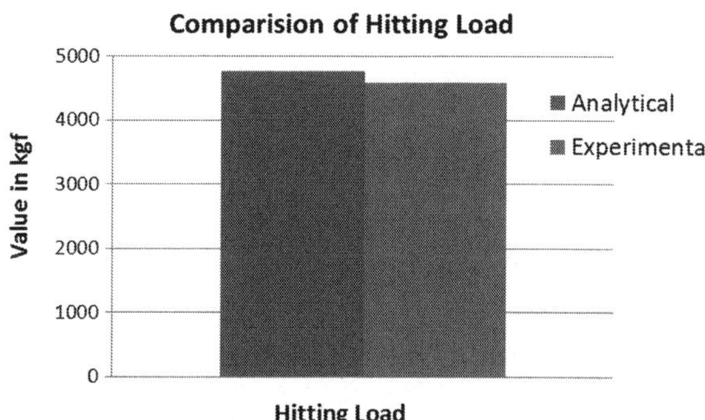

Fig. 6 Comparison of stopper hitting load values

4 Conclusions

The analytical methodology is developed to predict the shock load due to hitting of the front chassis stopper on the rear chassis stopper during steering operation in an articulated earth moving equipment. The analytical approach is verified using experiments conducted in the laboratory and field. It is found that a very good correlation exists between the analytical and experimental values of the stopper load. Hence the developed methodology shall be used to predict the shock loads in a similar kind of applications in order to design the structure to overcome failures.

Acknowledgments Authors wish to thank the management of Larsen & Toubro for the use of their articulated earth moving vehicle data and granting permission to publish this article. They also thank their colleagues for their assistance in carrying out the experimental work.

References

1. Martin PH, Tammy RE, Grenier SG (2010) Whole-body vibration experienced by haulage truck operators in surface mining operations: a comparison of various analysis methods utilized in the prediction of health risks. Appl Ergon 41:763–770
2. Milosavljevic S, Bergman F, Rehn B, Carman AB (2010) All-terrain vehicle use in agriculture: exposure to while body vibration and mechanical shock. Appl Ergon 41:530–535
3. Hostens I, Ramon H (2003) Descriptive analysis of combine cabin vibrations and their effect on the human body. J Sound Vib 266:453–464
4. Ericsson A, Slattengren J (2000) A model for predicting digging forces when working in gravel or other granulated material. 15th ADAMS European users conference
5. Sadak P (2000) Test of impact effect on an airplane's LANCAIR landing gear shock absorber. 15th ADAMS European users conference
6. Andersson K, Sellgren U (2005) Reality-driven virtual wheel loader operation. Proceedings of virtual concept, vol 45
7. Rehnberg A (2008) Vehicle dynamic analysis of wheel loaders. Thesis, Royal Institute of Technology (KTH)

A Simple Portable Cable Way for Agricultural Resource Collection

Shankar Krishnapillai and T. N. Sivasubramanian

Abstract A significant problem for Indian farmers today is the difficulty of obtaining labor for farm operations. The most labor intensive operation is the post-harvest resource collection. A simple, economical, compact cable way system has been designed, developed and fabricated to haul sugarcane loads from the field. The cable way is made up of simple collapsible steel 'A' posts, which can be assembled on-site from ready-made frames. The sugarcane loads are slung on trolleys which move on a steel cable passing over the top of the frames, the cable being tightened with a chain pulley block. The trolleys are pulled along by a recirculating rope operated by a winch. The empty trolley is pulled along by the same rope along a cable way running near the bottom of the 'A' post. Several trolleys can move simultaneously along the cable way. With a 2 HP motor the trolley speed was noticed to be about 0.3 m/s and the overall performance of the prototype was satisfactory. The entire setup is fabricated from easy available and low cost components.

Keywords Material handling · Cable way · Rural technology · Socially relevant

S. Krishnapillai (✉)
Department of Mechanical Engineering, IIT Madras, Chennai 600036, India
e-mail: skris@iitm.ac.in

T. N. Sivasubramanian
Pothu Vivasayee Sangam, Thottakurichi, Tamilnadu 639113, India

A. Chakrabarti and R. V. Prakash (eds.), *ICoRD'13*, Lecture Notes in Mechanical Engineering, DOI: 10.1007/978-81-322-1050-4_81, © Springer India 2013

1 Introduction

There various problems faced by the Indian farmer in the 21st century, chief among them is a critical shortage of manpower for all sorts of farming activities [1]. The typical Indian farm is small and restricted to a few acres; hence large scale mechanization as in the West is not very feasible [2] and the labor crisis is severe. Lack of insufficient manpower is especially severe in the post-harvest period, when significant manpower is required to transport agricultural produce from the field to the collection point outside. Thus in recent times there are severe labor problems in the collection of agricultural resources which discourages the farming profession in the younger generations [3]. The aim of the cable way is to provide an economical and simple solution to this problem.

Cableway (or ropeway) systems are used for material handling or transportation of passengers. Cableways are particularly useful in rugged terrain and in environmentally sensitive areas. The cableway is possibly the most efficient of all modes of transport. The cableway technology is quite basic and it has many advantages, which are presented by Dwyer [4]. First, it can cover steep, rugged and otherwise inaccessible ground. Being capable of moving goods in a straight line, it can easily reduce road haul distances by a sizeable factor. Another advantage of the cableway is its 'non-intrusive' nature i.e., minimal environmental impact. This is emphasized in the environmental assessment report of ropeway across Guwahati river in Assam State, India by the Pollution Control Board of Assam [5]. The area requirements of cableway are minimum with a small cleared pathway under the posts, which are widely separated. These environmental concerns are highly important in Pugalur where some sugarcane fields are wetlands criss-crossed with canals and cable way is the only method to access the fields with minimum environmental changes. The third advantage of the ropeway is the moderate energy consumption and minimum labor requirements as pointed out by Wuschek [6] in his study of energy saving potential of cableways. Various cableway systems are discussed by Edward et.al. [7] which include mono-cableway and bi-cable way and their many variations. The bi-cable way system is adopted here, where the load moves along on a cable on rollers and is hauled by a moving rope clamped to the load. The current cable way system was implemented with an NGO for the benefit of sugarcane farmers in the Karur District in Tamilnadu State, India.

2 Objective and Challenges

After interactions with the farmers, it was decided that the cableway system and the steel posts must be portable and economical to fabricate and operate. Farmers do not opt for a permanent cable way since such facilities are needed for only about 10–15 days in a year. The main components are (1) steel posts to carry the cableway (2) winch and motor unit and (3) chain pulley block and (4) steel cables and (5) recirculating rope. Several farmers could share the portable cable way. Given the budgetary constraints of the project, it was decided the prototype cableway could be fabricated and tested for 100 m. This distance could be readily extended with more posts. It should be able to continuously haul several head loads of sugarcane (each 25 kg) over this distance at a moderate speed; the speed could be readily increased with a more powerful winch. The components must be cheap and economically purchased and fabricated.

3 Description of the Portable Cable Way

Figure 1 shows the photograph of the portable cable way in operation. It shows a typical cableway post made of easily available tubular steel pipes. The pipes are bent into two trapezium frames, which are hinged at the top and opened at the bottom and held thus with tie rods to form a sturdy ‘A’ shaped post when viewed transverse to the cable way (this assembly is shown in Fig. 2). Also shown is a trolley (with sugarcane load in sling) fitted with two nylon wheels moving along the top cable, and the recirculating rope which pulls it.

More technical details are shown in the next schematic diagram (Fig. 3). There are many ‘A’ posts which are assembled on-site from the portable frames. There

Fig. 1 Photograph of portable cable way

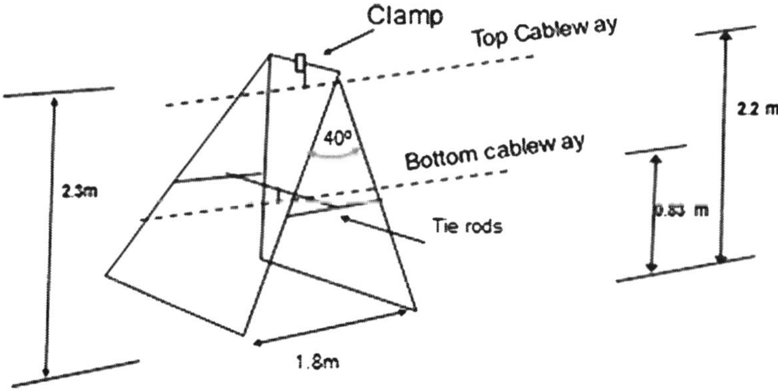

Fig. 2 Assembly of a cable way post

are two end posts which are especially sturdy, and a number of weaker intermediate posts (20 were used in the project). The 'A' post opens at an angle of about 40° (see Fig. 2) and is 2.3 m high with a base of 1.8 m (measured parallel to the cableway) and 1.78 m (transverse to cable way) forming a sturdy structure resistant to compressive and sidewise forces. The frames are held in place with tie rods (shown in Fig. 2). These tie rods also support the bottom cable. Usually there is a spacing of about 5–7 m between two posts.

A 12 mm steel cable passes over the top of the posts. Both ends of this cable are buried 1.5 m in the ground and tightened with a 2 tonne chain pulley block. A recirculating rope provides the motive power to pull the loaded trolleys along the top cable and the unloaded trolleys along the bottom cableway (in opposite directions). This rope is wound round a winch drum driven by a 2 HP motor; the loads moving at a walking speed. The upper cable is tightened using the chain pulley block. A turnbuckle suffices for the lower cable way which carries lighter

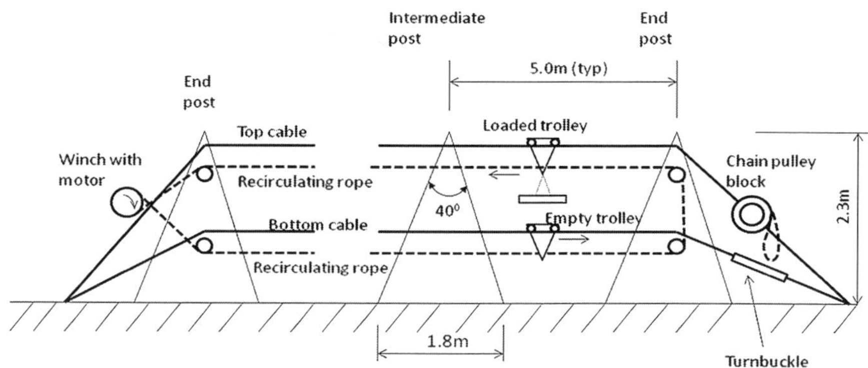

Fig. 3 Schematic diagram of cable way

Fig. 4 Chain pulley block to tighten cable

loads (it is also possible to connect both cables to the chain pulley block, with a slight modification).

The cable tension when calculated for a single concentrated load of 50 kg at mid-span for a permissible sag [8] of 3 % of span (span being 7 m) is 500 kgf. The downward compressive force on the end posts was 500 kgf and in the intermediate posts was considerably smaller. Hence the end posts have large tubular cross sections (38 mm OD, 2 mm thick) and intermediate posts have smaller cross sections(32 and 2 mm). The diameter of the cable was finalized as 12 mm (with about 7 tonne breaking capacity) because it has to be sufficiently broad for the trolley wheels to roll on.

Figure 4 shows a photograph of the chain pulley block, usually used for hoisting loads, now adapted to tightening the cable. It is capable of considerable mechanical advantage in lifting heavy objects. The 'hook' end of the device, which is normally connected to the load to be hoisted is now attached to the end of the cable coming from an end post. The other end of the chain is wrapped on a concrete block and buried about 1.5 m under the ground. By pulling the side loop chain, the operator tightens the cable with ease to the desired tension. Thus by using a simple chain pulley block, tightening is accomplished without recourse to more expensive hydraulic and pneumatic devices usually seen in cableways.

Figure 5 shows the photograph of the winch with 2 HP motor with a rated speed of 1,500 rpm. A standard reduction gear box is used to connect the motor to the winch drum. The drum is about 0.34 m in diameter with a corrugated surface for better grip. An 8 mm diameter nylon rope made endless is wrapped around on the drum. This rope (see Fig. 3) pulls both the top and bottom trolleys in opposite directions.

Among the other components of less significance is the trolley in the photograph (Fig. 6) which is made of a triangular steel plate (with about 0.18 m sides).

Fig. 5 Winch to pull the load

Two nylon rollers attached to this plate roll over the 12 mm diameter cable. The trolley can be clamped or unclamped to the recirculating rope. There is a simple sling to put the sugarcane load. Once the loaded trolley reaches the destination, it is unclamped from the recirculating rope. It is lifted off and put on the lower cable way and again clamped to the rope. When the cable way is in operation, up to five loaded trolleys were simultaneously hauled before deterioration of speed and performance took place.

4 Economy and Labor

A brief cost comparison is attempted here between the traditional head-load method and the cable way. The total cost of cableway development and fabrication (including 20 posts and one 2 kVA generator) was Indian Rupees 2,50,000. This device can be shared by many farmers. In the absence of a cable way, 10 headload workers per day are usually engaged in small farms, at a labor cost of Rs 250 per head per day. Ten head-load workers carrying about 25 kg head-load of cane per person, do not exceed 5 t of load transferred per day (it is variable, depending on terrain, weather etc.). The cable way operating at a modest speed of 0.3 m/s, can take a fresh load every 30 s (i.e., the time taken for the previous load to cover the initial span of 5 m). Thus a cable way working 8 h a day can transfer 24 t per day, with only 3 operators required to operate the cable way. However, it may be required to engage four workers for half a day to install the cable way, and also incur a transportation charge of about Rs 500 to carry the posts to the field. Thus it is seen that the cable way transportation rate is about 5 times that of manual transfer, and only 3 workers are required per day as opposed to 10.

Fig. 6 Photograph of trolley with clamp and sling

It may be noted in the above analysis that the number of workmen and time required for transportation increase with obstacles such as small trenches, furrows, water ways, irregular surface and also the weather, and more workers may be required in such cases. With the portable cable way such obstacles do not matter once the cable way is erected and in continuous operation. In a cable way of the current design only 3 workers are required to continuously operate the cableway irrespective of its length and the effort required of them is less demanding than head-load operations.

5 Conclusions

A simple, economical, compact portable cableway has been developed, fabricated and tested for transportation of sugarcane loads from farm to collection point. The cable is tightened by a chain pulley block. Using the concept of recirculating rope, and by means of upper and lower cable ways contained in the post, there is provision for two way travel of trolleys (loaded and unloaded). Within the budget constraints of the project, the prototype cableway has been tested for a distance of 100 m and works well. There are significant economies in labor charges by using this device.

Acknowledgments The authors thank the IIT Madras Alumni (Class of '81) for funding this project.

References

1. Mahesh R (2004) Labour mobility and paradox of rural unemployment—farm labour shortage: a micro level study. Indian J Labour Econ 47(1):25–37
2. Majumdar A, Karmakar S, Gupta JP (2004) Status of farm mechanization in Bihar, India. J Interacademicia 4(4):21–30
3. Sharma A (2007) The changing agricultural demography of India: evidences from a rural youth perception survey. Int J Rural Manag 3(1):27–41
4. Dwyer CF (1975) Aerial tramways, ski lifts, and tows. Forest Service, U.S. Department of Agriculture
5. Wuschek M (1982) Ropeways-economical and energy-saving. Int Seilbahn Rundschau J 3:118–222
6. State Pollution Control Board (2009) Environmental impact assessment report. Guwahati Ropeway Project, Assam
7. Neumann ES, Bonasso S, Dede ADI (1985) Modern material ropeway capabilities and characteristics. J Transp Eng 111:651–653
8. Brockenbrough RL, Merritt FS (1999) Structural steel designers handbook, 3rd edn. McGraw Hill, New York

Bio Inspired Motion Dynamics: Designing Efficient Mission Adaptive Aero Structures

Tony Thomas

Abstract Birds gave humans the idea of flight. Detailed study of bird flight reveals the complexity of nature's engineering. Birds are highly adaptive in flight. Blending man made technologies with nature's natural engineering designs help in improving efficiency and operational flexibility of flying machines. In-flight alterations corresponding to flight characteristics are crucial for better operational efficiency and flexibility. The paper discusses the research and design methodology used to develop small Nature (Bio) Inspired Mission Adaptive Aero Structure models. Conventional aircraft frame and structures are re-engineered to seamlessly respond to instantaneous flight requirements and adapt suitably. The two primary objectives of this study are to design aircraft structures for

- Time critical mission scenarios (military/counter terrorism/law and order)
- Efficient and comfort flying (civil aviation/commercial aircrafts).

Keywords Aircraft dynamics · Nature engineering · Flexible airframes · Adaptive flying structures · Next generation aircraft]

1 Introduction

The earth has been a living planet for nearly four billion years. During this time nature had to solve many problems of life. Later humans became part of this breath taking diversity. We humans with our large brains set us apart from the natural

T. Thomas (✉)
Mechanical Engineering Division, I.E.I Kochi Centre, Homeo Hospital Road, Near Pullepady ROB, Kochi, Kerala 682 017, India
e-mail: tony-thomas@in.com

A. Chakrabarti and R. V. Prakash (eds.), *ICoRD'13*, Lecture Notes in Mechanical Engineering, DOI: 10.1007/978-81-322-1050-4_82, © Springer India 2013

world to an artificial one by using man made technologies to solve the various problems of life. This alienation has made us successful to a great extent, but at a greater cost to our planet Earth. Now it is high time to start looking at the nature with new eyes. At these starting years of the third millennium, we stand in the brink of a new revolution; a world where technology and nature stay hand in hand. There are answers already in nature, and we need nature and technology to work together to solve problems efficiently.

1.1 Flight in Nature

In nature, the creatures need to move long distances as economically as possible. It is clearly understood that no creature that wastes energy will survive the unforgiving hand of natural selection. It was birds that first gave humans the idea of flying. Dreaming of flight and actually taking to the air are two very different and complex things. Around hundred years ago this dream finally took off. Well now the basics of flight are known to many of us, but birds are actually more complicated. An example; a flapping wing produce forward thrust; in short the birds' wings are both 'engine' and 'wings'.

A simple example—House flies

It might hold the secrets to new spy vehicles or aerial search and rescue equipment. Normally it is considered as a germ carrying nuisance, but it is really a marvelous flying machine. It is extremely complex and difficult to engineer and recreate one. Understanding the insect flight in high detail will help us in building smaller and smarter flying machines.

Understanding the basic principles behind nature's designs is the key to successful 'Bio-Inspired' thinking. Natural systems rarely break down. This is because nature goes in for a lot of redundancy i.e. there is always a backup system.

2 Mission Adaptive Aero Structures

Mission Adaptive Aero Structures (MAAS) refer to technologies introduced to the aircraft frame and structure enabling them to change or adapt themselves (or with manual assistance) to instantaneous flight conditions or operational requirements. In short these are flying machines that can morph seamlessly in-flight.

2.1 MAAS: The Civil Aviation Perspective

Albatross, a bird with the largest wingspan on earth and possess very high thermal efficiency; can remain airborne for months at a time. It is one of the most perfect flying machine ever created. When coming to aircraft flight, there is something that pilots and aviators refer as 'the real joy of flying'. It is not easy to recreate this 'flying like a bird' experience in aircrafts. MAAS technology in commercial aviation perspective focuses on 'designing an artificial albatross', a humble approach to be part of a new era in civil aviation with next generation aircrafts capable of in-flight morphing for comfort and efficient air travel.

2.2 *MAAS: The Military Perspective*

In a multi threat global environment, there is an unprecedented need for operational flexibility and responsiveness. In future battle fields wide variety of aerial vehicles deployed in various combinations are required. But huge costs involved hinder such diversity. The solution is a single aircraft capable of morphing instantaneously to multiple operational requirements.

3 The MAAS Technology

Conventional aircraft wing frames and structures are re-engineered to obtain flexible but strong sections that can undergo profile changes in-flight. These profile alterations change the aerodynamic properties of the whole aircraft depending on the situational/mission requirements. Since these shape changes are on a large scale (depending on aircraft size), developing and controlling them are complex tasks. Numerous modifications and the use of many advanced materials are required to develop it. Carbon fiber wing skin, network of integrated sensors, flexible strong airframe and micro actuators are some of the essential technologies required. The network of sensors detects the force, loading, temperatures and other parameters at different regions/sections on the wings and structural frame. This data is combined with the flight path and profile requirements, so that alterations can be made to the wing structure to change the aircraft's aerodynamic property instantaneously in flight. The most important design concern is to imply and maintain large seamless shape change on the aircraft's wing.

The basic design of MAAS consists of an internal lightweight Aluminum-Carbon composite octagonal frame (Octocomb) structure with three dimensional flexibility. This enables smooth alterations in wing structure and geometry. The wing is covered with thin and high strength carbon composite interweaved skin. The wing consists of multiple sections with individual motion and flexibility. The longitudinal and lateral deformations induced in the Octocomb frame change the camber and chord of the wing's aero foil. The wing profile can be altered by contracting or expanding the wing sections. The wing skin is designed according to the sectional arrangement corresponding to that of the frame. An additional layer of elastic support is provided between the sectional joint regions under the

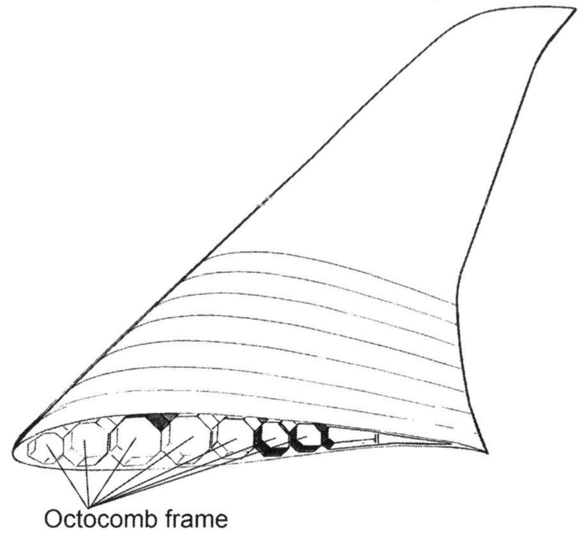

Fig. 1 Wing cross-section in default configuration

skin in order to impart more flexibility and strength. This avoids failure as these regions experience more mechanical load during shape change. The wings are designed to be lighter and smoother in-order to maintain a laminar airflow over a large portion of the wing. This is an important feature, as it can reduce drag that in-turn reduces fuel consumption. The wing has to be rigid as well as flexible (Figs. 1, 2).

3.1 The Design Features

The important aspects taken into account while designing are

- Silent and eco-friendly flying.
- Lightweight structure.
- Integrated airframe and wings.
- Less pollutive propulsion systems.

The propulsion systems, structure and airframe integration of the aircraft is an important area that requires fine tuning. The integration of body and wings into a "single" flying wing profile, otherwise called the Blended Wing Body possess many advantages. Both the body and wings provide lift, allowing a slower approach and takeoff, which would reduce noise as well as improves fuel efficiency. The flaps, or hinged rear sections on each wing (as seen on conventional designs) can be eliminated. These components are a major source of airframe noise when an aircraft takes off and lands. Engines can be embedded in the aircraft fuselage with air intakes on the upper side of the plane rather than underneath,

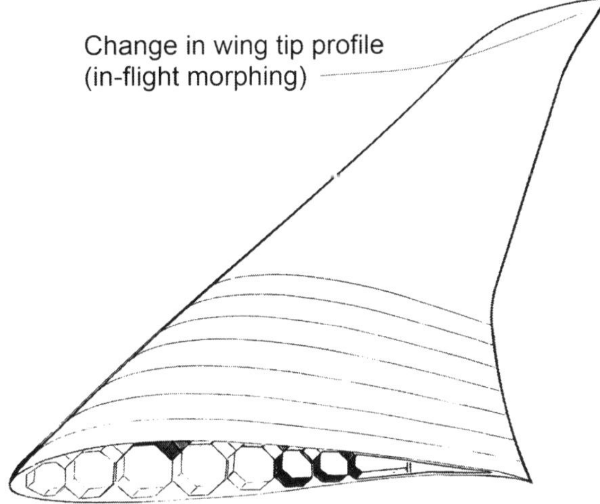

Fig. 2 Wing tip morphing (*seamless shape* change of the wing)

which screens much of the noise from the ground. This design is different from conventional ones. The engine air intakes are designed to provide a smooth airflow. Bringing together such modifications from conventional designs along with the incorporation of mission adaptive structures can produce an aircraft with numerous advantages.

4 Model 04Ax and 05Ax

The Models 04Ax and 05Ax are basically conceptual designs developed to analyze some unique aircraft features. Two palm sized models are currently used in the study. The swift and eagle are the two birds that were closely studied to develop the 04Ax and 05Ax models.

In case of eagles, when they are high in the sky they soar and their wings are fully extended. They glide to increase lift and reduce drag. This helps them to glide effortlessly and navigate for long durations in their search for a prey. To catch a prey, they fold their wings (fast attack profile) and dive in.

Swift, a bird whose wing-morphing ability makes it an exceptionally versatile flyer and allows it to eat, sleeps and mates in air. It spends almost its entire life in the air. During flight, it continually changes the shape of its wings. When they fly slowly and straight on, their extended wings carry them 1.5 times farther and keep them airborne twice as long. To fly fast, it sweeps back its wings to gain a similar advantage (Figs. 3, 4).

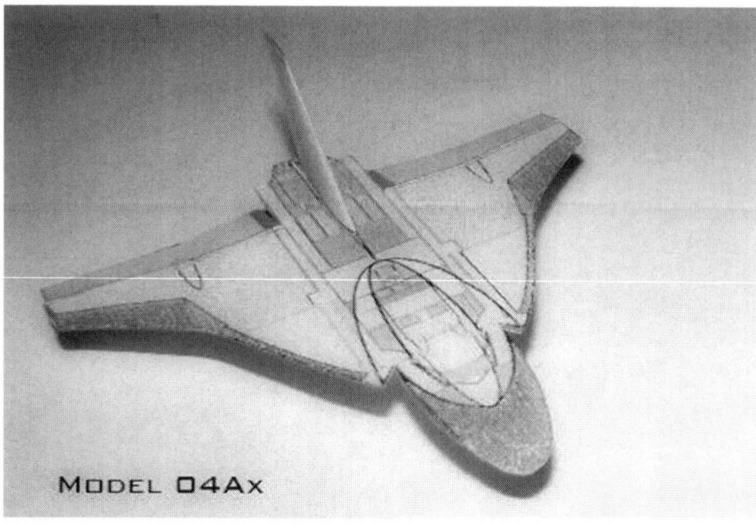

Fig. 3 Model 04Ax

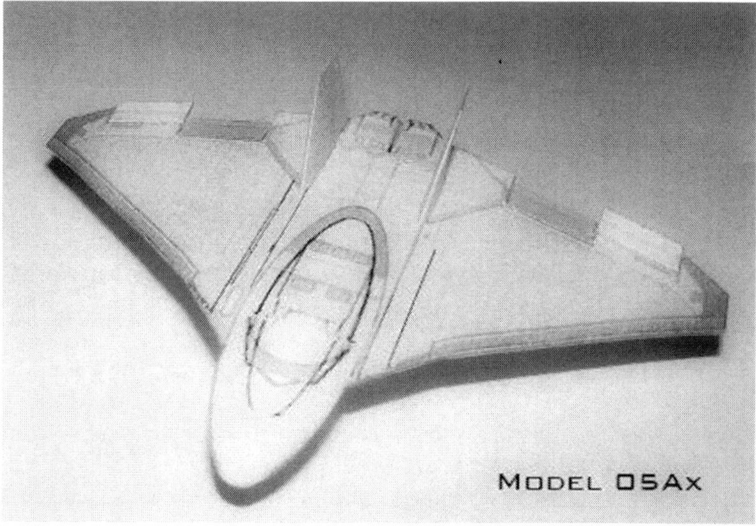

Fig. 4 Model 05Ax

5 Intelligent Micro Air Vehicles

Intelligent Micro Air Vehicles (IMAVs) refer to micro aerobots that are extremely light, palmtop flyers powered by small power cells. Their small size is apt for spying, flying through windows, finding and tracking insurgents, search and reconnaissance inside buildings, confined spaces and disaster struck areas.

For example, consider a typical mission profile; A human controller launching an IMAV from his palm on a mission to fly around a building, take images and transmit them live back to a small control console with him. The craft can be guided to enter the building through a window or hover outside undetected. Real time information can be transmitted from the IMAV to the controller and decisions can be taken without delay.

The IMAVs with MAAS can be highly adaptable and flexible as mission changes. A single aircraft can perform different missions. Conventional aircrafts can be optimized for a maximum of perhaps two flight characteristics like high-speed flight, maneuverability, or for range. The requirement of a whole fleet of different types of planes can be eliminated once such micro aerial vehicles (MAV) are in operation.

6 Conclusion

Mission Adaptive Aero Structures can provide high levels of in-flight adaptability combined with multi-operational flexibility and responsiveness unmatched with any other conventional aircraft designs. The models 04Ax and 05Ax are conceptual designs developed to analyze and study various flight-structural dynamics related to MAAS on a small scale.

The advantages of MAAS technology are:

- Development of reduced cost multipurpose & multi-mission aircraft.
- Higher levels of adaptability, operational flexibility and responsiveness than conventional designs.
- Silent, Eco-friendly and Energy efficient design.
- Reduce defence costs and aircrew losses in combat.
- Development of automated, stealth platform for Intelligence Surveillance and Reconnaissance (ISR) missions as well as combat (especially time critical targets) missions.

Acknowledgments I wish to thank Er. V. Vijayachandran and Dr. P.M Radhakrishnan for sharing their views and suggestions on various aspects of this project.

References

1. Skillen MD, Crossley WA (2007) Modeling and optimization for morphing wing concept generation, NASA/CR-2007-214860, Purdue University, Indiana
2. Weisshaar TA (2006) Morphing aircraft technology: new shapes for aircraft design, RTO-MP-AVT-141 (http://www.rto.nato.int/abstracts.asp), Aeronautics and astronautics department, Purdue University
3. Biologically inspired technologies in NASA's morphing project keynote address on 'smart structures and materials'; vol 5061, 2003
4. Gonzalez L (2005) Morphing wing using shape memory alloy: a concept proposal, University of Puerto Rico, Mayaguez
5. Majji M, Rediniotis OK, Junkins JL (2007) Design of a morphing wing: modeling and experiments, Texas A&M University, College Station, TX, 77843-3141
6. Bliss TK, Bart-Smith H (2005) Morphing structures technology and its application to flight control, Department of mechanical and aerospace engineering, University of Virginia
7. Iannucci L, Fontanazza A (2008) Design of morphing wing structures. In: 3rd SEAS DTC technical conference, Imperial College, Edinburgh
8. Smith K, Butt J, von Spakovsky MR (2007) A study of the benefits of using morphing wing technology in fighter aircraft systems. In: 39th AIAA thermophysics conference (AIAA 2007-4616)
9. Center for energy systems research, department of mechanical engineering, Virginia Polytechnic Institute and State University, Blacksburg, 24061, David Moorhouse, US AFRL, Wright-Patterson AFB, OH 45433

External Barriers to User-Centred Development of Bespoke Medical Devices in the UK

Ariana Mihoc and Andrew Walters

Abstract It is widely accepted that user-centred approaches to the development process produces better products and therefore brings commercial rewards. Despite such acceptance, the majority of manufacturers of medical devices fail to adopt such development principles. This paper will examine cases of manufacturers of bespoke medical devices, where one might perceive that the engagement with the end-user throughout the development process is critical to product quality. In a previous study undertaken by the author, interviews with manufacturers of bespoke medical devices indicated a perception that three external stakeholders to present barriers to the application of a user-centred design approach. This paper reports on a follow up study to understand the practice and agendas of the three external stakeholders, in order to draw a comparison with the manufacturers' views. The findings revealed mismatch between the product development process that manufacturers of medical devices are encouraged to apply and the practicalities of complying with the needs of the identified stakeholders.

Keywords User-centred design · Medical device · Product development process

A. Mihoc (✉) · A. Walters
The National Centre for Product Design and Development Research,
Cardiff Metropolitan University, Western Avenue, Cardiff CF5 2YB, UK
e-mail: armihoc@cardiffmet.ac.uk

A. Walters
e-mail: atwalters@cardiffmet.ac.uk

A. Chakrabarti and R. V. Prakash (eds.), *ICoRD'13*, Lecture Notes in Mechanical Engineering, DOI: 10.1007/978-81-322-1050-4_83, © Springer India 2013

1 Background

The challenge of making better products has pushed the design process from a linear approach, to an iterative one, placing the end-user at its heart [1]. User-centred design processes have proved to increase the usability and accessibility of products [1–4]. These characteristics are especially important in the case of medical devices, where products' misuse can have devastating effects [1, 5, 6]. Despite the support from research, legislation and standards, manufacturers of medical devices have proved to be reluctant to engage end-users in their product development process or to follow specific usability tests [6–9]. The causes for this have been identified as: the perceived high cost and delays from adopting a user-centred design approach, the lack of tools and awareness for such approaches and, many manufacturers perceiving a little value in engaging with end-users [6, 9].

This paper examines the case of manufacturers of bespoke medical devices e.g. artificial limbs, facial prosthetics etc., where the engagement with the end-user throughout the development process is believed to be critical to product quality. In this study, bespoke medical devices have been defined as medical products that are based upon the specific geometrical measurements of individual end-users.

In a previous study undertaken by the author, interviews with manufacturers of bespoke medical devices indicated three external stakeholders that were perceived to be barriers to the application of a user centred approach [10]. The identified stakeholders where:

- Research Ethics Committees (REC): through the long and complex mechanism of obtaining approval to access the patients;
- National Healthcare Services (NHS): through their predominantly cost based purchasing decisions that exclude the views' of the products' end-users;
- Standard Agencies, namely the British Standard Institution (BSI): through the lack of harmony amongst the standards of quality across international regions and, the lack of specialised standards for this type of medical device.

2 Aim

This paper reports on a follow-up study to understand the practice and agendas of the three external stakeholders in order to draw a comparison with the manufacturers' views [10]. This will help clarify the external barriers that the manufacturers face in the development of bespoke medical devices, the environment in which they operate and, how the climate can be improved.

3 Method

The research method chosen for this study has been developed with consideration to the complexity and roles of the three external stakeholders. For this reason, there have been employed two different methods for running the study: a literature review followed by a qualitative study. Due to the nature of the stakeholders running a quantitative study would have been difficult. The study started with reading and analysing literature regarding the three stakeholders. This offered insights into how REC, NHS procurement department and BSI are organized, how they function, what are their roles and, who is governing them. Studying the literature had some limitations; it offered only general information about the stakeholders and their practice in theory. The literature did not provide answers regarding their interaction with manufacturers of medical devices, and, it did not always make clear how development happens in practice. Thus, a plan was developed to contact these stakeholders directly, to arrange face-to-face interviews. Survey research was restricted due to the exploratory nature of the questioning. Additionally, as the aim is to develop depth of understanding, a focussed engagement with a small sample group was deemed more appropriate. Because all three bodies have been appointed by the government and, act according to a precise code of practice and regulations, it was sensible to interview only one key stakeholder from each body and to exclude the use of quantitative research. Semi-structured interviews have been undertaken with key people from each stakeholder. The transcribed interviews were analysed and provided valuable insights into how these bodies function, what is their role and what regulations apply to them. The findings have been compared with the manufacturers' views as revealed in the previous qualitative study [10]. Figure 1 illustrates the method used in the previous study and this one.

Each interview was tailored for the particular stakeholder and the participants' selection was undertaken with consideration for whom the manufactures of bespoke medical devices are most likely to interact.

The interviewee from the REC was the chair of a committee that is flagged for medical devices applications (including bespoke medical devices). The interview covered subjects regarding the application process, guidance in making the application, timeline in obtaining the approval, reasons for refusing an application, considerations in limiting the access to end-users.

In the case of NHS, accessing people from procurement department has proved to be difficult. Despite repeated attempts to contact key stakeholders with purchasing decision powers from NHS, over the 3 months allocated for this study, nobody has accepted to take part in an interview. In order to overcome limitations that these refusals have imposed, a semi-structured interview was carried out with the director of NHS product testing laboratory, a key player in the procurement process within NHS Wales. The interviewee is employed by NHS to run tests on mass consumption products that will be bought by NHS Wales. His role is to inform the procurement department prior to purchasing decision meetings on the

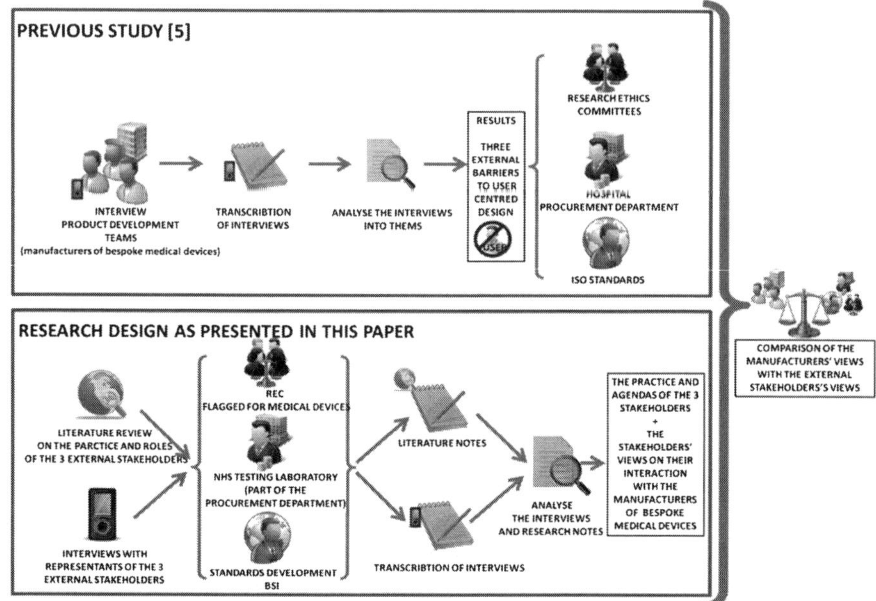

Fig. 1 Research method used in the two studies

results of the tests. The organisation independently tests and reviews the quality of products. Although the interviewee had no purchasing decisions in NHS, through his involvement in the procurement department he was able to offer direct insights on the activities and practice of the department. The interview revealed the impact that cost has on the purchasing decisions and the differences in procurement practice between NHS Wales and England.

BSI is the national standards body. They are appointed to develop new standards and publish European standards in the UK. As the manufacturers indicated that a lack of standards harmony and a lack of detailed guidance for the development of bespoke medical devices is a particular barrier, it was relevant to interview a key stakeholder from BSI. The interviewee was working in the BSI's department that develops standards. This interview covered the role and practice of BSI, the method of developing standards, harmonisation of standards and, issues regarding standards for medical devices.

This study presents some limitations. Although in the first study the interviewed manufacturers were both British and international companies that provide bespoke medical devices in the UK, the second study has been limited to the UK stakeholders. This is partly because the study is concentrating on the UK context and also because of the limited time for delivering this study (3 months). However the interviewed manufacturers are developing products outside UK and are dealing with the equivalent of other's countries health services, standards and regulations bodies and research ethics committees that might behave similar to the British ones.

4 Results

The results indicate mismatch between the product development process that manufacturers of bespoke medical devices are encouraged to apply [11, 12] and the practicalities of complying with the identified stakeholders.

The user-centred design process proposed by the literature derived from ISO 13407 on Human Centred Design Process (Fig. 2) is composed of 4 iterative steps [11, 12]:

- Analyse: the product development team identifies who are the end users of the product, how, when, where and for what will they be using the product. This stage is characterised by direct observation of the end user with the aim of correct identification of their needs.
- Design: the identified needs of the end users are translated into requirements for the new product. The development team generates concepts for the product based on the resultant requirements.
- Implement: the generated concepts from the previous stage are implemented in prototypes.
- Validate: the prototypes are tested with the end users. If the requirements have been met, the prototype is validated and can go onto the launching stage. If not all the requirements are satisfied, the prototype is refined until validation. The refinement of the prototype is done via the iteration of the whole process.

The interviews with the manufacturers indicated a perception that the three external stakeholders created barriers to the adoption of a user-centred-design process [10]. As a consequence the development process adopted by the manufacturers is markedly different from Fig. 2, the one proposed in the literature.

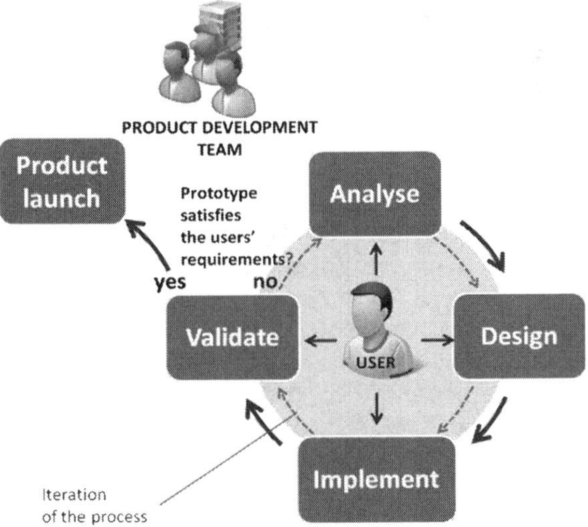

Fig. 2 The user-centred design process proposed by the literature. *Source* [11, 12]

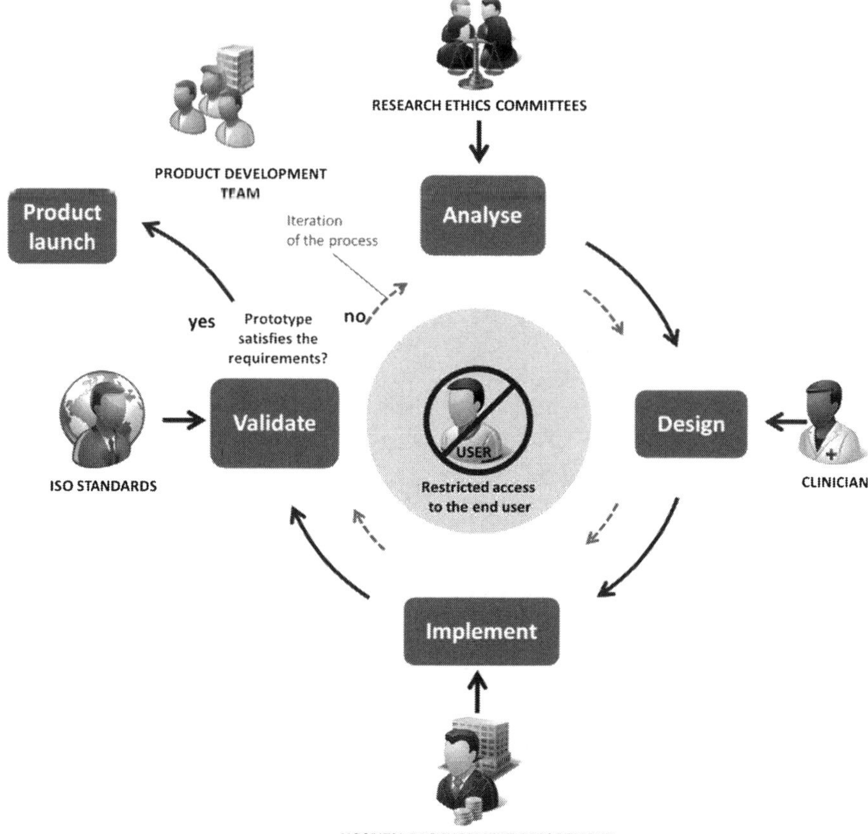

Fig. 3 User-centred design in the context of bespoke medical devices. *Source* [10]

Figure 3 illustrates the influence of the three external stakeholders at every stage of the development process.

The long and complicated application process for research ethics encourages manufacturers to find other sources of obtaining data with regards to the end-user. Most of the manufacturers will turn towards the clinicians who are perceived to have expert user knowledge and might also play a key role in the purchasing decisions made by hospitals. As a result the design of products is often done according to a clinician's requirements [10]. During the implementation stage, the use of prototypes and the material choice will depend on the budget imposed by the client which in most cases is represented by hospitals. The purchasing department within hospitals are perceived to base most of their decisions on cost and as a result force manufacturers to create low cost products [10]. Additionally, manufacturers perceive that currently there are insufficient standards of quality that would offer clear guidance with regards to the development and validation of these devices [10].

This study compares the manufacturers' views [10], with the views of the three external stakeholders. Interviewing and analysing the practice of these stakeholders has validated many of the manufacturers' statements, and revealed further issues that could raise concerns in the development of bespoke medical devices. Table 1 draws parallels between some of the manufacturers' statements [10] and the statements of the external stakeholders interviewed in this second study.

4.1 Research Ethics Committee Study

Interviewing members of the REC, flagged for medical devices applications has indicated the following issues.

The application process that manufacturers have to go through is still long—at least 2 months, and it necessitate a long and detailed preparation prior to the submission of the application.

Although the website of the REC offers detailed guidance, most applicants still need the help of an advisor in order to complete an application.

Interviewed members of the committee have admitted that although the website has been updated, it still needs improvements so that it is easier for people to prepare applications.

The manufacturers must predict all the tests and changes that they will require for each prototype at the time of the application. Any change not considered in the application, necessitates a new application. This final point is where the development process adopted by manufacturers clashes with the research review mechanism. It is difficult, and in some cases impossible, to predict the results of prototype tests and the modifications that the design team will need to make. Further, depending on the solution that is adopted, new tests might need to be defined in order to refine and validate the prototype. The potential requirement to go through the application for research review every time an unforeseen modification arises, discourages manufacturers from applying due to the imposed delays, thereby actively discouraging user engagement.

4.2 NHS Procurement Department Study

Stakeholders from the procurement department within NHS have been interviewed. The NHS England makes much of their purchasing decisions on a commercial basis that is due to the cost conscious economy. As a result the price set by manufacturers represents a major criterion in the purchasing decision. In Wales, NHS has a testing laboratory for the basic products that hospitals buy. When a contract needs renewing, the laboratory will test the products offered by participating manufacturers. The laboratory submits the products to tests according to both the standards and the requirements expressed by the users. However, this

Table 1 Comparison of statements from the two studies

Manufacturers' statements	Stakeholder statement	Insights
"Ethical approval is very hard. It takes us 5–6 months to get it. This is affecting us very much because for instance when we wanted to do tests in the summer because of the hot days, we ended up having to do them in the winter so the results were futile for some of them. This was a sweat management test"(Design Director at a UK Prosthetic Company)	"In the application process the most difficult part to complete is the IRES form definitely… The guidance [for completing the application] is on the website, is not always easy to use but the information is there and the coordinators will always help"	-The research ethics approval application process is difficult -From the manufacturers perspective it can cause great delays
"The vast majority of the NHS procurement teams base all their orders, their procurement pretty much on cost. Most of them have a scoring system. You have to submit your cost for products and then that makes 70 % of your overall score for the contract… We are planning to apply for a new contract with the NHS. We are looking now to see where we can cut more the cost of our products. This might mean using a lower quality material for our devices" (Clinician at an International Bespoke Medical Devices Company)	"Our funding from the Welsh Government is to test medical devices on behalf of the Welsh NHS and to provide technical advice on medical device related issues" "Because the NHS is trying to get value for money all the time, is driving prices down. The way some manufacturers and suppliers can get prices down is by sourcing pour quality" (Director of NHS Wales, product testing laboratory)	-NHS procurement method constraints manufacturers on reducing the cost of products. - The cost reduction has a high impact on the product's quality
"Standards are needed because they are there to reduce risks" "The problems arise where the people that wrote the regulation or the requirement are not directly involved with the end-users. They are either politicians or administrators, or they write the things generally, for a different purpose". (Project Manager at a U.K. company)	"If you are thinking of a medical device, the directive pretty much sais it's got to be safe. (…)there will be a standard that supports it, how can you manufacture a product in accordance to that directive. But actually within each country it will be up to the government to determine how that directive comes about". (BSI representative)	-Manufacturers need standards that specially developed for bespoke medical devices -The development of standards depends on government's agenda

laboratory does not have the capability to test complex products. The laboratory will also ensure the maintenance of quality of the stock delivered by manufacturers throughout the duration of the contract. The purchasing department within the NHS Wales makes much of their purchasing decisions based on the recommendations of this laboratory in order to ensure good quality products. But the final decision is still being done on a value for money basis.

In the UK, an independent organisation exists that has the capability to test all medical devices. However they can only issue reports with regards to the quality of the products tested. It is then up to individual hospitals if they wish to refer to these reports when they make purchasing decisions. Further, this body has the power to issue warnings if one of the products fails the quality standards.

The model adopted in Wales helps the manufacturers to improve their products, and, it encourages them to include the end-users in their design process. The model followed by England and Scotland imposes greater cost constraints on the manufacturers and as a result reducing cost becomes a priority in the design process. Furthermore, not including the views of the end-users of the product will turn manufacturers towards satisfying the requirements of the purchasing body and not the real user.

4.3 BSI Study

The interview of the representative from the BSI has revealed the following:

The British Standards are generic guidelines for manufacturers to follow in order to improve the quality of the products.

The standards of quality in the area of medical devices are based on the Medical Device Directive adopted by the European Union. Every country has translated this directive into standards of quality or regulations. Although the aim is to achieve harmonisation between standards at an international level, this is not yet the case. Thus, manufacturers of medical devices are forced to decide on the standard of quality to follow depending on the countries in which the product will be commercialized.

The lack of detail and specifications for the particular case of bespoke medical devices makes it even more difficult for manufacturers to follow specific and relevant standards.

5 Conclusion

The results of the interviews and analysis of the three stakeholders confirm the barriers to user-centred design indicated by the manufacturers of bespoke medical devices. This study reveals how external stakeholders influence the development process of bespoke medical devices. This provides an important contribution to the field of design. It makes apparent where the functioning of these bodies is at odds with the design and development process. Thus the study is the first step towards the development of solutions to overcome theses barriers.

It is evident that REC, NHS procurement department and BSI have become barriers to a user-centred design due to their practices. A better climate for the application of a user-centred design might be created if each of these stakeholders reviewed their methods. Listed below are the proposals and future research themes to begin to overcome the barriers.

REC has been criticised by the interviewed manufacturers for the rigidity, complexity and timeline of the application process. Adjusting their application process to the current commercial environment is a critical step in not only maintaining the ethics of the research but also encouraging research overall.

NHS has been criticised during the first study for its money driven approach. Concentrating on people can be much more cost effective than providing them with the bare minimum. A user-centred design approach would determine that purchasing bespoke medical devices should be done with the end user in mind.

BSI has been criticised in the first study for the lack of harmony in standards at international level and lack of guidance for the case of bespoke medical devices. The harmonisation of standards is a problem that needs to be reviewed at international level. As the main provider of guidance in product development, BSI plays an important role in assuring a high quality of bespoke medical devices. The lack of standards for such devices needs to be filled with specific guidance that has been developed through thorough research and tests in collaboration with the industry. However, it must be recognised that REC, NHS and BSI have developed their practices to safeguard risk and exploitation of users. Although it may seem ironic that their current practice is not serving users of bespoke medical devices well, creating new suitable and robust procedures is surely to be a significant challenge.

The study raises awareness over barriers that commercial organisations face, that might not otherwise be considered in the academic environment. Therefore, the research provides a practical resource for the translation of design theory to practice.

6 Future Study

The first two studies offered insights from a macro level in the development of bespoke medical devices. The research will continue to identify barriers to a user-centred design process at a micro level, through analysing the manufacturing companies that are producing such products in more depth. Further analyses of needs and attitudes of end-users towards their inclusion in the development process will also be undertaken. This will present a fuller picture of the issues that developers of bespoke medical devices have in engaging the end-user within their processes. The overall aim will be the generation of a user-centred design approach for the development of bespoke medical devices that will be applicable in the commercial environment, together with recommendations, tools and techniques for its application.

References

1. Thimbleby H (2008) Ignorance of interaction programming is killing people. Interact, Sept/Oct 2008, 15:52–57
2. National Patient Safety Agency (2010) Design for patient safety: user testing in the development of medical devices. NPSA and MATCH, Mar 2010
3. Nordic Innovation Centre (2011) New methods for user driven innovation in the health care sector. Available at http://www.nordicinnovation.net/prosjekt.cfm?Id=3-4415-244. Accessed 25 May 2011
4. Ulrich KT, Eppinger SD (1999) Product design and development, 3rd edn. McGraw-Hill, London International Edition, pp 53–68, 187–207
5. Kyberd P, Wartenberg C, Sandsjö L, Jönsson S, Gow D, Frid J, Almström C, Sperling L (2007) Survey of upper-extremity prosthesis users in Sweden and the United Kingdom. J Prosthetics Orthot 19(2):55–62
6. Winters JM, Story MF (2006) Medical instrumentation: accessibility and usability considerations, CRC Press, New York (31 Oct 2006)
7. Council Directive (1993) 93/42/EEC of 14 June 1993 concerning medical devices. Official J Eur Communities, L169
8. Lemke MR, Winters JM (2008) Removing barriers to medical devices for users with impairments. Ergon Des: Q Human Factors Appl 6(3):18–25 (20 June 2008)
9. Money AG, Barnett J, Kuljis J, Craven MP, Martin JL, Young T (2011) The role of the user within the medical device design and development process: medical device manufacturers' perspective. (BioMedCentral), BMC Med Inform Decis Making 11:15
10. Mihoc A, Walters A, Eggbeer D, Gill S (2012) Barriers to user-centred design in the development of bespoke medical devices: a manufacturers' view.In 13th national conference on rapid design, prototyping and manufacturing, Lancaster, 22 June 2012
11. Medical Safety Design (2011) User-oriented interface design for medical products. Available at http://www.medical-safety-design.de/en/medical-safety-design/user-centered-interface-design/. Accessed 12/03/2011
12. Usability Professionals' Association (2011) What is user-centred design? Available at http://www.upassoc.org/usability_resources/about_usability/what_is_ucd.html. Accessed 20/09/2011

Autonomous Movement of Kinetic Cladding Components in Building Facades

Yasha Jacob Grobman and Tatyana Pankratov Yekutiel

Abstract Movement of building façade cladding is used to control buildings' exposure to environmental conditions such as direct sunlight, noise and wind. Until recently, technology and cost constraints allowed for limited instances of movement of facade cladding. One of the main restrictions had to do with the limitations that architects face in designing and controlling movement scenarios in which each façade or cladding element moves autonomously. The introduction of parametric design tools for architectural design, combined with advent of inexpensive sensor/actuator microcontrollers, made it possible to explore ways to overcome this limitation. The paper presents an ongoing research that examines the potential of autonomous movement of façade cladding elements. It defines types of autonomous movement strategies and compares the advantages of these strategies over those of traditional methods of centrally controlled movement. Finally, it presents and discusses several case studies systems in which autonomous movement for building cladding elements is implemented.

Keywords Kinetic cladding components · Responsiveness · Interactive · Decentralized control · Arduino

Y. J. Grobman (✉) · T. P. Yekutiel
Architecture and Town Planning, Technion Isreali Institute of Technology, Haifa, Israel
e-mail: yasha@technion.ac.il

T. P. Yekutiel
e-mail: tanyapankratov@gmail.com

A. Chakrabarti and R. V. Prakash (eds.), *ICoRD'13*, Lecture Notes in Mechanical Engineering, DOI: 10.1007/978-81-322-1050-4_84, © Springer India 2013

1 Introduction

The notion of kinetic architecture can be traced back as far as the Roman Empire with legendary tales describing revolving rooms in Emperor Nero's palace [1]. Although the modern fascination with motion is associated with movements like Futurism in the early 20th century [2], research on kinetic architecture dates back to the 1960s [3]. From its earliest days, the research in this field encompassed a wide range of directions in various scales: plug-in cities by Archigram on an urban scale; the moving roof elements in Santiago Calatrava's Milwaukee Art Museum on a building scale; moving or folding walls in the Rietveld Schroder house or the moving office space in the OMA Bordeaux house on an interior design scale; moving structures as the Strandbeest by Theo Jansen or the Crate House by Alan Wexler on a furniture scale [4].

Starting from the early 1990s, when the immense implications of the information technology (IT) revolution in architectural design became evident, new types of computer-controlled movements in architectural design have emerged (mainly physical but also virtual). These new types were based on ideas such as interactive design and responsive environments [5–7].

One of the building industry's leading trends in both research and actual implementation is kinetic building facades, or more precisely, kinetic building cladding systems [8]. Indeed, the increasing focus on green architecture and especially on the environmental behavior of buildings has greatly increased the interest in high-performance facades. Until recently, technology and cost constraints allowed for only limited centrally controlled scenarios for movement of building facades or facade cladding. Centralized control over the facade elements limits the amount and type of movements they can perform to simple operations that are usually executed simultaneously by all the elements. The introduction of parametric design tools for architectural design, combined with advent of inexpensive sensor/actuator microcontrollers, makes it possible to examine different and more complex types of control over the building facade elements.

The following paper presents an ongoing research study that examines the potential of employing decentralized control of building facade kinetic elements. It begins with a brief discussion of the function of building facades and the types of kinetic movement of facade elements. It then looks at possible scenarios in which decentralized control could be employed. Finally, it presents and discusses two different case studies that examine a framework for decentralized control of cladding elements.

2 Controlled Elements/Mechanisms in Building Facades

Building facades have numerous distinct functions. Hutcheon [9] organized these functions into two groups with a total of 11 functional requirements. The first group consists of the items that relate to the facade as a barrier for the control of heat flow; air flow; water vapor flow; rain penetration; light, solar and other radiation; noise; and fire. The second group consists of overall requirements, such as providing strength and rigidity; being durable; being ecstatically pleasing; and being economical. Another aspect, which was not mentioned by Hutcheon and has become increasingly important, is visibility or visual exposure (both from inside outward and from outside inward) [10].

Introduction of kinetic elements influencing the facades performance may be done into main aspects; the first will influence the parameters of the openings of the facade itself. The second will influence the geometry of the facades or of an external cladding layer that is part of the facade. This research concentrates on the second group.

Numerous kinetic facade cladding systems have been developed in each of these aspects over the last 15 years. Loonen [11] presents in his thesis an overview of 100 different systems that control the performance of parameters from both groups. The control systems that actuate those kinetic façade mechanisms or systems can be differentiated into three main types, which also correspond to the evolution of these systems. The first type consists of elements that are actuated directly by a manual switch. In the second type, sensors are introduced. The information from the sensors is used to actuate the kinetic cladding elements. This type, which makes up the majority of present-day kinetic façade cladding systems, relies on centralized control, in which actuation is controlled through a central unit (computer).

The third type, which is the focus of this research, introduces the idea of decentralized control. The idea is based on the use of newly developed tools in parametric design, combined with sensor/actuator microcontrollers such as Arduino (an open-source single-board microcontroller, able to process signals passed by sensors to the microcontroller and translated through code into physical actions such as lighting or activating engines [12]). It refers to autonomous and direct actuation of kinetic facade elements by sensor/actuation units.

The following diagram describes the evolution of the types of control and the differences between them (Fig. 1).

3 Implications of Decentralized Control Over Building Elements

The idea of a centralized control over kinetic facade elements is easy to understand: It is meant to provide a single solution to a change in the basic conditions. For example, it is clear that once an environmental condition such as sunlight has changed, the control device must change the position of the cladding elements so that more or less light can penetrate the building.

Decentralized control is more complex. By definition, it's meant to handle multiple conditions and generate various responses. It is based on local, cheap and

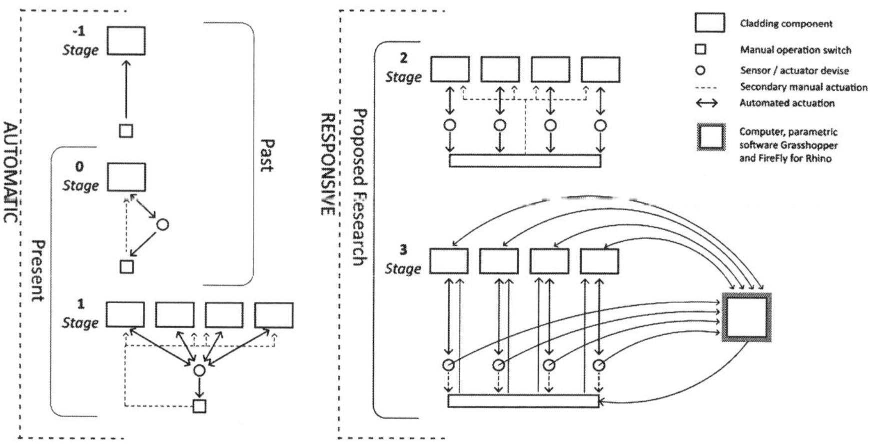

Stage (-1): Cladding component actuated by manual switch. *Stage (0):* Cladding components are connected to a sensor/actuator device and can adapt automatically to changes in environmental conditions. *Stage (1):* Similar to stage (0) but with a central control unit. *Stages (2) and (3):* The building facade consists of several small-scale cladding components, each of which is connected to a local sensor/actuator and can perform automated adaptation. There is communication between the components. In Stage (3), the introduction of a central computer enables the accumulation and processing of data to facilitate better adaptation of the system over time.

Fig. 1 Evolution of types of control over kinetic facade cladding elements

less powerful [than a central control computer (PC)] computers or microprocessors, which are connected to the kinetic elements.

The following advantages were identified to decentralized control in relation to traditional centralized control:

1. Efficiently—the possibility of insulated response to local environmental conditions emerging in a particular part of the façade. As opposed to traditional method where the adaptation can only be carried out by the whole facade
2. Redundancy—in multi connection system each component can substitute for its neighbor
3. Low cost—each component is relatively cheap and does not require high cost central operation system
4. Calculation time—the microprocessor located in each component will perform only the basic calculations and thus increase the efficiency of the system rather than performing the calculations of the data within the hole facade
5. Functional and compositional freedom.

The move form centralized control to decentralized control could be seen as three steps in terms of the relationships between the micro controllers and a central control unit. In the first level (centralized control) each cladding component is connected to a sensor devise and collects information about its immediate environmental conditions. The collected data from the entire facade is transferred to the main computer where the data processing takes place; as a result the facade will perform the necessary kinetic adaptation. The kinetic adaptation may be

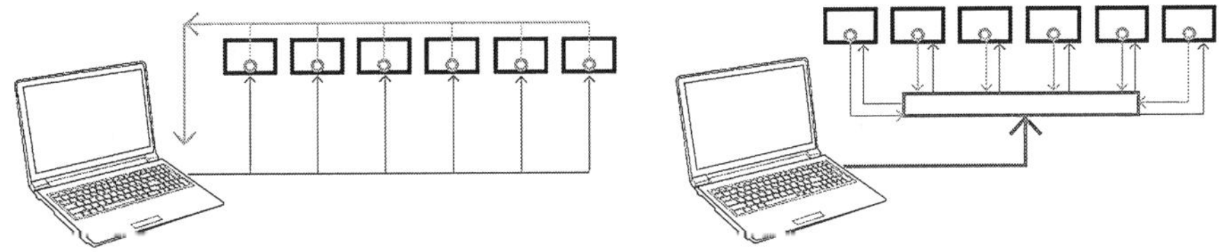

Fig. 2 Centralized control information flow (*left*). Decentralized control information flow through an "information hub" or a "bus systems" (*right*)

different for each cladding component but will be based on data processing in the main computer (see Fig. 2).

The second level preserves the data collecting principals of the previous level but disconnects the actuation process from the main computer and introduces an information hub or a "Bus system" for information navigation with in the facade. Each component will receive data from the environment and translate it to actuation. The information hub shares the information each unit has with all the units (Fig. 2).

The third level introduces the autonomous decentralized operation. Each component will receive data from the environmental conditions and data inputs from the neighboring components as well. The first evaluation will be done between the external and internal sensors of each unit, at the next stage the data from the neighboring component will be added as secondary input in order to preserve the systems equilibrium as whole. In this setting the components will influence each other and the kinetic adaptation of the facade will be performed by methods of flock behavior. The connection to the main computer in this case is not necessary and could serve as user interface for maintenance and occupants control device (Fig. 3).

Before developing a decentralized control mechanism for a cladding system, one must define scenarios in which this type of control could have potential either to handle situations that could not have been handled before or to introduce better results than those produced by the current method of centralized control. The following subsections presents trajectories that were identified as possessing this kind of potential.

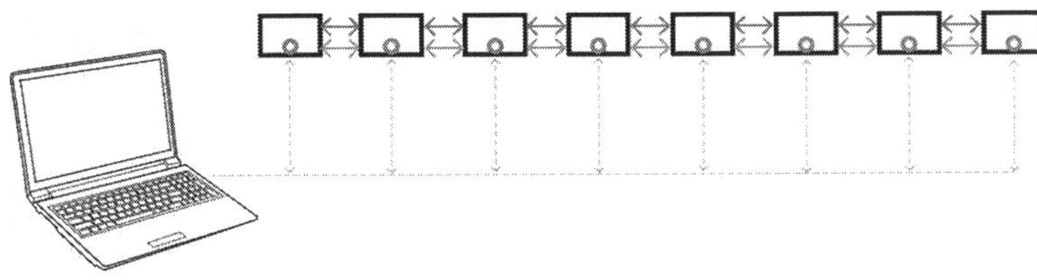

Fig. 3 Autonomous decentralized control information flow

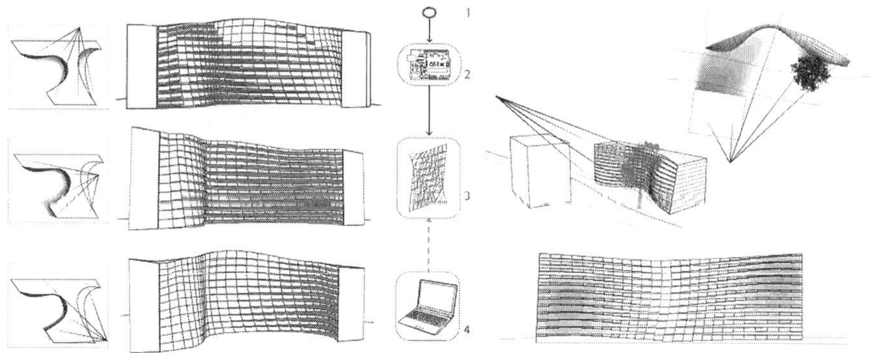

Fig. 4 Evolution of types of control over kinetic facade cladding elements

3.1 Handle Changing Local External Conditions

Partial shading by building, building elements or external elements—change the local position of kinetic shading elements according to the shade/reflection created by, inter alia, neighboring buildings, balconies and trees (see Fig. 4).

Visual exposure control—change the visual exposure of designated areas of a facade (see Fig. 5).

3.2 Handle Local Internal Changing Conditions

This possibility deals with internal changing conditions that require different facade settings, such as an increase in the number of people within a space that requires more natural ventilation, a change in the room's functions (see Fig. 6a, b).

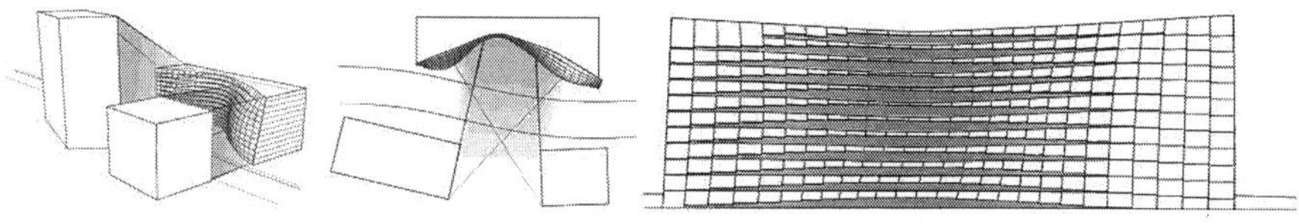

Fig. 5 The complexity of urban layout may cause undesirable privacy issues to building's occupants

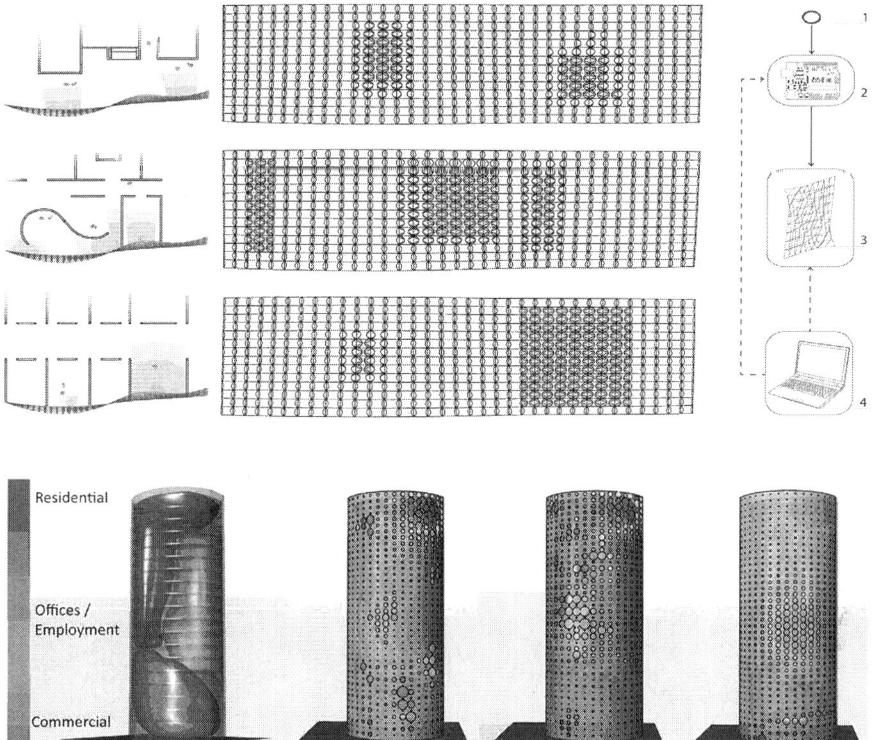

Fig. 6 **a** Programmatic flexibility. **b** Building's functional zones

3.3 Preceding Reaction or Flock Behavior

Actuating elements before their sensor feels the change based on information coming from the sensors of neighboring facade cladding elements (see Fig. 7).

4 Design Experiments

To examine the potential of the trajectories mentioned earlier, several design experiments were performed. In these initial experiments, light was chosen as a performance criterion for the actuation of the kinetic reaction The experiments' main aims, at this stage of the research, were to examine the efficiency and suitability of the kinetic mechanisms and the possibility of activating decentralized control with Arduino.

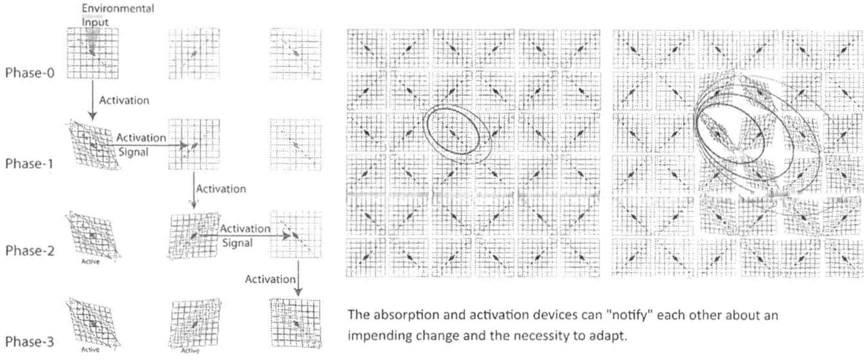

Fig. 7 Communication between the components

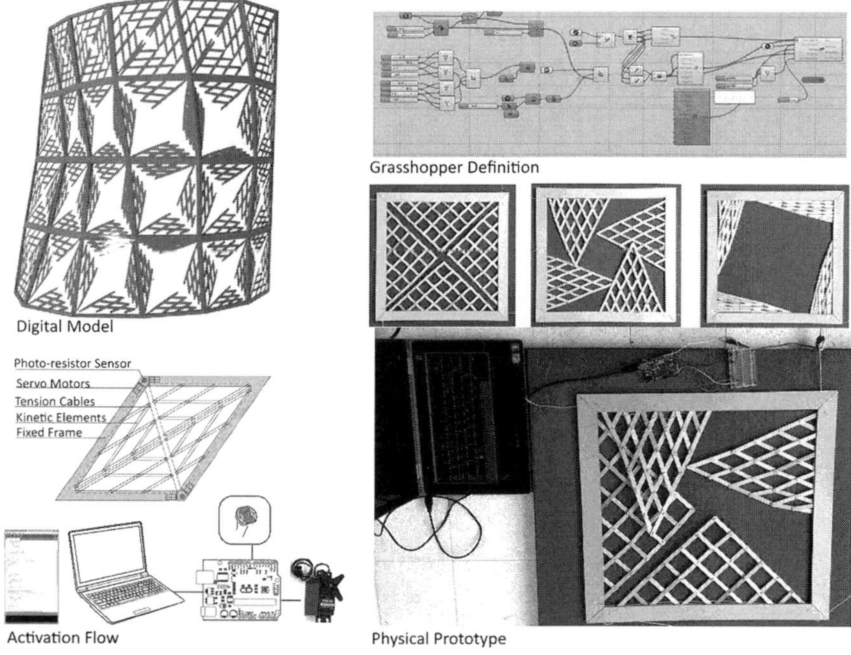

1. A photoresistor sensor measures the light levels for each component.
2. The data received by the sensor are transferred to the Arduino microcontroller.
3. The mechanism is activated by the servo motor according to code that defines the ratio between the light levels and the opening of the component.

Fig. 8 Design experiment no. 1

4.1 Design Experiment No. 1: Cladding Component System Based on a Pantograph Principle

Based on a pantograph principal, the components' mechanism is able to fold and expand. The cladding components are divided into triangular elements connected to light-measuring sensors and servo motors, which operate the mechanism through a net of cables.

An algorithm (in Grasshopper for Rhino) connects between the components' geometry and the various parts of the kinetic mechanism. Each component can move both autonomously (activated by the sensor and controlled by the Arduino microprocessor) and via instruction from the main computer. A physical working example of this experiment was developed. The model is described in Fig. 8.

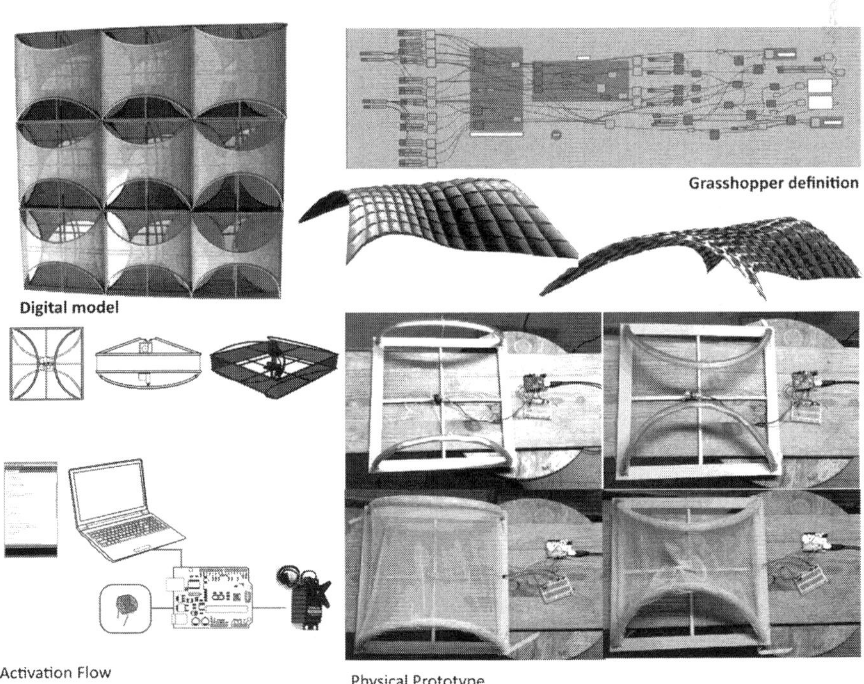

Fig. 9 Experiments no. 2

4.2 Design Experiment No. 2: Telescopic—Tension Components

The operating mechanism in this experiment combines two complementary principles. The main functional element consists of telescopic arches that are stabilized by a net of cables operated by servo motors. The main component has two perpendicular layers providing independent control over the different climatic parameters and allowing various degrees of shading and ventilation. The components' geometry is based on a rectangle and can be applied to facades that are based on quadrilateral geometry. The cladding movement is based on an algorithm (similar to the one in previous case study) and can vary in size and angles, making it suitable for application to complex geometry. A physical working model was developed for this experiment. The model is described in Fig. 9.

5 Conclusions and Future Research

Decentralized control over building facade cladding systems generates a new level of architectural complexity, one in which the designer cannot entirely control all of the design's architectural aspects [13] or to a certain extent needs to internalize how to allow the "human eye" to lose control over the design [14].

The design experiments and the preliminary research presented in this text are obviously only the initial steps of the research in this realm. Nevertheless, they seem to show both the plausibility and some unique possible capabilities of the proposed approach. The next stage of the research will concentrate on a comparative performance examination throw working prototype of various scenarios for decentralized control over kinetic cladding components in relation to traditional kinetic cladding systems.

References

1. Apollonio U (ed) (1973) Documents of 20th century art: futurist manifestos. Viking Press, New York, pp 19–24
2. Randl (2008) Revolving architecture: a history of buildings that rotate, swivel, and pivot, 1st edn. Princeton Architectural Press, New York
3. Zuk W (1970) Kinetic Architecture. Van Nostrand Reinhold, New York
4. Kronenburg R (2007) Flexible: architecture that responds to change, 1st edn. Laurence King Publishers, London
5. Kroner WM (1997) An intelligent and responsive architecture. Autom Constr 6(5–6):381–393
6. Fox M, Kemp M (2009) Interactive architecture, 1st edn. Princeton Architectural Press, New York
7. Oosterhuis K (2002) Kas oosterhuis: programmable architecture. L'Arcaedizioni, Milan

8. Moloney J (2011) Designing kinetics for architectural facades: state change, 1st edn. Routledge, Londan
9. Hutcheon N (1963) CBD-48 Requirements for exterior walls—IRC—NRC-CNRC. Can Build Dig 48
10. Shach-Pinsly D, Fisher-Gewirtzman D, Burt M (2011) Visual exposure and visual openness: an integrated approach and comparative evaluation. J Urban Des 16(2):233–256
11. Loonen R (2010) Climate adaptive building shells. Faculty of Architecture, Building and Planning. Eindhoven Univ Technol
12. Banzi M (2012) "Arduino tinkering," decoded conference. Available http://vimeo.com/24489835. Accessed 31 Mar 2012
13. Saggio A (2005) Interactivity at the centre of avant-garde architectural research. Architect Des 75(1):23–29
14. Eisenman P (1992) Visions unfolding: architecture in the age of electronic media. Domus 734:17–21

Design, Development and Analysis of Press Tool for Hook Hood Lock Auxiliary Catch

Chithajalu Kiran Sagar, B. W. Shivraj and H. N. Narasimha Murthy

Abstract The use of hook hood lock auxiliary catch is vital to all automotive cars. This paper deals with the design, development and analysis of new hook hood lock auxiliary catch and design the press tool to reduce number of stages of operations to manufacture hook hood component. The study of existing hook hood lock auxiliary catch revealed that there was development of potential crack on hook side and to overcome this, new hook hood auxiliary catch was designed with embossed profile located on the center of hook side and bend leg side. Stress analysis was performed, results revealed that new hook hood design had no potential cracks and was able to with stand higher stress value of 7–17 MPa. In design of press one stage of operation was reduced. Misalignment in embossed profile was 200 microns and was reduced to 20 microns. Further press tool can be analyzed for nonlinear and fatigue analysis.

Keywords Press tool · Embossing · Potential crack

1 Introduction

The design of press tools and its manufacturing functions are highly specialized and knowledge in nature of stampings parts, which are cut and formed from sheet materials. Bang and Thomas [1], reported a concept design of safety hook latch system was used in an weight reduction and design optimization process. When

C. K. Sagar (✉) · B. W. Shivraj · H. N. Narasimha Murthy
Department of Mechanical Engineering, R. V. College of Engineering, Bangalore, Karnataka 560059, India
e-mail: ckiransagar058@gmail.com

A. Chakrabarti and R. V. Prakash (eds.), *ICoRD'13*, Lecture Notes in Mechanical Engineering, DOI: 10.1007/978-81-322-1050-4_85, © Springer India 2013

force of 2,700 N was applied to safety hook there was development of potential crack. To overcome this, new design was development with lesser weight material SAPH440 steel and this was analyzed using Abaqus Software. Author reported that new design had no potential crack, with higher load bearing capacity of 3,300 N and weight of safety hook was reduced.

Nye [2], had developed an algorithm for orienting the part on the strip to maximize material utilization. The algorithm optimally nests convex or non-convex blanks on a strip and predicts both the orientation and strip width that minimize material usage. Based on minkowski sum concept, strip orientation was described by sweep line vector and strip width for any blank orientation were described by maximum perpendicular distance between the strip longitudinal axis and perimeter of component, using these results material utilization has been calculated and minkowski sum was generated. Venkata Rao [3], Presented analytic hierarchy process (AHP). This helped in selection of a suitable strip layout from amongst a large number of available strip layouts for a given metal stamping operation. This methodology was capable of taking important requirements of metal stampings. By this author had obtained logical and rational method of strip layout evaluation and selection. Shailendra [4], presented an intelligent system for selection of materials of press tool components for sheet metal work. Knowledge for the development of the system was acquired, analyzed, tabulated and incorporated into a set of production rules of IF-THEN variety. The system was coded in the Auto LISP language and loaded into the prompt area of AutoCAD. The output of this system gave easily available material for press tool and was useful for preparing bill of material. Eugen and Gavril [5], this paper presents analysis behavior of deformed sheet material. The analysis method, used here, was Finite Element Analysis, on ANSYS software. Author had considered that the work piece was pre-deformed, using a conventional method, by deep forming process. Author reported that most deformed zones, and most exposed ones by considering stress and strain state, von mises equivalent stress and strain and buckling region were studied.

The objective of the present work is to develop new hook hood lock auxiliary catch design with embossed profile and no weight reduction has been made. Design of press tool was to reduce one stage of operation in progressive press tool by using support punch concept in first stage of profile piercing operation. Press tool 2D&3D assembly designed, manufactured and tested for defects. Redesign of press tool to overcome defects.

CRCA is selected for hook hood lock auxiliary catch and D2 is selected for punches and dies. Material properties are shown in Table 1.

Table 1 Material property as per ASTM

CRCA (cold rolled close annealed steel) ASTM-A336	
Yield strength	270 MPa
Young's modulus	2.05×10^5 MPa
Poisson's ratio	0.28
D2 (high carbon high chromium steel) ASTM-A681	
Yield strength	2,150 MPa
Young's modulus	2.1×10^5 MPa
Poisson's ratio	0.28

*A MISUMI CORP. Standard component for press dies catalog

2 Study and Development of New Hook Hood Lock Auxiliary Catch Design

The study of existing hook hood design has intricate profile at center of hook, bend leg side and two piercing holes as shown in Fig. 1. New hook hood design was developed with embossed profile on hook side and bend leg side as shown in Fig. 2.

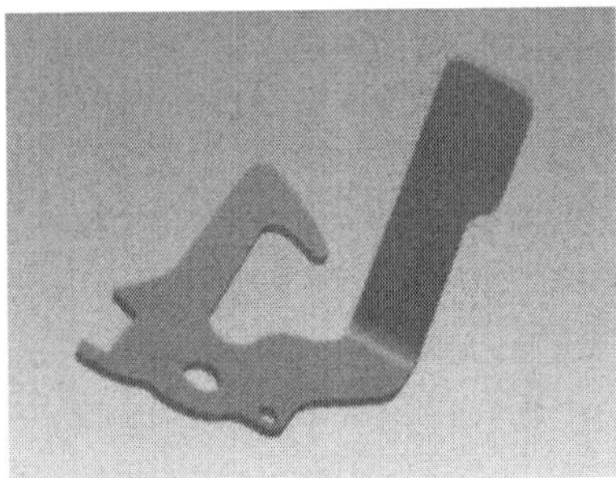

Fig. 1 Existing hook hood design

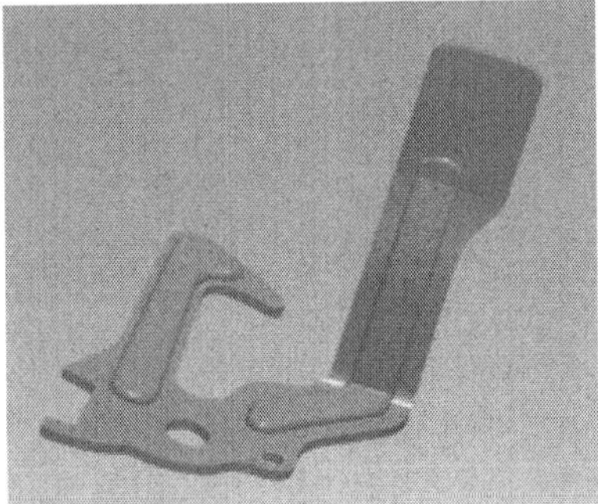

Fig. 2 New hook hood design

2.1 Stress and Total Deformation Analysis of Hook Hood Lock Auxiliary Catch

Meshed model shown in Fig. 3. Initially mid-surface was taken and meshed with 2D elements using four nodded quad, chosen element type shell 181, it contains 847 number of elements and 964 number of nodes and meshed using hyper mesh software [5].

Meshed model shown in Fig. 4 meshed with 2D elements and shell 181 element type was chosen, it contains 4,345 number of elements and 4,555 number of nodes [5].

Equivalent stress induced in existing hook hood was 7 MPa, when a load of 2,700 N was applied and there was development of potential cracks near hook side as shown in Fig. 5. Equivalent stress induced in new hook hood was 17 MPa, for same load of 2,700 N and there was no potential cracks developed as shown in Fig. 6.

Total deformation in existing hook hood was 9 microns, but there was development of potential crack by this we can predict that for every cycle of loading of 2,700 N there was plastic deformation of 9 microns on hook side as shown in Fig. 7. Total deformation in new hook hood was 17 microns and there is no potential crack or deformed regions as shown in Fig. 8.

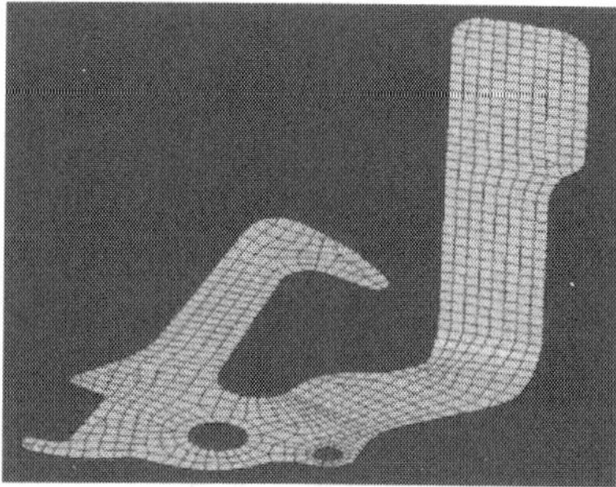

Fig. 3 Meshed existing hook hood

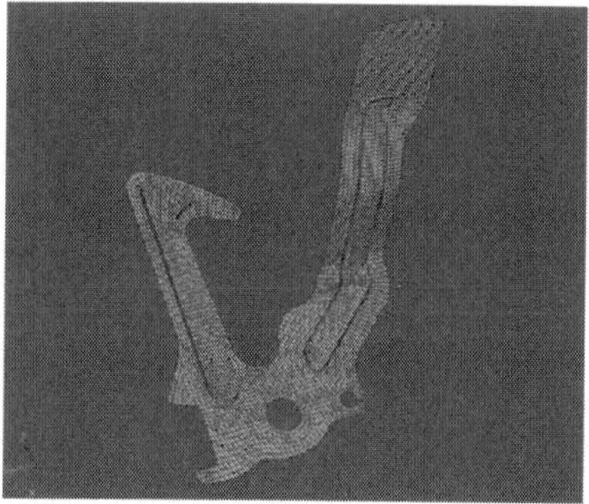

Fig. 4 Meshed new hook hood

Fig. 5 Equivalent stress of existing hook hood

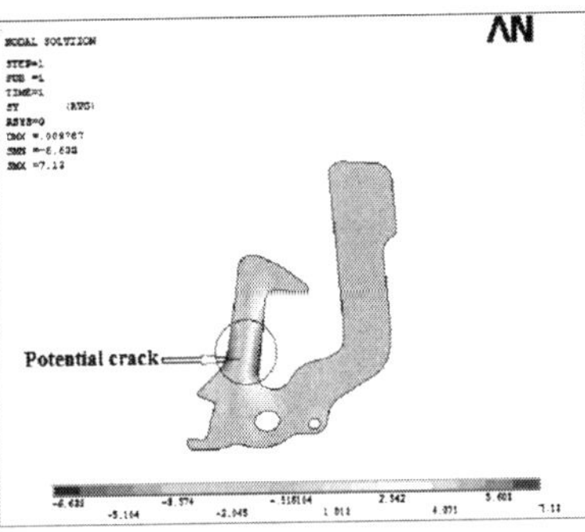

Fig. 6 Equivalent stress of new hook hood

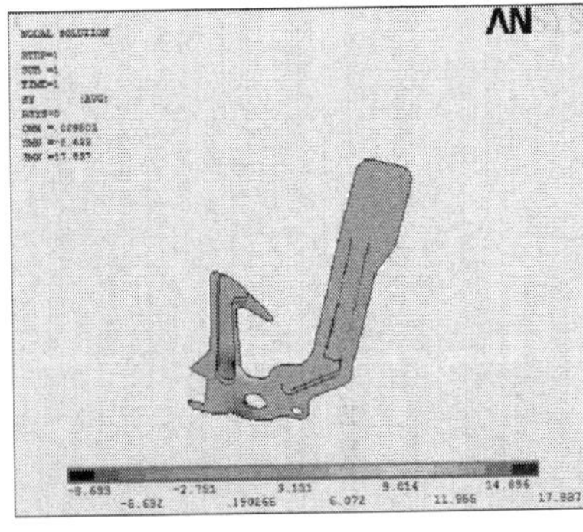

Fig. 7 Total deformation of existing hook hood

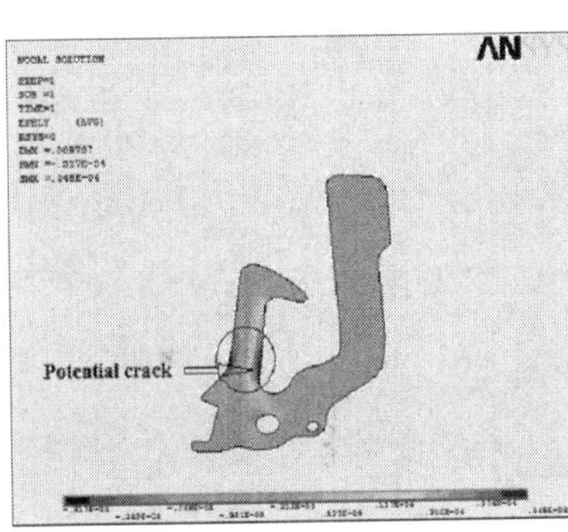

Fig. 8 Total deformation of new hook hood

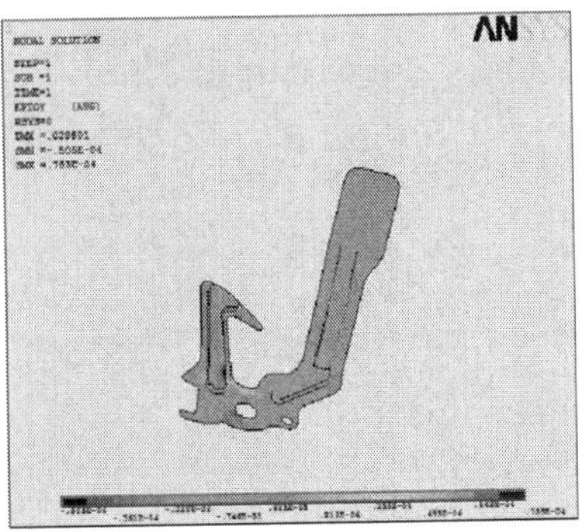

3 Press Tonnage Calculation for Progressive and Stage Tool

Shearing force:

$$F_{sh} = SLt \tag{1}$$

S Shear strength of the material, N/mm^2
L Shear length, mm
T Material thickness, mm

3.1 Press Tonnage Calculation for Progressive Tool

3.1.1 Stage 1: Piercing

$F_{sh1} = 360 \times 115.32 \times 2 = 830,304\,N,\ F_{sh1} = 830,304/(9.81 \times 1,000) = 8.46\,tons.$
$F_{sh2} = 360 \times 110.38 \times 2 = 79,473.6\,N,\ F_{sh2} = 79,473.6/(9.81 \times 1,000) = 8.10\,tons.$

3.1.2 Stage 2: Forming Piercing

$F_{sh1} = 360 \times 75.4 \times 2 = 54,288\,N,\ F_{sh1} = 54,288/(9.81 \times 1,000) = 5.53$ tons.
$F_{sh2} = 360 \times 289.75 \times 2 = 208,620\,N,\ F_{sh2} = 208,620/(9.81 \times 1,000) = 21.26 \approx 22$ tons.

For embossing operation take 50 % of shear force

$$F_{sh2} = 0.5 \times 21.26 = 10.63 \text{ tons.}$$

3.1.3 Stage 3: Piloting and Coining

$F_{sh} = 360 \times 21.99 \times 2 = 15,832\,N,\ F_{sh1} = 15,832/(9.81 \times 1,000) = 1.61$ tons.
For coining operation take 20 % of shear force

$$F_{sh} = 0.2 \times 1.61 = 0.33 \text{ tons.}$$

3.1.4 Stage 4: Piercing

$F_{sh} = 360 \times 37.7 \times 2 = 27,144\,N,\ F_{sh1} = 27,144/(9.81 \times 1,000) = 2.76$ tons.

3.1.5 Stage 5: Piloting and Blanking

$$F_{sh} = 360 \times 388.56 \times 2 = 279,763.2\,\text{N},\ F_{sh1} = 279,763.2/(9.81 \times 1,000) = 28.51 \text{ tons.}$$

Total tonnage = 33 + 32 = 65 × 1.5 – 97.5 ≈ 100 tons, based on standard machine available.

3.2 Press Tonnage Calculation for Stage Tool

3.2.1 Stage 1: Coining Operation

$$F_s = 360 \times 162.09 \times 2 = 116,704.8 = 11.89 \text{ tons}$$

3.2.2 Stage 2: Bending Force

$$F_b = 60\,\% \text{ of the shear force of blanking operation in N} \tag{2}$$

$F_b = 0.6 \times 279,763.2 = 167,857.92\,\text{N},\ F_b = 167,857.92/(9.81 \times 1,000) =$ 18 tons Total tonnage = 12 + 18 = 30 × 1.5 = 45 tons ≈ 63 tons, based on standard machine available.

$$\text{Cutting Clearance} : C = 0.005\ t\ \sqrt{Fs} \tag{3}$$

$C = 0.005 \times 2 \times \sqrt{50} = 0.07$ mm per side, for progressive and stage tool.

2D detail drawing was drawn using Mechanical desktop. Based on above calculations shut height, plate size, die block thickness, punch and die sizes were decided and based on this sectional and top assembly views were drawn for progressive and stage tool shown in Figs. 9, 10, 11, and 12.

4 Analysis of Punches and Dies

Embossing punch and die were 3D meshed using solid 92 tetrahedron element with element size 4 mm and fine meshing were made near loading region i.e. punch and die segment as shown in Figs. 13 and 14. A load of 208,620 N was applied and stress induced in punch was 128.7 MPa and die was 162.16 Mpa which was lesser, when compared with the yield strength of 2,150 MPa. By this we can predict that design is safe.

Bending punch and die were 3D meshed using solid 92 tetrahedron element with element size 2 mm and fine meshing were made near loading region. A load of 167,857 N was applied and stress induced in punch was 2,388.2 MPa which

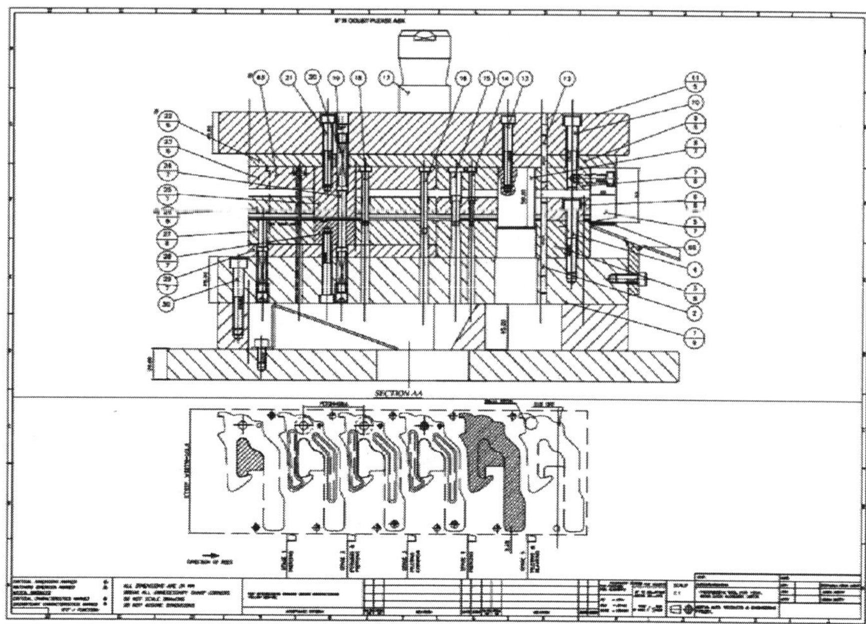

Fig. 9 2D sectional view of progressive tool assembly drawing

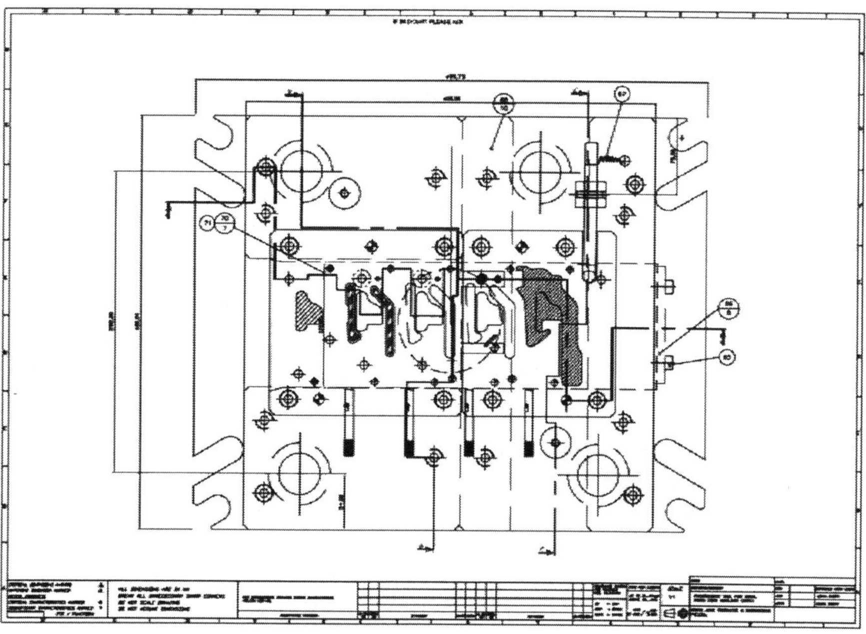

Fig. 10 2D Top assembly view of progressive tool

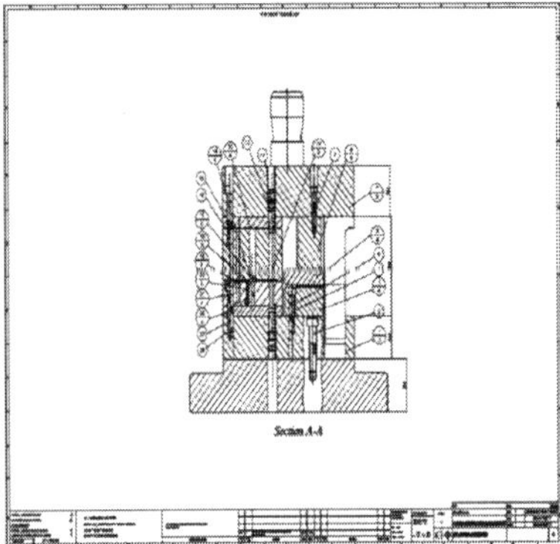

Fig. 11 Sectional view of stage tool assembly

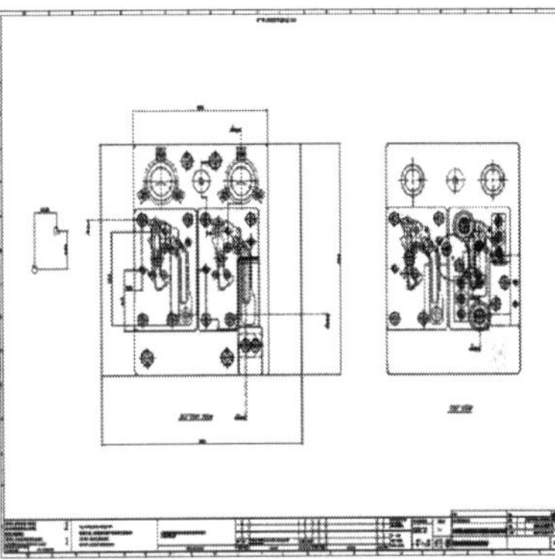

Fig. 12 Top assembly view of stage tool

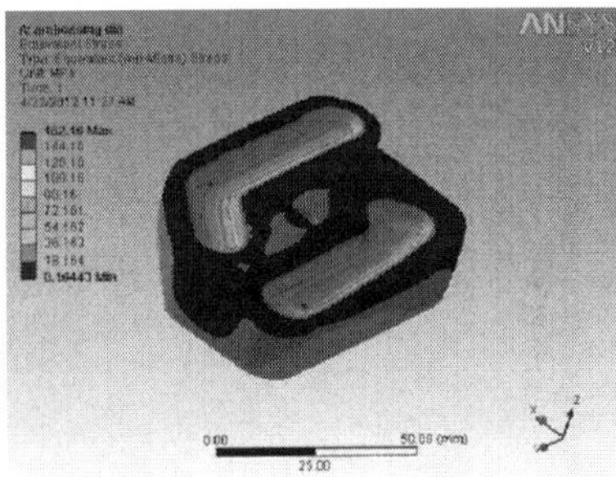

Fig. 13 Equivalent stress on embossing punch

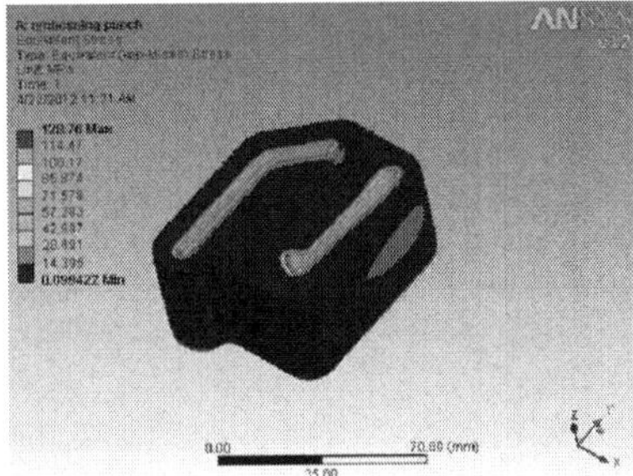

Fig. 14 Equivalent stress on embossing die

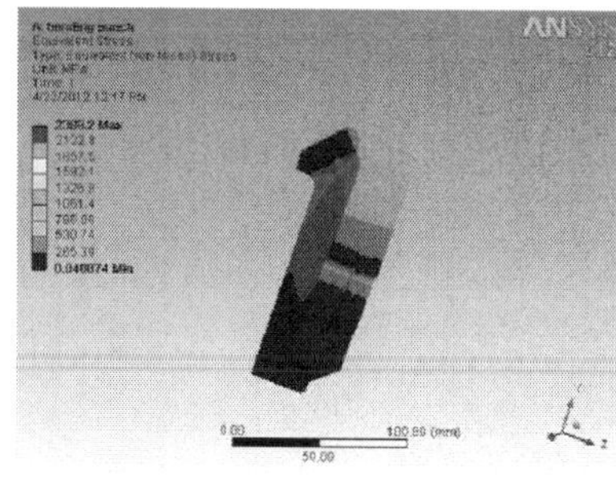

Fig. 15 Equivalent stress on bending punch

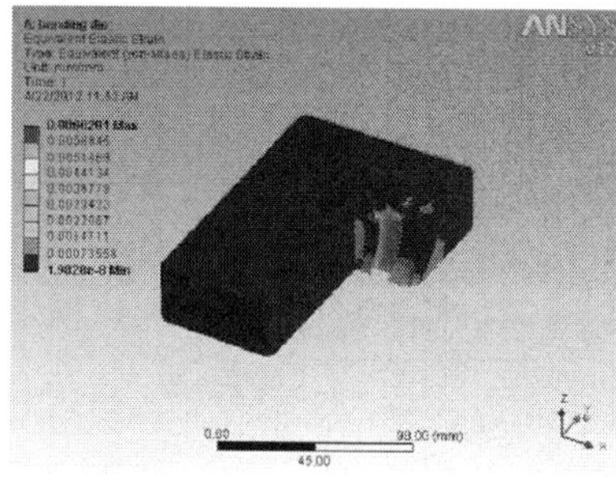

Fig. 16 Equivalent stress on bending die

Fig. 17 3D model of progressive tool assembly

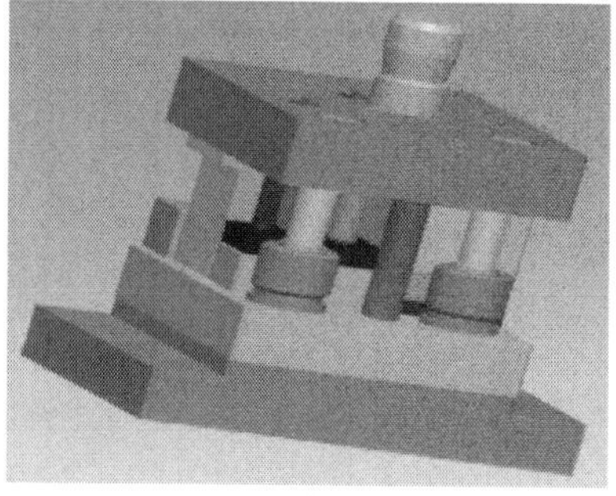

Fig. 18 3D model of stage tool assembly

Fig. 19 Manufactured progressive tool assembly

Fig. 20 Manufactured stage tool assembly

exceeds yield strength value of 2,150 MPa, so to overcome this problem, top plate thickness of stage tool was increased to 2 mm and stress induced in die was 1,390.2 MPa which was within the yield strength value and by this we can predict that design is safe shown in Figs. 15 and 16.

3D assembly model of progressive and stage tool were design to check misalignment between punches and dies as shown in Figs. 17 and 18. Later Press tool was manufactured for progressive and stage tool as shown in Figs. 19 and 20.

In initial strip layout a support punch concept was introduced in first stage of profile piercing operation and in second stage pilot piercing operation was carried along with embossing operation as shown in Fig. 21. Later tool test was carried out for progressive tool, result revealed that there was a misalignment of 200 microns in embossed profile due to improper locating operation carried in second stage of embossing, so to overcome this pilot piercing operation was adapted in first stage itself, so that proper locating operation will be made in second stage, again tool has been redesigned and tested. Result reveal that misalignment in embossed profile was reduced to 20 microns as shown in Fig. 22.

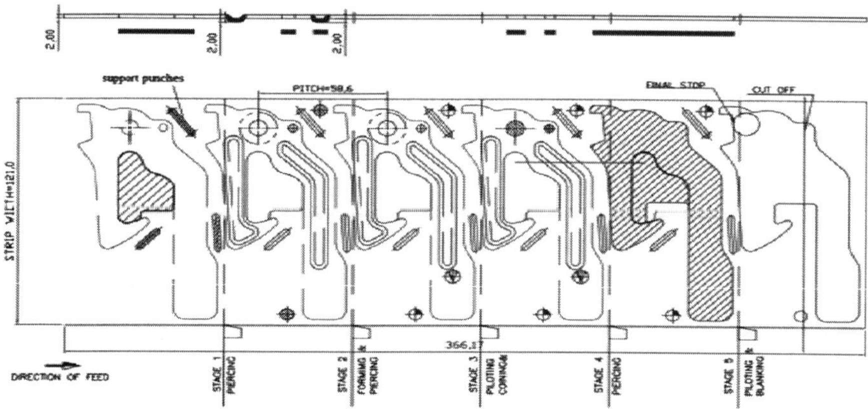

Fig. 21 Initial strip layout design of progressive tool

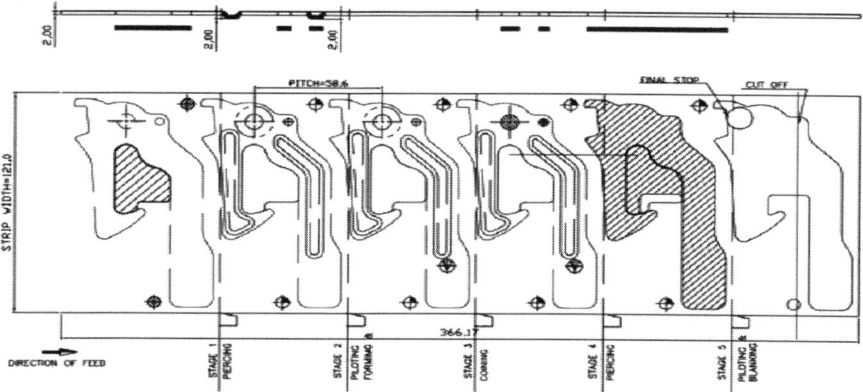

Fig. 22 Redesign strip layout of progressive tool

5 Conclusion

Design of new hook hood lock auxiliary catch with embossed profile was able to withstand higher stress value with no potential cracks when compared with flat existing hook hood lock auxiliary catch.

Use of support punch concept had reduced one stage of operation in progressive press tool, but test result revealed there was 200 microns misalignment. So use of pilot piercing operation instead of support punches in first stage and results reveal that misalignment was reduced to 20 microns. By this we can conclude that design of hook hood and press tool were optimized and was cost effective.

Acknowledgments Thanks are addressed to R.V. College of Engineering and Aditya Auto Products and Engineering PVT. LTD. faculty for its support.

References

1. Bang J, Thomas JC (2008) Optimization of a hood latch system. Bachelor of Science thesis, Worcester Polytechnic Institute, pp 46–52
2. Nye TJ (2004) Stamping strip layout for optimal raw material utilization. J Manuf Syst 19(4):239–242
3. Venkata Rao R (2004) Evaluation of metal stamping layouts using an analytic hierarchy process method. J Mater Process Technol 152:71–76
4. Shailendra K (2011) An intelligent system for selection of materials for press tool components. J Eng Res Stud 2(2):119–130
5. Eugen RI and Gavril G (2010) Finite element analysis of a forming process, using static structural (ANSYS) and ANSYS LS-DYNA. Fascicle Manage Technol Eng, NR2 9(19):1–177

Design of a Support Structure: Mechanism for Automated Tracking of 1 kWe Solar PV Power System

Pravimal Abhishek, A. S. Sekhar and K. S. Reddy

Abstract In the current scenario of increasing environmental problems and their global effects, there is a pertinent need to explore research in renewable energy technologies that do not affect the environment adversely, yet sustain the progress of mankind's growth and development. This paper presents the work intended to address the aforementioned needs by designing cost-effective solar tracking units for improving the solar power generation. Literature review reveals that tracking the Sun in both East–West direction and North–South direction can improve the power output by 25–30 %. This improvement can particularly play a significant role in the scenario of large scale grid connected solar power generation. Within this frame work, the present study attempts to develop a solar tracking system that can enable the photovoltaic solar panels to effectively trap the available solar energy by tracking the Sun. The system is designed such that it can withstand the fluctuating wind loads, remain safe even in the most adverse weather conditions, and possess the quality of frugality, ease of maintenance and repair.

Keywords Solar energy · Photovoltaic power system · Solar tracking · Bi-axial rotation mechanism

P. Abhishek · A. S. Sekhar (✉)
Mechanical Engineering/Machine Design Section, Indian Institute of Technology Madras, Chennai, India
e-mail: as_sekhar@iitm.ac.in

P. Abhishek
e-mail: pravimalabhishek@gmail.com

K. S. Reddy
Mechanical Engineering/Heat Transfer and Thermal Power Laboratory, Indian Institute of Technology Madras, Chennai, India
e-mail: ksreddy@iitm.ac.in

A. Chakrabarti and R. V. Prakash (eds.), *ICoRD'13*, Lecture Notes in Mechanical Engineering, DOI: 10.1007/978-81-322-1050-4_86, © Springer India 2013

1 Introduction

It is quite conspicuous that increasing environmental problems and their global effects are mounting a pressing need to push the frontiers of research in renewable energy technologies which do not manifest environmental adversities, yet sustain the progress of mankind's growth and development. Among such renewable energy sources, Solar Energy is a promising option. Its affluence, scalability and ease of access, makes it one of the most promising renewable energy resources.

In order to tap the full potential of such energy resource, one must design a system which can effectively capture the available solar energy and efficiently convert the captured solar energy into electrical energy. This paper is a definitive attempt to explore the opportunities in effectively capturing the solar energy by designing an optimized mechanical system and a support structure to support and rotate a set of solar photo-voltaic modules which are capable of generating 1 kWe power for about 8–10 h per day at Chennai.

1.1 An Insight into Solar Photovoltaic Technology

Solar Photovoltaic collectors, available in the market, can be broadly classified into two types, based on the nature of the collection of Solar Energy—Flat Plate collectors and Parabolic Dish collectors [1].

Flat-plate collectors are the more commonly used type of collector today. They are arrays of solar panels arranged in a simple plane. They can either be fixed in a static position, or rotated continuously to track the Sun. Tracking the Sun using automated machinery that keeps the Solar Panels facing the sun throughout the day, results in the better effectiveness in capturing the solar energy. The additional energy they take is due to the correction of facing more than compensates for the energy needed to drive the extra machinery.

Focusing collectors are essentially parabolic-dish collectors with optical devices arranged to maximize the radiation falling on the focus of the collector. These are currently used only in a few scattered areas. Solar furnaces are examples of this type of collector. However, one problem with focusing collectors in general is due to temperature. The fragile silicon components that absorb the incoming radiation lose efficiency at high temperatures, and if they get too hot they can even be permanently damaged. The focusing collectors by their very nature can create much higher temperatures and need more safeguards to protect their silicon components [1]. The present study deals with the flat-plat solar collectors.

1.2 Effect of Tracking on Performance of PV System

Solar Energy Conversion depends on two factors—Efficiency and Effectiveness. While efficiency deals with the ability in converting one form of energy into another, effectiveness deals with the ability to collect the available energy. Both these factors determine the power generation capacity of solar photovoltaic cell. At present, the efficiency levels of solar photovoltaic cells are quite limited. Therefore, tapping as much energy as possible will improve the power generation capacity significantly. In other words, for a given value of efficiency of a Solar Photovoltaic cell—the greater the effectiveness of tapping the solar energy the better the power generation capacity, irrespective of the degree of efficiency. Literature review on this subject reveals that a tracking solar photovoltaic cell can generate 25–30 % [2] more power than the static one. Achieving that increment in power production in a cost effective way is the primary goal of the present study.

Challenges in pursuit of greater effectiveness are multitude ranging from complexity in design, cost-effectiveness, exposure of mechanisms to adverse atmospheric conditions, accuracy in tracking, social acceptability etc. All the possible challenges were comprehensively taken into cognizance alongside an explicit outlay of the need matrix to embrace an 'inclusive design approach' for accomplishing the established goals.

2 Methodology

1. The methodology of the study is to perform a comprehensive design analysis of an existing solar tracking system. Computational techniques like Computer Aided Design (CAD) and Finite Element Analysis (FEA) were used to meet these goals.
2. The second step following the design analysis is to develop a multi-dimensional need matrix and a product function addressing the technical requirements and latent needs of the end users.
3. Based on the information acquired from the above processes, the final step is to conceptualize a set of feasible design solutions, deploy a decision making criteria and select the most optimal solution that can fetch maximum results and fulfill all the needs of the users. Standard principles of product design were followed in the second and third steps.

2.1 Organization of the Work Elements

The layout of the study is broadly discussed in three sections. A thorough study on the various aspects of solar tracking was conducted on a single axis tracking system. This is outlined in the Sect. 3. Following this comprehensive study, a detailed need analysis is presented in Sect. 4. With the study results and the need analysis at hand, standard principles of product design were deployed to make a conceptual design of a solar tracking system in Sect. 5. Approximate estimates of the cost of such a system were outlined in Sect. 6. Summary and important conclusions of the entire study are presented in a nutshell in Sect. 7.

3 Design Analysis of a Single Axis Solar Tracking System

3.1 Modeling of the Existing Single Axis Solar Tracking System

The existing single axis solar tracking system is configured to generate 1 kWe power for about 8–10 h per day at Chennai. It consists of six nos. of the solar modules with a capacity of 175 W each. The configuration is be made in two rows of three modules each. The base mount is made in the shape of a 'Y' and the solar panels along with the support structure are mounted on the base frame using knuckle joints. The gear mechanism and the motor are placed at the center of the 'Y' frame which connects to the shaft that rotates the solar panels. The details of the system are modeled in Solid Work and are shown in Fig. 1a and b.

The analysis involves a three stage process. The first stage comprises of evaluating all kinds of loads that are acting on the system. The second stage includes the development of solid 3D model using a CAD tool. The third stage involves a comprehensive stress analysis using an FEA tool to generate the detailed information of the stresses induced. Due attention was given to consider the use of standard procedures so that the obtained results are reliable and repeatable. The outcomes of these processes form the foundation for re-design of system taking into consideration all the engineering requirements, manufacturing and assembly aspects, reliability, frugality, safety, ergonomics and ease of maintenance and reparability.

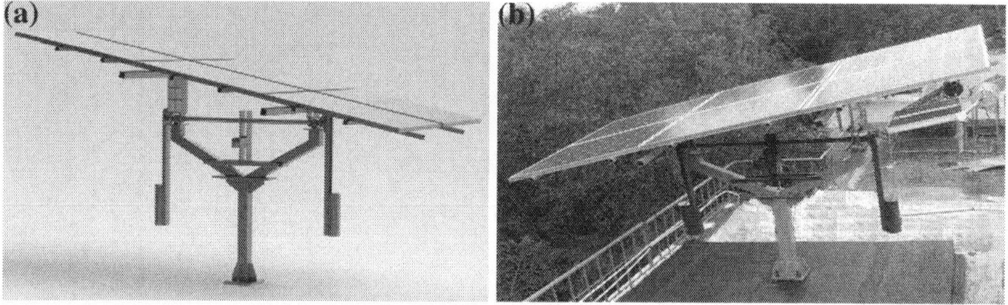

Fig. 1 1 kWe solar PV system with tracking at IIT Madras. **a** 3-D model. **b** Photograph

Table 1 Typical dimensions of a 175 W solar PV module

Parameter	Value
Power output	175 W
Length	1318 mm
Width	994 mm
Depth	46 mm
Weight	16 kg

3.2 Types of Loads

The loadings were calculated on the basis of standards given in ANSI/ASCE [3]. According to the given standards, there are five basic types of loadings on the structure.

1. Dead Loads (DL): Self-weight of the structure
2. Wind Loads (WL): Lift and drag due to flow of wind over the structure
3. Imposed Loads (IL): Sum of all the loads like rainfall, sand etc
4. Snow Loads (SL): Loads due to snow falling on the structure
5. Special Loads (SPL): Loads which are unpredictable, and impulsive

Among all the loads the most important ones considered for the analysis were the dead loads and wind loads. Since the loads caused by the other three are not quantifiable, a sufficient factor of safety was considered to account for them. Typical dimensions of a 175 W Solar PV Module are shown in Table 1.

3.3 Estimation of Loads

Six solar panels are mounted in 3×2 matrix each panel with a capacity of 175 W. The panels were supported using a grid type steel structure. Each panel weighs 16 kg and the grid frame along with the connected cables weighs around 14 kg. Therefore, the total dead load on the structure is 110 kg. While the dead loads are always constant, wind loads create a random time-varying loading on the structure. The amount of load subjected on the structure depends on the velocity of the wind, and the angle of attack of the wind, inclination of the solar panels, structural design, geographical factor and proximity with neighborhood structures. It also depends on the density of the air, but the variations due to these factors are assumed to be negligible. The flow of wind over the structure is a typical fluid mechanics problem that deals with flow of fluids over a body. Two kinds of forces are generated in such a situation—the drag force which acts along the flow of the wind and the lift force which acts on the body perpendicular to the flow of wind. These forces are calculated for a particular wind speed with varying inclination of the solar panels. The highest values of the forces generated are calculated and the structure was designed to with stand the highest load to ensure high degree of safety and reliability.

In order to calculate the wind speeds and to estimate various correction factors, standards were used based on IS 875: 1979 [4]. The structure falls within the scope of the standards and they are aptly applicable to the current analysis. Chennai belongs to the wind zone 5 of India. Basic wind speed is the based on the peak gust speed averaged over a time period of round about 3 s and corresponds to 10 m height above the mean sea level in an open terrain. The details of the winds speeds are as below:

Designed wind speed is given as:

$$(V_Z) = V_b \times k_1 \times k_2 \times k_3 \tag{1}$$

Location: IIT Madras, Chennai, Tamilnadu (13° N, 82° E)
Wind Zone: 5
Basic wind speed (V_b) = 44 m/s
k_1 = probability factor; the existing system was designed for 25 years. Thus $k_2 = 0.9$
k_2 = terrain height and height factor; the terrain of IIT Madras belongs to category 3. This category includes terrains with numerous closely placed structures. The structure is placed on the roof top whose total is round about 15 m. k_2 is determined based on both height and terrain characteristics. The above conditions suit Class A criteria and thus $k_2 = 0.9$
k_3 = This factor deals with the local topographic features like hill, cliffs, valleys and escarpments etc. which can significantly affect the wind speed. Since the location has no such features in the vicinity, the factor is chosen to be one ($k_3 = 1$).

Using the above factors, the design wind speed is calculated based on the basic wind speed. The design wind speed so arrived is 35 m/s.

The speed data of the above standards is used to ensure the structures are stable even when there is a significant increase of the wind velocity for a short time. However, the average wind speed at the ground level in Chennai is round about 12 m/s. The analysis on the existing design was done both at the 12 and 35 m/s and the stress values have been computed. The proposed new design was conceptualized to ensure maximum stability with minimum amount of material.

FEA analysis was done for the current solar tracking system to investigate the maximum induced stress, and it was further elaborated in the design of the new system for a different configuration of the solar panels. The temperature of air was set to be 25 °C and the density of the air was chosen to be 1.17 kg/m^3. The values of the lift and drag forces acting on the solar panels were computed and the results are shown in Table 2. Pressure and velocity contours and the lift and drag forces acting on the solar panels were obtained from the above analysis. Typical velocity contour and pressure contours are shown in Fig. 2a and b.

Table 2 Lift and drag forces due to flow of air on the solar panels

Angle (θ)	Lift (N)		Drag (N)	
	35 m/s	12 m/s	35 m/s	12 m/s
0°	92.45	NA	776.9	NA
30°	−8,935	−717	5,678	456
45°	−10,832	−826	11,564	859
60°	−10,020.2	−662	17,940	1,189
80°	−5,080	−248	19,780	1,478

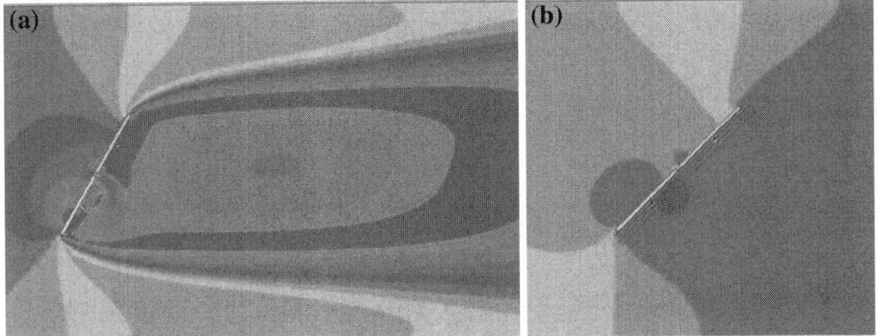

Fig. 2 FEA analysis of the solar PV system. **a** Velocity contour, **b** Pressure contour

Stress Analysis for the above structure was done using a FEA Tool. The material used for the structural support is AISI 1020 structural steel. Since it is a ductile material and the situations involve combined tensile and shear stresses acting on the same shape, defining an effective stress that create equivalent amount of distortion would be easier to conceive. Therefore, Von-Mises effective stress was used to determine the maximum stress induced in the structure. The maximum value of von-mises equivalent stress at a wind speed of 12 m/s is 81.65 MPa, while it is 220 MPa when the wind speed is 35 m/s. These values should be compared against the yield strength of the material which is 280 MPa. Therefore, the structure is under critical loading when the wind speed is around 35 m/s.

4 Need Analysis and Product Function

A complete identification of the need analysis is imperative prior to the design process. It would provide the designer detailed information of what to develop and also serves as resource material for future reference and research on improvisation of the system. Requirements for the system are summarized in Table 3 with their relative importance.

Table 3 Need Analysis Matrix

Need	Description	Rating
Automated E-W tracking	The system has to track the Sun every day from east to west for about 12 h	10
North–south tracking	Rotation along the North South direction across 1 year to account for the change in solar angle of incidence	10
Ease of use	System should be easily operable, maintainable and repairable	9
Structural stability	Should withstand the loads caused by dead weights and the wind loads. The structure should be stable at least for a standard design speed of wind	10
Dual RPM rotation	The system should have an incorporated dual RPM rotation in order to facilitate quick return motion	10
Frugality	The design should be frugal to ensure that the capital cost invested in solar power plants is the least	9
Scalability	The design should be scalable to be used for smaller are larger systems of similar configuration	8
Flexibility	The system should be flexible that it can cater to the needs of any solar module purchased from the market	9

Based on the above requirements, a set of functions were listed out which should be performed by the tracking system. These functions form the production function for development of new design. Various types of methods in which these functions could be performed were explored. Out of all the functions, those set of functions which could be feasible for further study were filtered. From the feasible domain, a single solution was selected based on relative advantages and the final functional configuration was developed by integrating the individual functional units [5].

The most basic function is to track the Sun in the East–West direction. In order to achieve this function, various actuation systems and mechanisms have been studied. The relative advantages and disadvantages are mapped in Table 4 for driving mechanisms and Table 5 for actuation systems. An appropriate choice of the systems was made based on needs. This exercise decision making was followed for all the other functions that need to be performed and a conceptual design was made incorporating all the solutions obtained in the decision making process. These conceptual sketches were further developed into CAD models and a detailed

Table 4 Relative advantages and disadvantages of various driving mechanisms

Mechanism	Advantages	Disadvantages
Power screws	Most important advantage is the ability of self-locking [9, 10] provides one step reduction in the RPM by serving as one of reduction pairs	Axial thrust forces induced on the shaft upon which the gear is mounted
Ratchet Mechanism	Reduction of the speed with least number of gears	Cannot provide a self-lock system. A separate mechanical system to lock the shaft at every position is required

Table 5 Comparison of actuation systems

Feasible actuation system	Advantages	Disadvantages
Piston	Eliminates the need for a multiple gear reductions as in the case of a motor driven mechanism. Improves structural simplicity	Cannot take the load of the solar panels. A supplementary support system needed. Fluid sensitivity to surrounding temperatures can cause inaccuracies. Pistons are costlier and their reliability is relatively low when compared to motors
Normal DC motor	Cheap and widely available. Ease of installation service and repair	Requires a series of reduction gears to reduce the speed to the required level. Different set of gears have to be used for providing dual RPM motion
Stepper motors	Very compact and a minute angular rotations can be controlled with high accuracy	Highly expensive

design was made in the later stages. The following section gives a brief overview of the new design features.

5 Design Features of the New Solar Tracking System

Taking into consideration the complexity and the cost benefit analysis, the design was equipped with a manual North–South Rotation instead of an automated rotation. The angle that needs to be rotated is quite and the cost-to-benefit ratio of such a mechanism is intuitively quite low. Therefore, the system has one automated East–West tracking mechanism and a provision for manually adjusting the North South inclination across the year. This mechanism is shown in Fig. 3.

A quick return mechanism is required to revert back the solar panels to the initial position to start tracking the sun for the next day. This rotation should be quicker than the regular tracking which is at an RPM of 0.00645. Since the driving mechanism is chosen to be power and the driver, a DC motor, a simple gear changing mechanism can provide dual RPM rotation to the worm. This is also shown in Fig. 3. The worm is attached with two spur gears at both the ends and motor can be housed inside the column and the worm can be rotated by two different gears trains to achieve dual RPM

Trusses were designed to bear the loads transferred from the solar panel frames. These Trusses are cheap and easy to make; most importantly, they give a great flexibility in varying height of the solar tracking system, as shown in Fig. 3. All the components of the new design were subjected to stress analysis using an FEA

Fig. 3 Design of the new solar tracking system. Inset—Bi axial rotation mechanism

tool. The yield strength of the material is around 384 MPa, and safety factor of 2.5 was chosen which will peg the upper limit of allowable stress at 153.6 MPa. Against this maximum allowable stress, the induced stress on the structure was 98.91 MPa which is well within the allowable stress limit. This brings the study to the last stage of the design process.

6 Economics of Solar Power

Compared to conventional power, (coal based thermal power) solar energy is four times more expensive [6]. Nevertheless, cost is not always the determining criteria. Solar power is much flexible and scalable which it more attractive in areas where access to conventional power sources is forbidden due to geographical impediments. Efforts are also being made to improve the efficiency of the solar panels. At the same time, national governments across the globe are assisting effective penetration of solar power by various market interventions. Currently, India is implementing the National Solar Mission, under the National Action Plan for Climate Change (NAPCC) which envisages installing 20,000 MW of solar power by 2022 [7]. Taking into consideration all the techno-commercial factors, the Central Electricity Regulatory Commission (CERC) in India has come up with a normative capital cost for Solar Photovoltaic power generation. It must be noted that the CERC normative costs do not distinguish tracking solar photovoltaic systems and static photovoltaic systems. For the FY 2011–2012, the normative capital cost was pegged at 1442 Lakh/MW [8]. The estimated cost of solar tracking system is round about 1,650 Lakh/MW which is a significantly high when compared to the normative limits. However, the increase in power production by 22.65 % can deliver an additional power generation of 2.05 kWhr power per day. The CERC estimates assume that the operational timer period of solar power generation system is 25 years. Therefore, the additional power harnessed by tracking systems across 25 years has a much greater value which can offset high capital cost. Also a single unit of solar tracking system can save 2,784 kg of carbon dioxide emissions per unit power generations, which can earn 2.78 carbon credits per unit power generation.

7 Summary and Conclusions

The primary objective of the study is to design a tracking system that can support solar panels which are capable of generating 1 kWh electrical energy. The methodology adopted for carrying out the study is to first perform a comprehensive design analysis of an existing structure, and then use the information to develop a new design by adopting standard principles of product design.

An improvised design was made catering to a multi-dimensional need matrix. The key features of the new design include a biaxial rotation of the solar panels, dual rpm drive for quick return, reduced stress levels in the structure along with reduced usage of the material and a robust safety mechanism to withstand adverse cyclonic conditions. 25 % material usage reduction was achieved from the previous design with a simultaneous increase in the ability to track the Sun and produce more power.

The support structure need not be necessarily a single column. Use of more legs might save more material. But the idea behind using a single leg is to ensure that the foot print of the structure is as low as possible on the ground. This is can allow the designer to increase the height of the column according to the needs. For instance, the height of the solar tracking system can be doubled from the ground so that more clearance is available at the ground level. Since, the footprint is least possible, more land would be available for any other productive usage instead of getting underutilized for solar power production.

References

1. Mousazadeh H (2009) A review of principles and sun tracking methods for maximizing solar outputs. Renew Sustain Energy Rev 13:1800–1818
2. Renewable Energy Report—Global Status (2011) Renewable Energy Policy Network for 21st Century

3. ANSI/ASCE 7-88 ACSE (1990) Minimum design loads for building and other structures
4. IS875:1979 Code of practice for design loads (other than earthquakes) for buildings and structures
5. Akiyama K (1991) Function analysis: systemic improvement of quality performance. Productivity Press, Portland
6. Avijik Nayak (2012) Cost economics of solar kWh. In: Energitica-India, vol 22. National Productivity Council, Kolkata (Jan/Feb 2012)
7. Jawaharlal Nehru National Solar Mission, Mission Document, Government of India
8. Renewable Energy Tariff for the FY 2011–2012, Order-256-2010, Central Electricity Regulatory Commission, Government of India
9. Norton RL (2007) Machine design: an integrated approach. Dorling Kindersley Publishers, India, ISBN 81-317-0533-1
10. Hudley DW (1984) Handbook of practical gear design. McGraw-Hill, New York, p 366

Automated Brain Monitoring Using GSM Module

M. K. Madhan Kumar

Abstract I propose to develop a project to simulate an abnormal bio-signal synthetically for analysis and testing of the device built for monitoring a person's brain status. This system acquires and analyzes neural signals with the goal of creating a communication between the brain and GSM mobile. It is a low cost solution for an automated emergency response system and also provides constant patient monitoring without affecting their day to day activities. This system ensures safety as well as mobility at all times.

Keywords Automated constant monitoring · Brain status · GSM mobile · Neural signals

1 Introduction

Monitoring the brain using **Electroencephalograph (EEG)** is usually done with maximum steps and a care taker need to monitor the patient's status consecutively. This system is used for acquiring normal brain activity and also for monitoring abnormal brain activity in patients.

For example, Coma patients in intensive care will be continuously monitored by **Health care taker (HCT)**. In this concept, the HCT can get earlier information after patient's brain decides. The brain is sensed and the brain status is informed to HCT through text (SMS) or call to their mobile which is called Biotelemetry system.

M. K. Madhan Kumar (✉)
Department of ECE, G.K.M College of Engineering and Technology, Chennai, India
e-mail: newtalantons@gmail.com

A. Chakrabarti and R. V. Prakash (eds.), *ICoRD'13*, Lecture Notes in Mechanical Engineering, DOI: 10.1007/978-81-322-1050-4_87, © Springer India 2013

Other than Coma patients, for mentally retarded persons even for senior citizens it will be helpful in finding out whether they are in hungry, tension and also whether in sleeping. For normal people it will help in knowing their situation like whether he/she is alcohol or drug addicted, attitude and behavior change and so on from any distance. Below, we made the Bridge to understand the relations between 'Brain activity' and GSM Mobile is provided by the EEG.

2 Innovations

Emergency response system helps us to identify the consciousness of the patient and the Medical history of the patients stored within the GSM module or ROM that helps with immediate and effective treatment. Integration of two sensors (brain and pulse) using a single processor to improve the accuracy of the measurements. The significance of this project is to continuously monitor the brain status via GSM module.

3 Salient Features

Multiple parameters such as Brain waves and pulse rate are sensed. The sensed report makes emergency response system and can be used for first aid info and patient medical history stored as SMS in the mobile phone at all timcs. The system is low power consumption with Battery powered and efficient power management system. It is the motion Capture of brain in continuous time.

4 Motivations

In most emergency situations we are lacking of quick access to hospitals/trained medical care and Sometimes, emergencies even go unnoticed for a long period. Need for Low cost portable solutions to enable even in small nursing homes, dispensaries and can be carried out everywhere. It increases the need of continuous in-house monitoring of patients without affecting mobility in hospitals. Availability of life saving first aid support cannot be expected in all situations, so a novel monitoring system addressing these issues is needed. Hence this system improves necessity for continuous monitoring of patients for neurodegenerative and neuropsychological disorders. The caretaker can monitor the patient's mind status via mobile phone (text message or call).

5 Block Diagram

5.1 Figures

The system will sense the Brain waves and pulse rate of patients using sensors placed in our body (Figs. 1, 2 and 3).

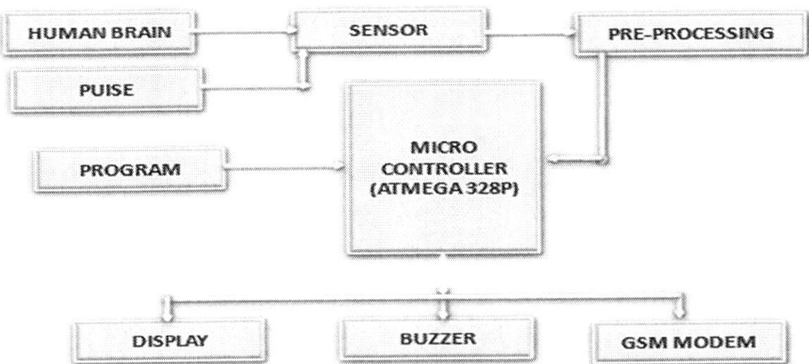

Fig. 1 General block diagram

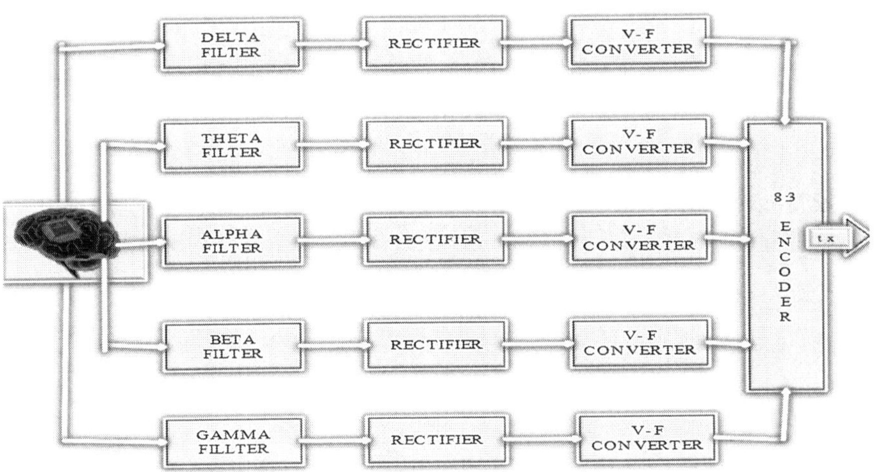

Fig. 2 Transmitter side

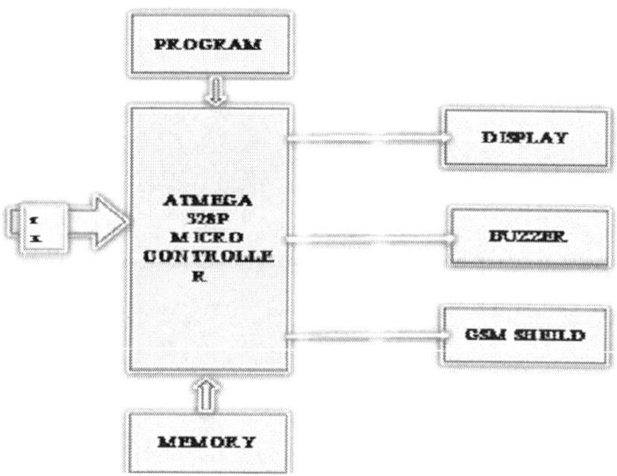

Fig. 3 Receiver side

5.2 In Case of an Emergency

The system will communicate with a GSM module and send the status to emergency (or pre-defined custom) numbers that can be along with the GPS co-ordinates for ease of locating the person.

Necessary information is displayed/played out loud in buzzer for assisting nearby people around, to provide immediate first aid.

5.3 In Case of Constant Monitoring

All measurements and diagnosed report will be transmitted to the paired mobile, where it is stored in a micro-SD card. Thus, these measurements keep a constant check on the patient's health conditions.

Previous medical histories of the patient are also stored in the mobile. This report would be very helpful in case of emergencies, for any physician to treat immediately.

6 Technologies

1. **EEG**: For monitoring Brain waveform.
2. **Induino board**: Evaluated board with ATmega328p controller.
3. **GSM Shield**: GSM modem for transmitting SMS (Bio telemetry).

Biosignal: Any signal measured and monitored from a biological being, although it is commonly used to refer to an electrical biosignal.

Neuro-Signal: Neuro means brain; therefore, 'neuro-signal' refers to a signal related to the brain.

6.1 What is EEG?

An electroencephalograph (EEG) is the recorded electrical activity generated by the brain. In general, EEG is obtained using electrodes placed on the scalp with a conductive gel. In the brain, there are millions of neurons, each of which generates small electric voltage fields.

The aggregate of these electric voltage fields create an electrical reading with which electrodes on the scalp are able to detect and record. Therefore, EEG is the superposition of many simpler signals. The amplitude of an EEG signal typically ranges from about 1–100 μV in a normal adult, and it is approximately 10–20 mV when measured with subdural electrodes such as needle electrodes.

The 10–20 international system is used as the standard naming and positioning scheme for EEG measurements (Figs. 4, 5).

The original 10–20 system included only 19 electrodes. Later on, extensions were made so that 70 electrodes could be placed in standard positions. Generally one of the electrodes is used as the reference position, often at the earlobe or mastoid location.

6.1.1 Placement of Electrodes

The electrodes placed in right side, numbers are even and left side numbers are odd.

Start: Nasion to Inion
Position: FP2—FZ—CZ—PZ—O2
Second: Ear to Ear

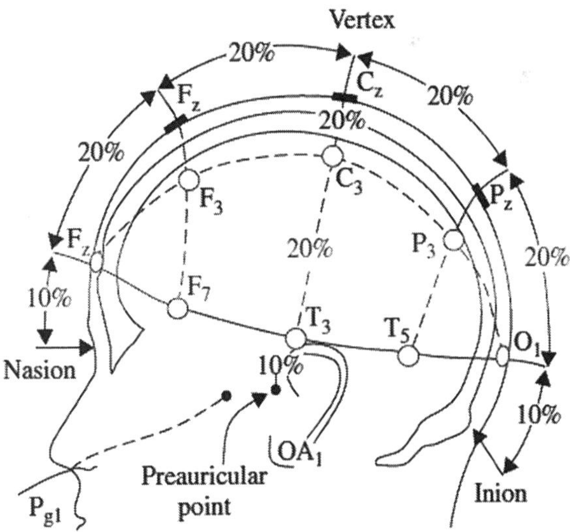

Fig. 4 Electrode placements-1 [15] (*side view of head*)

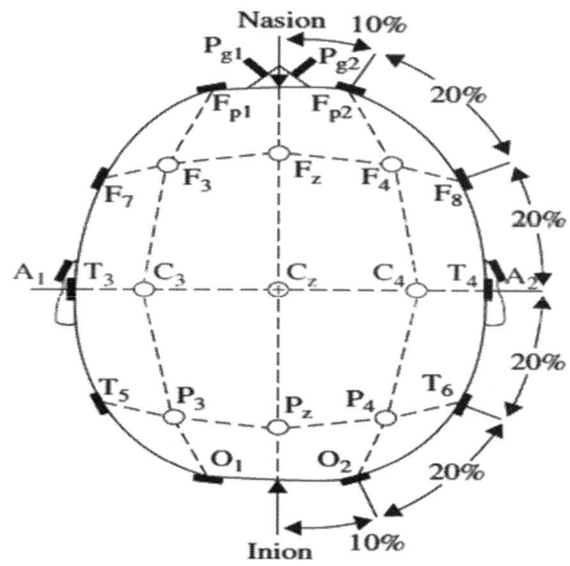

Fig. 5 Electrode placement-2 [15] (*top view of head*)

Position: T3—C3—C2—C4—T4
Third: Full circumference around
Position: Starts from FP2
Left side ≫ FP1—F7—T5—O1
Right side ≫ FP2—F8—T6—O2
Last: Marks between located
Position: F3—F4—P3—P4

6.1.2 Brain Waveforms

EEG is generally described in terms of its frequency band. The amplitude of the EEG shows a great deal of variability depending on external stimulation as well as internal mental states. Delta, theta, alpha, beta and gamma are the names of the different EEG frequency bands which relate to various brain states.

The intensities of the brain waves on the surface of the brain may be 10 mV; where as those recorded from the scalp have smaller amplitude of approximately 100 μV. The frequency of these brains ranges high in Hz but due to skull we can obtain the range of 0.5–100 Hz. The character of the brain wave is highly dependent on the degree of activity of the cerebral cortex.

Table 1 3 Brainwave frequency (HERTZ) and mental states listing

S. no	Brain wave type	Frequency range (Hz)	Mental status and conditions	Waveform
1	Delta	0.1–3	Deep, dreamless sleep, non-REM sleep, unconscious	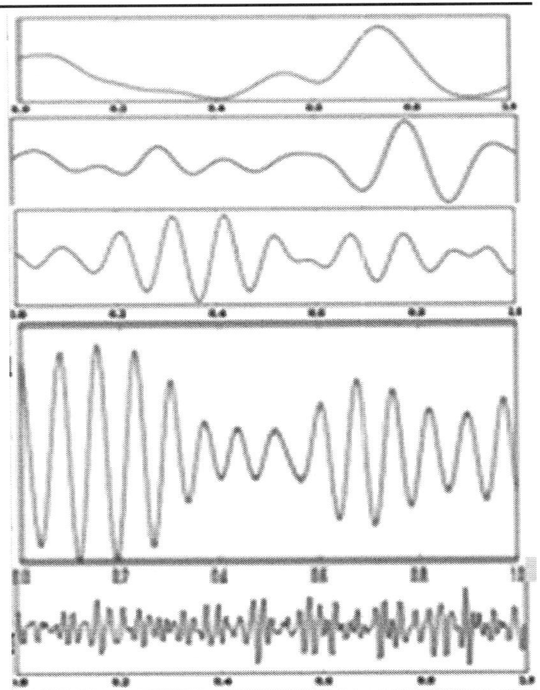
2	Theta	4–7	Creative, recall, fantasy, imaginary, dream	
3	Alpha	8–12	Relaxed, but not drowsy, conscious	
4	Low beta	13–15	Relaxed yet focused, integrated works	
5	Midrange beta	16–20	Thinking, aware of self and surroundings	
6	High beta	21–30	Alertness, agitation	
7	Gamma	30–100	Motor Functions, higher mental activity	

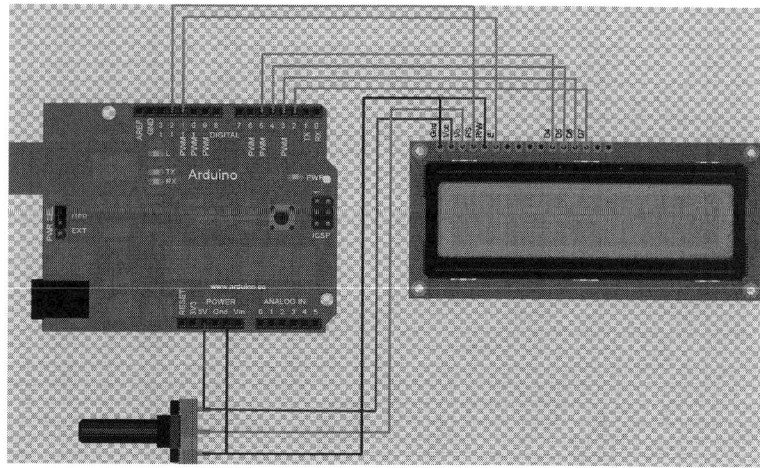

Fig. 6 Induino board with LCD display

6.1.3 Brainwave Frequency (HERTZ) and Mental States Listing

These frequencies are of all types; light, sound, electrical, etc. The two- or three-character source codes after each frequency [16] (Table 1).

6.2 Induino Board

The InduinoX is an Indian made clone of the Arduino Duemilanove with a host of added features. It is inbuilt with ATmega168 microcontroller. It has 14 digital input/output pins (of which 6 can be used as PWM outputs), 6 analog inputs, a 16 MHz crystal oscillator, a USB connection, an ICSP header, and a reset button (Fig. 6).

6.2.1 Specifications

Microcontroller	ATmega328
Operating Voltage	5 V
Digital I/O Pins	14 (of which 6 provide PWM output)
Analog Input Pins	6

Fig. 7 GSM modem

6.3 *SIM900 GSM/GPRS Shield for Arduino–IComSat v1.0*

IComsat is a GSM/GPRS shield for Arduino and based on the SIM900 Quad-band GSM/GPRS module. It is controlled via AT commands (GSM 07.07, 07.05 and SIMCOM enhanced AT Commands), and fully compatible with Arduino (Fig. 7).

7 Prototypes

Instead of EEG machine, we are using Brain Wave frequency oscillator to show immediate action where it finds easier to explain. It produces the frequency of 0–100 Hz. Voltage is of up to 3 V that supports Microprocessor (Figs. 8, 9).

7.1 *Results and Observation*

By varying the frequency (brain) in the prototype, the mental status which is coded with respect to frequency is received by the HCT. We have tried with patient in: (1) coma, (2) retrieved from coma, (3) person in hunger, (4) tension and (5) in sleep and got successful result by text format in mobile.

8 Future Works

The Entire kit can be made compact inside a wrist watch. An additional SD card storage can be provided in the wrist watch to store all the relevant brain activity, which can be used for future references. Neuro sky can be used to detect brain activity. The cost of this project will be around 10,000 (Figs. 10, 11).

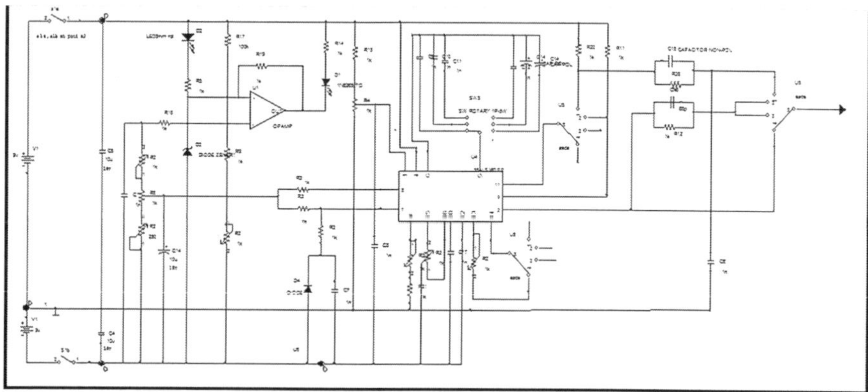

Fig. 8 Circuit diagram of brain frequency oscillator

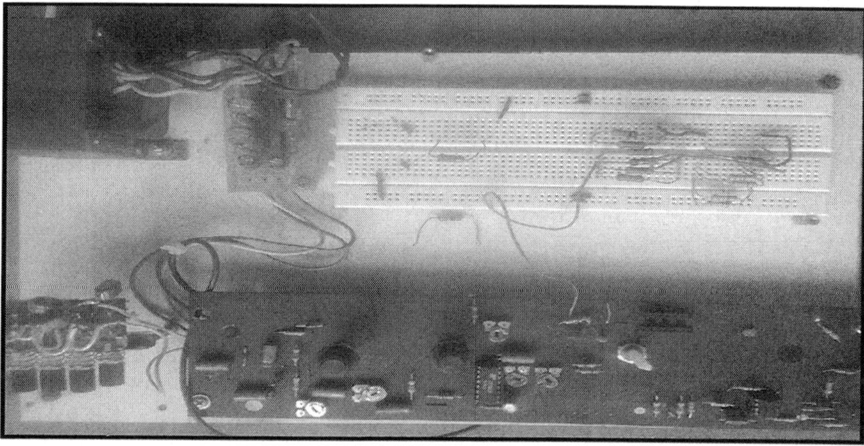

Fig. 9 Snap shot of brain frequency oscillator

9 Difficulties Faced

1. Minute variations in frequencies are complex to handle for the prototype.
2. Brain frequency oscillator produced voltage of maximum 8 V where micro controller limit is up to 5 V.
3. Access to brain functioning and the properties of waves were hard due to the multi disciplinary nature of the project.

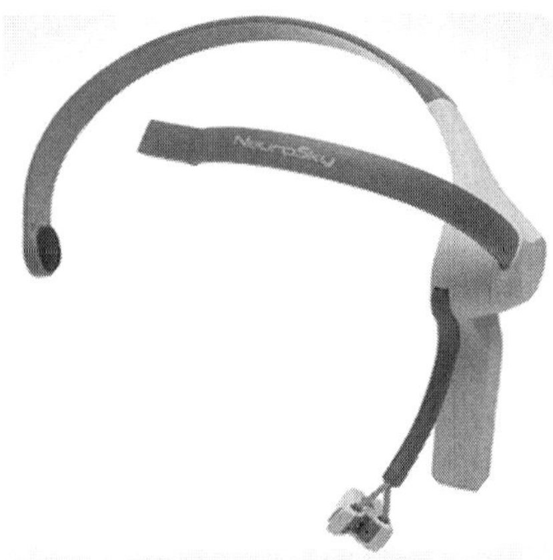

Fig. 10 Neuro sky

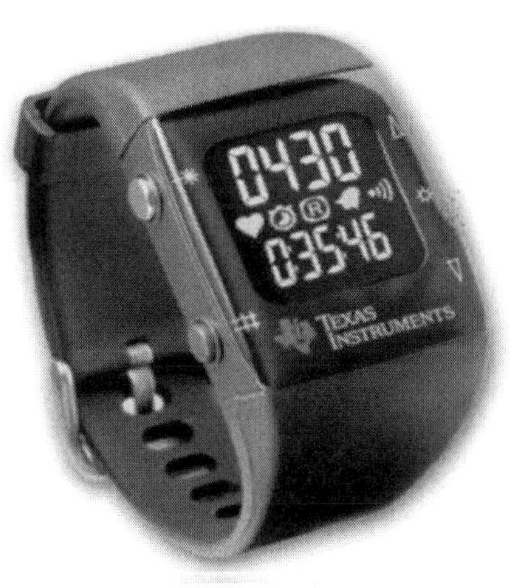

Fig. 11 T.I. wrist watch

10 Solutions to Problems Faced

1. We found that the Brain frequency will be constant for short duration.
2. Limited the voltage using Voltage rectifier and made 3 V to pass on micro controller.
3. Queried with doctors, research students and Bio medical dept. engineering college students and professors of various colleges for collecting data's.

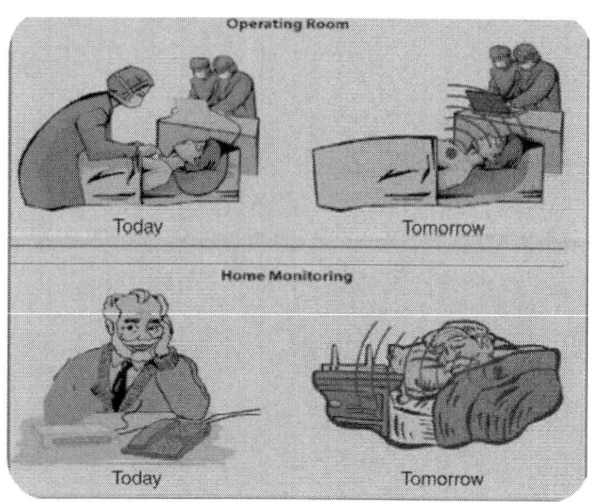

Fig. 12 Precaution is better than curing

11 Advantages

- Patients can be monitored via mobile phones.
- This can be utilized by the senior citizens who feel hard to communicate.
- Storing of Neural signals can be used for Investigation and for future medical affairs (Fig. 12).

12 Applications

For Monitoring

- Mentally retarded person
- Coma patients
- Senior citizens
- Defense purpose

13 Conclusions

In summary, it is possible to interlink brain and a mobile phone with an EEG machine. Monitoring the brain is usual but notifying instantaneous brain status or mental activity like hunger, anger and tension is a new thing we have implemented with biotelemetry. Thus it can be concluded that it is the immediate system that can be more comfortable for patients as well as for care takers in emergency situation.

Acknowledgments This project prototype has been done and success fully submitted to final phase of ***PDMA India innovation contest*** (***National level student design contest*** **2011–2012**). Through PDMA, ***LAP Lambert Academic Publishing, Germany*** asked this project for *publishing in their magazine*. This project idea has discussed with Dr. Premnath Kishan, PhD in ***National Institute of Mental and Neuro Science***, Bangalore. The overall project activity is monitored by Dr. D. Balasubramaniam, PhD, *HOD of GKM college of engineering* and Mrs. Su. Suganthi, *Asst. Professor of GKM college of engineering*. B. Ananda Narayanan, *M. Tech. IIT,* R. Anand final *year ECE from RMK Engineering College* and R. Sujitha *final year ECE from Jerusalem college of Engineering* supported well to finish this project.

References

1. Parelus P, Principe J, Rajasekaran S (eds) (2001) Making sense of brain waves: the most baffling frontier in neuroscience. Biocomputing. Kluver, New York. Chap. 3, pp 33–55. Lecture: "International conference on biocomputing" at the University of Florida, Gainesville FL 25–27 February 2001 Organizers: Parelus P, Principe J, Rajasekaran S
2. Harper R, Rodden T, Rogers Y, Sellen A (eds) Being human: human-computer interaction in the year 2020. Microsoft Research Ltd, Cambridge, ISBN: 978-0-9554761-1-2
3. Brain fingerprinting From Wikipedia, the free encyclopedia
4. Brain Wave Signal (EEG) of NeuroSky, Inc. December 15, 2009
5. http://www.youtube.com/watch?v=HVGlfcP3ATI&feature=fvwrel
6. http://www.youtube.com/watch?v=yI_8cx4m0Is&feature=related
7. Langleben DD et al (2005) Telling truth from lie in individual subjects with fast event-related fMRI. Hum Brain Mapp 26:262–272
8. Freudenrich C, Boyd R How Your Brain Works
9. Figueiredo N, Silva F, Georgieva P, Tomé A. Advances in non-invasive brain-computer interfaces for control and biometry. Department of Electronics, Telecommunications and Informatics Institute of Electronic Engineering and Telematics of Aveiro (IEETA) University of Aveiro Portugal
10. Makeig S Mind monitoring via mobile brain-body imaging. Swartz Center for Computational Neuroscience, Institute for Neural Computation, University of California, San Diego
11. The effect of high frequency radio waves on human brain activity: an EEG study Ke Wu, 1 Amirsaman Sajad, 2 Syed A. A. Omar, 3 and William MacKay 4 1Third Year Undergraduate, Human Biology Specialist, Genes, Genetics and Biotechnology. 2Fourth Year Undergraduate, Physics and Human Biology Double Major, University of Toronto. 3Fourth Year Undergraduate, Human Biology specialist in Genes, Genetics and Biotechnology, University of Toronto. 4Professor at Department of Physiology, University of Toronto; email: william.mackay@utoronto.ca. Corresponding author: Ke Wu, ke.wu@utoronto.ca
12. Baporikar V et al Wireless sensor network for brain computer interface. Int J Adv Eng Sci Technol 8(1):75–79
13. IEEE/Bio-medical papers of EEG
14. Sanei S, Chambers JA EEG Signal Processing. Centre of Digital Signal Processing Cardiff University, UK, p 17

15. FBA Fundacio de Biofisica Aplicada, Material de apoyo, Brainwave Frequency Listing
16. Ghiyasvand MS, Guha SK, Anand S, Deepak KK A new EEG signal processing technique for discrimination of eyes close and eyes open. Centre for Biomedical Engineering, Indian Institute of Technology New Delhi, India. Department of Physiology, All India institute of Medical sciences
17. Anderson A III, Wegener C, Thome R, Shikha, Lorenz M (2005) Portable electroencephalogram biofeedback device. Final Report Update: 16 Dec 2005 BME 200

Part X
Design Training and Education

Mapping Design Curriculum in Schools of Design and Schools of Engineering: Where do the Twains Meet?

Peer M. Sathikh

Abstract Schools of engineering in established universities such as Purdue offer courses in engineering design which, '… allows (students) to apply the fundamentals of engineering and science to solve open-ended design problems' and to learn a structured problem-solving process in the broad context of product design, considering marketing, problem definition, conceptual design, design evaluation, detailed design, manufacturability, and economic feasibility. To anyone teaching in a design school, it sounds like industrial design taught in a school of design. If both the schools proclaim that they teach 'design', then what are they teaching? Where is the difference? Where do they overlap? Where are the commonalities? This paper sets about mapping the typical courses taught at schools of design and schools of engineering offering undergraduate degrees in industrial design and mechanical engineering respectively. Through this the author highlights the possibility of a hybrid programme, which may represent the real world nature of design in the built environment today.

Keywords Engineering design · Industrial design · Education

P. M. Sathikh (✉)
School of Art, Design and Media, Nanyang Technological University,
81 Nanyang Drive, Nanyang 637458, Singapore
e-mail: peersathikh@ntu.edu.sg

A. Chakrabarti and R. V. Prakash (eds.), *ICoRD'13*, Lecture Notes in Mechanical Engineering, DOI: 10.1007/978-81-322-1050-4_88, © Springer India 2013

1 Introduction

Design, together with creativity and innovation, has become a common word in the Vocabulary of the 21st century, seen and heard more often, be it in the newspapers or social media or television, etc. As an industrial designer the author is left wondering what 'design' means to different people. Is it the shape; the function; the software, the performance? As an educator, the author is left with even more intriguing question, 'what do we mean by *teaching design*?' These questions became more apparent with more and more students facing the dilemma faced by students considering design education options between an engineering school and a design school? The world over, design (industrial design or product design) is a four year degree programme.[1] Product design is also taught as a specialisation or as a part of mechanical engineering in four-year engineering degree programmes in most of the countries. Which way should one go?

What, is the difference between design as taught in design schools and engineering schools? Where lie the focus and the difference? Is there an overlapping area that can be exploited to develop an undergraduate programme that can produce designers who are can fit into a 'total product development' environment.

The objectives of this paper are:

- To look at the curriculum, syllabus, aims and objectives of design courses/ programmes at undergraduate level offered in design schools and engineering schools, taking examples of each.
- To identify the differences between these two.
- To present a visual map which shows the elements of the two courses within the same spectrum.
- To explore the possibility of a curriculum that can combine elements of both which could be relevant to the future direction of the globalised world.

2 Curriculum, Syllabus, Course Aims and Objectives (Outcomes)

Heywood [1] defines curriculum to be *the formal mechanism through which intended educational aims are achieved.* Curriculum pertains to the entire length of study in a programme and consists of a set of courses, which allows an institution to achieve the overall aims and objectives of a particular programme and/or course as

[1] While most offer four year degree programmes (BFA, BID or BDes), some schools, especially in the United Kingdom offer a three year BA programme in design which usually requires students to have completed a one year foundation in art and design which then adds up to a four year equivalent.

it was originally intended. Syllabus, on the other hand could be defined as an outline or overview of a particular course setting out the topics, reading materials, time lines, expectations of class engagement by the lecturer, expected learning outcome (objectives) from the student at the completion of that course. In essence, studying the curriculum, syllabus and the aims and objectives of individual courses of an institution could be an indicator as to what the students may turn out to be at the end of the studies. A study of the course of aims of design programmes in engineering schools against those in design schools could then lead to a better understanding of 'design' as taught in engineering schools and design schools.

3 Aims and Outcomes for Engineering Design in Some Engineering Schools

The following have been taken from either the brochure/prospectus or the website of the respective universities mentioned. The highlighting in bold letters is by the author.

3.1 Purdue University

Students in the School of Mechanical Engineering at Purdue University have many opportunities to participate in design projects, which allow them to apply the fundamentals of engineering and science to solve open-ended design problems. Sophomores in the Cornerstone design course learn a structured problem-solving process in the broad context of **product design, considering marketing, problem definition, conceptual design, design evaluation, detailed design, manufacturability, and economic feasibility**. Seniors in the Capstone design course bring together design process knowledge with technical analysis capabilities that they have learned in the core curriculum to **develop their own conceptual designs into working prototypes**.

3.2 Nanyang Technological University, Singapore

This flagship four year Mechanical Engineering degree programme has been meticulously tailored to meet the needs of the local economy and beyond…The programme has three streams to cater to our students' differing needs and interests. Most students will enrol in the mainstream while approximately **20 % of the students in each cohort may choose to pursue an in-depth specialisation in either Design or Mechatronics** starting from their second year of study…The Design Stream's emphasis (is) on creativity, technology and design methodology.

Trains you to be an innovative design professional, competent in engineering. Courses you can expect to explore:

- Creative Thinking and Design
- Product Presentation.

4 Engineering Education Accreditation Body's Requirement for Engineering Design Curriculum

ABET, Inc., is the recognised accreditor for college and university programmes in applied science, computing, engineering and technology in the USA. The design-related requirements that ABET places on US engineering programmes for accreditation states that a curriculum must include most of the following features:

- development of student creativity;
- use of open-ended problems;
- development and use of modern design theory and methodology;
- formulation of design problem statements and specifications;
- consideration of alternative solutions;
- feasibility considerations;
- production processes;
- concurrent engineering design; and detailed system descriptions;

ABET seems to specify what are the necessary requirements but **does not mention the essential requirements** that is necessary in order to produce an engineering designer (http://www.abet.org). The course examples shown in Sect. 3 of this paper shows a level of adherence to ABET's requirements for accreditation of an engineering design course.

5 Aims and Outcomes for Product Design in Some Design Schools

The following have been taken from either the brochure/prospectus or the website of the respective universities mentioned. The highlighting in bold letters is by the author.

5.1 Nanyang Technological University, Singapore

The BFA in Product Design at the School of Art, Design and Media, offers a **curriculum in design methodologies and an environment conducive to**

innovative thinking. Beyond problem-solving skills and in-depth analysis of users, markets and cultural values, students also learn to redefine problems and question traditional methods while developing **new means of seeing and thinking**. Through close interaction with faculty, in small studios and classes, and with a dynamic laboratory and workshop for producing new forms, students gain the experience, knowledge and vision to create innovative work of strong conceptual value.

5.2 Purdue University, USA

The undergraduate programme (in industrial design) is a four-year degree with an emphasis in form giving for manufactured goods. Students' graduate with the ability to be **innovative problem solvers** and create **aesthetically appropriate forms** that can be **manufacture** by industry.

6 Design Accreditation Body's Requirement for Product/ Industrial Design Curriculum

National Association of Schools of Art and Design (NASAD), (http://nasad.arts-accredit.org/) is the recognised accreditor for college and university programmes in art and design in the USA. NASAD's competency requirements for recognising an industrial design programme in US as well as in other countries are:

- A foundational understanding of how products work; how products can be made to work better for people; what makes a product useful, usable, and desirable; how products are manufactured; and how ideas can be presented using state-of-the-art tools.
- Knowledge of computer-aided drafting (CAD), computer-aided industrial design (CAID), and appropriate two-dimensional and three-dimensional graphic software.
- Understanding of the history of industrial design.
- Functional knowledge of basic business and professional practice.
- The ability to investigate and synthesise the needs of marketing, sales, engineering, manufacturing, servicing, and ecological responsibility and to reconcile these needs with those of the user in terms of satisfaction, value, aesthetics, and safety. To do this, industrial designers must be able to define problems, variables and requirements; conceptualise and evaluate alternative; and test and refine solutions.
- The ability to communicate concepts and requirements to other designers and colleagues who work with them; to clients and employers; and to prospective clients and employers. This need to communicate draws upon verbal and written

forms, two-dimensional and three-dimensional media, and levels of detailing ranging from sketch or abstract to detailed and specific.
- Studies related to end-user psychology, human factors and user interface.
- Opportunities for advanced undergraduate study in areas which intensify skills and concepts already developed, and which broaden knowledge of the profession of industrial design. Studies might be drawn from such areas as engineering, business, the practice and history of visual art and design, and technology, or interdisciplinary programs related to industrial design.
- Easy access to computer facilities; woodworking, metalworking, and plastics laboratories; libraries with relevant industrial design materials; and appropriate other work facilities related to the major.
- Opportunities for internships, collaborative programs, and other field experiences with industry groups.
- Participation in multidisciplinary team projects.

7 Differences Between the Two Schools

NASAD defines industrial design as 'the professional service of **creating and developing concepts and specifications that optimise the function, value, and appearance of products and systems for the mutual benefit of both user and manufacturer**. Industrial design involves the combination of the visual arts disciplines and technology, utilising problem-solving and communication skills'. This definition is the same as that adopted by the Industrial Designers Association of America, [2].

ABET defines engineering design as **'**the process of **devising a system, component, or process to meet desired needs**'. This could be elaborated as 'the process of ensuring, through systematic analysis, simulation and testing, that a(product) design functions as intended and that it can be produced at acceptable cost' as stated by Templemann and Pilot [3].

7.1 Clear Differentiation on the Accent of Design Education

Based on the definition of design by NASAD and abet the main differences between design in design schools and engineering schools can be summarised as below:

- Industrial design's emphasis is on use and function is focused towards the users, leading to a user centred design approach to design.
- Industrial design also gives importance to the aesthetics of the final result (product or object)

- Engineering design's emphasis is on use from the point of view of function drivers (mechanical, electrical and electronics, structural, etc.)
- Engineering design education lays importance on optimising the function drivers to efficiency, manufacture, cost.

Womarld [4], Eggink [5] and many others tend to show a clear difference in industrial design education requirements from engineering design education as well.

7.2 *Anomaly of Sorts in Practice*

In studying the curriculum of the design programmes engineering schools in many parts of the world, one could believe that principles and processes of industrial design would also be taught. Where does this lead to in terms of the industry understanding of design? How does one understand relationship between 'fuzzy front end' aspect represented by industrial design and the 'analysis, synthesis and optimisation' approach of engineering design?

8 Mapping the Curriculum

Table 1 compares the curriculum between typical industrial design and engineering curriculum (nearest to engineering) placing five attributes namely, core skills, core knowledge, core studio/hands on, additional knowledge, capabilities.

It is apparent from Table 1 that for either programme to offer attributes from the other, some of the important core elements of the original programme need to be dropped. This leads us to the question of how rigorous would the integrated 'industrial/product design engineering' course be, in terms of justifying a curriculum that has both the accents of design? While a direct answer may not be possible, an indication of the challenges of such integration maybe possible through the visual mapping of typical curriculum based on Table 1, as shown in Fig. 1.

Figure 1 visually positions courses along subject areas in the X axis and focus (or intensity) of the courses on the Y axis as proposed by the author. Engineering courses are depicted as having a breadth ranging from science to application while design courses have a breadth ranging from art to application. One can clearly see from this map that engineering design harnesses science through technology to apply to real world situations while design harnesses art through humanities to apply to real world situations. Both the schools seem to overlap at the application level, with engineering reaching out to humanities and design reaching out to technology. As per this model, engineering courses are centred on technology while design courses are centred on humanities, and perhaps, social sciences. The challenge in developing any hybrid programme, then, is to find ways to bring

Table 1 Comparison of typical curriculum

Attribute	Industrial design	Engineering (related to design)
Core skills	Sketching/freehand drawing	Detailed level technical drawing CAD/CAE
	Model making	
	Visual communication techniques	
	Technical drawing/CAD/CAID Photoshop, illustrator skills, etc.	
Core knowledge	History of art, design and society	Math/physics/chemistry
		Mechanics of solids/machines
		Fluid mechanics/hydraulics
	Theory of design/design methodology/design research	Thermodynamics/thermal engineering
	Design elements/aesthetics	Machine element design/systems Design/part and component design
	Human factors/ergonomics	Analysis/optimisation
	Interaction/user experience	Materials/manufacture
Core studio/hands on additional capabilities	Product design (up to 6 courses)	Intermediate projects
		Capstone/final year project
	Conceptual design	Industrial engineering
		Operation management
	Form studies	Project management
	Minor projects	Project report, presentation project report viva voce
	Final year project	
	Marketing/product planning	
	Project management	
	Project report, multimedia presentation	

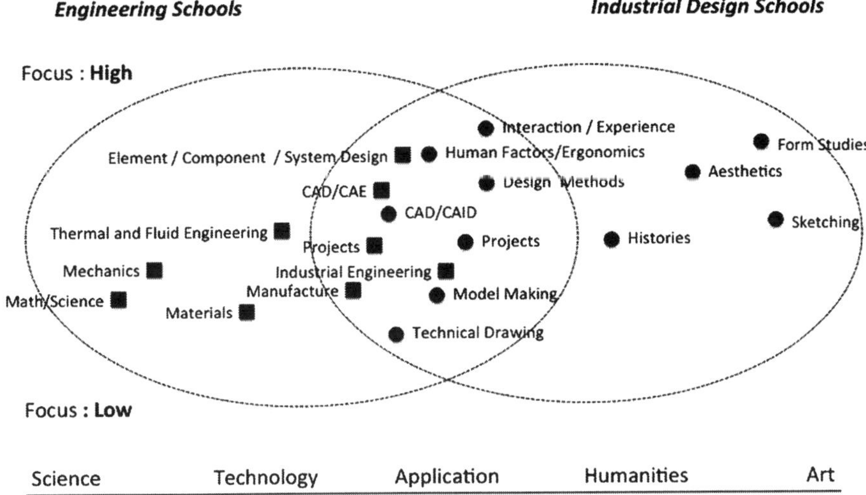

Fig. 1 Visual mapping of curriculum. ■ Engineering courses. ● Design courses

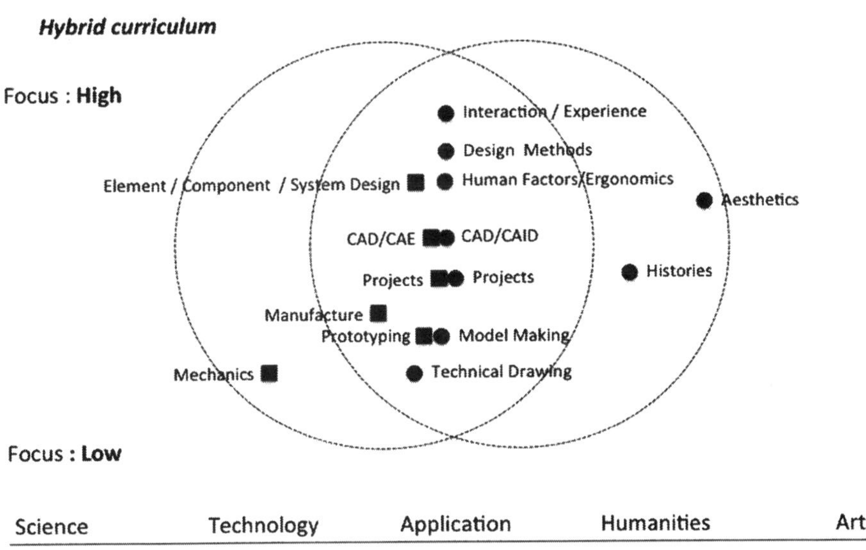

Fig. 2 Visual mapping of a possible hybrid curriculum. ■ Engineering courses. ● Design courses

subjects related to both together without losing the overall relevance of the courses.

One possible scenario for a hybrid curriculum is shown in Fig. 2, which depicts a four-year professional (hybrid) design programme. This curriculum compromises on two important elements of a typical design curriculum, namely sketching

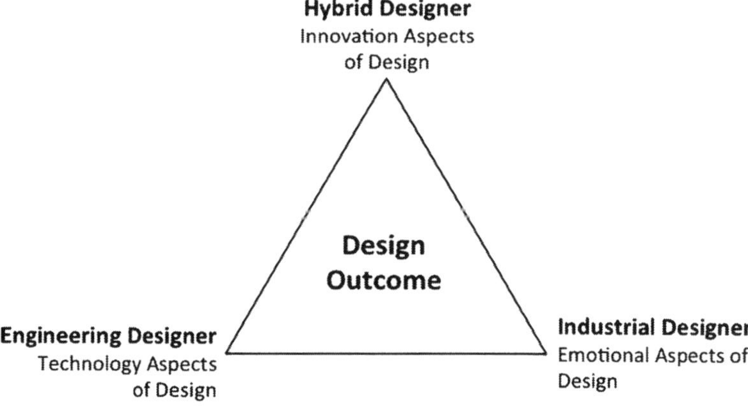

Fig. 3 Contribution zones of designers

and form studies as core subjects. This also has taken out engineering fundamentals in math/science and technology courses on thermal and fluid engineering. In other words, subjects that are fundamental/foundation to both have been removed. While it can be argued that basics of these subjects can be introduced under subjects termed foundation studies or electives, what is missed out is the rigour at which they were taught under traditional/classical curriculum of engineering or design. The end result of such hybrid curriculum could be graduates who have solid process and project knowledge who may not have sufficient skill or knowledge in making decisions on form and other aesthetics related issues as well as not be able to fully utilise the domain knowledge required in mechanics or thermal, fluid engineering, etc. The positive result of such hybrid curriculum would be the designers who will excel in managing projects whose fundamentals of engineering and art have been clearly defined. These hybrid designer-engineers could fit well into situations where decisions on design have to be made 'on the fly' at mass manufacturing centres which are far away from the origin of design, such as in China or Vietnam. They would have strong application knowledge and project management skills together with sufficient peripheral knowledge in areas related to art and engineering to make such decisions.

Is hybrid curriculum, then, be the new trend in design education? While the hybrid designer-engineer will figure prominently in the design of new products, the author feels that fundamental decisions related to the emotional side of a product (which require art, humanities and social sciences) will still be influenced by designers from the classical design schools, while fundamental issues related to engineering and technology applications will still come from design engineers from the traditional school of engineering. Figure 3 depicts possible zones of contribution by graduates/designers from the three curriculums. In presenting this model of contribution, the author's intention is to point out different products require different levels of focus. Furniture, ceramics and personal products such as watches, spectacles and jewellery require higher level of involvement in art and emotion. In such instances designers from a classical school of design would play dominant roles in the determining the final outcome of the product. Similarly, products such as precision instruments, engines and motors require engineering designers to play a dominant role in the design of such products. Products requiring new paradigm in user experience and application of technology such as smart phones and tablet computers benefit from inputs from designers originating from hybrid curriculum. There would be exceptions to this, especially in smaller companies perhaps, where one designer plays all three roles.

9 Discussions

The intention of this paper was to understand the curriculum of design degree programmes offered by schools of engineering and schools of design before visually mapping the two curriculums to compare them. This map gives a clear idea of the focus of each of these curriculum and allows for identifying possibility of establishing a hybrid curriculum, as has been established by several institutions around the world. The author highlights through this paper, that such hybrid programmes may compromise on the subjects/topics which are fundamental to engineering courses and which form the foundation for design course. In doing so design outcome may lose on fundamental principles of engineering and attention to visual aspects of design. At the same time, hybrid designers could play a prominent role as 'innovators' especially in products requiring new paradigms in user experience and technologies.

This paper, by outlining the three possible curriculum that produce designers with different focus and accent to their dominant roles in design, presents the importance of consideration of the human (designer) aspect in design research. It may be stated that designer, environment and process form the eco-system for effective design outcomes that show relevance through their success in the market place. The designers, in turn, are products of the different curriculum that produced them in the first place.

Acknowledgments This paper has drawn many of the curriculum and subject information from the websites of the various universities and institutions mentioned. Steps have been taken, as much as possible, to present information, which are up to date. The author wishes to place on record, his gratitude, for the availability of such information, to those universities and institutions that are mentioned in this paper, as well as those which are not mentioned in this paper whose information has been important in presenting ideas depicted herein.

References

1. DymHeywood J (2005) Engineering education: research and development in curriculum and instruction. Wiley, USA. ISBN 13 978-0-471-74111-4, ISBN-10 04-471-74111-6
2. Industrial Design Society of America (IDSA). (2010) ID defined. http://www.idsa.org/

3. Templeman E, Pilot A (2011) Strengthening the link between theory and practice in teaching design engineering: an empirical study on a new approach. Int J Des Educ 21:261–275
4. Womarld PW (2011) Positioning industrial design students to operate at the 'fuzzy front end': investigating a new arena of university education. Int J Des Educ 21:424–447
5. Eggink W (2009) A practical approach to teaching abstract product design issues. J Eng Des 20(5):511–521

A National Academic-Industrial Research Program with an Integrated Graduate Research School

Göran Gustafsson and Lars Frenning

Abstract ProViking® is a successful Swedish twelve-year research program in product development and production. Rather than to stimulate the emergence of new companies, it aims at strengthening the existing Swedish industry by producing scientific results at the highest international level and Ph.D.'s for work in the industry. ProViking® comprises a large number of research projects, jointly run by universities and industrial companies, and a national graduate school which provides courses for the Ph.D. students that work in the projects. The total ProViking® budget is close to 110 M€, of which 43 M€ is supplied by the Swedish Foundation for Strategic Research and the rest by the industrial project partners. Since the program started in 2002, forty-one different research projects have so far resulted in several hundred conferences and journal papers. Some fifteen patents have either been granted or have applications pending as a result of the research, and almost one hundred Ph.D.'s have graduated.

Keywords Research · Program · Industrial · Graduate · School

G. Gustafsson (✉)
Department of Product and Production Development, Chalmers University of Technology, SE-412 96 Gothenburg, Sweden
e-mail: gorang@chalmers.se

L. Frenning
PPU-ProViking, Chalmers University of Technology, SE-412 96 Gothenburg, Sweden
e-mail: lars.frenning@proviking.se

A. Chakrabarti and R. V. Prakash (eds.), *ICoRD'13*, Lecture Notes in Mechanical Engineering, DOI: 10.1007/978-81-322-1050-4_89, © Springer India 2013

1 Introduction

Sweden is a small but highly industrialized country with many successful international companies. The increasing global competition in the last few decades in particular has however forced an increased emphasis on the development of new and better products as well as faster, cheaper and more efficient and environmentally friendly methods to conceive of, develop, design and produce them. In addition, the vast majority of the Swedish international companies were founded on the basis of inventions and innovations in the late 1800s and the early 1900s, so there was and still is also a strong consensus across the national political borders that Sweden needs to nurture its small and medium-sized enterprises (SMEs) in order to create conditions which helps more of them to grow into large companies.

The background for what will be presented in this paper was therefore a strong desire to stimulate in particular the development of Swedish SMEs, and to do this through academic-industrial cooperation which would even better than before utilize the knowledge and competence in the domestic universities to further improve Sweden's international industrial competitiveness.

The ProViking® research program [1], which was launched in 2002, is an effort to strengthen the Swedish industry through a large number of research projects which involve academic as well as industrial partners with the intention that they shall produce useful scientific results at the highest international level. In addition, an integrated graduate school shall educate Ph.D.'s that are interested in and have backgrounds that are suitable for industrial rather than academic careers.

Previous papers on ProViking [2, 3] have mainly discussed aspects of ProViking's national graduate research school, and how that institution has developed over the years. This paper presents both parts of the ProViking program, i.e. the research projects and the graduate school. The first author is the Director of Studies of the ProViking National Graduate Research School and the second author is the Program Director of the entire ProViking research program. They have both held their positions since the start of the program.

2 The ProViking Research Program

2.1 Two Phases

ProViking is the result of expressed Swedish industrial needs for long-term research in the area of product development and production. Before and shortly after the turn of the last century there existed several different national research programs with or without integrated graduate research schools which aimed at strengthening various parts of the Swedish industry, or support it in several ways. Two of them were the Swedish Engineering Design Research and Education Agenda, ENDREA [4], and PROPER. Both included graduate schools. ENDREA focused on engineering management and product development while PROPER was dedicated to production research.

In the early 2000s however, the Royal Swedish Academy of Engineering Sciences, IVA [5], assembled the most important domestic stakeholders in the entire field of product realization to talk about what ought to follow ENDREA, PROPER and some other programs. The hearings and intense discussions with both industry and academia resulted in a new approach to the national industrial research: a broad program dedicated both to product development and production research, and with an integrated graduate research school, aimed at increasing the competitiveness of the Swedish engineering industry in general. Universities and companies should work together in the projects and produce research results at the highest international level. The doctors from the graduate research school should have knowledge and experiences that would make them suitable for industrial careers. Finally, ProViking, as the program was later named, should boost existing companies rather than stimulate the advent of new enterprises.

The ProViking program was launched in 2002. With its very wide coverage it has come to incorporate not only ENDREAs and PROPERs fields, but it extends beyond them. ProViking has from the start been financially supported by the Swedish Foundation for Strategic Research, SSF [6], and the total budget for its twelve years is around 110 M€. Of this sum, SSF directly supports the participating universities with 43 M€ while the remaining and larger part of the funding is provided by the industrial project partners in the form of cash, in kind (work) and other resources.

ProViking was initially planned as a six-year effort, and SSF established close coordination with VINNOVA [7] and the Knowledge Foundation [8], two other domestic research financers in the same field. However, towards the end of that period, in 2007, ProViking's scientific achievements were evaluated by four invited academic specialists from Canada, Denmark, Germany and Hungary, and their report showed such good results from both the research projects (see Table 1) and the integrated graduate school that SSF decided to extend ProViking by another six years. This was another sign of success since SSF normally only finances efforts that are limited in time to six-year periods. The two ProViking periods are somewhat overlapping and according to the plan the extended program expires at the end of 2013.

Table 1 Success of the ProViking1 research projects on a five-point scale, as assessed by the ProViking Scientific Council in 2007. (The non-integer numbers are due to split scores.)

Score		Number of projects
5	Excellent	3,5
4	Very good	7
3	Good	7
2	Fair	0,5
1	Poor	0
Total number of projects		18

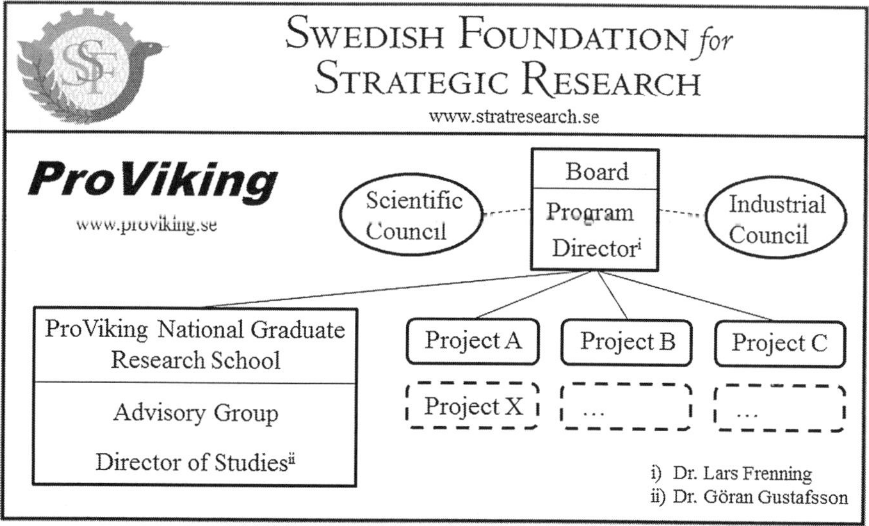

Fig. 1 Organization of the ProViking research program

The ProViking research program is governed by a Board with representatives from universities as well as industry. To assist them and the Program Director there is a Scientific Council with researchers from international universities which assesses applications for new projects, and an Industrial Council which provides advice on the industrial relevance of ideas and proposals for research projects. The integrated ProViking National Graduate Research School which will be described later in the paper has its own governing board in the form of an Advisory Group that supports the Director of Studies. Figure 1 shows the organizational structure of the ProViking program which has remained almost unchanged for a decade.

ProViking is also one of the founders of The Swedish Production Academy (Sw. Svenska Produktionsakademien), an organization of academics with professional interests in production. It supports cooperation in production research and production education and has an ambition to be an influential partner in this area. There are plans to create a corresponding organization for product development professionals. With an aim to boost the production knowledge in the domestic industry as well as in academia, ProViking has also both headed and largely funded an IVA project titled Production for competitiveness.

2.2 The Research Projects

The first phase of the program, ProViking1, in 2002–2007, involved eighteen different research projects. The second phase, ProViking2, which runs in 2008–2013, comprises twenty-three projects. All of these have and have had a focus on industrial development and/or manufacturing in Sweden in order to produce new concepts, theories, methods, tools and work routines that can strengthen the Swedish industry. The areas of interest are ideas, product development, design, production, product support, maintenance, end use and recycling, all in a life-cycle perspective. Some projects can be considered as commercially high-risk ones, since they have involved very novel ideas and complex technology.

The ProViking projects can all be categorized as applied research projects in the sense that the research questions that they deal with have been suggested as well as formulated by the industry, and not by university researchers. They vary in both length and amount of money allocated to them, but they all involve academic *and* industrial partners/institutions. Although SMEs are of particular interest, also large companies take part. Most of the projects have far more industrial participants than the minimum two, which of course reduces the cost for each of them. From the universities senior researchers as well as Ph.D. students take part, and foreign post docs have also been involved in some cases.

In addition to the several hundred conference and journal articles that have been published to report the scientific results from the ProViking research projects, the outcomes are also presented at yearly Result Days in selected cities all over Sweden. These are open to everyone interested since the policy of the Swedish Foundation for Strategic Research is to make public all results from the research projects that it finances. The idea behind the Result Days is also to give other companies than those directly involved in the ProViking projects opportunities to meet the researchers in person and discuss with them. As a complement to the Result Days, and as an alternative for those interested but who cannot attend, ProViking has also in collaboration with several other actors developed the Result Center™ [9]. This is a meeting place on the web for all those involved in Swedish product development and production research which contains information about companies and institutions as well as individual researchers and ongoing research projects, ProViking's and others.

Since some of them are still running, there has yet been no evaluation of the ProViking2 projects. In all about fifty different companies and organizations were involved in ProViking1 though, of which twelve were SMEs. Of the in total eighteen ProViking1 projects, three focused on management, eight on production and seven on design, and the success rate of them is shown in Table 1. It is believed that the comparatively high scores, with part of one project rated fair and the rest better or *considerably* better, is due to the fact that each problem is worked on by people with different knowledge and background from industry and academia in cooperation. This creates very productive and intellectually challenging environments where the results exceed the sums of the individual contributions. On the educational side this is of course very positive also for the Ph.D. students who do their research in ProViking projects. They get the best of both industry and academia, and they develop an understanding of industrial research which makes it natural for them to look for careers not only in universities but also in companies.

As examples of very concrete and useful outcomes from the ProViking program it can be mentioned that during ProViking1, six patents came out of the eighteen projects. In ProViking2, five patents have so far been granted on basis of research results from the twenty-three projects in this phase, and another four patent applications are pending decision. It should then be noted that a Swedish patent is only granted for an invention that has a certain “innovation height”, which means that it is substantially different from and better than anything that is hitherto known, and that it is also useful in practice.

Some ProViking projects have resulted in very interesting and novel applications. One example is a project to develop a new way to build an electric motor. The magnetic parts of today’s standard motors consist of stacks of thin steel plates with layers of insulation between them, but with the new method, one such part is instead cast in one piece made of a soft magnetic moldable composite. The new material is a good magnetic conductor at the same time as it has the mechanical properties of a very strong polymer. Motors which are made in this way are not only cheaper to make than the standard ones, they also have higher efficiencies than what comparable existing designs have.

Another project is devoted to the problem of corroding surfaces in fuel cells. One way of protecting these surfaces is to gold plate them, which is obviously a very costly technique. The project studies how a new material, which is not only cheaper but also increases the life of the plates as well as the efficiency of the fuel cell, can be applied to reach the best overall result.

In a third ProViking project a method has been developed for automated disassembly of scrapped LCD screens, the yearly number of which is on the rise. Instead of incineration, which is the present and from an environmental point of view far from ideal alternative ending for these devices, the process makes it possible to recover not only harmful substances from them but also components which can be reused to build new and different products.

Efficient induction heating is the subject of a fourth project, which has demonstrated that remarkable energy savings are possible at the same time as the cycle time can be reduced, and several other projects are devoted to the development of various simulation techniques. Besides these there are more than 30 other finished or ongoing projects which deal with many other problems.

Although the intention behind ProViking is and has always been to strengthen *existing* domestic companies, as a bonus some new enterprises have also emerged based on the results from a few of the research projects.

2.3 The ProViking National Graduate Research School

Already at the beginning of the first ProViking phase a graduate school was set up to organize courses for all Ph.D. students who were somehow to become engaged in the ProViking research projects. The purpose was and still is to produce researchers with industrial as well as academic competence for work in the industry. This is in contrast to traditional Swedish graduate education, which has rather tended to prepare the students for academic research careers.

All Ph.D. students who do research in and are financed by ProViking research projects, also part-timers, are members of the ProViking National Graduate Research School. Besides these, the school has also attracted a large number of other Ph.D. students who want to take part in its activities. If a student is doing research in a subject that fits in with ProVikings profile, he/she is usually admitted. A clear sign that the school is very attractive is the fact that of today's about 105 member students, the vast majority belongs to the second category, i.e., they are not associated with ProViking projects and have themselves asked to become members. 25 % of the students in the school are industrial Ph.D. students.

It should be mentioned that it is the students' home universities which confer the Ph.D. degrees on them. ProViking only provides projects to do research in and its graduate school offers courses that the students need to take to be able to graduate. At the time of writing this article, 97 students have received their Ph.D.'s during the time when they belonged to the ProViking National Graduate Research School.

The graduate school has an Advisory Board which consists of representatives of the major research universities (senior researchers/teachers), the industry and the Ph.D. students. The board assists and advices the Director of Studies on the course program and the other activities in the school.

2.3.1 The ProViking Course Program

A Swedish Ph.D. program is four years long, full time, and of that time the student typically devotes one year to course work and spends three years doing research and writing the thesis. (This balance can vary somewhat between different universities, and between different research subjects in the same university). Not long ago Swedish Ph.D. students usually took their courses at their home university, but during the last two decades this has changed so that many students now also take courses at other universities, often within the framework of some type of graduate school.

The ProViking National Graduate Research School organizes and finances courses at all major Swedish research institutions. The exact number of courses as well as their topics is constantly discussed. Both vary over time to cater to the needs of the students, which are in constant change as students graduate and leave the school and new students with other research interests enter. Courses are not duplicated, i.e. there are no two ProViking courses with the same content at two different universities, so students who need to take a particular course therefore do so at the university which offers it, which may not necessarily be their own. It has therefore become completely natural for the students during their Ph.D. programs to take courses at several universities, for some even in other countries.

The present course program contains three courses which all students have to take. The survey course in product realization, which covers all stages from perceived problem or product idea via design, production and use to retirement, serves the purpose to give all students a common view of the process that a product undergoes during its life, and of how industry works in these stages. The other two mandatory courses are on the theory of science and oral presentation and scientific writing, respectively. Besides these there are five different themes, each with a number of different courses. Four of the themes are design theory and development methodology, leadership and organization, materials science and production processes and systems, and the fifth comprises courses that do not fit into any of the other four themes. The students choose at least two courses among those in the themes.

The courses cover a wide area of subjects, which is both natural and necessary since the ProViking projects extend across so many different technical subjects. As mentioned before, the graduate school caters not only to the students which work in the ProViking projects and consequently have very different backgrounds and research interests, but also to many other engineering Ph.D. students who have joined at their own initiative. The latter tend to be an even more heterogeneous group than that which consists of students in ProViking projects.

With a geographical distance of 1,200 km between the universities with ProViking students that are the farthest apart, the ProViking courses cannot be organized and run in the same manner as local courses usually are, i.e. with lectures a couple of hours every week. Instead they are organized as 3–4 gatherings of several days in a row, with periods in between when the students, as they do most of the time, study at their home universities.

This way of organizing courses is advantageous in at least three ways:

- Each course is run by the university which is best at it, so the students perceive a course quality which is in general higher than it would otherwise have been
- Specialization and shared work among the universities (since essentially the same course is given at one institution instead of repeated at several) reduces the total national cost for education
- The gatherings give the students tremendous opportunities to meet and learn to know colleagues from other universities and develop personal networks, which is expected to lead to more cooperation and business when in the future the students will have assumed influential positions in the industry.

The ProViking course work for an individual student amounts to about half of the total course work needed for the student to get his/her Ph.D. The exact requirements vary not only between universities, but can differ also between scientific subjects in the same university. The total work for a Ph.D. in terms of hours is nationally standardized though, so a university or scientific subject which requires a lower amount of course work for a Ph.D. has correspondingly higher expectations on the research work and the doctoral thesis.

All ProViking courses are free of charge for the ProViking students themselves since the graduate school pays the universities for the education that they provide. The courses are however open also to engineers from the industry, and *they* have to pay fees if the course-giving university so decides. Professional engineers are very welcome to take the ProViking courses, not primarily because they improve the course economy but since our experience is that the course quality increases with their presence. Interviews with students and engineers who have taken the same course together confirm that both groups benefit from studying together with each other. The students are in general more updated on modern theory than the engineers are, which sometimes in practice makes them somewhat of assistant teachers in the courses. The engineers, on the other hand, have a vast collective practical experience that the students lack, for obvious reasons. This is of course also of value to the country. Not only do in this way the time and money spent produce a greater amount of collective learning than would otherwise have been the case, but engineers and students also get to know each other, which stimulates future contacts and cooperation between them.

The ProViking course with the largest proportion of engineers is one in Lean Product Development, a philosophy for how to do PD work which is presently very popular both in universities and in the industry. This was originally a course for professional engineers organized by the Chalmers School of Continuing and Professional Studies, which was later, and very successfully, opened also to Ph.D. students.

2.3.2 Other Activities in the Graduate School

The graduate school has a yearly three-day gathering for the Ph.D. students, when among other activities the group visits several industrial companies. The last gathering in May 2012 took place in a very SME-dense part of Sweden, where eight ongoing ProViking projects were presented to invited local industrialists. This was a success, since the immediate reaction from the industry people was that the results from four of the projects would be directly applicable in local producing factories.

The ProViking students can apply for scholarships from the graduate school to conduct part of their Ph.D. program at a university or other qualified research institution abroad. The length of a stay can vary from about a month up to a full academic year, and the intention is that it should widen the student's horizons by giving him/her the opportunity to study and do research in an environment which is different from the one at home. Some twenty students have so far applied and received scholarships to go to as different countries as England, France, Germany, Switzerland, China, Australia, Brazil, Canada and the United States. Most of the students have been doing research at universities, but quite a number also in industrial companies. After having returned to Sweden, they have written reports on their stays which have been published on the graduate school's homepage to stimulate other students to also consider spending some study time abroad.

3 Discussion

It is apparent from the "lessons learned" in the ENDREA program, as stated in its final report [4], that ProViking is reaping the fruits of having been preceded by other efforts at establishing national graduate schools. It was not easy in the beginning for ENDREA to get accepted by the academic advisors, but that has not been a problem at all for ProViking.

One advantage of having a graduate school of ProViking's size is that it is relatively easy to assemble large enough groups of students to actually be able to run courses. The inevitably heterogeneous background of the students—in relative terms, of course—sometimes works the other way too, so that there are fairly few students who are interested in some of the courses. But if they had not taken part in a national school they would have belonged to even smaller groups at their home universities.

What is seemingly most appreciated by the students is the opportunities they get to meet fellow students from other universities and to develop their own networks. They are also positive to the frequent contacts with engineers from industry that they get in courses, in projects and during other activities like Result Days and the yearly three-day gatherings.

The companies who take part in the ProViking projects are in general very satisfied with the research results that they get, and they also appreciate the opportunities to work closely together with Ph.D. students in the projects, which is a good way of getting to know prospective employees. ProViking has a good reputation in the industry and is always welcome, which is in turn a great asset to the school when it comes to organizing study visits in courses and during the yearly gatherings.

4 Conclusions

Since there is no previous domestic and national effort like ProViking, neither in scope nor in terms of budget, it is difficult to compare it with other programs. In some ways it resembles ENDREA, but it has a much wider scientific scope, larger budget and longer time frame. ProViking's industrial results in terms of faster, better and more economic products and processes and more competitive companies are also largely, for commercial reasons, confidential information. However, judged from the volume of published research results from the projects, the number of patents and produced Ph.D.'s, as well as the expressed appreciation of the industrial partners and the positive outcome of the assessments, it seems fair to say that ProViking, with its integration of research and education and wide scope, is and has been a very successful research program which can be a model for future efforts.

5 The Future

The second and, as it seems, final phase of the ProViking program is now nearing its end. In 2012 the Swedish Foundation for Strategic Research adopted a new strategy which during the foreseeable future excludes further financing of technical research of the type that is carried out in ProViking, as well as financing of national graduate research schools. At the time of writing this paper, the possibilities for a second extension of the research program by SSF therefore look bleak. Many ProViking projects are already closed and according to the present schedule the last courses in the graduate school will be run in late 2013. The universities in ProViking have therefore commenced discussions on how to act if the whole research program comes to an end then. If that is the case, there is a readiness and strong interest among them to try to get support from some other financer to allow at least the activities in the graduate school to continue. Now, when already several generations of Ph.D.'s have graduated from national research schools like ENDREA, PROPER and ProViking, it seems completely natural to everyone involved that this is how the national graduate education in engineering subjects should be conducted also in the future.

Acknowledgments The support from the Swedish Foundation for Strategic Research and its encouragement to publish results and experiences from the ProViking program's research projects and national graduate research school is greatly appreciated.

References

1. ProViking® (2012) www.chalmers.se/hosted/proviking-en/. Accessed 6 Oct 2012
2. Frenning L, Gustafsson G (2007) A national graduate research school in product realization. In: International conference on design education (ConnectED 2007), University of New South Wales, Sydney, Australia, 9–12 July 2007
3. Gustafsson G, Frenning L (2010) A national graduate research school with close industrial collaboration. In: The 6th international CDIO conference, École Polytechnique, Montréal, Canada, 15–18 June 2010
4. ENDREA Final Report (2012) Swedish Foundation for Strategic Research (SSF), 8 Mar 2004
5. The Royal Swedish Academy of Engineering Sciences (2012) www.iva.se/en/. Accessed 6 Oct 2012
6. Swedish Foundation for Strategic Research (SSF) (2012) www.stratresearch.se/en/. Accessed 6 Oct 2012
7. VINNOVA (2012) www.vinnova.se/en/. Accessed 6 Oct 2012
8. The Knowledge Foundation (2012) www.kks.se/om/SitePages/In%20English.aspx. Accessed 6 Oct 2012
9. Result Center™ (2012) www.resultcenter.com/?lang=en. Accessed 6 Oct 2012

Future Proof: A New Educational Model to Last?

Mark O'Brien

Abstract This paper will discuss the research, introduction, development and delivery of a recently established Masters of Arts suite of programmes. This ambitious Cross-disciplinary programme now comprises of 22 named Masters awards, taught in mixed disciplinary cohorts with multiple entry points. It draws students from across the world and represents a unique structural format within the UK Art and Design education sector. The paper will examine the experience of the delivery to the first cohorts and consider the strengths and weaknesses of the model suggesting recommendations for continuing improvement in the provision of highly flexible cross-disciplinary taught Masters provision.

Keywords Curriculum development · Cross-disciplinary · International

1 Introduction

The School of Art, Design and Architecture (SADA) at the University of Huddersfield (UoH) comprises of approximately 2,000 students and staff across 3 Departments, Art and Communication, Fashion and Textiles and 3D Design and Architecture. It is a well equipped and resourced School occupying 3 prominent locations on the central campus in Huddersfield. Historically the majority of provision has been and remains at Undergraduate level, however, in the last 4–5 years there has been a significant increase in the provision of both Postgraduate Taught (PGT) and Postgraduate Research (PGR) students as well as significant

M. O'Brien (✉)
The University of Huddersfield, Queensgate, Huddersfield HD1 3HD, UK
e-mail: m.a.obrien@hud.ac.uk

A. Chakrabarti and R. V. Prakash (eds.), *ICoRD'13*, Lecture Notes in Mechanical Engineering, DOI: 10.1007/978-81-322-1050-4_90, © Springer India 2013

developments in the general research profile and capacity within the School. Although PGT provision had existed within the School at various times during the previous two decades this had diminished in size to only 2 small courses and some franchised provision overseas by 2007/2008. A recently recruited senior management team for the School embarked on a series of new initiatives including the development of PGT Masters provision. The research and documentation was developed by a team of subject specialist staff drawn from across the School, which was validated to commence in January 2009. In September 2008 a Postgraduate Academic Leader was appointed with a variety of duties including the establishment and management the new provision.

2 UK Background and Context

The interest in and examples of cross-disciplinary initiatives have been widespread over the last decades, however, there has been a particularly keen interest within the design and creative industries sectors in recent years. One of the main catalysts for this interest was the publication of Sir George Cox's *Review of Creativity in Business* [1], known as the Cox Review. This set out the need to embed creative capabilities at the heart of the UK's competitive positioning within a global economy.

He called for:

> Business people who understand creativity...and who can manage innovation; creative specialists who understand the environment in which their talents will be used and who can talk the same language as their clients and business colleagues; and engineers and technologists who understand the design process and can talk the language of business.

The report encouraged universities to create 'centres of excellence [...] that specialise in multi-disciplinary programmes encompassing both postgraduate teaching and research.' The focus would be on Masters level programmes which would 'bring together the different elements of creativity, technology and business', enabling students from different backgrounds and with varying levels of industrial experience to work together.

The outcome, said Cox, would be:

> executives who better understand how to exploit creativity and manage innovation, creative specialists better able to apply their skills (and manage creative businesses) and more engineers and scientists destined for the boardroom.

Many universities in the sector had already been engaging in the development of new courses and research centres with varying levels of collaboration with other disciplines. Over 30 of these were brought together in 2006 when the Design Council set up a Multi-disciplinary Design Network, supported by the Higher Education Funding Council for England (HEFCE) and the National Endowment for Science, Technology and the Arts (NESTA), which 'aimed to facilitate the sharing of knowledge and best practice across universities, to improve curriculum design and assess the impact of these new programmes' [2].

The network was never intended to be exhaustive and it recognised that other similar activities were taking place within the sector (such as at Huddersfield), it did however, include the highest profile and best resourced examples of collaborative multi-disciplinary activity. Eight of these were brought together as a series of case studies [3]. They included probably the two most reported: Design London, a collaboration between Imperial College Business School, Imperial College Faculty of Engineering and the Royal College of Art and the Centre for Creative Design (C4D) a partnership between Cranfield University and the London College of Communication, University of the Arts London. Both programmes were listed in the Business week top 30 global Design courses (2009).

2.1 Definitions and Terminology

Various terminology has been used to describe cross-disciplinary practice in UK universities without reaching any agreed definitions. As the authors of the Design Council reports [2] note 'despite more than 40 years of cross-disciplinary practice in universities there is still a lack of precision about what the terms 'inter-disciplinarity', 'multi-disciplinarity' and 'trans-disciplinarity' actually mean'.

Citing earlier work (McEwenet al. 2008; Lawrence and Despres 2004) the authors use the term 'multi-disciplinarity' to describe a situation involving the co-operation of disciplines in which the disciplines themselves remain unchanged. Inter-disciplinarity they suggest would involve an attempt to 'integrate or synthesise perspectives from several disciplines' [2] and trans-disciplinarity would transcend or involve a transgression of disciplinary norms.

The courses devised at UoH are probably closer thought not wholly consistent to the definition Inter-disciplinary. The courses although mostly of an Art and Design nature draw content and some students (MA Fashion Management, MA International Design, Marketing and Communication) from non-design backgrounds, and within the module content there is material from other academic disciplines although delivered in and Art and Design specific manner. The chosen term used at UoH has always been 'Cross-Disciplinary' representing the disciplines within Art and Design: Fashion, Graphics, Interiors etc. and recognizing the changing nature of these disciplines and the benefits of learning from practices elsewhere, as well as providing input, content and thinking from across other academic disciplines.

3 Background and Course Development

The research and development of the original courses took place during 2007/2008 The, then new MA developments emerged from the existing MA culture in the School of Art, Design and Architecture and were evolutionary in nature, building on the experience of previous MA's at Huddersfield. Academic and industry professionals were widely consulted spanning the art and design industry, educational models both current and historic were referenced internationally to inform the rationale.

The following issues and objectives were identified in 2007/2008:

- The School of Art, Design and Architecture has grown considerably during the last 12 years from approx 450 students to now approx 2,000.
- During this period we have developed clusters of degrees in 10 subject areas and we have positioned ourselves as a school at the business/professional end of the spectrum.
- We have established a solid foundation of collaborative programmes and centres which over the years have been more academically and financially beneficial to the school.
- However, we have made little effort to establish a significant strong international recruitment profile.
- It is now an opportune moment to build upon the solid undergraduate base and move more coherently and strategically into postgraduate provision.
- (Internal report).

3.1 A New Art and Design Thinking

The original ethos of the courses was to 'promote a 'new art and design thinking' and the emergence of new art and design professionals whereby students will develop heightened creative analytical skills combined with business skills,' 'Key among these is 'employer engagement' [4]. It was and still is felt that, 'Masters students now need to develop a much wider skills set including: business acumen, art and design management skills, multi-disciplinary skills, art and design initiative and leadership, alongside philosophical and intellectual attributes'. The original courses aimed 'to offer an environment, which celebrates and interrogates the blurring of subject boundaries where for example fine artists can influence and work with fashion designers and vice versa.' (Internal report).

The courses were designed to respond to growing needs of internationalization of the curricula, global markets and the importance of creativity aligned to commerce. The future of postgraduate art and design education was seen to be moving to such commonalities within multi-disciplinary approaches. The MA programmes constitute the need for a common cross-disciplinary postgraduate delivery and as such has been specified to create common research collaborations. This was not dissimilar to PGT developments elsewhere in the sector (Nottingham Trent University, Manchester Metropolitan University, University of Northumbria) many Universities have promoted collaborative and shared provision within Art and Design often from a philosophical and pedagogical position allied as in the case at Huddersfield with a pragmatic response to reduced recruitment, a withdrawal of external funding (European Social Fund, ESF) and the need to provide efficient delivery models.

3.2 Initial Provision

The first cohort of students (Circa 20) were recruited in January 2009 to the original 8 programmes in a suite named the, 'Cross-disciplinary Masters in Art and Design'.

- MA International Graphic Design Practice
- MA International Design Marketing and Communication
- MA International Fashion Design
- MA Textiles
- MA Spatial Design
- MA 3D Digital Design
- MA Fine Art
- MA Digital Media

The 8 courses shared a series of generic educational aims in addition to three course specific aims for each named award and represented an ambitious spread of disciplines to be taught in a cross-disciplinary, collaborative format.

The common aims were:

- To enable graduates and professionals from varied and international backgrounds to further develop, demonstrate and apply autonomous skills in research, analysis and practice through a structured programme, which integrates scholarly, business, creative perspectives and processes.
- To provide students a unique cross-disciplinary, cross culturally and collaborative flexible environment in which personal and professional ambitions can be achieved through independent and teamwork.
- To provide the opportunity for students to develop a perceptive and acute appreciation and application of the importance and significance of innovation and entrepreneurship by becoming art and design practitioners who also have the ability to challenge and innovate at a business and society level.
- The stimulation of an inquiring, imaginative, analytical, intellectual and creative approach encouraging independent judgement, strategic decision making and

critical self-awareness leading to advanced knowledge and practice within art and design.
- To promote the relationships between theory and practice as fundamental for lifelong skills together with knowledge transfer and a creative application to further practice or scholarship.

Towards the end of the first delivery a significant number of changes were made to the documentation however, this related mainly to the assessable submissions and team provision as well as a move to more generic module specifications within the specialist provision. The original ethos, aims, learning outcomes and philosophy of the courses was retained and 3 new awards were added:

- MA International Fashion Management
- MA International Fashion Promotion
- MA Costume

3.3 Impetus for Change

The original courses commenced in January 2009 as the delivery to the first cohort. The rationale in choosing January over September was largely due to pragmatic considerations related to the initial validation schedule. It was felt that student demand was greater for September start programmes, the vast majority of Art and Design Masters provision is September start. However, over the two initial intakes (January 09 and 10) a good relationship had started to develop with International agents and it was felt that a move to September only start would damage the continuity of these evolving relationships. Also it was felt that the overall market would be greater with 2 potential intakes. Hence the decision was taken to accept Cohort 3 in September 2010 and cohort 4 in January 2011. Cohorts 3 and 4 were taught in entirely separate groups whilst research was undertaken to look at models of delivery that might enhance efficiency and provide alternatives in terms of student choice.

Research in the sector showed that there were only 27 taught Masters programmes nationally in Art and Design with dual or multiple-entry including the 11 at UoH. The only other major provider was Northumbria University with 10 courses all the others being dispersed at a variety of institutions frequently involving elements of Distance Learning. A variety of outcomes were considered, the final model chosen included two distinct routes which involved joint teaching of cohorts as mixed groups: a 12 month September start course which commenced in September 2011 (cohort 5) and a 16 month January version commencing in January 2012 (cohort 6).

3.4 Designed Outcomes

3.4.1 Full-Time 12 Month (September Start)

The September start version of the courses resembles the standard taught format within the sector, 3 terms of study ending in a 60 credit major project. On closer examination however, it has some divergence from conventional programmes. As cohorts from each entry point are taught together (5 with 6, 6 with 7 etc. see Fig. 1) the modules taken by September start students in their terms 1 and 2 have to be non-sequential in nature as January start students will undertake these in the reverse order. To facilitate this there is an extended and intensive Induction module for the first four weeks of the courses.

3.4.2 Full-Time 16 Month with Professional Engagement (January Start)

As can be seen from the model (Fig. 1), the January start version of the courses is undertaken over 4 terms (16 months). Students complete the same 180 credits as the September start students with the addition of a 20 credit 'Professional engagement' module to be completed at some point between Easter and October for a minimum of 5 weeks equivalent study. The module can be undertaken by

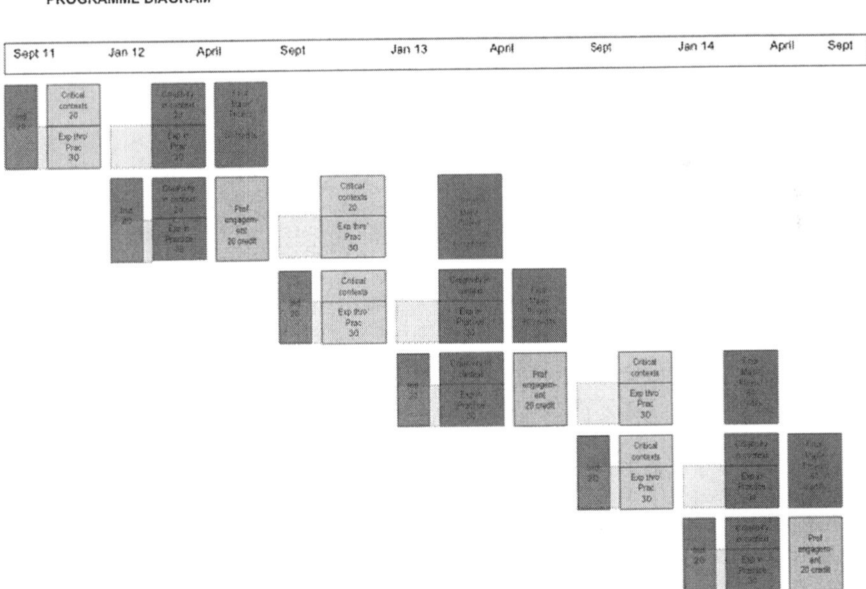

Fig. 1 Course diagram with multiple entry points

project or by placement/internship however, extensive efforts have been made to ensure that it cannot be offered or 'sold' as a guaranteed placement. The option is proving popular in International markets and also provides differentiation from the September version. Students passing the module in addition to the other modules receive a 200 credit 'with Professional engagement' degree award.

4 Delivery Experience

The experience of Full-time delivery is limited but positive. Students have reacted well to integration with new cohorts and have found the experience of meeting new cohorts as refreshing and rewarding. The student mix and interaction is positive as the courses always have 'old' students whenever a new cohort enters. There are therefore excellent opportunities for both formal and informal student mentoring.

4.1 Lessons, Experience and Recommendations

4.1.1 Non-sequential Delivery

The course consists of two main types of modules, as can be seen from Fig. 1, there are 2 'Context' modules and 2 'Exploration' modules. The 'Context' modules are primarily theory based, taught in mixed cohorts with some team elements. These can be delivered in a non-sequential manner, they contain different content from each other and students have become familiarised with many of the learning and assessment strategies from the Induction module.

The 'Exploration' modules are the 'subject' modules taught largely on a 1:1 or in small groups by Course Leaders and other subject specialist staff. The modules are generic across the 11/22 courses. The non- sequential nature of this learning does vary by course, in some courses it is sequential, students build from one project to the next, leading to their final Major Project, however, they do benefit from the interaction with students on 'other' parts of the same course. Because much of the delivery is 1:1 it is possible for students to build their learning in a sequential manner however, the longer term efficiencies required suggest that this is an issue that will need to be revisited.

4.2 Cross-Disciplinary Delivery

The courses are ambitious in the range of provision, from theory (Fashion Management) to Fine Art, some are quite technical (3D Digital Design) some very specialist (Costume). As a spread it is on occasion difficult to provide material that is of interest to all and also material that is essential to some but only very small numbers. A range of materials have been developed to enhance this cross-disciplinary delivery and have been largely well received however, there is feedback from both students and staff about relevance to individual requirements and the efficiency of delivery of specialist teaching to occasionally very small groups.

Much of the shared delivery and assessment is by team projects, this has evolved during the experience of delivering the courses, initially projects were fully cross-disciplinary with no two students in a group being on the same course. However, whilst this is still generally the case in the Induction module, for subsequent modules students have been clustered with students on related courses which has proven to be more successful and popular. As numbers increase (currently 35) increased clustering of delivery and material will continue. Modules are also supported by cross-disciplinary seminars, these are intended to be discursive and participatory in nature. These have been successful with the initial cohorts however, as the courses continue to develop this is evolving in response to student requests, seminars are a balance of delivered content as well as discussion.

4.2.1 Part-Time Delivery

Initial cohorts did contain part-time students and although delivery was never simple, it was possible to integrate students attending one day per week and completing the course over a 28 month period. Although numbers were small there were instances of high student achievement and satisfaction. However, once the current version of the courses was introduced in September 2011 the PT experience has suffered. The block delivery of the Induction module and subsequent impact on other modules has made part-time delivery very difficult and the decision has been made to withdraw provision. Delivery had always been challenging and economically questionable however, the PT students did bring fresh perspective and professional experience to the student body. Students interested in part-time study are now encouraged to apply to MA/MSc's by Research and are still welcome to attend taught sessions across all courses.

4.2.2 Integration of International Students

The nature of the student body has evolved during the evolution of the courses, becoming increasingly International in nature, drawing students from all the major continents of the world. This adds an interesting and welcome dynamic. The courses have been designed to provide an International perspective and to represent the realities of a global economy. One of the greatest assets of the courses in this respect is the range of international and cultural experience within the student body itself. The benefits of bringing students together into a larger community also enhance this, in a mixed group of 35 or more students there will usually be over 10 nationalities represented.

5 Conclusion

Although the courses have only recruited students for just under 4 years, this has involved delivery to 8 separate previous and ongoing cohorts (by January 13) and hence a notable learning experience. The courses have undergone significant amendments during this period and have been revalidated externally (December 2011) and subject reviewed externally (February 2012). It is felt that the courses have now developed a degree of maturity and stability; a model has been developed to facilitate the efficient delivery of multiple courses across a wide range of disciplines within the School with dual entry points. The central lessons to be drawn from this experience, of interest to academics and institutions considering cross-disciplinary and/or multiple entry delivery are manifold. Central amongst them however, would be to: consider with care the blend of programmes to be clustered, to examine the feasibility of further 'sub' clustering and the balance between the 'sub' and the main cluster. Multiple entry points can bring enhanced benefits to both recruitment and the student experience, although the associated complexities raised above do need to be acknowledged.

This is a unique model within UK Masters provision in Art and Design. Although not without challenges given the complexity of the provision, the Course team recognise that the structure and nature of provision provides a flexible, rich and rewarding learning experience for students.

References

1. Cox G (2005) "Review of creativity and design in UK business" HMT, London
2. Design Council (2010) Multi-disciplinary design education in the UK. Report and recommendations
3. Design Council (2010) Multi-disciplinary design education in the UK. Eight case studies
4. Clews D (2008) Future proof—new learning in the creative and cultural industries, ADM HE

Talking Architecture: Language and Its Roles in the Architectural Design Process

Yonni Avidan and Gabriela Goldschmidt

Abstract Architects use language intensively along the design process. Students are often asked to talk about their concepts, which sets the verbal language as the main tool used by students for communicating information, in spite of the fact that the architectural act is conceived as visual/spatial. The study challenges the notion that language is inferior compared to visual representation, and places the verbal expression as an essential part of the design process. The study follows architecture students, whose verbal concepts during one semester were mapped in terms of consistency, variability and development. A correlation was found between the percentage of evolving concepts in the process and the final studio grade. Analyzing semantic networks of design processes showed a higher number of links between concepts for students with higher grades, supporting the argument that language has an important role alongside graphic products in the architectural studio.

Keywords Design process · Design education · Language · Verbal concepts

---

Y. Avidan (✉) · G. Goldschmidt
Technion—Israel Institute of Technology, Haifa 32000, Israel
e-mail: yonniavidan@gmail.com

G. Goldschmidt
e-mail: gabig@technion.ac.il

A. Chakrabarti and R. V. Prakash (eds.), *ICoRD'13*, Lecture Notes in Mechanical Engineering, DOI: 10.1007/978-81-322-1050-4_91, © Springer India 2013

1 Introduction

1.1 Language and Communication

Language is a means of communication which is used to transfer knowledge, ideas, feelings and other information through different sounds and signals that give them meaning [1]. In the human language there is a limited number of vowels and constants, which constitute all the words and determine their pronunciation. In linguistics it is customary to attribute the term language to a system of signs and symbols that are subjected to syntax. The symbols themselves are meaningless; the syntax is what gives them meaning when they are in a certain context [2]. "The meaning of each expression does not constitute another simple projection of the author's intent, instead it is comprehended in the relation between the creator of the expression, the language it is stated in, the discourse it takes part of, and the understandings of the readers/listeners who interpret it" [3, pp. 20–21]. According to the authors, between the subject of the expression and the recipient there are at least two mediating factors—the language as a collection of meaningful symbols and the discourse as a social action of exchange of words and the creating of meaning. This idea is identical in essence to the ideas of the philosopher and architect Wittgenstein [4] who sees the language as an assortment of tools which work in different ways. The meaning of a word is set according to the way it is used and its location in a sentence. Therefore, it is possible to sort a group of words in new ways according to new sets of rules, thus gaining new readings of the same words, which receive different meanings [5]. This concept is most relevant to the learning process in the architectural studio, where the design language is built through concepts that have certain initial meanings butare given a new expression. As a result their meaning changes [6].

1.2 Language and the Design Process

Architects use language intensively along the design process, both as students and as experts [7]. Words help architects to express and explain their ideas to others, which is essential in teamwork, common among designers [8]. Students are often asked to talk about their concepts in design reviews, which sets verbal language as the main tool used by students for communicating information, in spite of the fact that the architectural act is, by definition, physical. What roles do words play in the architectural design process? Are there special elements in the spoken language that place it as a design tool along with graphic representations? What is the point of talking about architecture?

The design process, much like language is "a way of communicating ideas. Both can be considered cultural expressions or cultural communication" [9, p. 224], hence the proposal to examine these two as sharing a common structure. Oak [10] claims that design is intertwined with dissociation and interaction between participants, who create together an understanding of the object. In speech, evaluation is also apparent and as a result a student can change his or her mind if he or she chooses to do so. According to Oak, the nature of a conversation is not only the creation of concepts for its participants for the purpose of understanding, appraisal or using what is said; the conversation effects directly the future appearance of the designed object.

Dong [11] defines design itself as a language. In his research "The enactment of design through language" he contemplates the true role of language in the design process. He takes issue with the popular opinion that the language is subjected to the visual representation and in most cases functions as an accessory, an addition to the design itself which assists solely in negotiation. In his opinion, the language has an essential role in portraying the design, beyond the representation or the verbal description of the architectural act. He defines three executive parameters, through which the language portrays design:

(1) *Aggregation*, an arrangement of concepts, accumulated ideas and experience which are built over time along with the designers' personal experience.
(2) *Accumulation* of concepts, creating new representations from the collection of knowledge items.
(3) *Appraisal* of concepts.

These parameters allow to the design process to run its course and be brought to fruition through the use of language. An example can be found in Schön [12, 13] who describes the teacher Quist and his student Petra achieving congruence in the process of creating the design concept through words. Reflective design as defined by Schön is a process where in the designer responds during reflection in action, by understanding the problem and disassembling it into action strategies. Dong explains this as progressing towards a clear design concept, which is built from "cycles of convergence and divergence" [11, p. 11].

Schön [12] claimed that language reveals the ways in which we think and perceive reality. One cannot think of a new concept, only create new compositions and new relationships; therefore developing concepts and replacing one concept with another is highly important.

The purpose of this research is to examine the ways in which architecture students integrate verbal concepts in the different stages of the design process. The study challenges the definition of language as subordinate to visual representations, and argues that language is an integral part of the design process and can reflect its potential alongside visual artifacts.

2 Methodology

In this study, two studio groups were selected randomly from a total of eight groups in the third semester of Architecture studies, a total of 23 students. This selection was intended to widen the sample and diversify the range of design processes; the process was the same in both groups. All participants were informed about the research goals and procedure in general terms only, in the first meeting of the semester. The students were observed at work on their individual projects for 14 weeks, from the beginning of the semester, in regular classes and presentations, including the final presentation that marks the end of the design exercise. Class routines were not interfered by the research, as researchers took part only as non-participating observers.

Most of the material for the analysis came from the conversations between the teachers and the students (desk critiques), which were recorded and transcribed in order to glean verbal concepts from. The researchers mapped the students' concepts over time, as expressed by them verbally, in terms of consistency, variability and development. Conceptual maps of their design processes were examined and correlated with the students' final studio grades, given by the studio teachers. Semantic networks [14, 15] were also generated, which allowed tracing the different connections among verbal concepts, and comparing the semantic networks of the different students.

3 Research Definitions

3.1 Consistent Concepts

Students often express repeatedly topics which they find exciting and inspiring to work with along the semester. Concepts that were mentioned more than once in a specific design process were defined as consistent concepts, and were marked as such in the concept charts that were generated for each of the research participants. Figure 1 shows an example of a construction of the concept chart of a single design process. Every new concept that is mentioned by the student is documented in a new row, under the column representing the sequential number of the class. In class 2 for example, the student discussed with the teacher three verbal concepts ("different Places", "layers" and "public path"). In the following lessons the student kept on mentioning two of those three concepts (verbal concepts no. 2 and 4), while the verbal concept "layers" (no. 3) was dropped and therefore it is not considered as a consistent concept.

← lesson ..	lesson 6	lesson 5	lesson 4	lesson 3	lesson 2	lesson 1	
						a stone wall	concept 1
		open places	playgrounds	open places	different places		concept 2
					layers		concept 3
	public trail	public trail	public path	public path	public path		concept 4
			library				concept 5
	shared balconies	a shared balcony	a shared balcony				concept 6
	light spiral stairs	spiral stairs	a circulation system				concept 7
							concept .. ↓

Fig. 1 An example of a concept chart (partial)

3.2 Evolving Concepts Versus Unchanged Concepts

Evolving concepts are verbal ideas that developed along the process. They are represented by gray bars (see Fig. 1), while black bars indicate unchanged verbal concepts–permanent ideas which have not evolved throughout the design process. For example, verbal concept no. 4 is an unchanged concept–one can notice that although the student has mentioned this concept constantly, it remained verbally unchanged as it was called "Public Path/Public Trail". On the other hand, verbal concept no. 2 was defined as an idea that has evolved from "Different Places" through "Open Places" and into "Playgrounds" (lessons 2, 3 and 4 respectively).

3.3 Semantic Networks and the Design Space

The design space in which the student's project is developed is a dynamic psychological-intellectual realm created by both the teacher and the student. "It is used to identify the design problem and to facilitate the development of the design project. It places the imagined environment in a workable context, linked in reality" [16, p. 325]. The language of the design space is assembled from both visual and verbal components. And defines a shared 'database' as an infrastructure for communication. Wendler and Rogers [16] define five acts that can create the design space:

(1) *Specifying*—identifying processes that will lead to intended outcomes to surface.

(2) *Witnessing*—recounting personal experiences and expectations in the studio environment.

(3) *Structuring*—organizing processes, goals, objectives, perceptions and expectations.

(4) *Visualizing*—visual information, written and verbal communication.

(5) *Sharing*—structuring information that is seen as relevant by all parties.

In this research, Semantic Networks which were assembled from the teacher-student conversation protocols during the semester represent the design space of each of the participants. The semantic network's structure is composed of Nodes—verbal concepts mentioned by the student, and Connections between them. Figure 2 is an example of a semantic network characterizing a design space of a single student's process. This semantic network differentiates among three types of verbal concepts: Main Verbal Concepts (see Fig. 2, legend) represent ideas that the student defined as primary, and that led to 4 or more other ideas. In the design process described in Fig. 2 the student had 3 main verbal Concepts. The second node type is Conceptual Ideas, the student's "design wishes" that do not achieve a physical (spatial/visual) expression, followed by the last component of the network—the Design Acts, an actual architectural elements designed by the

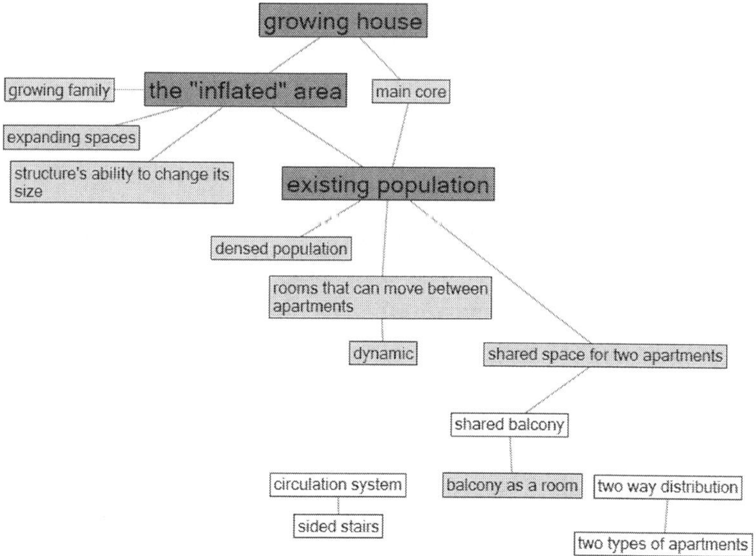

Fig. 2 Semantic network as a design space. *Legend dark gray* background: main concepts; *light gray* conceptual ideas; *white* design acts

student. In the example given in Fig. 2 the student decides on several Design Acts: shared balcony, two types of apartments, etc.

The semantic networks draw the boundaries of the space in which the student chooses to work, while the teacher can broaden, challenge and change those boundaries, allowing the student to use this new information first as a new analysis area, and afterwards as an integral part of his or her personal design space.

4 Results and Discussion

The study found an r = 0.537 correlation between the proportion of consistent concepts and the studio final grade, meaning that the more students adhered to their ideas along the process, their accomplishments in the studio were higher, as shown in Fig. 3. Figure 4 shows that the lack of ability to develop concepts throughout the process led to poor results. In addition, an r = 0.626 correlation was found between the final studio grade and the proportion of evolving concepts—ideas that developed along the process (see gray bars in Figs. 3 and 4). Similar to Schön's proposition [12], design processes wherein one concept was transformed into another by developing and enhancing existing verbal concepts, as well as continuous concepts along the design process, were rated higher at the end of the semester. As a result from these findings, an r = 0.704 correlation was found between the proportion of the combined total of consistent concepts and

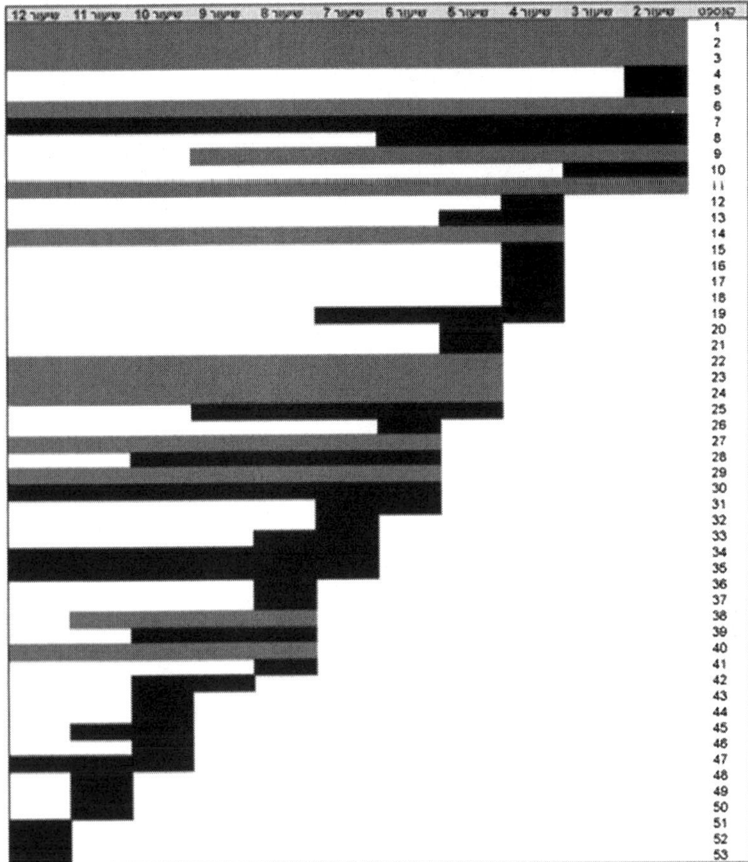

Fig. 3 Concept chart of a student graded 95. *Legend gray bars* evolving concepts, *black bars* unchanged concepts

evolving concepts. This indicates that designers who entertained a wide range of ideas during the semester, could develop them into new verbal concepts, which expanded and enriched their design process.

The charts generated from the students' design processes differ from one another not only in the number of consistent or evolving concepts. The charts' graphic characteristics such as length and slope can also shed some light on the quality of students' processes. In this research no correlation was found between the total number of the verbal concepts (represented by the charts' lengths) and the students' studio grades. Although in the examples given above the concept chart of a student graded higher is in fact longer, Fig. 5 shows a design process with a large number of verbal concepts, double the average number of verbal concepts of the entire group examined. In spite of the fact that the student had many ideas, most of them did not evolve nor were they carried over to the next class. As a result, the

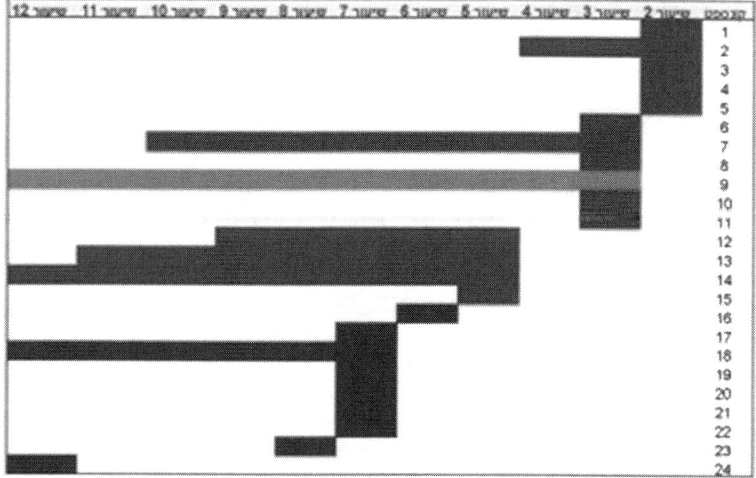

Fig. 4 Concept chart of a student graded 70. *Legend gray bars* evolving concepts, *black bars* unchanged concepts

student's grade, 70, was the same as the grade awarded to the student whose concept chart, shown in Fig. 4, includes significantly fewer ideas.

In addition, Fig. 5 demonstrates a chart with an acute slope, a prevalent quality in the design process, defining a process with clusters of Local Ideas that do not continue to the following class (see Fig. 6). On the contrary, a moderate slope of a chart stands for a design process that is rich with continuous concepts which, as mentioned earlier, this research takes to be a more successful design process (See Fig. 3).

As mentioned earlier, a semantic network analysis enabled to examine the connections between verbal concepts in the design process, as well as to define different kinds of concepts.

Comparing networks showed a higher number of links among concepts and therefore a more solid framework for students with higher studio grades, while low-graded students were characterized by a more disassembled structure of their networks (See Figs. 7 and 8, respectively).

Additional research findings shows an $r = 0.874$ correlation between the proportion of conceptual ideas at the beginning of the semester (class 2, top of a network) and the verbal concepts that were added in the same class. This reinforces the understandable notion that at the beginning of the semester architecture students explore and build their primary conceptual inventory. Referring to this data, an $r = 0.469$ correlation was found between the proportion of conceptual ideas in the final presentation, at the end of the design process (bottom of a network) and the studio final grade. These findings strengthen the importance of the presence of a solid and varied conceptual inventory for design that is being held until the later phases of the process.

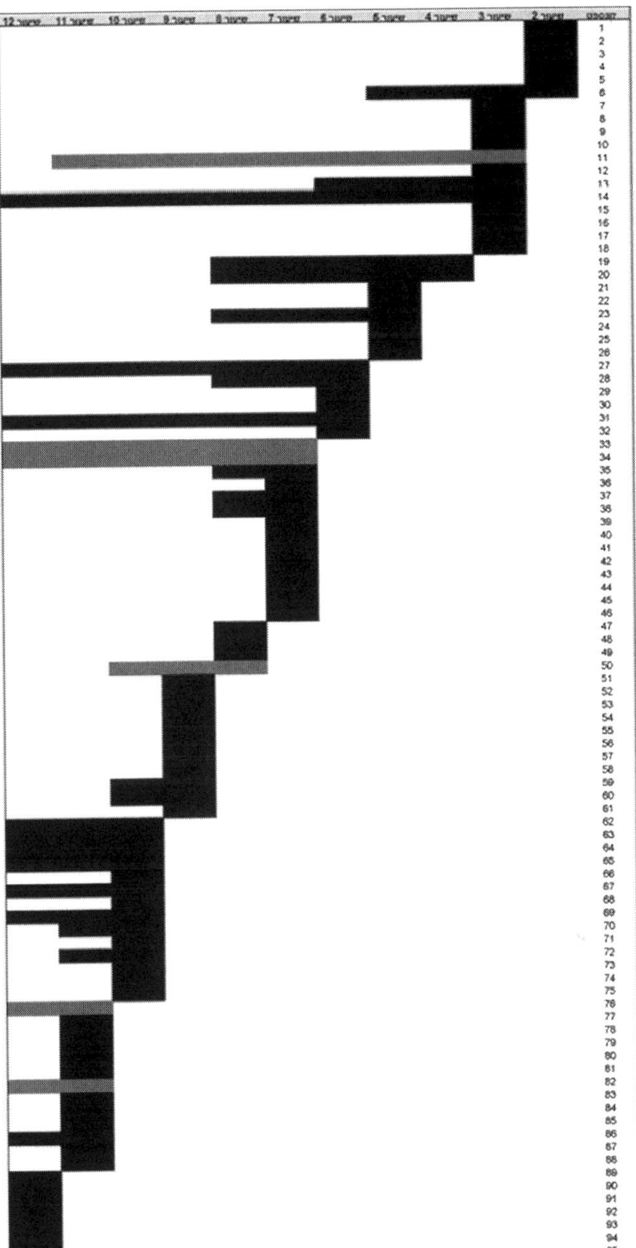

Fig. 5 Concept chart with an acute slope

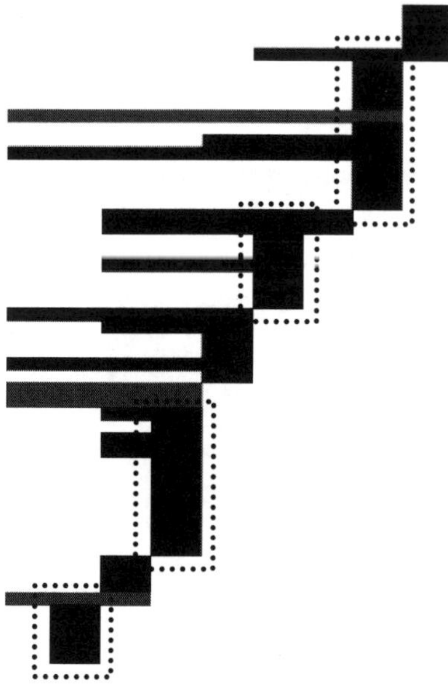

Fig. 6 Detail of the chart demonstrating clusters of local ideas

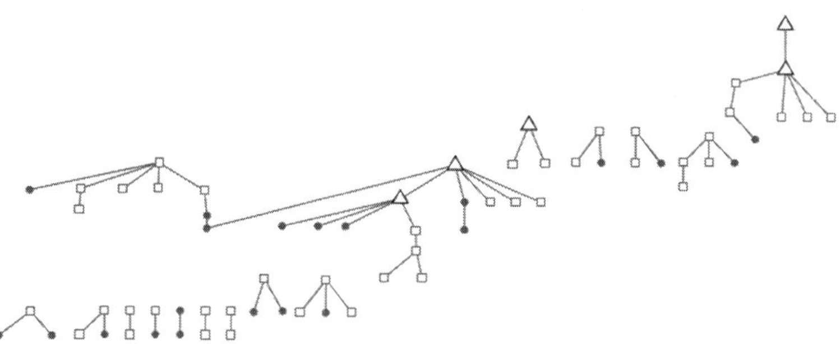

Fig. 7 Semantic network of a student's process graded high. *Legend triangle* main verbal concepts; *square* conceptual ideas; *black circle* design acts

Fig. 8 Semantic network of a student's process graded low. *Legend triangle* main verbal concepts; *square* conceptual ideas; *black circle* design acts

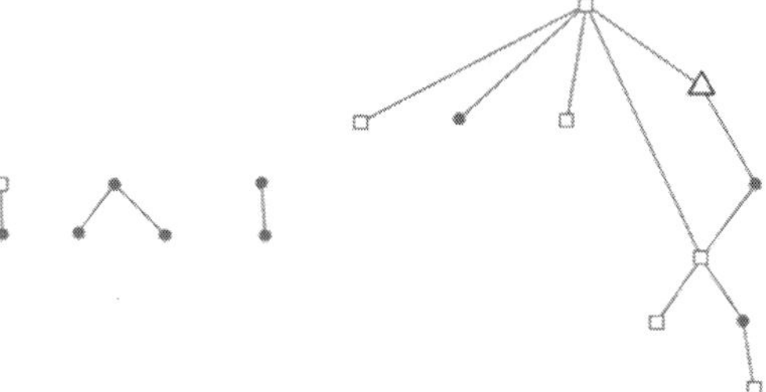

5 In Conclusion

The findings of this study support the claim that verbal language has an important role alongside graphic products in the architectural studio. Although this research followed only the verbal concepts that were obtained from the students, a clear connection between the verbal 'database' of a specific design process to its quality has been shown. A correlation between these two indicates that even though the teacher does not follow the students' verbal ideas consistently, his or her general assessment of the students' achievements also include their verbal choices.

Pinker [17] argues that the idea of which language is identical to thought is a common absurd: one knows the familiar feeling of searching for the right word for a specific thought or an idea. For this to happen, a conceptual realm between internal and external representation (expression) has to exist. This conceptual space also contains the endless possibilities of one thought as it transforms and changes into a chosen word. Most designers execute these bidirectional movements between internal and external representations hastily, as they make progress in the design process. It is proposed that meticulous choices of words defining the designers' wishes may lead to a more satisfying result. Furthermore, the progression of a search for accuracy will construct and expand the design space. Understanding its dimensions can "promote effective communication, so that each individual can better "see" what another "sees"" [16, p. 333].

Verbal concepts are varied and unlimited. This research distinguishes between two types of verbal concepts, characterizing the design process; while a conceptual idea can be used as an endless platform for developing other ideas, a design act is intended as an examination of the original idea and stands as its physical representation, which can result in another conceptual idea or design act. Other categorizations of verbal concepts can be established in order to reveal other design process structures, inferred from verbal concepts.

The architectural studio teaches its students design methods by using professional language and visual tools, but it cannot explore and review all verbal concepts, leave alone determine which ones have the ability to lead to a better design product. Nevertheless, developing and enhancing verbal motivators among architecture students may provide them with new insights that would refresh their processes and, ultimately, lead to better results.

Acknowledgments The Authors would like to thank the students and teachers in the faculty of Architecture and Town Planning at the Technion who participated in the research for their appreciated cooperation.

References

1. Ruse C, Reif JA, Levy Y (eds) (1988) Oxford students' Dictionary. Kernerman Publishing Ltd and Lonnie Kahn & Co. Ltd, Tel Aviv
2. Goldschmidt G, Litan-Sever A (2011) Inspiring design ideas with texts. Des Stud 32(2):139–155

3. Shalsky S, Alpert B (2007) Ways of writing qualitative research: from deconstructing reality to its construction as a text. Mofet, Tel Aviv
4. Wittgenstein L (1953) Philosophical investigations. Basil Blackwell Ltd, Oxford
5. Chomsky N (1978) On the biological basis of language capacities. In: Miller GA (ed) Psychology and biology of language and thought. Academic, NY
6. Vygotsky L (1997) Thought and language. MIT Press, Cambridge MA
7. Wong J (2010) The text of free form architecture: qualitative study of the discourse of four architects. Des Stud 31(3):237–267
8. Poggenpohl S (2004) Language definition and its role in developing a design discourse. Des Stud 25(6):579–605
9. Swierczek DJ (1985) Style as a language of design. Des Methods Theor 19(1):218–225
10. Oak A (2011) What can talk tell us about design? analyzing conversation to understand practice. Des Stud 32(3):211–234
11. Dong A (2007) The enactment of design through language. Des Stud 28(1):5–21
12. Schön D (1983) The reflective practitioner: how professionals think in action. Basic Books, NY
13. Schön D, Wiggins G (1992) Kinds of seeing and their functions in designing. Des Stu 13(2):135–156
14. Sowa JF (1991) Principles of semantic networks: explorations in the representation of knowledge. Morgan Kaufmann Publishers, San Mateo CA
15. Jackendoff R (1990) Semantic structures. MIT Press, Cambridge, MA
16. Wendler V, Rogers J (1995) The design life space: verbal communication in the architectural design studio. J Architectural Plann Res 12(4):319–335
17. Pinker S (2011) The language instinct: how the mind creates language. Shalem, Jerusalem

Cross-Disciplinary Approaches: Indications of a Student Design Project

H. Hashemi Farzaneh, Maria Katharina Kaiser, Torsten Metzler and Udo Lindemann

Abstract Cross-disciplinary approaches are adopted in technical product development for a number of reasons, including the improvement of the product quality and the reduction of time to market. However, the positive and negative effects of cross-disciplinary approaches such as cross-disciplinary teams or biomimetics are controversially discussed. In this work, we perform a case study with architecture and mechanical engineering students using biomimetics to gain insights to effects in a threefold cross-disciplinary project. The results indicate possibilities for improving cross-disciplinary team projects.

Keywords Cross-disciplinary team work · Biomimetics · Collaborative design

1 Introduction

Adopting a cross-disciplinary approach in technical product development is supposed to have a number of positive effects: a more profound problem understanding, higher quality of solutions and a shorter time to market are examples [1, 2].

H. H. Farzaneh (✉) · M. K. Kaiser · T. Metzler · U. Lindemann
Institute of Product Development, Technische Universität München,
Boltzmannstraße 15, 85748 Garching, Germany
e-mail: hashemi@pe.mw.tum.de

M. K. Kaiser
e-mail: kaiser@pe.mw.tum.de

T. Metzler
e-mail: metzler@pe.mw.tum.de

U. Lindemann
e-mail: lindemann@pe.mw.tum.de

A. Chakrabarti and R. V. Prakash (eds.), *ICoRD'13*, Lecture Notes in Mechanical Engineering, DOI: 10.1007/978-81-322-1050-4_92, © Springer India 2013

A cross-disciplinary approach can be conducted in different ways. One approach is to set up cross-disciplinary product development teams in order to develop a product for a task related to both disciplines. An example is a product for heating, ventilation, air conditioning and refrigeration (HVACR): both architects and engineers can contribute with their discipline-specific knowledge to the development. Another approach is to use information from different disciplines for inspiration. This is the case in biomimetics: the designer uses nature or results from biological research as inspiration for solving a technical task.

What are the effects if both these approaches are combined, i.e. a cross-disciplinary team uses information from another discipline? Understanding the effects of this combined constellation can give implications for supporting cross-disciplinary teams working on a cross-disciplinary project.

In this work, we explore the effects of cross-disciplinary approaches in a case study conducted with five cross-disciplinary teams consisting of 23 students of mechanical engineering and architecture collaborating in a biomimetic product development project. A particular focus lies on the comparison between the internal views of the participating students and the achieved outcome of the team work. To start with, this paper gives an introduction to literature on cross-disciplinary teams and biomimetics. Then, we describe the detailed proceeding of the case study. In the following section the internal views are presented. They are then compared to the outcome of the teamwork. In conclusion, this exploratory study shows positive and negative effects of cross-disciplinary approaches.

2 Literature Survey: Cross-Disciplinary Approaches

This section gives an introduction to literature on cross-disciplinary teams and on biomimicry, an approach to use information from biology to develop technical products.

2.1 Cross-Disciplinary Teams

From an industrial perspective, working in teams aims at synergy effects and information exchange to enhance productivity. In this context, teams are defined as temporary work groups solving problems, developing solutions or fulfilling tasks within the framework of a superordinate target [3]. Cross-disciplinary-teams consist of individuals possessing knowledge from different disciplines. According to the above understanding that one goal of teams is to "exchange information", the individuals can contribute with their heterogeneous information achieving a higher productivity. However, research on diverse teams including cross-disciplinary teams has resulted in contradicting conclusions. Mannix and Neale [2] reviewed psychological research on diversity in teams and found that the negative effects prevail in the majority of research contributions. They propose theories such as the self-and social categorisation approach. According to this theory individuals categorise others and have expectations based on this categorisation. This increases the tendency to develop stereotypes about individuals belonging to a different "category" [2]. In contrast, conflicts and confrontation can also have positive effects. Stempfle and Badke-Schaub [4] state that cognitive confrontation is necessary for creativity. Kurtzberg [5] observed that diverse team develop a higher number of ideas even though the individual team members feel less creative.

2.1.1 Biomimetics: Using Information from Different Disciplines

Engineers as well as architects are continuously searching for new solutions for their technical and design tasks in order to develop new, creative solutions. Nature offers a large repository of biological systems which can provide analogies or inspiration. Therefore, biomimetics are recommended as a creativity method [3]. Accordingly, Nachtigall [6] defines biomimicry as "learning from the design-, process- and development principles of nature". Still, applying biomimetics can pose a number of challenges due to the cross-disciplinary nature of the approach. Coming from different disciplines, mechanical engineers, architects and biologists use different models and terminologies [7]. This entails challenges for the search for biological inspirations and analogies as well as for their transfer to technical and design solutions. A number of researchers have focused on these challenges and developed approaches to support the biomimetic search and transfer: as to biomimetic search, databases of biological systems have been built [8–12]. Another research focus is on natural language analysis to map biological and technical terms [7, 13, 14]. The transfer of biological analogies to engineering and architecture is also addressed by Sartori et al. [8] using the SAPPhIRE approach to model both biological and technical systems. Other researchers propose development procedures specifically designed for biomimetics to facilitate the transfer [9, 10].

3 Combined Cross-Disciplinary Approach

In this work, we study a combined cross-disciplinary approach in product development involving three disciplines: a team consisting of individuals from two disciplines and a task focusing on these two disciplines for which information from a third discipline is required. The aim is to integrate information from the three disciplines to improve the development of a product. This threefold cross-disciplinary constellation discloses a number of questions:

- What are the effects of the cross-disciplinary team and how do they use information from the other discipline?
- What is the impact on the outcome of the project? What is the contribution of the three disciplines to the outcome?
- Which indications for a support of a project in this threefold constellation can be deduced?

We approach these questions with the case study described in the following section.

4 Case Study: Cross-Disciplinary Student Teams Developing a Biomimetic Concept for a Shell Construction

In this case study, teams consisting of students of architecture and mechanical engineering develop a biomimetic concept for a shell construction. 23 students participate in 5 cross-disciplinary teams. The teams consist of four to six students of which one or two are mechanical engineering students. This team constellation is due to the task *developing a shell construction,* which is considered mainly architectural but requires knowledge from mechanical engineering to ensure the technical functionality.

The students were guided and supervised jointly by members of the Institute of Shell Constructions from the Faculty of Architecture and the Institute of Product Development from the Faculty of Mechanical Engineering. Lectures focusing on shell construction, technical product development and biomimetics introduced the students to the project. In the first 2 weeks, the teams performed a literature research on biological systems which were used as inspiration for shell constructions. Then, within 7 weeks, they developed concepts for shell constructions addressing a chosen issue such as ventilation or lighting conditions. They tested and presented their concepts by models and prototypes. There was a mid-term presentation after 4 weeks and a final presentation after 7 weeks.

To capture the internal view of the students, they filled out six questionnaires. The questionnaire at the beginning of the project was aimed at capturing the students' previous experiences and their expectations. During the project there were four questionnaires to record the development of the teams: before and after the literature presentation, at mid-term and before the final presentation. After the final presentation a last detailed questionnaire was used for a retrospective view on the project. To deepen the understanding gained from the questionnaires, semi-structured interviews with five of the students were conducted. For the interviews we chose two teams and interviewed one mechanical engineering student and one or two architecture students per team.

As to the outcome of the project, we regard the final presentation of the concepts. The focus lies on the contribution of each discipline to the outcome.

The internal observations are compared to the outcome and discussed to analyse effects of the cross-disciplinary product development teams in combination with the use of information from another discipline. We conclude with the indications of the case study with regards to the questions presented in Sect. 3.

5 Internal View of the Students

We use the questionnaires and interviews with the participating students to analyse their internal view on the cross-disciplinary team work (Sect. 5.1) and the biomimetic approach (Sect. 5.2). In the following, due to limited space, we present only the most distinct results of the analysis of the questionnaires and additional insights gained in the interviews.

5.1 Cross-Disciplinary Team Work

The results are divided into the topics: *team-performance* and its development during the project, *positive and negative aspects of the cross-disciplinary team constellation*, *tasks of the individual students* and the perceived *importance of the cross-disciplinary team constellation.*

5.1.1 Team-Performance

Figure 1 shows the students' evaluation of their team's performance and its development during the project. This evaluation was designed according to Metzler and Shea [15]. The students could evaluate on a scale between a *dysfunctional team* (0) to a *functional team* (4) to a *high performance team* (8). Figure 1 displays the average values per team. Since some students did not fill out all questionnaires, the number of students varied from 16 to 23.

The first time the students evaluated their team's performance was in the second questionnaire before the presentation of the literature search (week 2). The other questionnaires were filled out after the presentation of the literature search (week 3), at mid-term (week 6), before the final presentation (week 9) and after the final presentation (week 10). The axis in Fig. 1 is therefore not proportional to the elapsed time. It can be noted that the average value per team varies from 3.75 to 7.25 throughout the whole project. It has to be added that the minimum evaluation by one student in one questionnaire is 2 (not displayed in Fig. 1). Still, it can be concluded that in general all teams considered their team functional or more. As to the development during the project, all teams except team 3 evaluated their team's performance higher at the end of the project than at the start. Team 3 evaluated their team's performance highest at the beginning of the project in comparison with the

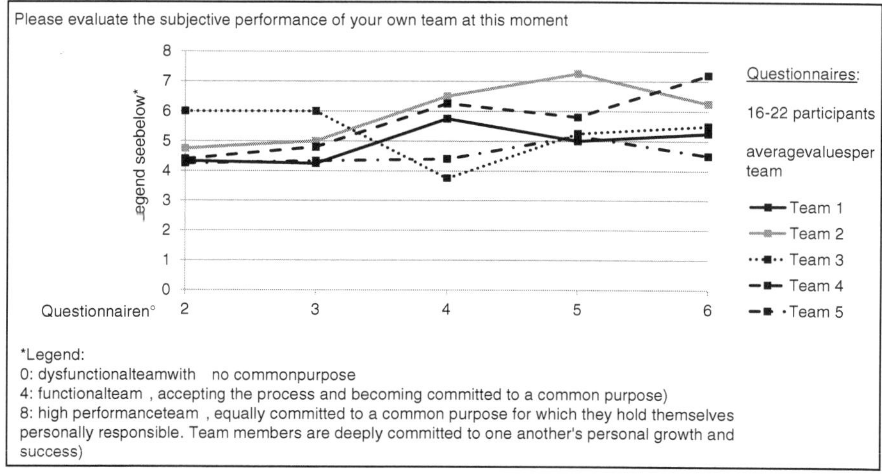

Fig. 1 Questionnaires: development of the team performance

other teams and lowest at mid-term. Towards the end of the project their evaluation of their team's performance increased.

5.1.2 Positive and Negative Aspects of the Cross-Disciplinary Team Constellation

In the questionnaires the students were asked for positive aspects in their team (*What contributes to the success of your team?*) and for negative aspects (*What causes difficulties in your team?*). The students could choose six options per question and suggest additional observations. Figure 2 shows the students' expectations at the start of the project (questionnaire 1) and their view at the end of the project, i.e. after the final presentation (questionnaire 6).

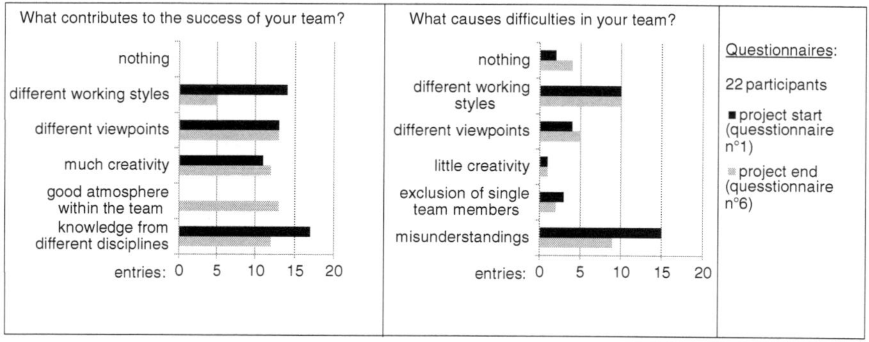

Fig. 2 Questionnaires: positive and negative aspects of the team work

As to the positive aspects, no student chose the option *nothing contributes to the success of the team*. Both at the start and at the end of the project, more than 10 students chose "different viewpoints" and "much creativity" as positive aspects. This positive view was sustained by a student in the interview stating that he gained awareness of the other discipline's viewpoint which caused him to leave his common patterns of thought.

At the start of the project, no student expected a "good atmosphere" as a positive aspect in a cross-disciplinary team. At the end of the project this option was chosen. 13 times which represents more than half of the students. They now perceived the good atmosphere as a factor in their team work even though they had not expected it.

At the start of the project the majority of the students (17) expected *knowledge from different disciplines* to be a positive aspect. At the end of the project the number of students choosing this option had declined to 12. This decline might be explained by statements from the interviews. In the interviews, three students stated that the technical ideas from the mechanical engineering students could not be pursued as far as they had wanted because of a lack of time. According to these three students the main contribution of the mechanical engineering students was their knowledge about systematic approaches in product development. On the other hand, they considered the architecture students more pragmatic, but less systematic.

Different working styles were expected to be a positive aspect by 14 students at the start of the project. At the end of the project solely five students chose that option.

With regards to negative effects, at the most five students chose *nothing*, *different viewpoints*, *little creativity* and *exclusion of single team members*.

Different working styles as a negative aspect was chosen by ten students at the start as well as at the end of the project. This view is sustained by two of the architecture students in the interviews. In their opinion, the architecture students were prepared to work more than the mechanical engineering students. One of them stated that "*architects accept iterations due to significant concept changes if the result can be improved*". The other one stated that "*when a model had to be finished, it was the architecture students who stayed and worked*".

Misunderstandings were expected at the start of the project by a majority of 15 students. At the end of the project the number of students choosing that option had declined to nine.

5.1.3 Tasks of the Individual Participants within the Team

Figure 3 shows the students' expectation about their individual tasks at the start of the project (questionnaire 1) and their view after the final presentation at the end of the project (questionnaire 6). The students could choose several options in the questionnaires. The answers of architecture and mechanical engineering students are shown separately to allow for a comparison.

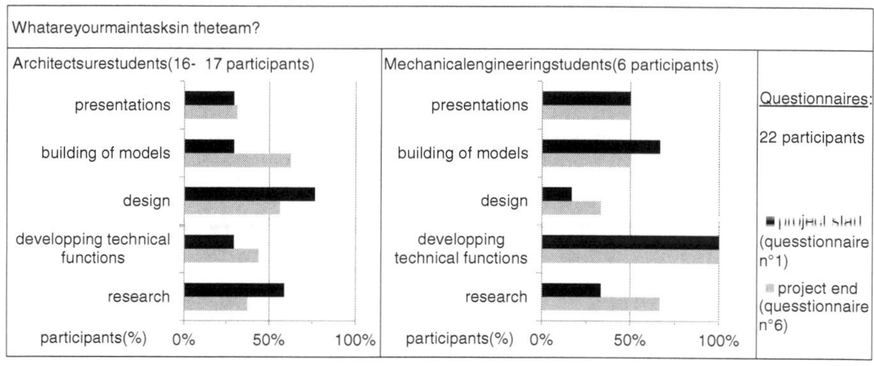

Fig. 3 Questionnaires: Tasks of the individual students within the team

At the start of the project more than 50 % of the architecture students chose *research* and *design* as their main tasks. The other tasks were chosen by about 25 % of the architecture students. At the end of the project, the rating of the task *building of models* had changed most: More than 60 % of the architecture students chose this option. In the interviews, one architecture students stated that he had expected the mechanical engineering students to be technical "tinkerers" who build a lot of models and prototypes, but was proved to be wrong during the project.

As to the mechanical engineering students, all of them chose *developing technical functions* to be one of their main tasks at the start and at the end of the project. Apparently, they felt this was their main responsibility, possibly because there were only one or two mechanical engineering students per team.

5.1.4 Importance of the Cross-Disciplinary Team Constellation

Figure 4 shows the degree of confirmation of two statements comparing the importance of the cross-disciplinary team constellation to the option to carry out the project with students from one of the disciplines. These statements were part of the questionnaire at the end of the project after the final presentation (questionnaire 6). The students could choose six options on a scale between one (not true) and six (very true). The options one to three therefore express declining disagreement with the statement, the options four to six show increasing agreement with the statement.

As can be seen in Fig. 4, the confirmation of the first statement (importance of cross-disciplinary team constellation) was high. The majority of both architecture and mechanical engineering students chose between four and six (very true) points.

With regards to the second statement, the result differs for the two disciplines: The mechanical engineering students all rather disagreed with the statement and

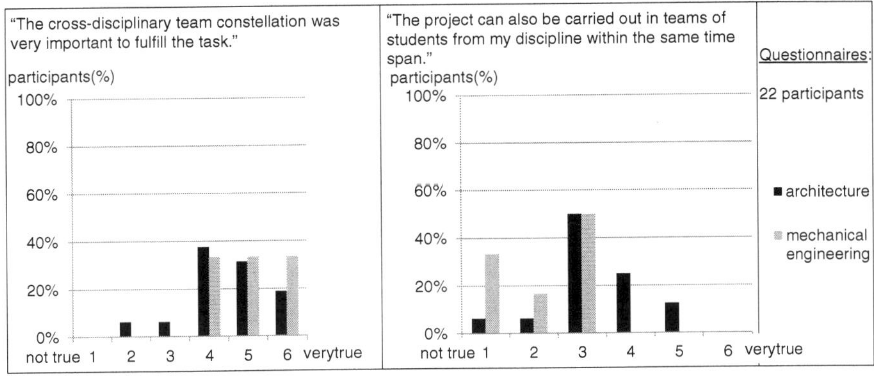

Fig. 4 Questionnaires: Importance of the cross-disciplinary team constellation

chose between one (not true) and three points. The majority of the architecture students chose between three and five points. This shows their tendency to see less importance in the collaboration with the mechanical engineering students.

5.2 Biomimetics

Figure 5 shows the evaluation of the students with regards to the influence of biomimetics on creativity and the importance of biomimetics for the task. This evaluation was part of the questionnaire at the end of the project after the final presentation (questionnaire 6).

As to the question "did you develop more creative ideas because of biomimetics?", students could choose on a scale between one (no more creative ideas) and six (much more creative ideas). About 60 % of the students choose five or six

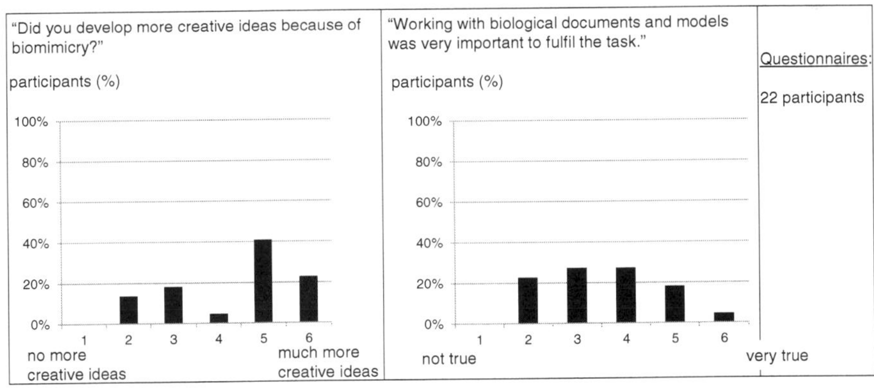

Fig. 5 Questionnaires: Biomimetics—creativity and importance

showing that the majority of the students believed to have more creative ideas due to biomimetics. This view was confirmed by four of the five interviewed students. One of the students stated that biomimetics had helped to disengage from existing solutions and to defer the feasibility of ideas. According to him, both fostered the development of more creative ideas.

To evaluate the statement *"working with biological documents and models was very important to fulfil the task"* students could choose on a scale similar to those in Sect. 5.1.4. About 20 % of the students evaluated this statement with two, three, four and five points respectively. This shows a moderate disagreement to a moderate agreement with the statement. Accordingly, in the interview one student stated that since the literature research on biological systems had been separated from the development of concepts they had not pursued ideas from that phase later in the project.

To sum up, a majority of the students were of the opinion that biomimetics triggered their creativity. On the other hand, almost half of the students did not think it was essential to fulfil the task. A possible reason for this is that they did not pursue a biomimetic idea until the end of the project, but solely used biology as an inspiration at the start of the project.

6 Comparison of the Internal Views with the Outcomes of the Project

For the final presentation, all teams prepared a presentation to show their final concept and its development. In addition they presented physical models of their preliminary and their final concept. We analysed this outcome of the project to assess the contribution of the three disciplines architecture, mechanical engineering and biology to the final concepts. Table 1 shows a short description of the final concepts and the contribution of the three disciplines. As can be taken from Table 1, all final concepts are predominantly architectural concepts as required by the task given to the student teams. In addition, all final concepts include elements from mechanical engineering. Either they are necessary elements to actuate the developed concepts (paraffin cylinders, electro-magnets, hydraulic cylinders) or additional elements to save or produce energy (water pumping systems, wind turbines). The third discipline, biology, on the other hand, is not directly visible in the final concepts, but the teams claim to have used biological systems as inspiration. Hill distinguishes between four different degrees of abstraction for biomimetics, of which using a biological principle as an inspiration is the most abstract one [9]. It can be concluded that in this case-study biology was abstracted to a high degree to serve as an inspiration.

In comparison with the internal views, the outcomes confirm the evaluation of the students to a high degree. As to the cross-disciplinary collaboration of architecture and engineering students, the final concepts confirm that both disciplines

Table 1 Final concepts of the teams

Team	Final concept	Architecture	Mech. eng.	Biology
1	Cooling and shading system for a building	Window blinds adaptable to different conditions during day and night	Water pumping system cooled during night-time in the window blinds through the building	The wings of beetles served as an inspiration for the folding of the window blinds
2	Shading system between the window panes	Ring elements connected via paraffin cylinders;they twist due to heat and change the shading conditions	Actuation of the elements via paraffin cylinders changing their length due to heat	Hairs of the old man cactus served as an inspiration to provide shading with thin elements that can be twisted to enhance their opacity
3	Cooling a building and producing energy	Façade leading the wind through the building (adaptive to the wind direction)	Wind turbines for the energy production	The den of the prairie dog served as an inspiration to use wind
4	Shading system between the window panes	(Un) folding elements to shade parts of the window	Electro-magnetic actuation of the window panes	Butterflies served as an inspiration to vary the geometry of elements
5	Shading system for a building	(Un) folding elements that are installed in front of the windows	Actuation of the elements via hydraulic cylinders	Mimosa serves as an inspiration for the folding mechanism

contributed and that the architectural part dominates. With regards to biomimetics, biology served as an inspiration, but the teams detached themselves from the biological systems when developing their concepts.

7 Conclusion, Discussion and Outlook

This work can provide a few starting points for further research. It cannot provide generally valid answers because of the limited number of teams and participants and the number of additional influences characteristic for such a case study. In the following, the indications of this work with regards to the questions of Sect. 3 are presented:

- What are the effects of the cross-disciplinary team and of the use of information from the other discipline?

As to the cross-disciplinary team, the positive aspects prevailed, as the students affirmed a positive influence of different knowledge and viewpoints. The main negative aspect was differing *working styles*. With regards to biomimetics, the teams perceived a positive effect on creativity but did not unanimously perceive a high importance for the project.

- What is the impact on the outcome of the project? What is the contribution of the three disciplines to the outcome?

Both architecture and mechanical engineering contributed to the outcome. Biological systems were not transferred or copied but abstracted and served as an inspiration.

- Which indications for a support of a project in this threefold constellation can be deduced?

To support the cross-disciplinary team work, this case-study indicates that a support of the teams to understand the working styles of the other discipline can be beneficial. In this case study teams used biomimetics on a highly abstract level. This is not necessarily negative, but if a more direct use of biomimetics is requested, measures for support can be beneficial. Possibilities are the inclusion of biologists or providing more support for the transfer of biological systems to architecture and mechanical engineering.

Acknowledgments The authors want to thank the participating students and Prof. Dr.-Ing. Tina Wolf and Dipl.-Ing. Philipp Molter from the Institute of Shell Constructions for their support.

References

1. Kleinsmann MS (2006) Understanding collaborative design. Ph.D. Thesis, Technical University Delft
2. Mannix E, Neale MA (2005) What differences make a difference? the promise and reality of diverse teams in organizations. Psychol Sci Public Interest 6(2):31–55
3. Lindemann U (2009) Methodische entwicklung technischer produkte, 3rd edn. Springer, Heidelberg
4. Stempfle J, Badke-Schaub P (2002) Thinking in design teams—an analysis of team communication. Des Stud 23(5):473–496
5. Kurtzberg TR (2005) Feeling creative, being creative: an empirical study of diversity and creativity in teams. Creativity Res 17:51–65
6. Nachtigall W (2002) Bionik—Grundlagen und Beispiele für Ingenieure und Naturwissenschaftler, 2nd edn. Springer, Heidelberg
7. Kaiser MK, Hashemi Farzaneh H, Lindemann U (2012) An approach to support searching for biomimetic solutions based on system characteristics and its environmental interactions. International design conference (Design2012), pp 969–978
8. Sartori J, Ujjwal P, Chakrabarti A (2010) A methodology for supporting "transfer" in biomimetic design. Artif Intell Eng Des Anal Manuf 24:483–505
9. Hill B (1997) Innovationsquelle natur: naturorientierte innovationsstrategie für entwickler, konstrukteure und designer. Shaker, Aachen
10. Gramann J (2004) Problemmodelle und bionik als methode. Ph.D. Thesis, Technical University Munich
11. Löffler S (2009) Anwenden bionischer konstruktionsprinzipe in der produktentwicklung. Ph.D. Thesis, Technical University Carolo-Wilhelmina
12. Asknature (2012) http://www.asknature.org, extracted 2012/07/20
13. Cheong H, Shu L, Stone RB (2008) Translating terms of the functional basis into biologically meaningful keywords. In: Proceedings of the ASME IDETC/CIE
14. Vandevenne D, Verhaegen P-A, Dewulf S, Duflou JR (2012) Automated classification into the biomimetics taxonomy. International design conference (Design2012), pp 1161–1165
15. Metzler T, Shea K (2011) Lessons learned from a project-based learning approach for teaching new cognitive product development to multi-disciplinary student teams. IDETC/CIE, Washington

Reflecting on the Future of Design Education in 21st Century India: Towards a Paradigm Shift in Design Foundation

Indrani De Parker

Abstract This research-in-progress is an attempt to establish the need for a paradigm shift in design education. The research also investigates aspects that need to be rooted and nurtured in the foundations of design education appropriate in the 21st century India. 'Design Foundation' or 'Basic Design', as it was referred to in early design education has come a long way since its origins at Bauhaus and its further evolution at Ulm and Basel. In the nascent period of design, which primarily involved industrial design, the work was focused on physical products including textiles and graphics. Today, however, to be relevant to contemporary society, designers need to work on complex issues that are interdisciplinary and much broader in scope. 21st century design education needs to be able to apply design and develop strategies to solve real issues and not just look at 'good form'. There is also visible shift from client-driven projects towards a more reflective 'issue based' design education that strives for more socially inclusive, locally/glocally/globally relevant solutions. It is becoming very important in design education to include political, social, economic and ecological discourses in a collaborative and trans/multidisciplinary way thus enabling a conceptual understanding of issues at stake as well as 'intangibles' like values, social responsibilities, empathy, humility and local/global relevance. Relevant design solutions seem to have shifted from 'Form Based' Design to 'Issue Based' Design. Design today is complex and large scale, and design education needs to address major issues. Design education needs to change, yet still retain its essential character. It needs to encourage trans-disciplinary thinking in students to better understand human beings and their needs, understand the economics underpinning issues and

I. De Parker (✉)
Industrial Design Centre, Indian Institute of Technology Bombay, Mumbai, India
e-mail: indranideparker@gmail.com

I. De Parker
366, Sector 'A' Pocket 'C', VasantKunj, New Delhi 110070, India

A. Chakrabarti and R. V. Prakash (eds.), *ICoRD'13*, Lecture Notes in Mechanical Engineering, DOI: 10.1007/978-81-322-1050-4_93, © Springer India 2013

the technological requirements of solving problems. An extensively documented case study was conducted to illustrate collaborative learning in design education for students of a Foundation Program, using urban–rural connections as an example. The study documented collaborative activities among educators, students, crafts persons, professionals business entrepreneurs and so on, in constant search for ways to improve learning, increase student involvement and maximize human interaction, establishing the rural context in design education.

Keywords Design foundation · Design education · Paradigm shift · Issue base learning · Collaborative learning · Rural–Urban connect

1 Background

Design Foundation or the Foundation Program of most design curricula has evolved from a need that was originally perceived at Bauhaus and Ulm as an introduction to 'elements and principles of design' and 'design thinking and action'. There is a need to revisit the traditions of design learning and try to understand the role played by basic design and see how it should be woven into the process of inducting new entrants into the realm of design thinking and action.

1.1 The Bauhaus and the HfG, Ulm Heritage

Modern design education originated during the industrial revolution where craft traditions and apprenticeship processes through which design used to be practiced, was steadily replaced by industrialization. The first school to formally create a series of assignments within a curriculum to introduce students to formal design education was the Bauhaus in Germany. Set up in 1919, post World War I, the Bauhaus was a creative center that was home to some of the greatest design thinkers of those times. The founders of the Bauhaus tradition identified those qualities that needed to be nurtured in an art and design student, both in the form of skills and sensibilities as well in their conceptual abilities and attitudes when dealing with materials and the real world of design action [1].

The Hochschule fur Gestaltung (HfG) Ulm, emerged as a continuation of the Bauhaus experiments in design education under its former students, Max Bill. However under the leadership of Tomas Maldonado, it's focus veered from a foundation in art to science and society. The faculty, comprising eminent teachers and thinkers across disciplines, experimented with design education and documented the results in a series of 21 journals published between 1958 and 1968. This research, theory building and sharing had a lasting impact on design education including design teachers in India [2].

The closing down of the HfG Ulm in 1968 saw the scattering of its faculty and students across the world, all steeped in the Ulm ideology of public good with design theory and action. This resulted in significant action on the ground in the form of new design education in Latin America by Gui Bonsiepe, in India by Sudhakar Nadkarni and H Kumar Vyas and in Japan by Kohei Suguira, besides the numerous other influences in Europe and the USA that continue to this day.

1.2 *Inherited HfG, Ulm Heritage and Influence of Pedagogy in India*

In India, modern design education began in the late nineteenth century with the opening of schools in architecture and art (commercial and fine art). On the request of the then Prime Minister Jawaharlal Nehru, Charles and Ray Eames' 'India Report' initiated Industrial Design practice and education in the post independence period. Charles Eames who had drafted the guidelines based on which the National Institute of Design (NID) was founded, had spent some time at HfG Ulm. In spite of the focus on Eames' report on Indian design tradition and sensibilities, the design education programs in India, like in many other countries, actually borrowed its pedagogy and thinking from Bauhaus as well as HfG Ulm school tradition.

This influence continued at Industrial Design Center (IDC), which was setup in 1969 at the Indian Institute of Technology Bombay (IITB) in Mumbai and later Department of Design (DOD) at IIT Guwahati, where the first and the only undergraduate program in design in India. Many early teachers at NID, IDC and DOD were trained in the same school. This deep-rooted connection between HfG Ulm and many design schools even today, influenced thoughts, ideas, philosophy and hence the Foundation Program.

Prof Trivedi [3, p. 9] notes in his article, Sarvodaya—Betterment of All, "One of the propositions put forward by the HfG, Ulm in its founding philosophy was that the quality of human life can be bettered by improving the quality of the man-made environment. But that alone would not be enough." He also quotes Tomas Maldonado who wrote in Ulm 2: "Man exists not only to utilize objects and even less—as they will make us believe nowadays—to consume products. But man will constantly be confronted with the intentional and unintentional demands of his consciousness. And these demands cannot be satisfied by soundly designed consumer goods alone".

2 Homogenization in Design Foundation Education

In September 2010, NID organized a conference in Kolkata (in collaboration with Goethe-Institut/Max Mueller Bhavan, HfG-Archive Ulm & IFA, Stuttgart (Institute for Foreign Cultural Relations, Germany), 'LOOK Back—LOOK Forward: HfG Ulm and Basic Design for India', where it shared, after fifty years, its curriculum through an extensive documentation of the work of the students of their Foundation Program. Surprisingly, students from different institutions, (such as National Institute of Fashion Technology (NIFT), Delhi; Pearl Academy of Fashion, Delhi; IILM School of Design, Gurgaon, Institute of Apparel Management (IAM), Gurgaon and Indian Institute of Craft Development (IICD), Jaipur), produced work similar to that of NID which has a rich heritage and infrastructure spanning five decades. This observation is based on an extensive photo-documentation of the works of the students of Foundation year of the mentioned design schools over a few years and demonstrates a certain homogenization that exists in design schools today. The profiles of the students defer significantly, but this similarity of work among students of across various design schools is apparent and could be attributed to the pedagogy followed, repeated and replicated over decades, which may, warrant a re-look at the current learning process.

3 The Current Design Paradigm

In the early days, the main of focus of industrial designers was form and function, materials and manufacturing. Today, design is the dutiful servant of technological, economical and political interest in almost every area of manufacturing and construction. This design paradigm includes not only products of consumption but also today's housing which is informed by and designed within a vision driven by short-term economic goals [4].

> Design was summoned to absorb the shock of industrialization, and to soften its devastating consequences upon the cultural web, in other words, to make industrial products culturally, socially, economically, symbolically and practically acceptable. Esthetics was then its privileged rhetorical tool, followed by ergonomics in the mid-twentieth century and semiotics in the late-twentieth century. But its most unique field of activity has remained the material product. Findeli [5, p. 15], emphasis as per original)

Most design schools in India have been stand-alone enterprises (or institutions). Classical industrial design is a form of applied art, which requires deep knowledge of forms and materials and skills in sketching, drawing, and rendering. The new areas, on the other hand, are more like applied social and behavioral sciences and require understanding of human cognition and emotion, sensory and motor systems, and sufficient knowledge of the scientific method, statistics and experimental design so that designers can perform valid, legitimate tests of their ideas before deploying them [6].

However, the issues are much more complex and challenging, in the current context. The current programs in India expect students to offer services to very diverse requirements. How does one train a student to design for global village as well as real villages in India as the demands of the global companies are very different from the demands of the villages in India?

4 The Paradigm Shift in Design

Findeli defines a paradigm as 'the shared beliefs according to which our educational, political, technological, scientific, legal and social systems function without these beliefs ever being questioned, or discussed, or even explicated [5].'

> The need to perceive concepts differently, to reframe our approach to complex systems, is a reality that we must reckon with and which requires new pedagogical methods. Rather than simply focus on passing on knowledge, then, it is necessary to develop thinking methods that will generate new knowledge. Moreover, these methods need to lead us to better solutions not only for business but for humanity and the planet as a whole. Peinado and Klose [7], emphasis as per original).

4.1 New Paradigms for Design Practice

The modern practice of design has been the model for design education since the days of the Bauhaus. Defined as an object centered process, the traditional goal of design has been to produce an artifact or environment that solves a problem.

> For academic programs arising from the arts, the beauty and humanity of such objects or environments are important. For programs arising from the sciences and engineering, usability and efficiency are paramount. And in between are the social sciences, where the issues of culture and social interaction reside. The distinctions within each of these disciplines are not simplistic, but the research paradigms they represent for producing objects and environments clearly have different value systems and methods, and historically, they have argued for very different curricular paths at the graduate level. The demands on design practice in the twenty-first century, however, are significantly different from those of the past, suggesting that these paradigms may require re-examination [8].

- Increasing complexity in the nature of design problems
- The transfer of control from designers to participants
- The rising importance of community
- The necessity of trans-disciplinary, collaborative work

The complex scale of problems, diversity of settings and participants, and demand for adaptable and adaptive technological systems argues for work being done by interdisciplinary teams composed of experts with very different modes of inquiry. How such experts collaborate as peers and the roles design can play in mediating collaboration present new opportunities for designers. The paradigm shift in the focus of the design process from objects to experiences demands new knowledge and methods to inform decision-making. It broadens the scope of investigation beyond people's immediate interactions with artifacts and includes the influence of design within larger and more complex social, cultural, physical, economic, and technological systems.

India, in this context, has a unique opportunity, to innovate a new kind of design education at the exact moment when four new NID (National Institute of Design) campuses have been announced. Thakara [9] elaborates, "India is not alone in needing to innovate new educational models. On every continent, outside its Big Tent—on the edge of the clearing—exotic new species of design and business education are emerging. These new schools and courses have names like Yestermorow School, Deep Springs College, Kaos Pilots, School of Everything, Social Edge, Deep Democracy, Center for Alternative Technology, Schumacher College, Living Routes, Gaia U, Crystal Waters, Horses Mouth, WOOF, The Art of Hosting. These 'outliers' (not mainstream universities) are where the real innovation is happening—in terms of content, form and business model. Few designers, few policymakers, and few entrepreneurs, have even heard of these places. But they are significant, for me, because they meet the requirements of these new times. They can be the competition—or the collaborators—for design education in India and beyond."

A trans-disciplinary approach broadens the 'objective' of design and rising complexity of contexts requires new multidisciplinary knowledge. Design students need to experience the benefits of working with others with the components of design being central.

4.2 Collaborative Learning in Design Education Today

Collaboration is a process in which two or more people or organizations work together in an intersection of common goals—by sharing knowledge, learning, and building consensus. Structured methods of collaboration encourage introspection behavior and communication. These methods specifically aim to increase the success of teams as they engage in collaborative problem solving.

Advocates of collaborative learning claim that the active exchange of ideas within small groups not only increases interest among the participants but also promotes critical thinking [10]. Quoting Johnson and Johnson [11], Gokhale argues that there is persuasive evidence that cooperative teams achieve at higher levels of thought and retain information longer than students who work quietly as individuals. The shared learning gives students an opportunity to engage in discussion, take responsibility for their own learning, and thus become critical thinkers.

Designers (i.e. the experts who have been specifically trained in design thinking and design knowledge) need to face systemic changes that are driven by a growing number of actors. These actors together can generate wide and flexible networks that can be collaboratively conceived, developed and generate sustainable solutions. The paradigm shift in design today changes the position and role of professional designers.

Traditionally, designers have been seen and have seen themselves as the only creative members of interdisciplinary design processes. In the emerging scenario this distinction blurs, and they become professional designers among many non-professional ones. However, this does not mean that the role of design experts is becoming less important. On the contrary, in this new context, design experts have the crucial function of bringing very specific design competence to these co-designing processes. That is, they become process facilitators who use specific design skills to enable the other actors to be good designers themselves. Manzini [12, p. 11], emphasis as per original)

Thus design schools can play a significant role in the emerging scenario and, generate new models and ideas in education to map the paradigm shift. Today's design problems exist at the level of systems and communities, and are too big and too complex for any single discipline to address. Collaborators need to be from fields as diverse as anthropology, cognitive psychology, computer science, business, social policy, etc. Current strategies of design education need to evolve to prepare students to address the interdisciplinary demands of complex, system-level problems.

> We are in the midst of a slow, but insistent shift in how we teach, assess, and organize our classrooms. After centuries of fixating on the solitary student's singular progress, we are currently experiencing the rise of a radical emphasis on collaborative, team-based learning. This is not just a slight course correction but has the potential to be a major shift of paradigm. This transformation poses profound challenges to the basic tenets of our educational system while it also forces teachers and administrators to invent new and novel ways to assess student progress and organize their curricula. We can no longer afford to wait for an Einstein to help us. Working effectively in groups and across disciplinary boundaries will be a key survival skill in the 21st century (James Hunt, Director, the Experimental Graduate Program in Transdisciplinary Design at Parsons)

[13].

5 Rural-Urban Synergy

It is more important now than ever before to make the connect between the rural and urban in design education. It is critical to create collaborations engaging in a more holistic approach to design, which includes all other actors and stakeholders both rural and urban, especially when today's design scenario exists at the scale of systems and communities; huge and complex.

> In India, on one hand, we are privileged to have a large rural base of people with agricultural and artisanal skills and a huge diversity of knowledge, tools materials and experiences. In the march towards a mostly Western, industrialized model of development much of this indigenous knowledge resource is being lost. Design skills could be used to trigger new imagination, propose daring new scenarios, which build on what people know and empower them to become partners in shaping their destinies. On the other hand, Indian industry and services are maturing rapidly. Indian corporations are becoming multinational. To remain competitive in the global marketplace, industry must respond to new sets of challenges from users who are seeking more than usefulness and usability. They are looking for emotional connectedness, commitment to green values, transparency, fair use of labor and so on. (Vision First, 2011, emphasized as per original).

Design discipline in India has been attempting to address the conflict between the need to rapidly modernize, need to promote economic development to tackle poverty and the need to minimize the effects of economic developments on traditional culture. Caught in this conflict, design schools in India have been walking a tightrope, balancing between international design approaches and those rooted in local issues and tradition of India. Globalization has exacerbated this conflict, forcing us to question the validity of the tightrope walk, particularly in design education [11].

5.1 Collaborative Connect Between the Rural and Urban in Design Education: A Case Study

An extensively documented case study was conducted to establish these complex urban–rural connections in design education. The key for designers today is to understand the stakeholders and how they are connected through the web of stakeholder networks. The study also documented collaborative activities among educators, students, crafts persons, professionals business entrepreneurs and so on, in constant search for ways to improve learning, increase student involvement and maximize human interaction, establishing the rural context in design education.

The underlying premise for collaborative and cooperative learning is founded in constructivist epistemology. Johnson et al. [16] have summarized these principles in their definition of a new paradigm of teaching.

> First, knowledge is constructed, discovered, and transformed by students. Faculty create the conditions within which students can construct meaning from the material studied by processing it through existing cognitive structures and then retaining it in long-term memory where it remains open to further processing and possible reconstruction. Second, students actively construct their own knowledge. Learning is conceived of, as something a learner does, not something that is done to the learner. Students do not passively accept knowledge from the teacher or curriculum. Third, faculty effort is aimed at developing students' competencies and talents. Fourth, education is a personal transaction among students and between the faculty and students as they work together. Fifth, all of the above can only take place within a cooperative context. Sixth, teaching is assumed to be a complex application of theory and research that requires considerable teacher training and continuous refinement of skills and procedures (p1: 6)

The action research method was adapted for this study. The objective was to integrate collaborative learning in design education, particularly for design foundation. This case study, presents the learning and experience of design students as a collaborative activity and show how it can happen effectively and successfully between the urban and the rural actors. Students were encouraged to develop the capacity to design for the unexpected.

The learning opportunities, which foster the above, were created in such a way that they operate on four levels, which are not discrete, linear or sequential. Taken together they enable experiences, which stimulate genuine understanding of complex relationships and processes.

The collaborative experiential learning layers are:

- Looking, listening and being
- Exploring, thinking and assimilating
- Questioning, experimenting, making and questioning
- Connecting, collaborating and co-creating.

The course 'Issues & Perspectives in the Craft Sector, is orchestrated by Ms Lakshmi Murthy. A professional Communication & Ceramic Designer, and Design Educator, she has been conducting this course since 2005 with refinements every year in context to the present needs and future projections. This module has been extensively documented for the past two years by the author, at the Indian Institute of Craft Design, Jaipur, Rajasthan, India, as part of the in progress research. The module stimulates the students's curiosity and encourages them to experience the synthesis within the craft sector, social sector, business and design. The Case Study is presented in form of a booklet.

6 Conclusion

India as an emerging market has grown rapidly, thus giving rise to both aspirations and anxieties about the potential socio-economic and environmental repercussions. This has thrown up new opportunities, both in the rural and urban spaces for designers, entrepreneurs, activists, policy-makers, investors, and so on. Design could focus on developing dynamic and flexible innovation systems, through which all actors collaborate to create and develop, options which encourage sustainable lifestyles and inclusive prosperity. Design students need to experience the benefits of working collaboratively with others with the components of design being central. A trans-disciplinary approach broadens the 'objective' of design and rising complexity of contexts requires new multidisciplinary knowledge.

Future design schools could follow a model of participatory and collaborative design education. The research attempts to establish that there has been a paradigm shift in design. Given this shift, it stresses the need for a concurrent and corresponding shift in the design pedagogy, most critically at the design foundation.

COLLABORATIVE CONNECT BETWEEN THE RURAL AND URBAN IN DESIGN EDUCATION: A CASE STUDY

References

1. Ranjan MP (2005) Lessons from Bauhaus, Ulm and NID: role of basic design in PG education. Design education: tradition and modernity. Scholastic Papers from the International Conference, DETM 2005. National Institute of Design. Ahmedabad
2. Ranjan MP (2011) Nature of design: the need for nurture in India Today. What design can do—Conference. Amsterdam
3. Trivedi K (2003) Sarvodaya Betterment of all. Page 9, ICSI Dnews
4. Fuad-Luke A (2002) Slow theory: a paradigm for living sustainably, design by development. MIT/Srishti School of Art and Design, Bangalore
5. Findeli A (2001) Rethinking design education for the 21st century: theoretical, methodological, and ethical discussion. Des Issues 17(1):5–17 (Winter 2001)
6. Norman D (2010) Why design education must change. Core 77 design magazine & resource. Viewed July 2012 http://www.core77.com/blog/columns/why_design_education_must_change_17993.asp
7. Peinado A, Klose S (2011) Design innovation: research-practice-strategy, researching design education, symposium proceedings, researching design education, 1st international symposium for design education researchers, Cumulus Association//DRS SIG on Design Pedagogy, Paris
8. Davis M (2008) Why do we need doctoral study in design? International Journal of Design vol. 2 (3). www.ijdesign.org
9. Thakara J (2011) Doors of perception: what kinds of seed. Viewed July 2012, http://wp.doorsofperception.com/next-skools/what-kinds-of-seeds/
10. Gokhale A A (1995) Collaborative learning enhances critical thinking. J Technol Edu 7(Fall):1
11. Johnson RT, Johnson DW (1986) Action research: cooperative learning in the science classroom. Sci Child 24:31–32
12. Manzini E (2011) Design schools as agents of (sustainable) change: a design labs network for an open design program, symposium proceedings, researching design education, 1st international symposium for design education researchers, CUMULUS Association//DRS SIG on Design Pedagogy, Paris
13. Hunt J (2012) Collaborative learning. Viewed July 2012, Posted: 05/24/2012 5:25 pm http://www.huffingtonpost.com/jamer-hunt/collaborative-learning_b_1543999.html
14. Vision First (2010) About creating design competencies in India.Viewed July 2012, http://issuu.com/visionfirst/docs/visionfirstproposal110212
15. Athavankar U (2005) Globalization and the new mantra of design education for India. Design education: tradition and modernity. Scholastic Papers from the International Conference, DETM 2005. National Institute of Design, Ahmedabad
16. Johnson DW, Johnson RT, Holubec EJ (1991) Cooperation in the classroom. Interaction Book Co, Edina

Design of Next Generation Products by Novice Designers Using Function Based Design Interpretation

Sangarappillai Sivaloganathan, Aisha Abdulrahman, Shaikha Al Dousari, Abeer Al Shamsi and Aysha Al Ameri

Abstract Several researches confirm that novice designers need to gain many abilities to perform like experts in the field. Design interpretation is a technique to analyze the working product and identify the functions performed by an artifact in the form of a function tree. This paper reports an investigation into the design of an Operating Table by novice designers and the use of Design Interpretation to gain abilities to perform like experts. The observations were made with respect to (1) Organized structure and cognitive action (2) Scoping and information gathering (3) Consideration of alternatives (4) Time spent on activities and tasks (5) gathering basic data (6) adaptive expertise and (7) procedural expertise. The results suggest that novice designers gain by carrying out design interpretation of the current generation when designing the next generation.

Keywords Design interpretation · Adaptive design · Next generation products

1 Introduction

A design expert is a person who has a comprehensive and authoritative knowledge or skill in the design area. A novice in the meantime is a person new to the field or activity. An expert therefore will normally have knowledge about the product, or at least sufficient ability to acquire the product specific knowledge needed, with ease. The novice on the other hand will have limitations in both product specific

S. Sivaloganathan (✉) · A. Abdulrahman ·
S. Al Dousari · A. Al Shamsi · A. Al Ameri
Department of Mechanical Engineering, United Arab Emirates University, 17555Al Ain, United Arab Emirates
e-mail: Sangarappillai@uaeu.ac.ae

A. Chakrabarti and R. V. Prakash (eds.), *ICoRD'13*, Lecture Notes in Mechanical Engineering, DOI: 10.1007/978-81-322-1050-4_94, © Springer India 2013

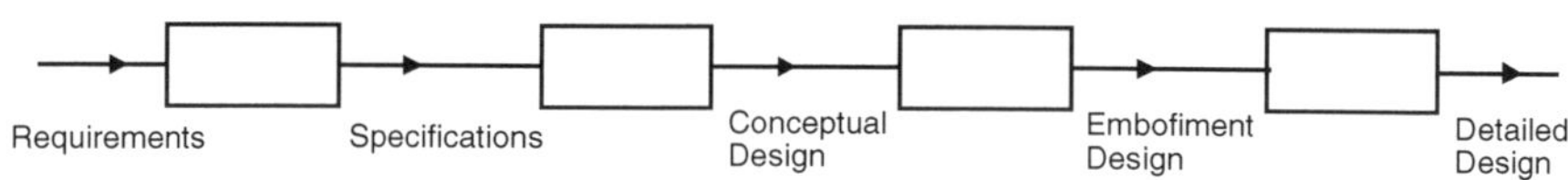

Fig. 1 Input output representation of systematic design process

and design domain knowledge. Ahmed et al. [1, 2] investigated the experienced and novice and conclude that (1) significant differences between them are present and the differences are noticeable in the early stages, and (2) supply of additional information expressed or used by the experienced designers is a credible way to support the novice designers.

Systematic product development starts with establishing the elements of a societal need called requirements and the process goes through the stages where specifications and conceptual, embodiment and detailed designs are being produced. From an input–output or black box visualization, requirements are the input and specifications are the output, specifications are the input and conceptual designs are the output and so on as shown in Fig. 1. Every product will have these in an explicit or implicit form.

Next generation products are improved variations of existing products. The variation may originate by having additional requirements, tighter specifications or/and better conceptual, embodiment and detailed part designs due the use of better insights and the usage of advanced technologies. The design of next generation products thus is not a new product development. Its development however has to follow a similar path of new product development from the point where the variation originates, but with better insight provided by the current generation.

Design Interpretation [3] is a technique where the embodiment of an existing artefact is extracted in the form of a parts tree and each operating condition is considered in turn to identify the functions performed by these parts or subassemblies. The functions are grouped into functional subsystems thus transferring the artefact from the physical domain to the function domain. The generated function structure of the existing product will give a better insight into the design of the current product. Whether Design Interpretation is a suitable means for novices to gather the initial knowledge needed to thrive in the design project is the research question and this paper reports an investigation into this question.

2 Literature Survey

Four aspects (1) expertise in design (2) comparison of novice and experienced in action (3) function based design and (4) design interpretation were seen as relevant and used in the investigation, and a brief review of the survey is given in the following sections.

2.1 Expertise in Design

If expert is an authoritative person in a subject, expertise is the basis of his credibility due to his or her study, training and experience in the subject matter. Cross [4] asserts that expertise is not simply a matter of possessing 'talent', but is the result of a dedicated application to a chosen field and without the dedicated application of the individual, levels of performance will remain modest. Chi [5] outlines, (1) excelling in generating the best solution (2) detecting and seeing features that novices cannot (3) spending a relatively more time analyzing a problem qualitatively, developing a problem representation by adding many domain-specific and general constraints to the problems in their domains of expertise (4) possessing more accurate self-monitoring skills in terms of their ability to detect errors and the status of their own comprehension (5) choosing appropriate strategies (6) using available resources efficiently and (7) retrieving relevant domain knowledge and strategies, as the characteristics of experts. Cross and Clay bourn Cross [6] note that experts take a systemic view of the design situation, choose to frame their view of the problem in a challenging way, and draw upon first principles to guide both their overall concept and detailed design.

2.2 Comparison of Novice and Experienced in Action

Several researchers have carried out comparative studies on how novice and experienced designers carry out design tasks. For instant Kavakli and Gero [7, 8] reports that the expert's cognitive actions are clearly organized and structured while there are many concurrent actions that are hard to categorize in the novice's protocol. They also found that the expert's cognitive activity and productivity in the design process were three times as high as the novice's. Based on these results, they present the view that the expert's structured and organized cognitive actions lead the expert to a more efficient performance than the novice.

Atman et al. [9] report their findings on comparison between student and experienced designers under the categories (1) problem scoping and information gathering (2) project realization (3) consideration of alternative solutions (4) total design time and transitions and (5) solution quality. They found that experts spent significantly more time on the task overall and in each stage of engineering design, including significantly more time in problem scoping. The experts also gathered significantly more information covering more categories. Results support the argument that in problem scoping and information gathering there are major differences between advanced engineers and students.

Ahmed, Wallace and Blessing [1, 2] in their research used observations, discourse analysis and interviews to identify differences between novice and experienced designers. Their findings from observations can be summarized as 'experienced designers consider several more related data and information when considering an issue than a novice'. In the discourses the novice designers' questions fall under the categories of (1) obtaining information (2) how to calculate (3) terminologies and (4) typical values. The supporting interviews among others identified that experienced designers knew which issues are important. They concluded that there is a significant difference between the experienced and novice and the difference is noticeable in the early stages. The main support for novices suggested by them is the additional information expressed or used by the experienced designers.

Robin et al. [10] report the work of Hatano and Inagaki which divides the expertise into (1) adaptive and (2) procedural types. Characteristics of adaptive expertise are inventing new procedures derived from expert knowledge to solve novel problems, a tolerance for ambiguity, fluidly adapting to new situations, performing minor variations in procedural skills and examining their effectiveness in a new context, engaging willingly in active experimentation and exploration, and being sensitive to internally generated feedback such as a surprise at a predictive failure or being perplexed by alternative explanations of a phenomenon. In contrast, characteristics of routine expertise include technical competence in solving familiar problems quickly and accurately yet only modest competency in solving novel problems, tendency to solve problems based on past solutions, highly standardized procedural skills, unwillingness to risk varying the skills, and having a preference for strategies that ensure quicker solutions over strategies that promote seeking alternative solutions.

2.3 Function Based Design

Establishing the function structure in systematic design process was first introduced by Pahl and Beitz [11] and the method has been adopted by other design text books as well [12]. Function Analysis has gone through several developments since then. For instance Ullman [13] outlines a hierarchical function structure as part of his design process and Andreasson [14] forwards Function Means tree as a method to establish conceptual designs. For this research it is sufficient to use the hierarchical function structure as outlined by Ullman.

2.4 Design Interpretation

Design Interpretation [3] is a technique aimed at establishing the function structure of an existing product in the absence of technical and design data of a product. It is achieved in three steps (1) process description (2) extraction of the parts tree and (3) establishment of the function structure. The first step is to operate the product and get familiar with the product to get a step by step description of the operation of the product. From the description the subassemblies and parts responsible for various actions are identified and a parts tree is established. The objective of the parts tree is the derivation of the function structure and as such an appropriate level of abstraction is decided. A similar technique called ‘Product Teardown’ is described by Otto and Wood [15]. It employs a procedure called Subtract and Operate Procedure, SOP for short, where the product is systematically dismantled, one by one, to identify the functions performed by each subsystem or part.

3 Formulation of the Study

The study was conducted by making observations on a group project lasting sixteen weeks by a group consisting of four undergraduate students in their final year. The students were expected to complete a conceptual design as their deliverable and in that sense it is a real life project and not a controlled experiment. Interim presentations, reports, log book and frequent interactions with the supervisor are the records that were developed and used in the study. The objective is mainly to understand the students’ thought processes and identify whether design interpretation is the right means to support them.

Findings from the literature survey which formed the basis for formulating the study is given in Table 1. It shows seven areas in which the experienced designers do much better than the novice. The transition of a novice into an expert needs development of the novice in these areas. The research question therefore is ‘Whether Design Interpretation assists novice designers to gather ability in (1) Organized structure and cognitive action (2) Scoping and information gathering (3) Consideration of alternatives (4) Time spent on activities and tasks (5) gathering basic data (6) adaptive expertise and (7) procedural expertise’.

4 Description of the Work

The task in this project is to develop a conceptual design of a surgeon’s operating table for use in operating theaters and the students were given a very brief description of the product. The first activity was to gather as much information as they can from the net and write details on a log book. Weekly meetings were held with the supervisor. After the first week and visiting a lot of sites the students managed to write only a couple of lines on the log book. In one student’s own words ‘I have visited several sites but I do not know what is important and what I have to write in the log book’. During the second week the students were introduced to the systematic product development process as outlined by Ulrich and Eppinger [12]. Contacts were made with a local hospital and a visit was made. Two engineers explained and demonstrated an operating table used in the hospital. The students were given ample opportunity to ask questions and to take as many photographs as they would like. The visit lasted about an hour and a half. The

Table 1 Basis of formulating the study from literature survey

Characteristic	Findings from the literature
Organized structure and cognitive action	Experts excel in this and get better results while the novices lack the ability and struggle
Scoping and information gathering	Experts spend a lot of effort in defining the scope of the problem and in gathering information
Consideration of alternatives	Experts explore alternative solutions or width first approach while novices take a depth first approach
Time spent on activities and tasks	Experts spend considerably more time in activities and tasks by considering different perspectives in detail while novices plunge into the details straight away
Basic data 1. Obtaining information 2. How to calculate 3. Terminologies 4. Typical values	Experts have standard methods and intuition to gather these data efficiently while this is the major constraint for novices
Adaptive expertise	Experts thrive in this due to their past case studies and the structured methodology while novices struggle a lot in this area
Procedural expertise	Experts again perform better in this because of their ability to choose right domain of knowledge for the task

numbers of questions asked were limited and excepting one student even the number of photographs taken was also small. The students could not even express their understandings among themselves with a standard terminology.

By this point design interpretation was introduced to the students and they carried out design interpretation of a hydraulic trolley to learn the technique. The photographs were systematically studied after that while trying to establish a function structure. It was felt that the details collected from the hospital were inadequate and a second visit was arranged. The visit lasted for about two hours and the students had more time on their own with the product and they took several photographs. There was only one engineer and the students asked several questions. Following the visits the students carried out systematic analysis of the photographs taken trying to understand the functionality of the parts and assemblies. In the meantime they became conversant with the systematic product development method by Ulrich and Eppinger [12]. They divided the operating table into ten sub-sections as shown in Fig. 2 together with two additional sub-sections the hydraulics and electricals.

With this division they could establish the functions performed by each of the subsystems and a function tree was formed as shown in Fig. 3.

By this time the students were comfortable with a standard terminology and design process model. Following the design process model outlined by Ulrich and Eppinger they established a Mission statement, recorded Customer Verbatim and formed target specifications. The group was ready for conceptual design. Then they realized that they could only think about the table for which they have carried out the design interpretation. A third visit to the hospital was arranged. There they

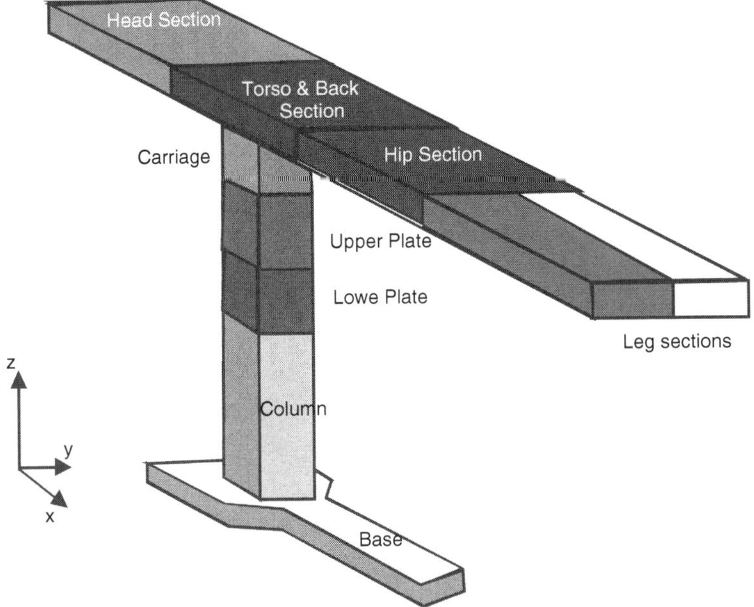

Fig. 2 Sub-sections of an operating table

saw different operating tables. The visit lasted for about two hours and the students were very busy discussing the functionality of the different tables with the engineer and were effectively carrying out design interpretations. Though no documents were made it was a thorough job with the usage of specific terms for sub-assemblies and parts. They could formulate a generic description for the operating table as 'an operating table is constituted with a flexible bed section carried on a shuttle which has the capabilities to (1) adjust the height (2) tilt the bed about the x axis (3) incline the bed about the y axis and (4) keep the various sections of the bed at different angles. It has minimum foot area so that other equipment and monitors can be brought to the vicinity without difficulty' [16]. With this definition and understanding the students were able to develop a CAD model of conceptual operating table as shown in Fig. 4.

5 Analysis and Discussion

Organized Structure and Cognitive Action: The project started with no structure and direction though the students knew about the generic design models and stages in the design process. Adopting the design model by Ulrich and Eppinger gave a definite direction. However the lack of product knowledge was evident in the early stages and in many meetings the common question was 'what shall we do next?'.

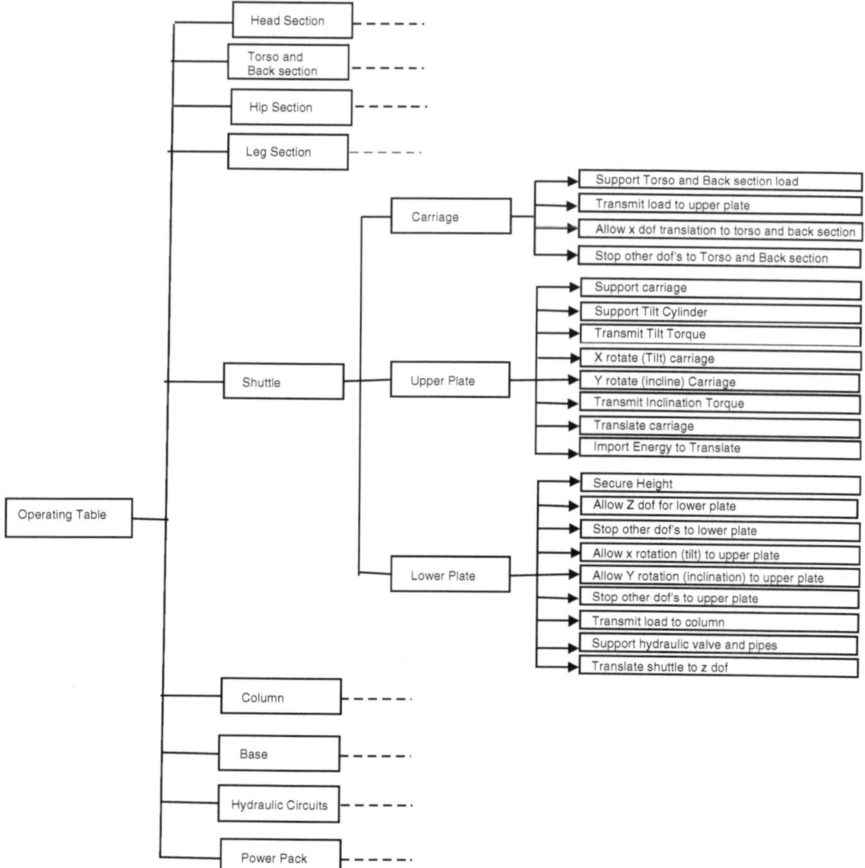

Fig. 3 Part of the function structure

The observations by Ahmed et al. were clearly evident. Design interpretation was not an easy task for them. But they had several pictures, especially those taken during the second visit. After working hard they established the Function Structure. With the function structure in place and the design model to follow the students started to progress well and this suggests that Design Interpretation helped them to structure the work.

Scoping and Information Gathering: Identifying 'what is needed and where to find them' was a difficult task for the students. Product descriptions, leaflets and manuals, health department directives, clinical requirements expressed by the engineers in the hospital and machine elements including the electronics, power and hydraulics were some of the areas visited by them. The situation changed considerably with the establishment of the function structure. It gave them a direction to narrow down and focus the search.

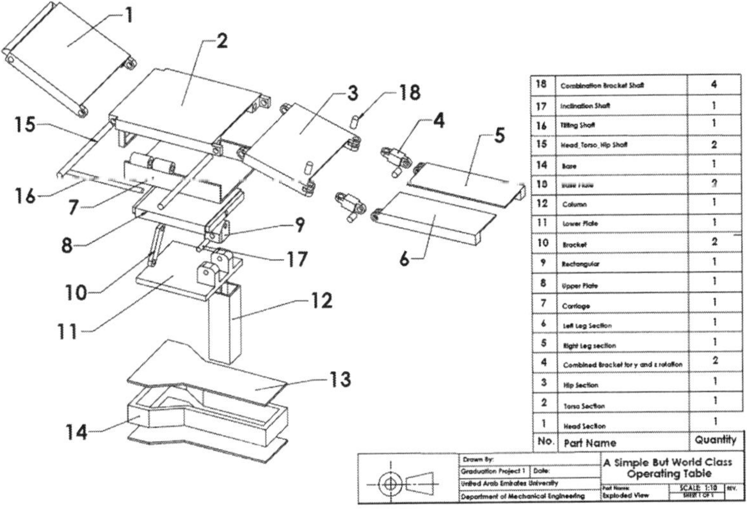

Fig. 4 CAD model of the conceptual operating table

Consideration of alternatives: This had two phases. After establishing the function structure the students tried to generate concepts and all what they could do was reproducing the table they analyzed supporting some kind of design fixation. However the situation changed dramatically when they analyzed different designs. The students were able to come up with different concepts. This is an area that needs further investigation.

Time spent: Time spent on the project increased as time went on. But it is difficult to say whether it was due the pressures exerted by the deadlines or by better insight into the problem.

Basic Data: Design Interpretation definitely had an effect on this area. The dominant one was the establishment of the standard terminology. This enabled them to have intellectual discussions with the engineers in the hospital during the third visit. They could obtain 'ball park values' for various parameters and identify information elements they should search for.

Adaptive Expertise: Though the mind's databases for past case studies of the students were limited, their interpretations of the functionality of the different tables during the third visit suggest that they recall the interpreted design of the first table. This suggests that design interpretation helps to structure designs for mind's database.

Procedural Expertise: No specific evidence was found to support or reject help from design interpretation. However it is worth pointing out that function structure helps to identify knowledge components needed for providing different functions.

5.1 Discussion

Design Interpretation is aimed at establishing the function structure of an existing product. In this project however its use as a tool to gain the abilities lacking in novice designers is being assessed. The observations were made passively while the students were working for a different target. In this sense it is an assessment of the technique on a real project but in an uncontrolled fashion. The results suggest that Design Interpretation helps novice designers to gain the abilities they lack when designing the next generation products. However the observation is based on a single project. Detailed investigation with more projects is needed to make a firm decision on usefulness of Design Interpretation.

6 Conclusions

Observations on the design and development of this moderately complex product suggest that Design Interpretation of the current generation product helps novice designers to gain the skills possessed by expert designers in the following areas, when designing the next generation product.

1. *Organized structure and cognitive action*: Product specific knowledge provided by Design Interpretation helped the students substantially to think of the product as a collection of ten subsystems and organize their thought process along those lines.
2. *Scoping and information gathering*: Design Interpretation broke the overall problem into subsystems and this enabled the search for the relevant information needed for the design for example the body mass distribution to design the bed sections.
3. *Consideration of alternatives*: This had two phases. In the first, design fixation played a major role. But after looking at different products, the fixation eased and they generated concepts where casting, welding and bolting were the main manufacturing processes.
4. *Time spent on activities and tasks*: Time spent on the work increased as time went on but it is not clear whether it is due pressure or due to the nature of design.
5. *Gathering basic data*: The main contribution by Design Interpretation in this area was the establishment of standard terminology. It made them confident and pro-active in collecting data and engaging in discussions with professionals in the area.
6. *Adaptive expertise*: Design Interpretation helps to add items to the mind's data base.

This paper describes the observations made on a single group of students in an uncontrolled setup. Though they provide interesting observations the conclusions can only help to formulate hypotheses. Further investigations on multiple projects in a controlled setting are needed to confirm this observation or otherwise.

Acknowledgments The authors wish to acknowledge their gratitude to the maintenance department of Tawam Hospital in UAE and Professor Yousef AlHaik and Dr AbdelHamid Mourad of UAEU for their intellectual contributions.

References

1. Ahmed S, Wallace KM, Blessing L, Moss M (2000) Identifying differences between novice and experienced designers, design for excellence. In: Sivaloganathan S, Andrews PTJ (eds) Proceedings of the engineering design conference 2000, professional engineering Publishing, Bury St Edmunds, Suffolk, pp 97–106
2. Ahmed S, Wallace KM, Blessing LTM (2003) Understanding the differences between how novice and experienced designers approach design tasks. Res Eng Des 14:1–11
3. Mendis MV, Sivaloganathan S (2000) Design interpretation: a methodology for adaptive design, the Journal of the institution of engineers Sri-Lanka. May 2000. Won the best paper award for the year 1999/2000
4. Cross N (2004) Expertise in design: an overview. Des Stud 25(5):427–441
5. Chi TH (2006) Two approaches to the study of experts' characteristics. The Cambridge handbook of expertise and expert performance, Cambridge, pp 21–29 (Chapter 2)
6. Cross N, Clayburn Cross A (1998) Expertise in engineering design. Res Eng Des 10:141–149
7. Kavakli M, Suwa M, Gero JS, Purcell T (1999) Sketching interpretation in novice and expert designers. In: Gero JS, Tversky B (eds) Visual and spatial reasoning in design, key centre of design computing and cognition, University of Sydney, Sydney, pp 209–219
8. Kavakli M, Gero JS (2001) The structure of concurrent cognitive actions: a case study on novice and expert designers. Des Stud 23(1):25–40
9. Atman CJ, Adams RS, Cardelia ME, Turns J, Mosborg S, Saleem J (2007) Engineering design processes: a comparison of students and expert practitioners. J Eng Educ 96(4):359–379
10. Adams RS, Turns J, Atman CJ (2003) What Could Design Learning Look Like? In: Design thinking research symposium VI, Sydney
11. Pahl G, Beitz W, Schulz H-J (2007) Engineering design: a systematic approach, 3rd edn. Springer, London
12. Ulrich K, Eppinger S (2011) Product design and development, 5th edn. McGraw Hill, New York
13. Ullman D (2003) The mechanical design process, 3rd edn. McGraw-Hill, New York
14. Andreassen MM (1998) Conceptual design capture, keynote paper. In: Sivaloganathan S, Shahin TMM (eds) Proceedings of the engineering design conference '98, Professional Engineering Publishing, Bury St Edmunds, Suffolk, pp 21–30
15. Otto K, Wood K (2006) Product design. Tata McGrawHill, India
16. Abeer A, Aisha A, Aysha A, Shaikha A (2012) Design and development of a simple but world class operating table, graduation project report, Department of Mechanical Engineering, United Arab Emirates University, UAE

Changing Landscapes in Interactive Media Design Education

Umut Burcu Tasa and Simge Esin Orhun

Abstract The recent advancements in networking technologies and tools are causing different transformations in the modes of communication and the design of products, which mostly is in direct relationship with the discipline of interactive media design. Having based its curriculum on project design courses for varying media, Department of Interactive Media Design in Yildiz Technical University was established in 1997 and gave its first graduates in the academic year of 2002/2003. Reflecting the effects of the spreading of media technologies in the way the students approach to design process and products, we believe that the graduation projects of the last decade have great potential to delve into the changes in the digital media landscapes. This research analyses this potential in parallel with the real-life design implementations in media technologies and a McLuhanian media theory, in terms of the dynamics of content flow and the reconfiguration of user relationship with the content.

Keywords Interactive media design · Media ecology · Design education

1 Introduction

"We shape our tools, and thereafter our tools shape us" [1].

The transformative effect of technologies and media has long been debated by scholars from different disciplines.

U. B. Tasa (✉) · S. E. Orhun
Department of Interactive Media Design, Yildiz Technical University, Istanbul, Turkey
e-mail: utasa@yildiz.edu.tr

S. E. Orhun
e-mail: sesin@yildiz.edu.tr

A. Chakrabarti and R. V. Prakash (eds.), *ICoRD'13*, Lecture Notes in Mechanical Engineering, DOI: 10.1007/978-81-322-1050-4_95, © Springer India 2013

Media ecologist Neal Postman drew attention to the ecological nature of technological change, underlying its power in changing the whole relationships and the root structure within the system, instead of just adding a new value to the system [2]. According to Lanier, a technologist himself, technologists, by changing the ways we connect to ourselves and to the others, directly manipulate our cognitive experiences; so it is impossible to work with computing technologies without getting involved in "social engineering" [3].

Activity theorists had long ago argued how consciousness of a human being is shaped through the "actions" that are being done [4]. If "we are what we do" as the activity theorists argue, we should consider the artifacts and tools all around us, by which our activities are mediated. In a recent article philosophy professor Andy Clark points exactly to this fact, as he argues embodied cognition and the fact that cognitive processes are products of not only neurological activities of brain, but also the complex interactions between the brain, the body and the "designed" environments that we live in [5].

While these arguments can be exemplified more, we argue that they become ever more crucial when it comes to the recent advancements in media and communication technologies. These ubiquitous and pervasive tools of technological design become more and more intimate to our daily lives and to our bodies day by day. And they not only "upgrade" our present activities with faster, more convenient and more efficient ones, but also create their own "needs" to the novel activities that they bring into our lives [6].

This invasion and transformation of new technologies have been occurring most intensely among young generation. That is why in this research, we intend to track how this invasion has transformed the mindset of our graduate-to-be students—both as the direct experiencer and the very designer-to-be of these media, in order to be able to take the picture of the change in the digital media "landscapes".

2 Background of the Analysis

In this research we analyzed 86 graduation projects developed, defended and succeeded in Interactive Media Design Department of Yildiz Technical University of Istanbul, from 2002/2003 to 2011/2012. In the analysis, we used Marshall McLuhan's forecasts on the changing trends due to the changes in media as a framework. We have not taken these forecasts as a priori true statements; rather the study has been conducted as a two-way test. On one hand it is a unique test of McLuhan's theories among a group of next generation of designers. On the other hand, McLuhan's theories provided us a framework to analyze the mindset of this group. The main purpose of the project has been to reveal possible future projections of how media has been changing and transforming.

2.1 Interactive Media Design Education

Interactive Media Design Department (IMD) of Yildiz Technical University of Istanbul was established in 1997 and began accepting undergraduate students in the academic year of 1998/1999. The department gave its first graduates in the spring semester of 2002/2003.

The focus of the department has changed from Communication Design to Interaction Design in due time, which affected the approach in graduation projects. In today's perspective, Interactive Media Design discipline and designers have great potential to set the basis for the new coming services and products. As the idea of innovation gets stronger, creativity becomes more valuable than ever. In the curriculum, the students are expected to work on a specific design problem and/or specific technology in each of the preceding project courses. In the graduation projects on the other hand, they are asked to generate a creative solution for a design problem that they have discovered themselves. While in the first years they were also expected to implement their design and they were motivated towards game design, since 2006 we try an alternative approach to generate more innovative works through the stages of the briefing, conceptualizing, planning and executing processes of the projects, with more emphasis on the conceptualization so that the practical impossibilities of execution shall not restrain innovative ideas from being developed [7].

2.2 Timeline of Interactive Media Industry

When we analyzed the projects, we came up with the clear observation that there is a paradigm shift in the graduation projects of the last decade around the year 2007. This is not only a projection of the change in educational approach at the department, but also due to the advancement in real-life media implementations. It should be noted that when in 1998, the first group of students was accepted to the department, only a minority of them had mobile phones, Windows 98 had just been released, and the only way to connect to the Internet was via dial-up.

During 2000s, we have witnessed a great leap not only in the pace and capabilities of these communication technologies that we use in daily life, but also in the spread of them among young people. However, it was especially after 2004 that the seeds of some groundbreaking media tools were planted. Facebook, Flickr and Google's Print Project were launched in 2004, followed by Youtube and novel Google services such as Google Earth and Google Books in 2005, and Twitter and Wii in 2006, the year when "You" were selected as the person of the year by Times magazine. With 2007, the most significant year in the short history of the projects, iPhone was out with built GPS inside. In 2008, the first Android powered phone was in market together with the street view display in Google Maps. These utmost mobile devices and services paved the way for millions of applications;

including location based ones as well as the use of touch screen and gestural technologies; and other hand-held devices like iPad, and gestural game consoles like Kinect for Xbox in 2010. The pace of the change itself has an ever increasing acceleration that makes it very difficult to keep track of the changing landscapes in the use of information technologies and communication media, however we can at least say that *gestural interfaces, ubiquitous connectivity, socializing through digital media, collaborative content creation, a shift of focus to user experience,* and being surrounded by an increasing *abundance of digital information* have been some of the "gifts" of these groundbreaking media of the last decade.

2.3 McLuhan's Projections Adapted

While it is rather clear and trivial to track the reflections of changes in the industry and in our educational approach on student projects, the changing landscape of the graduation projects has the potential to reveal more for the future, which we intend to uncover through a reading and comparison of McLuhan's visionary projections on media.

Philosopher and media theorist Marshall McLuhan is mostly well known for coining the statement "the medium is the message" and the term "global village". The theories that he developed on media and communication especially in 1960s have been among the most debated and referred theories in the field of media studies. While his "prophetic" visions of "Electric Age", may be argued to be controversial from many aspects, a specific reading on his media approach turned out to be very inspiring and promising for this research.

McLuhan evaluates the history of humankind in three ages namely Tribal, Mechanical and Electric Age. These ages corresponds to specific mentalities, rather than being chronological and successive periods. The mentalities shaping these ages are determined by the way people relate to the media and technologies, which are extensions of men that which result in a distinctive understanding and experience of life. In this manner McLuhan contemplates on a significant number of media from language to TV (which was the "new" media of his time) and on how these media have changed the human perception through altering "sense-ratios" and how even further transformed the experience of life. The first great transformation is the de-tribalization of Tribal Man by the 3000 years of mechanical and fragmentary media and technologies of Mechanical Age, such like phonetic writing, print, clock, assembly line, car and so on. This is a transformation that has already taken place in Western world and has given birth to the "modern man". And it is a transformation that has been carried over to tribal men of the other worlds via colonization and modernization. As the second collective transformation, he prophesies the "re-tribalization" of modern life by Electric media and technology, which is introduced with the electricity and the following communications media from telegraph and telephone to TV [1]. During this re-tribalization process, McLuhan suggests many paradigm shifts to occur presented in Table 1.

3 Methodology

Through a list of qualities as a checklist, each graduation project has been investigated in a historical perspective according to the paradigm shifts presented in Table 1.

These qualities questioned the existence or non-existence of customization; the structure, the nature and organization of the content, of the media, of the tasks, and of the activated senses; the existence or non-existence of social interactions, social responsibility and use of social media. Due to format restrictions we cannot present here further details but only a limited number of examples from student projects.

It should be noted that our intention has been revealing the possible trends underlying media innovations and opening these to discussion, rather than proving a quantitative hypothesis. In that manner, considering the highly qualitative nature of the data, any statistical data are only used as a clue to detect and present the changing trends.

4 Results and Discussion

Through the analysis of the projects, which we have accomplished utilizing -but not limited by—McLuhan's forecasts listed in Table 1 as a framework, we have

Table 1 The Process of Re-tribalization in the transition from Mechanical Age to Electric Age

Mechanical age	➔	Electric age
Mass-production	Re-tribalization	Customization
Mechanization		Information
Typographical and mechanical		Metaphorical and iconographical
Pyramidal hierarchy		Mesh structure
Centralization		De-centralization
Lineal and deterministic		Non-lineal and in-deterministic
Division and specialization of tasks		Integration and unification
Homogeneous and uniform		Heterogeneous mosaic mesh
The dominance of visual sense		Synesthetic and all-inclusive
Visual and abstract media		Tangible and organic media
Isolated individuality		Connected collectivity
Abstraction of space		Contextual spatiality
Fragmentation		De-fragmentation

reached some trend changes clearly manifest and in parallel with the milestones of the media industry and our educational approach.

These changes and discussions among them have been compiled in six compact arguments; namely "A shift to subjectivity", "Kingdom of information", "Dissolution of rational structures", "Back to heterogeneous nature", "Appraisal of synesthesia and tactility", and "Global embrace versus fragmentation".

4.1 A Shift to Subjectivity

Freeing the students from the technological capabilities of present day and encouraging innovative thinking, contributed to the development of more sophisticated and creative projects during the second half of the decade. Before 2006, almost all the projects with minor exceptions are Game design projects implemented in Macromedia Flash for PC platform. With 2006, the proportion of Tool/Service design type of projects which are task-specific assistants and/or guides, based on a variety of technologies and platforms, starts to increase each term, taking over the majority by 2010, most of which are customizable according to user preferences.

Beginning from 2007, whether Tool/Service or Game design, the projects start to reflect a "media consciousness" and a focus on experience design. That is to say, they consciously use and reflect on the media, considering the possibilities of the used media in designing the user experience. By 2008, the proportion of these experience-centric projects reach majority. As a comparative example in games, the left screenshot in Fig. 1 is from the *Çukursaray Hidden Room* (2005) project by Seren Köroğlu, which is a typical Flash based survival horror/puzzle game. As for the *Bunker* (2011) project by Mert Uzer in the right screenshot of Fig. 1, as a more current game project, it is a location-aware augmented reality multiplayer game application designed for smart phones, where the social media (Twitter and Face book) activity of the user is an input to the game play and the user may come across with other users in physical reality. The designer of this project underlines

Fig. 1 Çukursaray Hidden Room (*left*) versus Bunker (*right*)

his intention to encourage people to engage in physical activity while staying connected to social media.

As another comparative example, in Fig. 2, the projects *Kördüğüm* (2012) ("Blind Knot" in Turkish, which is a phrase used for "complicated situations") by Cansu Kaykaç, and *Finkles* (2011) by Cansu Taştan, respectively, are both examples of experience design projects. *Finkles* as a tangible medium prints a 3D tangible and organic form as output, which is generated by the random motions of the users during their play with the medium. The purpose is to give inspiration to designers. *Kördüğüm* as another tangible project on the other hand is a Tool/ Service, as well as an Experience design, which utilizes use of (textile) threads embedded with new technology as a novel interaction medium for visually impaired people. The aim of the project is to provide the visually impaired people a memorandum book where they can record and recall their memories. There is a shift from mass-produced "experiences" designed for all to customizable and personal experience centeredness, which was also foreseen by McLuhan. We can observe an increase in the subjectivity of both the designer in design process, and the user as s/he experiences the design.

4.2 Kingdom of Information

Although it was not McLuhan who coined the term "Information Age", it was he who claimed that everything, including ourselves, is being translated into the form of "information" [1] in this age.

Since the invention of telegraph, the increasing abundance of information has started to become a part of the human life. That is why crossword puzzles and knowledge contests entered into our lives: to create contexts that all this information abundance makes sense [5]. The massive amount of applications being implemented for iPhone, iPad and other mobile devices is the continuation of this process. As for the student projects, beginning from 2007, information—be it design, transference, or processing of information—starts to become a major element in their design. Besides, a switch from the universal and abstract language of typography to the story-telling and contextual way of metaphorical and iconographical communication, as another projection of McLuhan, is also manifested especially in the design of this information in these student projects.

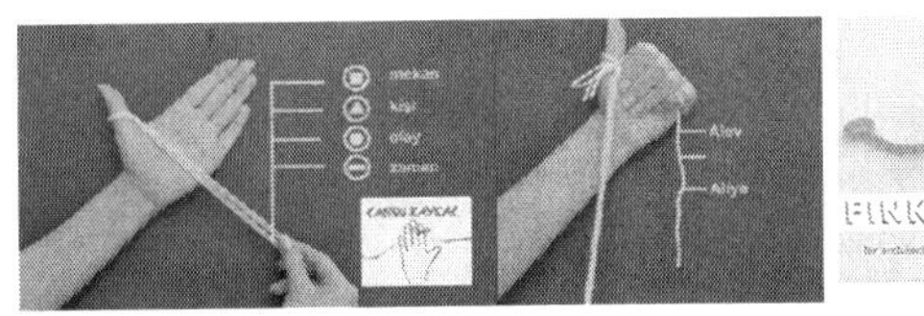

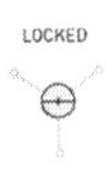

Fig. 2 Kördüğüm (*left*) versus Finkles (*right*)

4.3 Dissolution of Rational Structures

According to McLuhan, a linear, continuous, and deterministic structure and a central and pyramidal hierarchy between everything are the very nature of Mechanical Age, which is subject to dissolutions with the coming of Electric Age [1].

When we look at the course of projects, one of the pioneers, which indicate the upcoming paradigm shift, is an "interactive TV" design project from 2007. Although TV media itself was acknowledged by McLuhan as an example of "mosaic mesh" media, we have seen even further de-centralization of this media in due time [1]. The *Live Video Share and Interaction Platform* (2010) by Ömer Çıracıoğlu, a later gestural TV project is an example of this change. In this project the structure of TV media is totally turned upside down, with the inclusion of audience participation in the programs by broadcasting their own live videos, giving live feedback to the programs and by communicating directly with other audience.

4.4 Back to Heterogeneity

Division of tasks in the cycle of production and specialization among disciplines are another elitist and fragmentary face of Mechanical Age and modernization, while the specialized tasks and any other subject are supposed to be homogenous and uniform in themselves [1]. McLuhan suggests a unification and integration among tasks and disciplines, especially those of production, consumption and learning in Electric Age, and a heterogeneous mosaic structure instead of the uniformity that is diminishing [1].

This integration of tasks and disciplines instead of division and specialization, and a heterogeneous nature in the use of media are manifest in the projects around since 2007. For example, both of the projects in Fig. 3, *Live Video Share* (2011) and *Vibe Sense* (2011), include a unification of consumption and production, by inviting users to be active content creators. *Vibe Sense* (2011) by Meir Benezra is a project designed to strengthen the communication between the DJ and the audience in live music performances. Proposing a conceptual interactive ceiling, lighting and floor structure, the environment responds to the gestures and dance movements of the audience and translates them to the DJ in an augmented reality display. The audience learns through experience how their body feedbacks are communicated and responded, and thus begins to use this language to consciously communicate their musical requests to the DJ. This project not only combines the production, consumption and learning processes, but also integrates many different media from lighting to tangible and spatial media or to augmented reality in a heterogeneous manner.

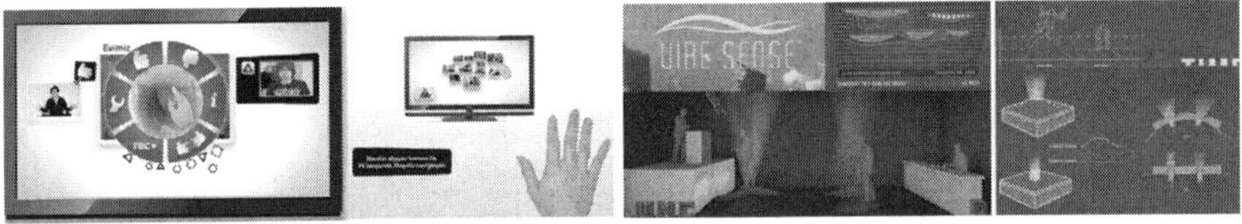

Fig. 3 Live Video Share and Interaction Platform (*left*) versus Vibe Sense (*right*)

4.5 Appraisal of Synesthesia and Tactility

The human senses and their "ratios" in the perception of reality is a discussion of great importance in McLuhan, who argues that the culture of Mechanical age is "visual", and sense of sight is a cool and detached sense [1]. Under the dominance of sight, the other senses like touch and smell, which are all-inclusive senses on the contrary to that of sight, and very crucial in the life of Tribal Man, loses their significance in Mechanical Age [1].

The prevision of the coming back of these underrated senses of tactile, audio and scent, in a synesthetic manner, are clearly manifest in many of the projects since 2007. While *Vibe Sense* (Fig. 3) translates motion into iconographical visual data, or *Kördüğüm* (Fig. 2) proposes sense of touch as a way to "write" down your memories, they both suggest all-inclusive and immersive experiences. Besides, in an increasing number of projects since 2007, we came across with the use of tactile and also "organic" material and media, such as ceramic or textile. For instance *Rina* (2012) by Özge Kantaroğlu (Fig. 4) is a conceptual interactive bond project, which is made from an electronics embedded textile, and is designed as an assistant for the traditional spring festival *Hıdırellez*. *Hıdırellez* is celebrated every May by thousands of people in a very crowded area where Balkan music and dances dominates the cultural atmosphere and many rituals from jumping over the fire to attaching tissues to the "Wish Trees" are performed collectively. Inspired by the handkerchiefs and neckerchiefs commonly used by Balkan musician and dancers and by the tissues attached to the Wish Tree, the student chose "textile" media as a communication, way finding, socializing and memorial tool. This project is an example of the tendency towards the use of tactile and also organic material in the overall projects.

Fig. 4 Rina: Interactive Cord

4.6 Global Embrace vs. Fragmentation?

Since 2008, the year when Facebook exploded in Turkey, the projects have started to consider social parameters increasingly. Some of the projects, like *Live Video Share*, *Vibe Sense* (Fig. 3) and *Rina* (Fig. 4), focused on designing social interactions. Some other projects like *Kördüğüm* (Fig. 2) started to consider social responsibility. And some projects went ahead and designed novel "social media" projects, such as *Peace Begins With You* (2008) by Ece Soner, a social media platform for peace activism on the web.

The clear shift from individual use to collaborative use among projects may also be regarded as a clue of the shift from the isolated individualism of Mechanical Age to the connected collectivity of Electrical Age [1]. However, this time, we abstain from claiming that this increase in the interconnectedness is a manifestation of the forthcoming "Global Embrace" that McLuhan anticipated.

"Fragmentation" or "segmentation" is the utmost critique directed towards not only to the Mechanical Age in McLuhan, but also to the whole modernization project in Deleuze [1, 8]. Whether the Electric/Information Age carries along this fragmentary nature of the Mechanical Age, or whether this nature is being transformed into a state of "Global Embrace" through new technologies, is an immediate question to answer.

On one hand, there are optimists such as artist and theorist Roy Ascott, seeing an embracing future in new technologies [9]. On the other hand other theorists, Sherry Turkle for instance, has strong claims that are based on interviews with hundreds of people, that new communication tools like social media and texting creates "connected but alone" people more than ever, along with the increasing risk of pathological communication disorders [10].

As it is not a trivial task to evaluate a design in terms of its "fragmentariness" in advance, we approached this problem indirectly in this research. The detachment of the human being from the rest of the life and from themselves, as a result of this fragmentation, begins with the abstraction of time and space [1, 5]. The notion of space as an abstract, detached, uniform, and quantifiable concept sets the very basics of de-tribalization of human [1]. "Contextual spatiality" thus points to an integrated relation of human to their spatiality. When we analyzed the projects, we see that beginning from 2007, an increasing number of them start to consider spatial information, through the use of location-aware, sensor, or ambient technologies. When we focus on the relationship between the user and the space, however, the observation is clear that it is never the *user* that is "location-aware", but the *media* itself. Space is also translated into the spiritual form of information, as McLuhan prescribes, and it is this information that the user is related to, not the space itself, which is most observable in augmented reality projects. Thus, we argue that, McLuhan's projections on Global Embrace, de-fragmentation and contextual spatiality, on the contrary to the previous projections, are not really coming into life yet.

5 Conclusion

Due to fast developments and adaptations that has been taking place in information technologies, there has been an increasing demand on the design of products, tools and applications, which however may not all be in favor of human beings. We believe that this changing landscape may be observed from the discipline of interactive media design as it provides a wide range of designs and innovations related with the ongoing advancements.

In this research we intended to delve into the mindset of gradute-to-be students of IMD of Yildiz Technical University, as they are potentially both the very subject and the future designers of the new media and technologies, which have a crucial transformational effect on the very experience of life. Thus our purpose has been not only to take a picture of the present situation of, but also to disclose and open to discussion any possible forecast for, the digital media landscapes.

The analysis not only showed an open connection between educational and industrial approaches and the course of the student projects, but also provided an experimentation of the paradigm shifts suggested by McLuhan. We concluded that there is a tendency towards *subjectivity*, *information hegemony*, *dissolution of rational structures* and *hierarchies of modernity* such as centralism and linearity, *a heterogeneous nature* where once divided subjects *converge and come together*, *end of uniformity*, and a *yearn for a synesthetic and tactile life experience*, which is in parallel with McLuhan's forecasts. The last observation is especially valuable, as it is different from the ongoing trend in industry. We came across with many projects in which the students consciously preferred not only tangible but also organic materials and media, which provide an all-inclusive sensual experience, having real textures and odor. This observation, we believe, can reveal two important outputs especially from the standpoint of product design. One is that, in the products that are being designed, technology should not diminish and substitute the organic and tactile practices of daily life, but should be integrated as an assistant to our daily and embodied experiences. That is why we believe the increasing research on embedding electronics media in organic material such as glass, textile and etc., will maintain a fruitful avenue.

There has been one remarkable output, which is in conflict with the McLuhan's theories, that we could not find true traces of the proposed shift from isolated individualism to "Global Embrace". While there is an ever increasing emphasis on social media, social interactions, social responsibility and inter-connectedness, among both real-life and student designs, there is also a strong debate going on whether this connectivity has been causing another type of detachment. We took the relation of human to their spatiality as the indicator in this debate, owing to the argument that the abstraction of space and losing direct contact with it is the first step in the process of fragmentation. We observed that although there is an increasing number of "spatial" projects, the spatiality in these are reduced to "information" again. If there is any awareness, it is not the *user*, but the *media* that is location-aware. In the demo video of the location-aware *Bunker* project in

Fig. 1, two "users" come across with each other within the game play but in a physical space, and they do not even greet each other.

We believe that, due to the great transformational effects of new media technologies, an ecological approach, which would consider all the related parties in their dynamic play holistically, is necessary both in the design of and the theory building on these media.

Acknowledgments We would like to thank our colleagues Adviye Ayça Ünlüer and Dr. Asım Evren Yantaç for their valuable support in collecting the necessary information from the departmental archive and to all graduates for their creative and wise works. This research would not be possible without them.

References

1. McLuhan M (1964) Understanding media. Routledge, London
2. Lanier J (2010) You are not a gadget: A manifesto. Knopf Doubleday Publishing Group, New York
3. Clark A (2012) Out of our brains. New York Times, New York
4. Norman D (2010) Technology first, needs last. ACM Interact 17:2
5. Postman N (1993) Technopoly: the surrender of culture to technology. Vintage, New York
6. Kaptelinin V, Nardi BA (2009) Acting with technology. The MIT Press, Cambridge
7. Brown T (2009) Change by design. Harpers Collins Publishers, India
8. Deleuze G, Guattari F (1987) A Thousand Plateaus: capitalism and schizophrenia, translated by brian massumi, University of Minnesota Press, Minneapolis
9. Ascott R (2003) Telematic embrace. EA Shanken (ed) University of California, California
10. Turkle S (2011) Alone together. Basic Books, New York

System Design for Community Healthcare Workers Using ICT

Vishwajit Mishra and Pradeep Yammiyavar

Abstract This paper presents a case study in the design of a mobile based community healthcare communication system within a larger health project named ASHA (Accredited Social Health Activists). The paper outlines the process of conceptualizing and modelling a health system involving community healthcare workers in a rural setting in India. The aim of the designed system is to enhance the efficiency of the field level community health workers in carrying out their prescribed work. The methodology involved, the identification of needs and problems as well as the development of the mobile based approach is outlined in this paper. It includes the initial design of the system up to the GUI and final working prototype run on a mobile. System analysis and information design principles were incorporated during the development process. Results of a limited testing of the design are also reported in the paper. The study reported includes findings from 8 village community healthcare workers and 3 doctors using the developed application. The targeted users were asked to evaluate the designed interfaces and information architecture in terms of their learn ability, ease of use, efficiency and adaptability. It was found that using the application reduced the cognitive load of the healthcare workers and assisted them in their work. The community healthcare workers were able to engage better with the rural people using this application.

Keywords Information system · Community system design · Qualitative research · Mobile based health monitoring system · GUI · Useability testing

V. Mishra (✉) · P. Yammiyavar
Department of Design, Indian Institute of Technology Guwahati, Guwahati, Assam, India
e-mail: vishwajitster@gmail.com

P. Yammiyavar
e-mail: pradeep@iitg.ernet.in

A. Chakrabarti and R. V. Prakash (eds.), *ICoRD'13*, Lecture Notes in Mechanical Engineering, DOI: 10.1007/978-81-322-1050-4_96, © Springer India 2013

1 Introduction

In the present age, the government is putting significant emphasis on fostering the health of people, especially in the rural areas. The rural health sector presents numerous diverse and challenging problems that are unheard of in the urban landscape [1–3]. Generally in developing economies like India, there is a scarcity of resources in terms of infrastructure as well as human resources. Most of the problems arise due to an acute lack of doctors and unawareness among people in rural India. This is clearly evident in the scenario considered in this paper [4–6].

From the field study, it was found that in the present context of North Guwahati region in Assam, there are just four doctors in the Primary Health Centre (PHC) at College Nagar to cater to a population of around one lakh spread over 19 villages. In such a scenario, empowering the community health workers becomes an imminent task at hand to bolster the efficacy of the many health schemes promoted by government. The members of ASHA have been empowered by government through means of limited training, free mobile phones, pamphlets, brochures, informative books, a free radio and a free cycle [7, 8]. In this paper, we have tried to understand the current system as functional in the North Guwahati area in our locality and have tried to propose design interventions for the broader system to plug the prevalent loopholes. The focus has been on empowering the ASHA members, Auxiliary Nurse Midwives (ANMs), Anganwadis, and other multi-purpose health workers in their work through the intervention of Information and Communication Technology. The paper intends to explore further how the ASHA members and other community health workers can be empowered and assisted in their daily work and be integrated with the central health care system through the means of an even efficient system and through the intervention of a mobile application utilizing the high penetration of mobile phones.

2 Current Scenario of Healthcare with Emphasis on North Guwahati, Assam

In the current situation, there is a District Health Centre in every district in the country. Besides this, there are some Primary Health Centres in each district, which come under the purview of the District Health Centre. To support these, there is one Health Sub-Centre in each village. The Sub-Centres come under the supervision of The Primary Health Centre (Fig. 1).

The focus of this study is the Primary Health Centre at College Nagar, in Kamrup District, and the Sub-Centres that come under its purview, especially the Moriyapati Sub-Centre at Amingaon, in North Guwahati, Assam. In the current system, there are 19 health sub-centres in the region between Amingaon and Bezara, in North Guwahati. There aren't any regular doctors assigned at the sub-centres, only community health workers work there, roughly 4 health workers at

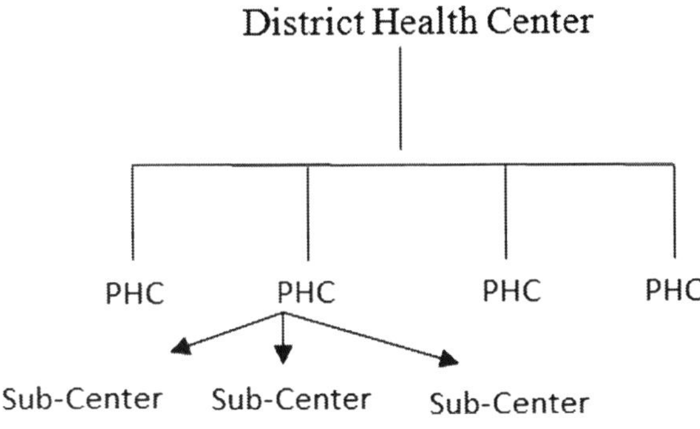

Fig. 1 Existing hierarchy: *PHC* primary health center

every Sub-Centre. Once a week, a doctor from Primary Health Centre visits the Sub-Centres. The Primary Health Centre (PHC), located at College Nagar, North Guwahati, monitors these Sub-Centres. The population of this region is approximately 1 lakh. There are 4 doctors assigned in the PHC at College Nagar. Besides them, there are around 15 other health workers to assist them in their work at PHC. The working hours of the doctors are between 9:30 AM to 2:30 PM from Monday to Friday. Approximately 180 people visit the Primary Health Centre at College Nagar every day during winters. The number rises significantly during summer months. On every Wednesday, the doctors have to go to the Sub-Centres on field visits and immunization. On every Friday, the doctors go to villages to spread awareness and look after people in their area. Thus, as can be analysed from the routines of the doctors, there is significantly great work load on them. Besides the doctors, there are many community healthcare workers and a chain of ASHA members, one ASHA for approximately every 1,000 rural people in villages. The ASHA members receive payment on performance basis and there isn't any monthly salary for them [9]. Their major duties are to spread the messages of government health schemes to rural people and bring about awareness in them. However, majority of the tasks that ASHAs perform are limited to the areas of maternal health, immunization, family planning and first aid. Besides the ASHAs, there are Anganwadi Workers (ANW), who cater to the cognitive and development aspects of children less than 5 years of age in villages [9]. The Anganwadis' receive a paltry monthly salary. Although the tasks of ASHAs and Anganwadis are demarcated, the roles performed by them in villages are generally blurred [10]. The government has given the Auxiliary Nurse Midwives (ANMs) and ASHAs' a free mobile phone [11]. Although there is an eligibility barrier of being class 10 pass for becoming an ASHA but generally the norms are relaxed [10]. Respected women in their middle ages and widows are mostly preferred for recruitment to the ASHA program. "The situation of healthcare was significantly grim before the inception of National Rural Health Mission (NRHM) in the year 2006. But after the inception of NRHM in 2006 and the Accredited Social Health Activists (ASHA) program in 2008, the efficacy of the intended health schemes has improved significantly. There has also been significant improvement in the ratio of the number of ASHA members to the number of rural people and it is reaching the desired ratio of 1 ASHA member for every 1000 people in the villages [9].

3 User Survey

This survey was conducted with the intention to understand the problems faced by community healthcare workers in their daily life while performing their task. The usage characteristics and importance of mobile phone in their daily lives were also studied. The sample sizes are indicated in the table mentioned below. The findings from the survey are reported in the paragraph that follows (Fig. 2).

3.1 Problems Faced by Community Healthcare Workers

After studying the current scenario and interacting with the stakeholders concerned, i.e. the healthcare workers, doctors, rural people, and assessing the information present in literature and other secondary information sources, the following issues and problems were considered to be the most concerning.

- Little Training (only 23 days training in 12 months followed by mere 2 days of follow up every 2 months) [9]. After the training, the health workers were left on their own, they did not know whom to approach for doubts on their lessons. Also, since the books and pamphlets that thcy rcceived after training sessions [9] weren't engaging and contained too much prose, the health workers generally did not read them.
- Inertia towards reading books and other written material. The Booklets and learning material given contains too much prose and is not engaging and one has to search sufficiently to gather information from books and there isn't any instant doubt removal facility.
- The Radio program that was aired for ASHA [12] members was too infrequent. It was aired only 4 times in a month. The health workers could not listen to the program again if they had missed a radio session once. Also, the members could not record the Radio sessions. Besides this, the information disseminated

Health Centre	District	State	Country	No. of Community Health Workers
Moriyapati Subcentre, Amingaon	Kamprup	Assam	India	6
Primary Health Centre, College Nagar	Kamrup	Assam	India	12

Fig. 2 User survey to asses needs of community health workers

through the ASHA Radio Program [12] was too generic in nature and only seldom catered to the demands of individual community health workers.

- The post cards given to ASHA members [12] was initially offering exciting prospects in terms of doubt clearance and query answering results, but this also reached stagnation since the answers through post cards took a lot of time to reach and till then the interest in the case was lost.
- The mobile phone given to the ASHA members and the free closed user group created by BSNL [11] is not used to its full capacity since the ASHA uses it only to make and receive calls. There isn't any integrated application to help them in their work.
- The village sessions that the ASHA organizes or the home visits made by them aren't very engaging. Besides, when an ASHA goes to check a particular patient, she generally does not know about the past patient history of that person. Also, even though most of the cases that come to ASHA concern maternal health, there isn't any checklist which ensures that the ASHA has covered all the necessary points while discussing it with a would-be mother.
- The health worker does not know what to do in case of a medical emergency as she has limited training. If the health worker can administer some basic first aid in cases of emergency, then it will be of great help in saving valuable lives.
- There are certain cases in which the ASHA goes to visit a particular patient and then she has a doubt in performing a particular task, say wrapping a band aid, or she is confused whether to feed the patient medicine A or medicine B for a particular ailment. In these situations, the ASHA cannot bring along her book with her and even if she brings, it will be a big embarrassment for her to look into it for solution. In these cases, the availability of a secondary learning resource to aid her was missing.

4 Context Diagram

Figure 3 shows all the entities involved to form the basic system and is termed as the Context Diagram [7]

A mobile service provider controlled by the health department will monitor the system. The rural people can send in their queries regarding health, appointments, etc. and the system will direct their queries to the community health worker in their locality. The health worker can register patients and send his/her data to the service provider, controlled by the health department. Health workers can also type in their queries or seek help in case of emergencies and the service provider can then connect the health worker to appropriate doctors who can assist them in their work. Health workers can also perform several other functions like learning their training lessons and taking self evaluatory tests sent by the health department.

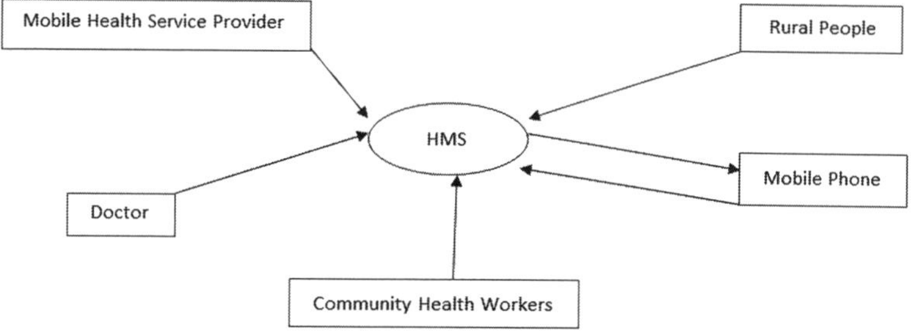

Fig. 3 Context diagram: *HMS* healthcare management system

They can send the result to the service provider, who can send it to doctors, who can monitor the community health workers progress over time.

5 System Modelling and Development

The detailed modelling of final designed system was done using Unified Modelling Language comprising of Use Case Diagram, Sequence, Relationship, and Activity diagrams among others [7, 13]. One of the several Use Case diagrams and Sequence diagrams is shown in Figs. 4 and 5. The sequence diagram depicts the sequence of interface interactions performed by a community health worker, given the tasks of selecting a mode of registration, searching for a patient who is already registered on the list, looking the patient's past history and finally counselling the patient.

5.1 Information Architecture

The information architecture (IA) of the application was designed keeping the system specifications in mind. The central aspects of the system, Registration, Learning Module, Notification, Self-Assessment tests, Important Contacts and Asking Doubts sections, were further detailed down. For example, within the registration section shown in the IA, one can choose from Children's Registration, Mother's Registration or General Registration, catering to other ailments which do not belong to the previously mentioned categories, like fever, diarrhoea, wounds, etc. This categorization was done since most of the patients that come to the health workers concern maternal and child health. Similarly, in the 'Learn' section, health worker has the option of learning lessons on various issues like Pregnancy, Family Planning, First Aid, Immunization, Hygiene, Nutrition, etc. In the notification section, the health worker can view messages and reminders sent to her from the health centre or rural people. The notifications from the health centre can be

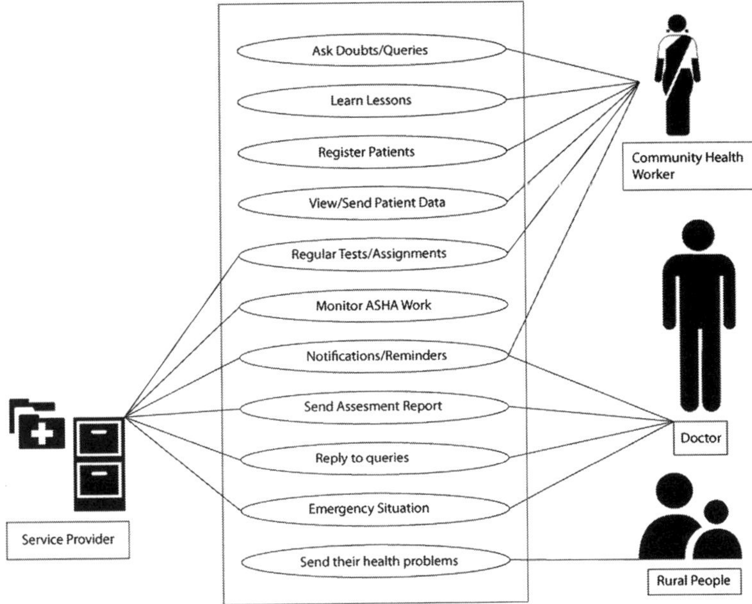

Fig. 4 Use case diagram

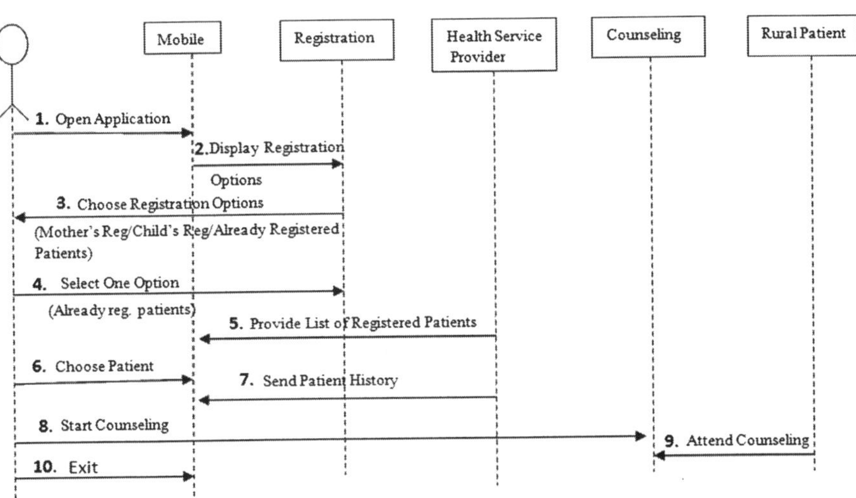

Fig. 5 Sequence diagram

regarding reminding the health worker about her home visits, or health check-up scheduled in the village, what items she should carry along with her, what lessons she has to revise in order to prepare herself for a particular task, etc. In the self-assessment or 'Test' section, the health worker can evaluate her knowledge of a

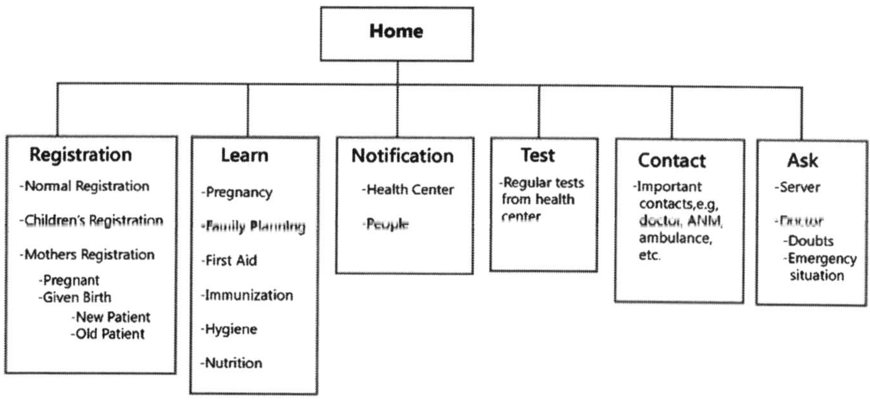

Fig. 6 Example of a basic information Architecture for the application

particular section by giving the predefined tests already loaded on her mobile application. The health centre monitoring the progress of ASHA can also send her tests to take. The answers to these tests are present in this section. In the 'Contacts' section, important contacts of all healthcare professionals and emergency numbers are provided. In the 'Ask' section, the health worker can send her doubts to the healthcare service provider, or in case of emergencies, for example, a large cut, snake bite, etc., she can send the photo of the affected part to a doctor, through the service provider, and slowly they can have a video conference where the doctor guides the health worker in emergency situations (Fig. 6).

5.2 *Functions in the Mobile Application*

Functions in the mobile application shown in Fig. 7.

6 Final Mobile User Interface Prototype Development

The prototype was developed using Adobe Flash CS 5.5 and Action Script 3. The specifications of the developed User Interface are as listed below:

Language Used: Hindi, Assamese
Voice Over: Hindi
Number of Icons: 6 on the first screen
Color Used: Color and Black and White both
Screen Size: 240 × 320 pixels
Software Used for Prototyping: Adobe Flash CS 5.5
Mobile Used for testing prototype: Dell XCD 28

Registration	• Registering a child→Options for registering a new born child or view patient data history of an already registered child • Registering a mother with maternal problems→Options for registering a new mother or view data of an already registered mother • General Registration for other illnesses like jaund ice, malaria, etc. Other options same as in the previous cases
Learning	• Pregnancy →Danger signs in pregnancy, correct breastfeeding techniques, nutrition, medication, etc. • First Aid→Large Cuts, Suffocation, Drowning, Snake Bite, etc. • Immunization • Family Planning • Cleanliness • Nutrition
Notification	• Notification from health center informing the health worker about village health day, her counselling sessions, patient visits, etc. and its time, venue and articles to be carried along with • Notification/requests from people. The mobile server will direct rural people's queries to the concerned ASHA in the village and then the health worker can look into the query
Evaluation/Tests	• Self-Assessment tests which the health worker can take to test her k nowledge on an any topic. For example, before an in house counselling session, the health worker may want to brush up her knowledge on danger signs in pregnancy and other maternal health topics • Tests sent from health center from time to time basis or before organizing a village health day/health camp. For e.g., before a malaria, pulse polio drive, etc., the health center can send some tests to the health worker over phone. The health worker in turn has to send her response and she can get a feedback. This way, she will be better prepared for her sessions and her efficiency will improve. • In case the health worker does not know the response to a question, she can discuss it with other health workers at sub center
Ask	• Send doubts/queries to mobile health se rver in case when she has a doubt or if she does not know about a health topic. • Send photo of affected area to doctor/ start tele-medicine, in case there is an emergency situation and doctor is not immediately available

Fig. 7 Table depicting functions in the mobile application available to the health worker

6.1 Features of the Developed Mobile User Interface

- Usable Information Architecture
- Easy Navigation
- Voice Over
- Notifications messages from health service provider
- Facilitates learning and self-assessment
- Easy registration of patients
- Enhanced engagement for both the health worker and her clients
- Checklist of important points to be followed

6.2 Final GUI on Mobile

Some screenshots of the final Graphical User Interface designed are shown below in Fig. 8. Wherever there is a voice over, the active loudspeaker button denotes it.

Fig. 8 Final GUIs

7 Testing

The first set of tests was done to understand the 'learnability' aspects of the proposed new GUI. Testing location was at North Guwahati, Assam, India. Total 8 community health workers were tested. The tests were carried out at health centres and at the residences of health workers. The user age group ranged between 37 to 46 years. Education level of the users was till class 10th. All the users had previous experience of mobile phone and they were using the free mobile phone that the government had provided them. Hindi language was used for testing purposes along with prompts in Assamese wherever required. All the users had undergone a course in Hindi language during their training program. Two tasks were given to be completed by the users. These tests were carried out on the final mobile prototype that was developed.

- Task 1: To go through the counselling section on her own. This section comprised a set of 12 messages. Each message asked a question which had to be replied in 'yes' or 'no'. The respondents also had to listen to each question with the voice over and then record their response to the question and move to the subsequent message.
- Task 2: To go to the notification section and read the message from either the health centre or rural people.
- Number of error/confusion spots denotes the number of instances when the user was confused or made an error while navigating through the screens.
- Mobile Phone: Dell XCD 28: Prototype in Hindi Language
- Stop watch was used to check the time taken to complete the task (Figs. 9 and 10).

7.1 Testing Methodology

Coaching method' was used to model the test [14]. Hindi and Assamese languages were used for giving instructions and clearing doubts of the users. Mobile phone was provided to the user to operate and time recorded to perform the task was

Fig. 9 Usability testing on the final mobile prototype (*extreme left*), and on paper prototypes (*middle* and *extreme right*)

Sr. No.	Educationin class passed	Agein Years	Time for T1in minutes	Number of Errors/Confusion	Time for T2 in seconds	Number of Errors/Confusion
1	9th	37	1:40	3	8:24	0
2	8th	45	1:48	2	6:23	0
3	8th	44	1:42	3	7:50	1
4	10th	42	1:50	4	8:24	0
5	9th	36	1:32	2	8:28	0
6	8th	44	1:38	2	7:82	0
7	10th	52	1:42	5	8:24	1
8	10th	46	1:42	3	7.88	1

Fig. 10 Table depicting time taken for the completion of tasks (*T1* Time taken for task 1, *T2* Time taken for task 2, Education-1st to 12th standard in a typical schooling system before entering into university. The table shows the data for the tests carried out on the final mobile prototype.)

noted using a stopwatch. The various screen configurations along with the purpose of the task were first explained, shown and demonstrated to the subjects to make them feel comfortable. The users were asked to use the application for three days. The tasks were then assigned and the test was conducted after this gap of three days. The time taken to perform the assigned tasks was an indicator of how quickly and easily the subjects could learn to operate the new GUI. The test also tested the learnability and the navigability (information architecture) of the information hierarchy of the application. The number of error/confusion spots would indicate the number of times the user were either unsure of what to perform or performed some task wrongly. This would be an indicator of the learnability as well as the cognitive load of the user while navigating through the information architecture.

8 Results

The test data indicates that the average time taken for Task 1, involving listening to the questions in messages under counselling section and then recording their responses, was 1 min and 42 s and the average number of errors faced while performing this task was 3. The error/confusion regions indicated that some regions in the interface were difficult for the health workers to comprehend. Interestingly, these error/confusion spots were uniform for all the health workers. The second task involving going to the notification section and reading a message took 7.8 s on an average and was performed with almost no errors.

9 Conclusion

Using the application ensured that the workers were able to remember and recall effectively, the lessons on general health taught to them during training. They were also able to clear their doubts and engage better with rural people. It saved them time and reduced cognitive load in carrying out their work. Cultural contexts of use need to be incorporated while designing information architecture [15]. The options/number of clicks possible on a screen have to kept as minimum as possible to reduce cognitive load. At every instance, the users must get appropriate feedback on the task they are performing and the number of tasks remaining to be completed. The insights from this project can be used to develop similar applications in agriculture, education, microfinance, insurance, etc. in developing countries.

References

1. Waterhouse Coopers P (2010) Emerging market report: healthcare in India 2007
2. Future of Health PSFK, Available at http://www.psfk.com/publishing/future-of-health. Accessed 2 Nov 2011
3. Srinivasan R (2005) Health care in India-vision 2020, issues and prospects. Planning commission, India
4. Chib A (2009) The acehbesar midwives with mobile phones program: design and evaluation perspectives using the information and communication technologies for healthcare model. In Mobile 2.0: Beyond Voice? ICA Pre conference
5. Pakenham-Walsh N, Priestly C, Smith R (1997) Meeting the information needs of health workers in developing countries. Br Med J 314(7074): 90
6. Kuriyan R, Toyama K, Ray I (2006) Integrating social development and financial sustainability: the challenges of rural computer kiosks in Kerala. In: Proceedings information and communication technologies for development
7. Van Dijack P (2003) Information architecture for designers. Rotavision Publication SA, UK
8. Inclusive Healthcare Management for Sustainable Development, Conference Background Note, Deloitte, Aug 2011
9. Ministry of Health and Family Welfare (MoHFW) (2007) Government of India, New Delhi. National Commission for Macroeconomics and Health Annual Report 2006–2007
10. Ramachandran D, Canny J, Das P, Curtell E (2010) Mobile-izing health workers in rural India. In: Proceeding CHI 2010, ACM Press
11. Mobile Phone to Sub Center ANM. Available at: http://www.nrhmassam.in/anm_mobile_content.php. Accessed 12 Feb 2012
12. ASHA Radio Programme under NRHM, Assam. Available at www.nrhmassam.in/asharadio.php. Accessed 21 Jan 2012
13. Yeates D, Shieldss M, Helmy D (1996) System analysis and design. Macmillan India Limited, India
14. Nielsen J (1993) Usability engineering. Morgan Kaufman Publication/Academic, USA
15. Kolko BE, (2002) International IT implementation projects: policy and cultural considerations. In Annual IEEE IPCC Conference, 352–359

Developing Young Thinkers: An Exploratory Experimental Study Aimed to Investigate Design Thinking and Performance in Children

Anisha Malhotra and Ravi Poovaiah

Abstract It is increasingly popular to teach creative thinking skills in schools. A diverse variety of programmes exist to support practitioners in this task, and some research has been gathered on the effectiveness of individual approaches. However, there is less research on the effect of such programmes in Indian schools and creative thinking. This study aimed to investigate the process of creative problem solving in children observes thinking strategies used by fifteen 11–13-year old children redesigning the traditional board game of Snakes and Ladders. There were three experimental conditions: individual, dyad and triad collaborative. The paper presents results where protocol analysis is used to investigate evolution of design ideas and various thinking strategies to analyze the levels of design thinking. The paper also examines the relationship between creative thinking and collaboration. Results demonstrated differences in levels of design thinking and performance gains for collaborative conditions both dyad and triad.

Keywords Creative problem solving · Design thinking strategies · Divergent thinking · Collaboration in children

1 Introduction

Problem solving has proven to be an effective way through which thinking can be practiced in schools. Conventional school assignments rarely give students the

A. Malhotra (✉) · R. Poovaiah
Industrial Design Centre, IIT Bombay, Powai, Mumbai, India
e-mail: malhotra.anisha@gmail.com

R. Poovaiah
e-mail: ravi@iitb.ac.in

A. Chakrabarti and R. V. Prakash (eds.), *ICoRD'13*, Lecture Notes in Mechanical Engineering, DOI: 10.1007/978-81-322-1050-4_97, © Springer India 2013

opportunity to work on challenging thinking problems which are open-ended. Hence, most students have little experience in skills such as planning, researching, problem solving, idea generation, reasoning and summarizing. Design problem solving encourages such practices where one has to continuously explore and think of multiple solutions for a single problem.

This paper presents findings from an on-going research aimed at supporting collaborative activities in face-to-face environments and developing thinking skills in creative contexts. One objective of the study presented here was to examine the ways in which children that are not trained in any kind of (game) design thinking approach a redesign task and to identify the kinds of strategies they use to accomplish the task. Fifteen children who volunteered for the study became game designers for an hour and worked on redesign of the game of Snakes and Ladders. Three conditions were investigated for the same task where children as individuals, dyads and triads solved the problem in three different sessions. The study presents a comparative difference and similarity in thinking strategies and design process for all three conditions.

The insights from this experiment could help understand a facilitator less situation and help improve design education or creative thinking skills programmes where the focus is on enhancing creative thinking skills in children.

1.1 Problem Solving: Convergent Versus Divergent Thinking

Problems vary in knowledge needed to solve them, the form they appear in, and the processes needed to solve them. The problems themselves also vary considerably, from simple procedural problems in elementary school to thinking of multiple ways to solve a problem with an unknown answer. If a problem is an unknown worth solving, then problem solving is "any goal directed sequence of cognitive operations" [1] directed at finding that unknown.

A critical attribute of problem solving is that the solution is not readily apparent or specified in the problem statement, so the learner must identify not only the nature of the problem, but also an acceptable solution, and a process for arriving at it. The approach to solve a problem may be convergent thinking (a single, known solution) or divergent thinking (one of several acceptable solutions). Well-structured problems with a single solution require the application of a limited and known number of concepts, rules, and principles being studied within a restricted domain. On the other hand, problems which are open ended look forward to solutions those are neither predictable nor convergent.

For a long time, psychologists believed that "in general, the processes used to solve unstructured problems are the same as those used to solve well structured problems" [2]. However, more recent research in creative problem solving in different contexts makes clear distinctions between thinking required to solve problems with convergent solutions and problems with divergent solutions.

1.2 Design and Problem Solving Approaches

Simon [2] points out, a unique feature of design problems is that they do not have a single right solution; there are always alternatives. Previous research on design identified different approaches of handling design problems and used bipolar descriptions, such a "top-down" versus "bottom's up" [3]. The view that has been called top-down tends to see the process of problem solving as one of breaking down a problem into more meaningful sub problems. Here, the designer maps out the context, content, and structure of a design at the beginning. The opposing view, called bottom-up, describes problem solving as a conversation with the situation, in which the design of the game emerges in the process of implementing it. These two views suggest that students may approach the design task from different positions, choose to emphasize different aspects of the design, and think about in different ways depending on their personal preferences.

2 Collaboration and Thinking Skills

The majority of theorists and thinking skills approaches actively encourage learners to work collaboratively (as quoted from [4]). Yet, to date, minimal research exists to endorse the benefits of collaborative learning when fostering thinking skills.

Wegerif [5] argues that collaborative learning improves children's ability to reason, and in general enhances performance on most activities. Wegerif and Mercer [6] coined the term "exploratory talk" to denote the ability to 'reason' through interaction and collaboration with others. Gokhale [7] conducted a study based on Johnson and Johnson's claim [8] that collaborative learning enhances children's critical thinking. There was no advantage of collaborative learning on factual knowledge. The collaborative learning condition experienced greater task enjoyment and were consequently more engaged and motivated. However, more research is needed to determine whether collaborative learning specifically enhances thinking skills.

Because a task division can hinder such a conceptually oriented interaction, we preferred to work with collaborative peer-work groups. In contrast with co-operative learning groups, students in collaborative peer-work groups try to reach a common goal and share both tools and activities [9]. According to Cohen [10], shared goals and tools can strengthen positive student interdependence.

3 Research Questions

We setup an experiment to investigate thinking strategies used by individuals, dyads and triads to solve a game redesign problem. In testing these conditions, the experiment also explores if such a situation can help design education as part of the school curriculum and practice oriented creative thinking which requires a redesign in the mode of instruction, motivation and methodology in a facilitator less environment. The methodology should help open doors to divergent thinking as against learning concepts with defined and absolute solution.

Our study aimed at answering the following questions:

1. In a no input and no facilitator setup, what are the thinking strategies used by children to solve a creative problem like redesigning a traditional game of Snakes and Ladders?
2. Where does the precedence lie for the redesign ideas?
3. What are the different approaches used or developed by individuals, dyads and triads to solve a complex task of redesigning a game?
4. Which group meets the criteria of redesigning the best by thinking of a number of qualitatively different ideas, individuals, dyads or triads?

The paper does not question the quality of the creative output of children in the process of redesigning the game of Snakes and Ladders. The paper concentrates on the nature of design process and various thinking strategies used by children. Later, in future experiments we observe how this ability may be enhanced through providing inputs, built on thinking strategies, to produce new ideas.

4 Method

4.1 Protocol Analysis- Use of Think Alouds for Investigating Thinking Process in Children

The term 'verbal protocols' is used to refer to human subjects' verbalization of their thoughts and successive behaviors while they perform cognitive tasks. The protocols are generally taken concurrently with the task performance but may also be taken in retrospection. Verbal reports, concurrent 'think-aloud' protocols, provide a valuable source of data about the sequence of events that occur while human subject is solving a problem or performing some other cognitive task.

As designers, when we frame a situation we create an initial design structure within which we begin to invent and implement solutions. Although this dialectic process is illustrated in think-aloud protocols collected extensively while adults attempted to solve a variety of problems; little is known about problem solving techniques and thinking processes when children solve a design problem.

4.2 Experiment Setup

The experiment was conducted with fifteen children of the age 11–13 in a controlled setup. The experiment was performed in a fairly large room which was designed especially for this experiment. The room was divided into four temporary sections (as shown in Fig. 1 below) to conduct parallel sessions. Visual barriers were created in the room so that participants are not able to see each other but are aware of other participant's presence. The barriers were a must so that participants do not get influenced by each other's thinking and concentrate without distractions. The first corner was used for pre-task warm-up session. The rest three corners were used by the participants to perform the task. Also, every session was video recorded separately with one camera per individual in the first session and per team for the next two sessions. A video camera was placed on a tripod at a distance from the participants to record their activities during the task.

The first session was for individual participants, second for 3 dyads and the third for 2 triad groups. Three sessions of fifty minutes each were conducted one after the other. The first twenty minutes included filling a short questionnaire and a short discussion on redesign. The next thirty minutes were given to the participants to work on the task. The experiment was planned in two levels for thirty minutes each- first for idea generation and second for designing and implementing the new game idea.

One researcher was present during all three sessions to conduct the warm-up session, observe and provide assistance whenever required during the task. The assistance did not include any kind of suggestions or guidance to help in thinking of a solution.

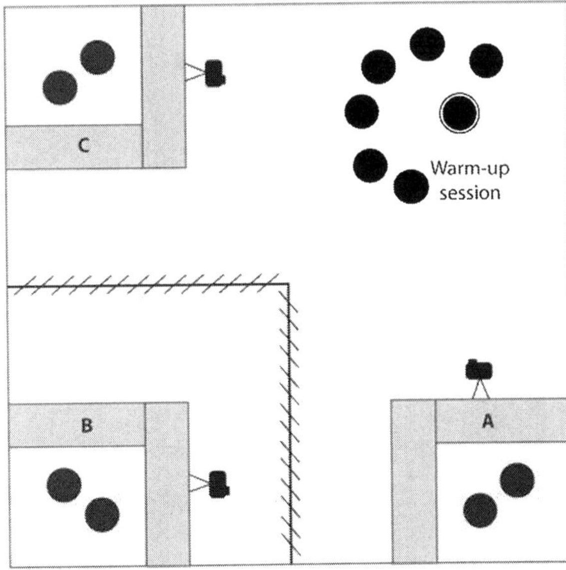

Fig. 1 Experiment room design

4.3 Procedure

Stage 1: A warm up session was carried out with children to break the initial inhibitions and get them to think and talk about games. This included filling up a short questionnaire individually and an informal discussion with the researcher on *what is redesign?* During this session we discussed how often do they play games and the kind of games they play?

Stage 2: The design problem was given to all the participants orally together in a group. The participants repeated the design problem before starting the task.

Stage 3: The researcher explained the process of performing the experiment. The participants were told to think aloud during the entire process of redesign. They were told to say 'I am done' when they felt they have no more ideas.

Stage 4: Participants started idea generation on the redesign of the board game.

Stage 5: Execute the 'new game' idea.

4.4 Participants

Fifteen different children participated from same socio-economic background for the experiment. Parents were informed of the study and gave their consent for children to volunteer and participate in the study. The participants were invited for voluntary participation for three separate sessions through posters and informal requests. None of the participants participated in multiple sessions. The mean age of the participants was 11.5. As shown in Fig. 2, three settings were investigated: (a) three children working individually (n = 3; 2 boys, 1 girl); (b) three groups working collaboratively in dyads (n = 6; 6 boys, 0 girls) and (c) two groups working collaboratively in triads (n = 6; 6 boys, 0 girls).

Randomly participants were selected for the sessions on a first come first serve basis and none of the participants participated in more than one session.

Fig. 2 Children solving the problem in each condition: (**a**) individual (**b**) dyad (**c**) triad

4.5 Material Provided

One A3 white sheet to draw, pencils, a ruler and an eraser were provided to each individual. The game of Snakes and ladders was made available only on request of the participants. No colors were provided in any of the sessions. For collaborative sessions, participants shared all resources except pencils.

4.6 Design Problem

The design task given to children was to 'Redesign the game of Snakes and Ladders to make it more interesting.' The focus and the only trigger to thinking that was provided were to focus on thinking of a number of different ideas to change the game and avoid making a 'single' new game.

4.7 Recording the Thinking Process: Logging of the Data

The verbal data, gestures, and the accompanying sketches were recorded on excel sheets by the team. The verbal protocol was divided into chunks of information for analysis in form of 'episodes and events'. The episodes are analyzed and insights are generated which may be useful both for theoretical and practical implications.

5 Results

Qualitative analyzes of transcripts, video recordings, sketches and background questionnaires revealed children used multiple strategies to solve the problem. Also, collaboration positively impacted children's engagement, participation, and enjoyment of the activity. Both content and frequency analysis was performed for the following three processes: Collaboration, Ideation and Execution separately and their combinations. This proved benefits of collaboration over individual problem solving.

5.1 Design Process

The individuals and teams had different processes to solve the problem where some took the approach of idea generation first and then execution and others thought while executing their ideas and the third category used a mix of both approaches. Though the approaches were different but like any other design problem solving activity, giving tentative ideas, keeping alternatives open and keeping some aspects vague were clearly visible in the protocols. A common step toward the redesign solution observed in all the groups except one individual was to draw the baseline grid. Later they used different strategies to think of hurdles and bonus points keeping the same aim of the game that of a race. The process of idea generation was observed to be different in individuals and in collaborative teams. Individual participants hardly questioned their solutions and fewer alternatives were thought of for the same idea. Whereas, in collaborative teams, almost every idea was questioned and improvised which led to a richer thought process and more number of ideas for the final solution.

5.2 Thinking Strategies

Children did not adopt one particular strategy but rather negotiated between skills and multiple thinking strategies. Carryover of elements from the existing game was observed as one of the strongest strategy used by children in the design process; where they replicated the most dominant elements of the existing game. Many visually dominant elements were borrowed either due to the presence of the game board with the participants or because of strong memory of a very often played game. Examples of this strategy can be seen in the Figs. 3, 4, 5 and 6. A square ten by ten grid with numbers 1–100 has been borrowed blindly and as a prerequisite for the redesign. All participants in all three conditions marked 'Start' and 'Home or Finish' on the grid making this again a 'race' where players chase each other to reach the final box. The movement on the new grid follows the existing left to right-up-left-up zigzag path. One long 'snake' like element and one long 'ladder' like element is also evident of in the new games. The redesigned games follow the rule of repetition of same hurdles and bonus points as in the existing game of Snakes and Ladders.

The common thinking strategies in all three conditions used at different stages of the design process are listed below:

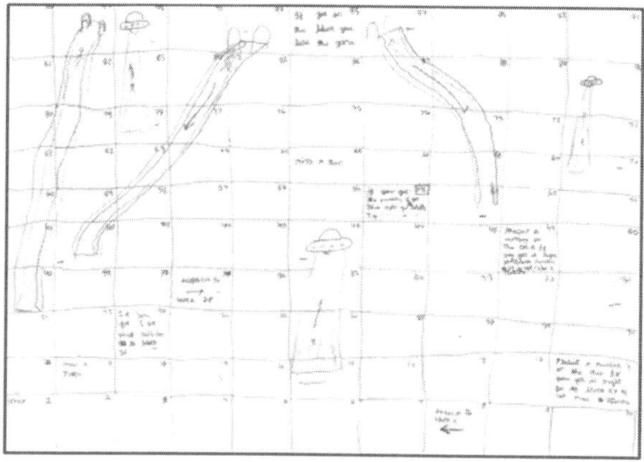

Fig. 3 Individual 1 (Individual B)

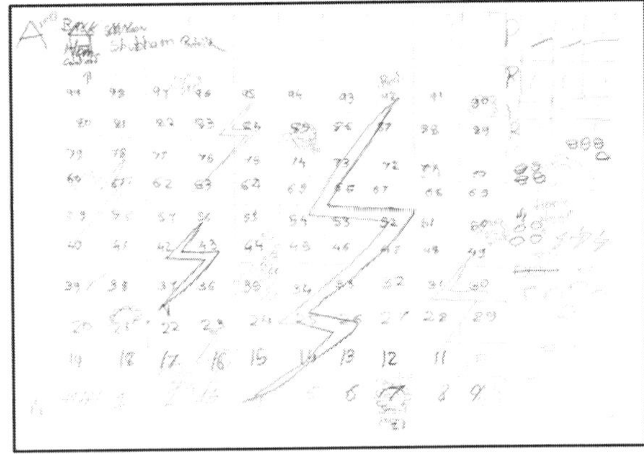

Fig. 4 Dyad 1 (group A)

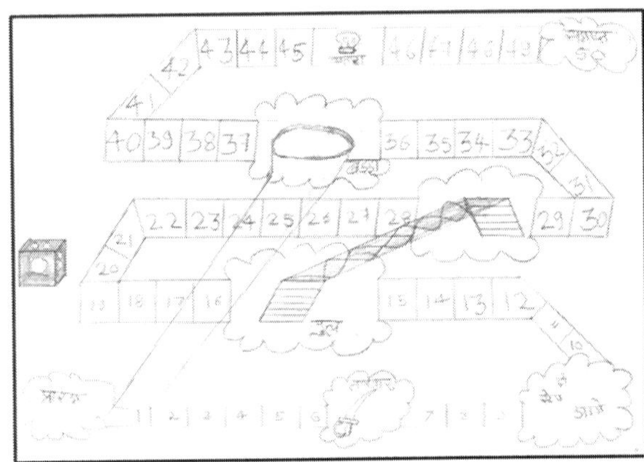

Fig. 5 Dyad 2 (group C)

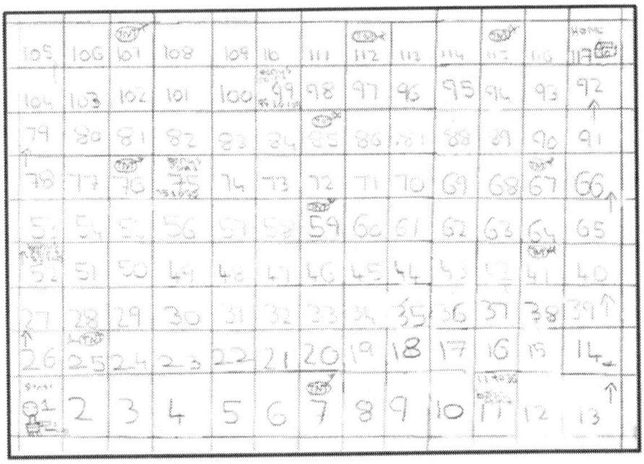

Fig. 6 Triad 1 (group A)

1. Carry over: At least one bonus point should lead you to the top row near the winning point.
2. Baseline for thought process: Drawing the grid, 'Start' and 'Home': Start is
3. Adaptation from other game rules memory: One obstacle in the first row which forces the player to restart the game. More hurdles than bonus points.
4. Analogy from real life situations: Snakes will be replaced by something which affects people in real life. For example: an electric shock, a sword, a pothole, a bomb. Ladders are replaced by objects that help people in real life to move ahead or forward. For example a tunnel, a bridge, a river and a boat, a UFO.
5. Role play: Thinking of real life situations like man going up in the UFO, sliding down, jumping on a boat to move ahead or using a tunnel. The idea of covering the bridge to give it a real look as the participant says the player will fall otherwise.
6. Visual simplification of snakes, ladders and other elements.

Replacement and use of analogy as a strategy was used excessively by the participants. For example:

"2: Where there are snakes we will make ladders and replace ladders by snakes." (Dyad)
"2: We will make a Man in place of Mickey mouse." "2: We will draw the man in the same position as Mickey Mouse." (Triad)
"3: Listen, we can have something in place of Snakes. Like cars or something." (Triad)
"2: Let us draw bombs and ladders now." (Triad)

The participants used either top-down or bottom-up approach to solve the problem. An example of use of top-down approach used by one of the dyad groups is shown below. The group makes a road map of their ideas and actions and later starts executing those ideas where they improvise some on the way.

"2: We will think first and then we will make. As and how we will think we will make."
"2: Here we will make start and there end"
"2: We will make the path like this. From here to here to here to here and end. Ok?"
"2: Later, we will put something like pot holes etc. etc. ok?"
"2: Here let's write FINISH"
"2: This. Start. Finish. Here we will go till 10 then we will have 20, 30 and 40 and finish."
"2: In between we will have some jackpot, then going ahead, then"

Real life situations and role play as a thinking strategy was also used by the participants where they thought of elements (hurdles and bonus) from real life which brings people harm and the ones that help them. Also, while executing a real person was always visualized on the board and design was improvised accordingly. For example:

"1: Draw something here or fill this with color, darken it.
2: Why?
1 (laughs): He will fall otherwise."
"2: Knife? Knife? A sword? When the player will come here he will be killed by a sword and he will come back to 'Start'. What say?"

Adaptation as a strategy was used by a few participants where they picked ideas from memory of the games they have played or play to make new changes to the existing game. Ideas like 'go back to start' or 'miss a turn' were taken and the idea of 'Snakes and Ladders of different genres'—Shooter, Adventure, Espionage and Sport seems to be an adaptation. These genres are most popular in Xbox 360, Wii and other online games.

5.3 Collaboration

Collaboration was an integral part of solving an unknown problem. Children without any training or forced collaboration setup were able to collaborate and work together comfortably towards accomplishing the goal. Many reasons were observed for collaboration among children. Children appeared to participate more actively when common resources were provided for multi–participant interaction. Children made verbal comments and physical gestures to provide input when they were not in control of the drawing sheet. Self initiated distribution of task lead to cooperation to execute the ideas.

For example:

"1: Hey, let us write numbers. You write one line and then I will write the next."
"3: You do the first line. I will do the second and let him do the third line."
"2: You write numbers and I will try drawing the man."

Combined thinking on alternatives and idea generation in a team was frequent. Example:

"2: what should we put here? We want something like..
1: a rat?
2: We don't want a rat. We want something..
1: Lizard?
2: King cobra?"

Implementation and difference in skill lead to interaction between the participants and hence the collaborated better and were found to be more engaged in the task. Example:

"1: Can you draw straight lines?
2: I don't know how to. Is this the way to draw the line? It is not coming straight."
"2: why are you drawing from here? Draw from bottom"
"2: Keep the pencil like this and draw straight."

The participants were constantly asking for each other's opinions and agreement especially in dyads where they worked closely together both on idea generation and execution. Phrases like "Ok?"; "You like it, right?"; "Understood?" were commonly used in their conversations which lead to better collaboration and combined responsibility.

Each transcript hence was coded for any of the three processes: collaboration, ideation or execution. A frequency analysis on occurrence of collaborative ideas in groups was performed on the content where collaboration was distinguished used for ideation or execution. For example, in dyad group 1, total number of collaborative ideation episodes were 16 (collaborative ideation = 9, collaborative execution = 7); for dyad group 2, total number of collaborative ideation episodes were 13 (collaborative ideation = 3, collaborative execution = 6, collaborative ideation for execution = 4); for dyad group 3, total number of collaborative ideation episodes were 17 (collaborative ideation = 6, collaborative execution = 9, collaborative ideation for execution = 2). The results show that there was collaborative ideation but there was more collaborative execution. This may be because of a redesign problem or because of lack of any design instruction given to the children.

5.3.1 Limitations to Designing

Skill proved to be a limitation to design and idea generation. Due to lack of knowledge or skill to draw children were unable to produce and collaborate.

"3: I am not able to make a man, how will we replace Mickey Mouse with a man?"

It was also observed that children were conscious and aware of the fact that their actions are being recorded. This awareness led to a forced collaboration for the camera to show participation. A few examples of that are mentioned:

"1: Let me also draw something"
"1: My photo is also coming let me also work." (Dyad)
"2: You are being watched by the camera. Beware!" (Dyad)
"3: Help please. We will not get anything. Camera is watching." (Triad)

6 Conclusions

This ongoing research study was conducted to investigate the need and benefits of collaboration in a facilitator (input) less setup, when children are involved in a creative problem solving task. Children did not adopt one particular strategy but rather negotiated between skills and multiple thinking strategies. Children used different approaches both top-down and bottom-up in order to solve the task in a successful manner. Design process followed by individuals was found to be linear whereas the groups followed a more iterative process where there was constant evaluation, feedback and improvisation.

The findings indicate that when children are solving an open ended problem with no external input or intervention, the collaborative condition experienced greater task enjoyment and were consequently more engaged and motivated than children solving the problem individually. The lack of skill proved to be a limitation to design and idea generation. Due to lack of knowledge or skill to draw children were unable to produce and collaborate. Otherwise, collaboration occurred at many instances during the problem solving session especially while implementing the redesigned game. It was observed that children probably need more motivation and slight direction to keep the process of thinking moving without disturbing the raw flow of ideas, especially in the case of individuals.

References

1. Anderson JR (1980) Cognitive psychology and its implications. Freeman, San Francisco
2. Simon HA (1996) The sciences of the artificial. MIT Press, Cambridge
3. Kafai YB, Resnick M (1996) Constructionism in practice: designing, thinking, and learning in a digital world. Routledge, London
4. Burke LA, Williams JM (2008) Developing young thinkers: an intervention aimed to enhance children's thinking skills. Think Skills Creativity 3:104–124
5. Wegrif, R. (2002). Literature review in thinking skills, technology and learning. Nest Futurelab Series, Open University
6. Wegrif R, Mercer N (1997) A dialogical framework for researching peer talk. In: Wegrif R, Scrimshaw P (eds) Computers and talk in the primary classroom. Multi-lingual Matters, Clevedon, pp 49–65
7. Gokhale AA (1995) Collaborative learning enhances critical thinking. J Technol Educ 7:1045–1064
8. Johnson RT, Johnson DW (1986) Action research: cooperative learning in the science classroom. Sci Child 24:31–32
9. Webb NM, Palincsar AS (1996) Group processes in the classroom. In: Berliner DC, Calfee RC (eds) Handbook of educational psychology. Simon and Schuster Macmillan, New York, pp 841–873
10. Cohen EG (1986) Designing group work: strategies for the heterogeneous classrooms. Teachers College Press, New York

Part XI
Posters

Learning from Nature for Global Product Development

Axel Thallemer and Martin Danzer

Abstract For both innovation and strategic design management education is shown via a case study how to teach through research in a multidisciplinary manner. Blurring boundaries of professional compartmentalisation and fragmentation of knowledge is leading towards a new era of innovation by not mimicking nature. In contrast to purely aesthetic design with its emphasis on subjective values, the focus of the innovation origination process here is on the rationalised formulation of functional shape in harmony with materials, production and environmental technologies. This is the opposite of prettifying or pure styling. It is also for this reason that there will be no dressing up of a predetermined technical package for the purpose of providing marketing or advertising with better sales or promotion arguments, but instead—and from the start—a concentration on the devising of analytic solution variants.

Keywords Methodology of fostering industrial innovation through observing nature followed by inductive reasoning · Studying live role models leading to purpose-driven design through scientific rationale vs. Subjective approach · New era of innovating by not mimicking nature

A. Thallemer (✉)
University of Art and Industrial Design, Chair of Industrial Design/Head of Scionic®
I.D.E.A.L., Hauptplatz 8, A-4010 Linz, Austria
e-mail: axel.thallemer@ufg.ac.at

M. Danzer
University of Art and Industrial Design, Head of Computer Aided Industrial Design/
Scionic® I.D.E.A.L., Hauptplatz 8, A-4010 Linz, Austria
e-mail: martin.danzer@ufg.ac.at

A. Chakrabarti and R. V. Prakash (eds.), *ICoRD'13*, Lecture Notes in Mechanical Engineering, DOI: 10.1007/978-81-322-1050-4_98, © Springer India 2013

1 Inspiration and Meta-Level

A humanoid industrial robot arm inspired by nature has been researched, designed, developed, fabricated and demonstrated by bachelor and master students of scionic® I.D.E.A.L. curricula. Both business and educational as well as academic research issues and contexts are being covered. The presentation spans from natural sciences in industrial design management, materials- and production technologies to purpose-driven Gestalt: scionic®

Visualisations clearly depict the design process, the managing of both the conceptual development and the alternative morphologies resulting in the final prototype in comparison to the common industrial solution.

Procedures: Taking inspiration from prototypes found in Nature, inductive reasoning will be used to open up a larger range of strategic solutions than would generally be available in the context of traditional engineering sciences, due to their compartmentalisation of specialised knowledge. With respect to suitable functional shapes, the wider range of deduced scenarios that thus arises, leads into the so-called morphology box. Working from the basis of the respective scientific aims, a combination of appropriate properties will be addressed on a multifunctional and interdisciplinary basis (Fig. 1).

The history of science also plays an ever more central role in the cultural history of humanity. The structuring of human artefacts existed long before the term "design". Coming as it does from the English-speaking world, it has only been since the 1970s that this term has come to be used to refer to the job description of those who were previously known as "Formgestalter" in the German-speaking world. The congeniality that existed in the Renaissance between art and science led to technological innovations. In Roman times, Vitruvius identified the quality characteristics of design as being "firmitas—utilitas—venustas" (stability, usefulness, beauty). In the wake of the increasing shift away from innovation, functionality and scientificality towards superficial prettification experienced in the design sector in the last decades, this natural sciences research and teaching approach is intended to effect a repositioning in the purpose-driven implementation of human artefacts, ranging from materials selection to associated fabrication technologies, taking into account sustainable environmental aspects. The research and development result is a robot arm with anthropofunctional motion patterns for both industrial automation applications and assisting humans directly due to its resilience.

1.1 The Design Process by Scionic® I.D.E.A.L.

Biologically inspired by analyses of lobsters' and grasshoppers' legs and of human pointing gestures, a two-segmented arm with an external skeleton has been developed with motion patterns similar to those of a human. Powered by artificial muscles, "AirArm" is capable of anthropofunctional motion while not looking humanoid (Fig. 2).

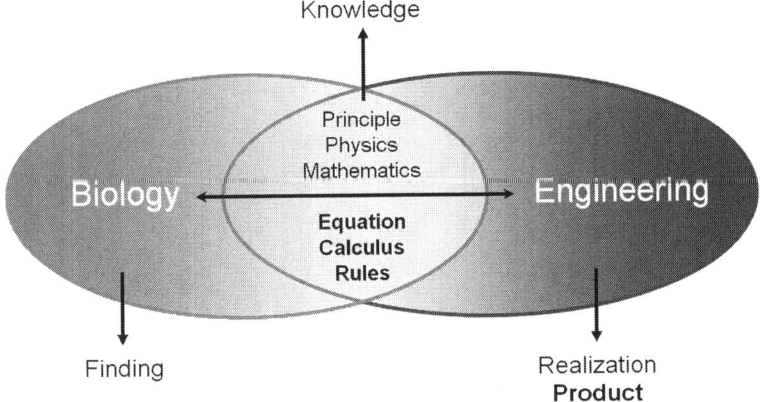

Fig. 1 Biological findings superimposed with engineering lead to extended scientific knowledge for the designer. *Source* Univ.-Prof. Dipl.-Ing. Dr. med. (habil.) Hartmut Witte

Only principles of properties and growth in nature were used in an abstract manner by inductive reasoning to achieve a purpose-driven form for the robotic arm without mimicking the real role models. This can be seen by no direct visual analogy between the life role models chosen and the derived human artefact. A scientific functional analysis of the human arm identified numerous opportunities for technical realisation. The technical purpose of the "arm" is seen as that of reaching as many remote points as possible within a hemispherical operating range from a specified point in space (like under Ref. [4]).

1.2 *The Computer Aided Industrial Design Process by Scionic® I.D.E.A.L.*

As this prospective robotic arm is representing a research project, which has been carried out for a large industrial enterprise, the challenge for us as an education faculty was to include the teaching by this research in the integrated virtual product development process of that named third-party funded project. Following the rules of the commercial world of professional practice, this study project was to be oriented towards large scale industrial production, enabling it to attain technical or functional superiority in the globally competitive environment.

Numerous IT applications, which together facilitate Concurrent Engineering, enable various departments within an organisation and also external service suppliers to have continuous access to product development data (Figs. 3, 4).

When starting the project, it was our explicit goal to implement a digital/virtual process in the product development—similar of what has been described as methodological principle for product development at large enterprises. To do so, we defined the 3D CAD data as pre-requisite for the complete development of AirArm.

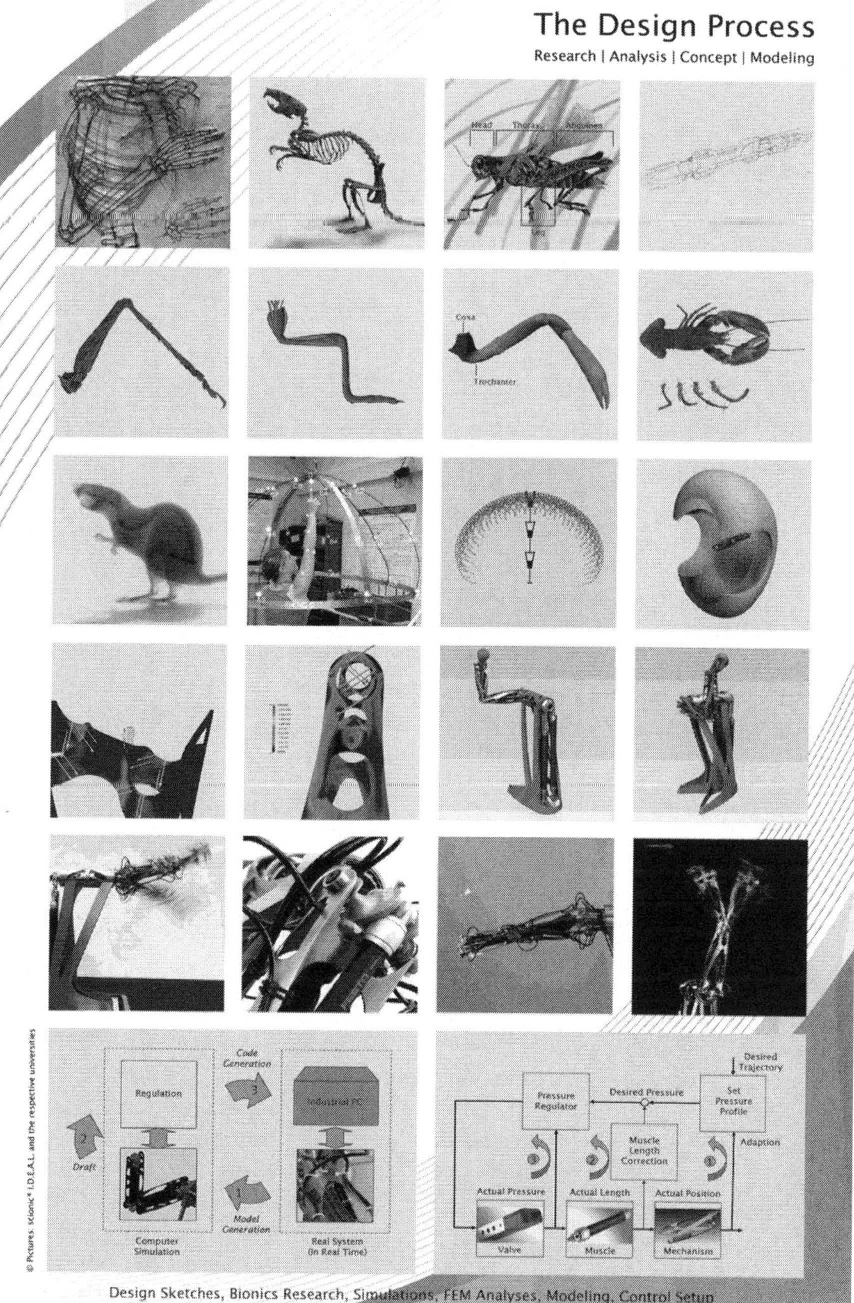

Fig. 2 Because mammal have about the same design from a meta-level, internal skeleton and proportionally scaled bones according to each species, arthropods have been analysed for a broader palette of design solutions as live role models from nature. *Source* scionic® I.D.E.A.L.

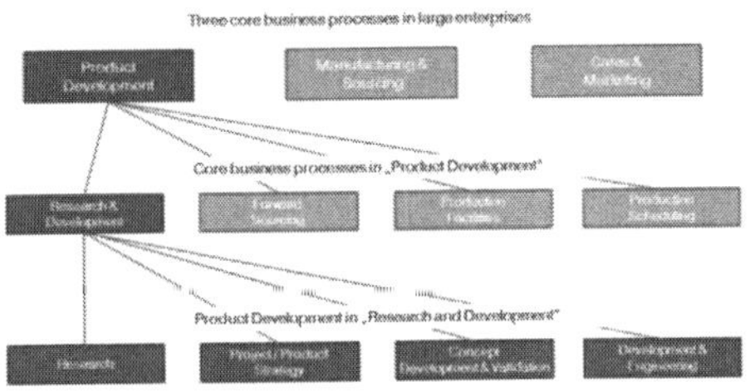

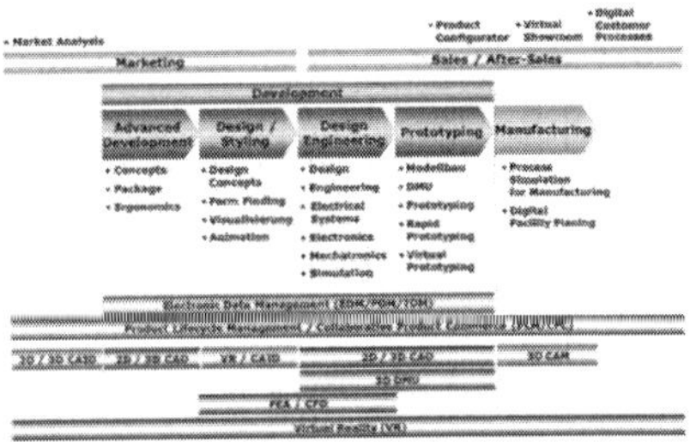

Figs. 3, 4 Research & Development as part of the three core business processes in large enterprises and Integrated virtual product development process. *Source* scionic® I.D.E.A.L.

Since the advent of CAID around 20 years ago, design solutions have involved the use of innovative information technologies within an integrated virtual process chain leading to a new product development process. At the end of the 1990s the first IT systems for 3D design and modelling tasks could be developed, again on the basis of newly defined algorithms, which made it possible to trace complex free form geometry by means of NURBS algorithms.

For security reasons and also because of the need for specialization, design and initial surface definition departments of an enterprise have tended to work separately from the development and engineering departments. This division, together with a certain degree of resistance towards computer technology, is the reason why, at the present time, there is no universal data-based communication throughout the field of virtual product development. This means that numerous 'gaps in the system' prevent the existence of an efficient, generally viable process chain such as Concurrent or Simultaneous Engineering. The status of a design and development process is checked and recorded by 'milestones'. As digital technology has advanced, two main scenarios for the (design) evaluation of projects have become established: digital reviews using visualisation based on real-time simulation and virtual reality and/or physical rendering of component parts based on digital input using the rapid prototyping procedure.

2 Purpose Driven Form Finding

As prerequisites for various possible approaches under the heading of "smart design", the following categories were outlined (like under Ref. [5]):

- Lightweight
- Flexibility
- Resilience
- Simplicity
- Robustness
- Adaptive control

2.1 Deriving Shape from Natural Role Models by Inductive Reasoning

In designing the joints, inspiration was derived from examples found in living nature, but not copied phenomenologically to display visually the live role models chosen. Via the joints of the grasshopper's leg located close to the body (in particular the coxa trochanter joint), the search led to the lobster's leg with its angularly displaced axes of motion. A two-segmented flexing system with muscles operating in contrary motion was chosen as the general principle for technical realisation. By analogy with the natural model, the design was to be kept as simple as possible, and the principles and structures were to be duplicated at various levels; this is known as self similarity. To ensure both lightness and robustness, triangulation of the arm modules was executed by analogy with the exterior skeleton of a grasshopper's leg (Figs. 5, 6). By crossing over the joint axes of the lobster's leg and adapting the segment lengths, a favourable compromise was achieved between simplicity and versatility for the reaching movements within the hemispherical range of operation. The pneumatic muscles as an antagonistic mechanism allow a high degree of yielding ability in combination with minimal expenditure of energy to remain stationary in a specified position (like under Ref. [6]).

Two-dimensional design and contour sketches were initially drawn up for all the functional components required for technical implementation. The three-dimensional realisation was effected using CAID, with verification of the datasets in design programs. The datasets generated by this means served as a basis for production of the functional components by 3D laser processing and CNC milling. In designing AirArm, care had to be taken to ensure that the various functional components could be readily produced from normal metal blanks. In view of AirArm's operation with the medium of water, it was manufactured in stainless steel. As bearings, standard production components were used, thus effecting the best compromise between weight and cost.

The dynamics of the arm system and the patterns of motion within its range of operation were already visualised by means of simulation and computer animation prior to production of the functional components, so that the kinetic characteristics could be analysed and critical system conditions identified at an early stage.

2.2 Transdisciplinary Verification Through Research in Teaching

Inspired by nature, all form finding was done by industrial design students of the author's research institute scionic® I.D.E.A.L., including 3D modelling, FEA analyses and inverse kinematics—the transdisciplinary verification took place both at Friedrich Schiller University of Jena, Germany, Institute of Systematic Zoology

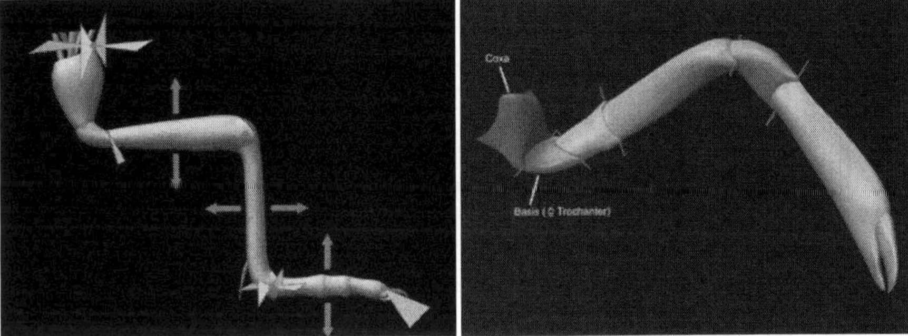

Figs. 5, 6 Degrees of freedom of a grasshopper leg (*left*) versus joint axes of a lobster leg (*right*). *Source* Univ.-Prof. Dr. Martin S. Fischer

and Evolutionary Biology and Phyletic Museum (high-speed HDTV X-Ray, infrared motion capture, Computer Tomography) and at Technical University of Ilmenau, Germany, Faculty of Mechanical Engineering, Department of Bio-mechatronics and Faculty of Computer Science Technology and Automation, Department of Systems Analysis. They furnished the scientific proof of this industrial designer's purpose-driven way of form giving by getting inspired through functional morphology as well as teaching through research.

The pneumatic muscles used as a drive mechanism have a very favourable ratio between the high mechanical forces attainable and their low weight. The advantages of AirArm thus come to the fore above all where rapid dynamic movement is required. On the other hand, the pneumatic muscles are also highly elastic, and the relations between their contraction, their pneumatic pressure and the forces produced are non-linear. As with its biological model, the technical system must also actively learn to deal with this situation—this is the task of AirArm's control unit!.

The initial biological inspiration for AirArm was provided by the legs of the grasshopper and the lobster—i.e., legs with internal muscles and an exoskeleton, as are typical of arthropods. However, since AirArm was to combine the structure of an arthropodic leg with the operating radius of the human arm, the axes of rotation and range of motion of the leg joints first had to be determined. It was found that unlike in the case of vertebrates, each successive joint of the arthropodic leg flexed at right angles to the preceding one. The central rotational movement of the arthropodic leg is the sum of the scope of movement of the four joints closest to the body. By contrast, the large scope of operation of the human arm is achieved through a high degree of manoeuvrability in the shoulder joint and muscle assisted flexibility of the shoulder blade (Table 1).

The participating students ideated and sketched about several dozen conceptual design solutions for the prospective arm only after a thorough analysis of most robots via the internet. Global findings were that most of the robots were intended for industrial use. Those could be clustered in six areas: axial, palletiser, scara, portal or gantry, hexapods or deltaic robots. These do all pretty much look the

Table 1 Manoeuvrability of a lobster's leg

Proportion of selected lobster leg segments to total leg length	Selected joint angles of a lobster's leg/Joint axis	Range of motion
Coxa 1	Thorax|coxa to the centre/to the side	Approx. 45°
Trochanter 1	Coxa|trochanter to the front/to the rear	Approx. 85°

Source Univ.-Prof. Dr. Martin S. Fischer

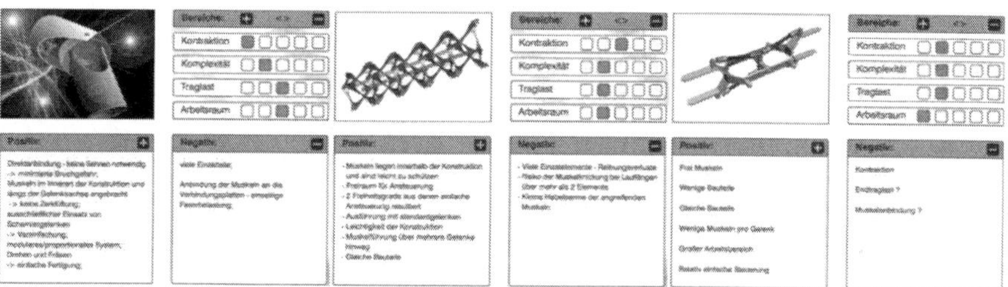

Figs. 7, 8, 9 The final three principles in the evaluated morphological box of about three dozen possible design concepts. The first and third were superimposed for the optimum of the final design concept. *Source* scionic® I.D.E.A.L.

same, no matter which make or company they are stemming from within each category, due to the special functionality those have to perform. Humanoid robots were only found in kinetic art, funfairs, toys, movie industry, laboratory research specimen and/or intended to interact with humans directly in the (far?) future as service robots. A special segment is prostheses. Static devices are recorded for some 3,500 years, mechanically moving ones for some 500 years. Industrial robots are for 50 years existing and operating (Figs. 7, 8, 9).

Digital product development allows photo-realistic designs to be visualized in real time, and subsequently analysed in Design Reviews using VR technology, such as CAVE environment. This offers great advantages compared to complicated and expensive hardware modelling. It is possible to check technical details at an earlier stage, thus anticipating the way the process will progress (front loading) and also improving and enhancing communication with other departments and outside service suppliers.

By means of finite element analysis (FEA) an optimized component shape and load capacity for all functional components was realized as well as sufficient safety tolerance ensured.By collecting the 3D CAD data of all mechanical components including air wiring, a complete assembly has been performed to check any design problems, also guaranteeing data accuracy. PDM systems help to improve design quality via digital mock-up which results in better manufacturing accuracy (Figs. 10, 11).

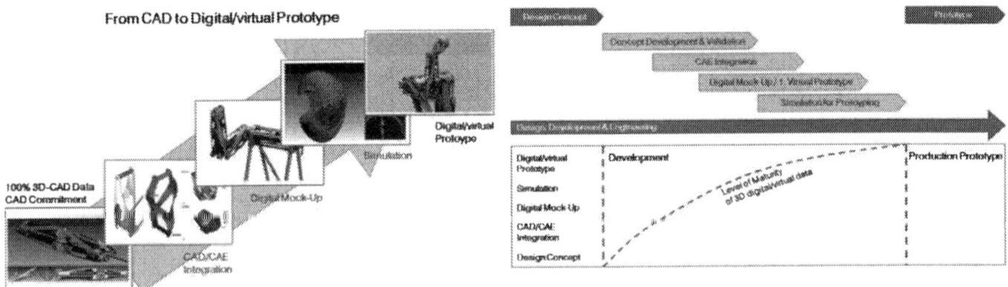

Figs. 10, 11 Increasing product definition and Level of maturity of 3D digital/virtual data during development lead-time. *Source* scionic® I.D.E.A.L.

3 Feasibility Studies on Anthropofunctional Motion

The movement of AirArm has already been simulated in the early design phase by generating two-dimensional scatterplots of 3–6-segmented arm modules. With the increasing definition of the 3D components and their interaction, the movement of the assembled component was simulated by animation SW using inverse kinematics features. This not only proofs the movements and interactions of the static mechanical components, but also included the air wiring and the respective movements.

The processes of reaching and grasping in mammal were investigated with rats, by means of a high-resolution X-ray camera with up to 1,000 HDTV images per second in two planes. This procedure highlighted the significance of the shoulder blade in executing these movements. Since this method is precluded with human subjects, surface measurements of the arm were used to determine the shoulder blade's role in human reaching movements (infrared motion capture). The results largely corresponded to those determined in the X-ray analysis of rats.

An anthropomorphic, i.e., human-oriented, approach was not adopted in the design of AirArm; rather, the principle of "rotation effected close to the body" was transferred to the use of linear instead of rotary drives (anthropofunctionality). After all, seeking inspiration from nature does not mean directly copying human arm movements (referred to as "biomimicry"), but entails the technical adaptation of human movement patterns, in this case with a lever system based on different lengths, proportions and modes of actuation. Reaching for points within the hemisphere was interpreted as the index finger "reaching" to points on a spherical surface. The movements of the human arm were registered with an infrared-aided motion analysis system. In evaluating the lines of movement, patterns of joint flexing were observed that could be summarised in the form of (apparently) simple rules. The structural solutions produced in the process of human evolution need not be copied; the significant requirement is that the appropriate lengths and forces are made available by technical means. It is not necessary to imitate the human shoulder joint. The evolutionary burden of man once being a primordial fish, going on land, transforming from quadrupedal to bipedal motion, can be seen in the

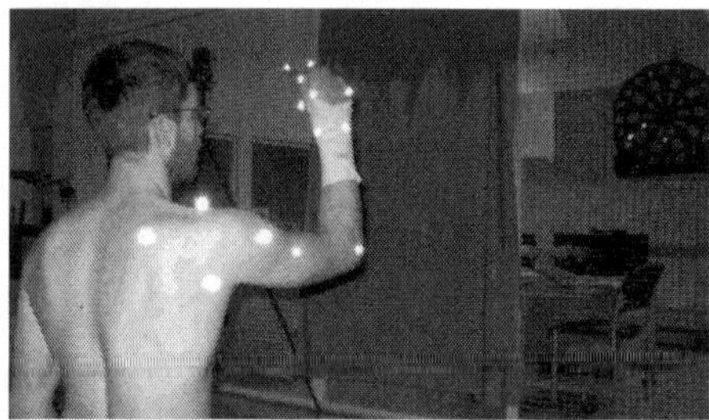

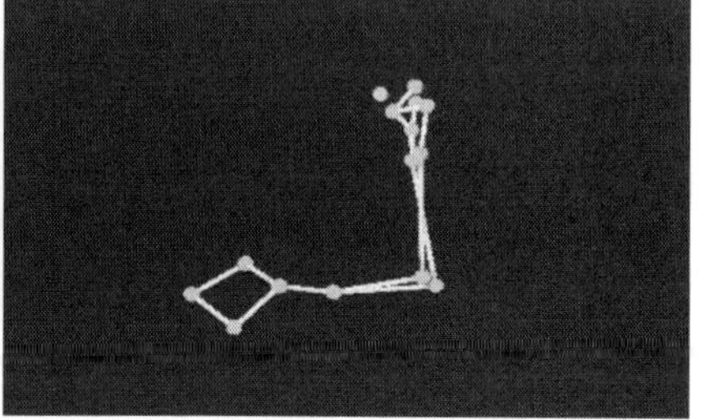

Figs. 12, 13 Human overshot dart throw with infrared motion capture. *Source* Univ.-Prof. Dipl.-Ing. Dr. med. (habil.) Hartmut Witte

morphology of our arm and hand. Once the breast flipper, the finrays are forming now our five fingers, the shoulder muscles were once pulling water through the gills, later the forelegs became arms.

In analysing the pointing movements, the coordinates of the reference points—marked in the form of reflective spots—and of the targets are registered by means of an infrared camera system generating up to 1,000 images per second. This tracking of coordinates as a function of time provides the basis of motion analysis (Figs. 12, 13).

The entire control system is designed on the basis of a model: the mathematical model of AirArm is first devised on a computer, on which the control system is then drafted and optimised. This system is then transferred via automatic code generation to an industrial PC, which controls AirArm on a real-time basis. AirArm is real-time controlled by means of three nested feedback loops (Figs. 14, 15, 16, 17, 18, 19).

4 Conclusion

The major difference to earlier contributions (like under References [1, 2, 3]) to that topic lies in the transdisciplinary interaction of evolutionary biology, zoologically functional morphology and biomechatronics as well as automation and systems engineering, systems analysis, where industrial design is in the centre of that innovation process described here. By teaching through research on real third-party funded projects and managing a cluster of different universities as well as their faculties and students at various locations, a new way of digital innovation process inspired by natural role models for purpose-driven industrial design could be proven academically. The transdisciplinary crossover between industrial design, engineering and natural sciences is exactly the scope of our bachelor and master curricula yielding to the university degree of (Dipl.-Ing.) "diploma engineer in industrial design"; a unique proposition in industrial design education at university level. The triangle of knowledge in industrial design education comprises purpose-driven form giving inspired from nature, materials-, manufacturing-

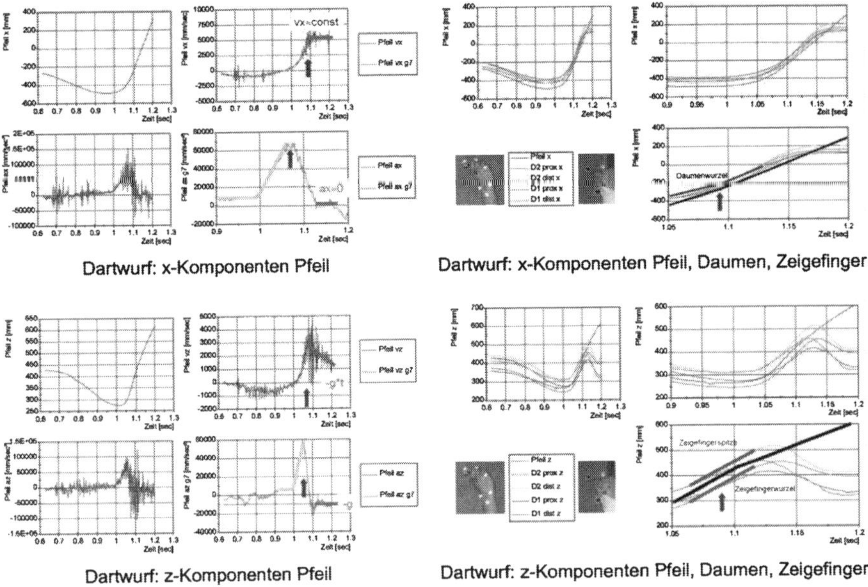

Figs. 14, 15, 16, 17 Motion tracking of airarm's overshot dart throw (*left side*) versus human one (*right side*). *Source* Univ.-Prof. Dipl.-Ing. Dr. med. (habil.) Hartmut Witte

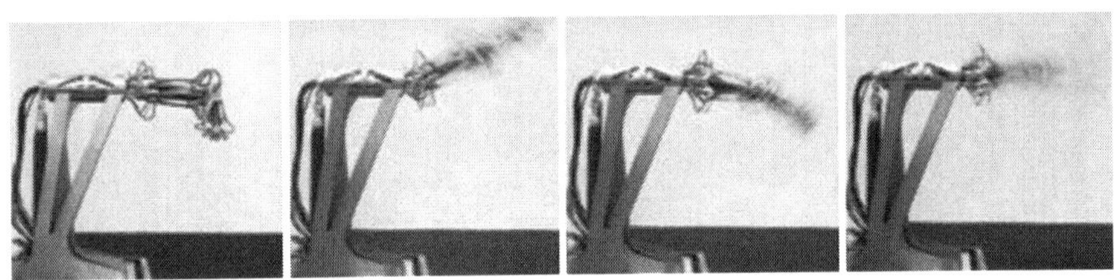

Fig. 18 Maximum acceleration 12 g, maximum velocity 11 m/s. *Source* scionic® I.D.E.A.L.

and environmental technologies combined with the digital, virtual development process. It is rarely the case, that industrial design ideation and conceptualisation comes first and engineering is following "simultaneously" through the process;—mostly even in industrial design technical packages are beautified and embellished for the mere sake of advertising and marketing only. The rationale here is the scientification of industrial design by teaching through research versus a subjective "I design" or autographic "designed by me" approach, so commonly known in the world of design. It is rare that design by learning from nature is not mimicking the natural role model visually or phenomenologically, our paper however, shows inspiration by inductive reasoning. The difference is that one cannot see the initial live role model chosen from nature visualized by the (industrial) design (like under Ref. [7]). One is tempted to state, the more similar the looks of the man-made

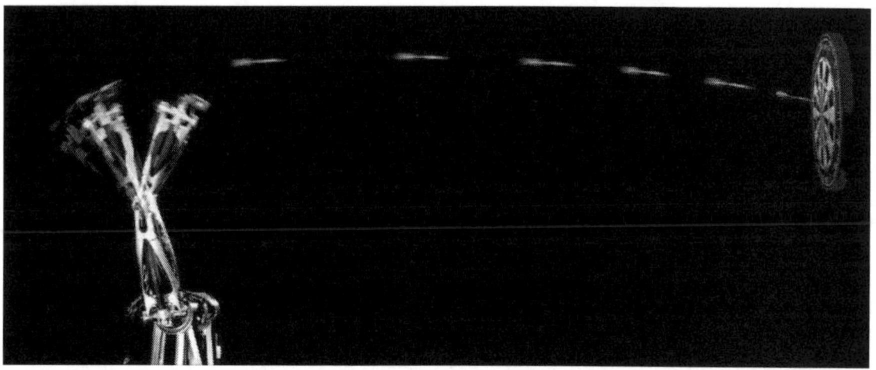

Fig. 19 Airarm throwing dart. Typical power consumption of 600 W, maximum 2,500 W. *Source* Univ.Prof. Dr.-Ing. (habil.) Christoph Ament

artefact as compared with the natural role model selected, the more likely a mere direct copy has been deduced. But copying nature is futile, because its materials and fabrication methods are completely different to man-made products as well as the reason.

References

1. Aschenbeck KS, Kern NI, Bachmann RJ, Quinn RD (2006) Design of a quadruped robot driven by air muscles. In: Proceedings of biomedical robotics and biomechatronics, 2006
2. Nelson N, Hanak T, Loewke K, Miller DN (2005) Modeling and implementation of McKibben actuators for a hopping robot. In: Proceedings in international conference on advanced robotics, 2005
3. van der Smagt P, Groen F, Schulten K (1996) Analysis and control of a rubbertuator arm. Biol Cybern 75(5):433–440
4. Edelmann TH (2008) An arm, inspired by two legs. Des Rep 4(08):52–53
5. Thallemer A, Danzer M, et al. Airarm—an anthropofunctional robot arm with inherent flexibility. In: 3rd mobiligence conference proceedings, 19–21 Nov, Awaji, Hyogo, Japan
6. Dale-Hampstead A (2001) The fluidic muscle by Axel Thallemer. Modern Classics, London
7. Thallemer A (2010) How bionic innovations redefine design tasks. In: Kronhagel C (ed) Mediatecture. Springer, New York

Design2go. How, Yes, No?

Nikola Vukašinović and Jože Duhovnik

Abstract This paper investigates and discusses the opportunities and possibilities of mobile and ubiquitous technologies in the NPD process. During the 2012 EGPR—internationally based NPD process in virtual environment, we made an analysis on technologies and services that were used for the purposes of the course. Our particular interest was, what platforms the students used for different tasks of the NPD process, and are there any mobile alternatives. Since mobile technologies rely more and more on cloud services this opens many other issues as well: intellectual properties rights, protection of personal information, availability of services and information for different participants, standardization of the protocols which should be well considered before any engineering process such as NPD. Our first observations showed that on one side student participants use more and more of different mobile and cloud technologies available, but on the other side there are situations where they still feel much more comfortable when using "old-fashion" technologies, especially when communicating. One interesting fact is also constantly growing wish of students to use the IT web services which they are familiar with despite all necessary IT infrastructure for their work is provided by the course organizers. This is especially important message for the organizers of such courses, to learn how to balance between accepting the opportunities of new internet tools and threats of privacy and control over the intellectual property.

Keywords Mobile technologies · NPD process · Internet services · Product design · Virtual team

N. Vukašinović (✉) · J. Duhovnik
LECAD Laboratory, Faculty of Mechanical Engineering, University of Ljubljana, Askerceva 6, Ljubljana, SI-1000Slovenia
e-mail: nikola.vukasinovic@lecad.fs.uni-lj.si

J. Duhovnik
e-mail: joze.duhovnik@lecad.fs.uni-lj.si

A. Chakrabarti and R. V. Prakash (eds.), *ICoRD'13*, Lecture Notes in Mechanical Engineering, DOI: 10.1007/978-81-322-1050-4_99, © Springer India 2013

1 Introduction

Design has never been so mobile. Smart phones, tablets, net-books, ultra-books, cloud computing, social networks, fast mobile networks, are allowing us to work, create, develop our ideas and share them with the others every second of our lives, no matter where are we or where are the others. But is that really a truth or only an urban myth, triggered by the sellers of mobile devices and media. We will try to answer this question in perspective of the new product design. The NPD process in virtual environment was chosen since it covers a broad range of different activities during different phases of the process, from marketing research to get insight into the project constraints to core engineering tasks which are necessary to finish the project successfully [1].

The question was, whether all product design and development process can really be done on the go with the support of ubiquitous mobile technologies, and there is no further need for an office space, discussion rooms and computer workstations, or maybe we still need them to complete the tasks successfully?

In this paper we will open several point of views for the discussion. What are the phases of product design, which of them could be done mobile and which could not, how we should be concerned about the safety of our intellectual properties and work when working mobile, there are several comfort and ergonomic topics we will touch as well as the meaning of mere human–human interactions.

All this issues have been observed and investigated for several months in real product development process during the last EGPR school process. This is the international school of anew global product design, which gives a participating company five working prototypes and a plenty of fresh ideas every year, and is a never ending source of research material for different design studies every year [2]. In the year 2012 the course connected 38 students from 5 European universities in virtual Multi-X environment, to develop an industrial problem from the idea to the working prototype for participating company from Croatia. Combining the work in virtual multi-X environment, which represents a great learning and realization challenge [3–5] and narrow time constraints which were less than 5 months for the whole NPD process, we got a stimulating conditions for the communication within the virtual teams [6].

The course was divided into four phases of development process which are logical frames of different NPD tasks. In the first—fuzzy front end—phase students gathered information about the company, market, existing products, etc., to set up a design vision which leads into a definition of a design problem to be solved in the following phases. This process consisted of internet data mining, literature overview, interviews and costumer surveys. These activities demanded a lot of work to be done outside the office as well as different ways of electronic communication within team members, industrial partners and external sources of information. The main form of the information during the first phase was digital text and graphics, while the voice communication was used mostly during VC team meetings and some face-to-face interviews.

The second phase is called creative phase, and its final goal is to generate several creative concepts of solutions for the design problem. This phase depends mostly on cooperation within the team members, who were allocated around Europe. Therefore, the main information stream connected different team members and consisted mostly of real time verbal communication (voice and text), text notes and graphical documentation in form of digital pictures and photos.

The third phase represents the detailed design of the selected concept, which consists of CAD modeling, analytical and numerical simulations and analysis and generation of technical documentation. Therefore, this phase demanded daily communication among team members and company representatives to coordinate the activities and exchange the information. Concurrently a comprehensive computer work had to be done for modeling and analyzes. For that reason the amount of digital information exchanged within the team increased respectively.

In the last, fourth phase, all members of virtual teams initiated the process of obtaining and production of components for the prototype realization. At the end of this phase they finally met in person for one week to assemble and present the working prototype. The first part of this phase consisted of communication with the part manufacturers and suppliers while the workshop part of this phase consisted mostly of local tasks, face-to-face communication among team members and local partners, while the exchange of digital information decreased compared to the previous phase.

Due to the virtual nature of the NPD teams most of the work process demanded various means of electronic communication. The methods of communication and the contents of the information to be shared within the teams were in a strong correlation with the phase of the NPD process and each of the tasks requires appropriate information and communication technologies (ICT) infrastructure [7]. However, the results of some researches show that just the availability of the ICTs does not necessary lead to use of them. Therefore, it is essential to establish standards for availability and acknowledgement of communication, which define how dispersed team member will be available for collaboration and how quickly they will respond to the messages [7].

These standards should be specified carefully since other studies showed that the frequency of the communication has a delicate influence on the creativity within the teams. Namely, there exist some optimal frequency of communication within the team, while too low or too high frequency has negative influence on the creativity [8].

However, during the EGPR course these standards were only vaguely specified by course organization—e.g., the use of VC equipment, formal weekly meetings—while the choice of other communication channels depended on team members—e.g., the service for file exchange and instant messaging programs.

Many studies also confirmed that different IT tools have different influence on the market performance, innovativeness and quality of a product, but mostly they foster the results [9]. For example, E-mail communication has been proved to be excellent tool for the engineering project management and information sharing, but it is not that useful as a problem solving tool [10, 11].

The last study [11], not only showed that the IT tools were less suitable for problem solving than for communication purposes, but showed, that web tools in general are more suitable for information sharing, project management and data mining and research than for the creative work. Creative work namely demand more complex services or programs, more computer power as well as the optimal rate of the filtered information flow [8], to establish best condition for creative process and good decision making.

2 Internet Services and Safety of Information

During the course we've been monitoring, what were the services that student used to fulfill different NPD tasks. As it has been mentioned in the introduction, the students had a lot of freedom to establish their own protocols and standards for the (synchronous and asynchronous) communication and document exchange and sharing. Hereby we have to mention that for the purposes for the course, we established the infrastructure for the file depository (FTP server) and teleconferencing equipment for regular VC meetings.

From the Fig. 1, which show the services used for the file exchange, we can see, that despite FTP server, which students used mostly for sending material to the coaches and company, they used Google Documents and e-mail services, which served mostly for their internal communication.

The reason for that is the experience of young generation with these services. They want to use the services which they are familiar with and they know how to use. Therefore, the young generation used third-party online services for the information exchange while many of the students still had to learn how to use the FTP despite the fact that this is an old and most common protocol for the file

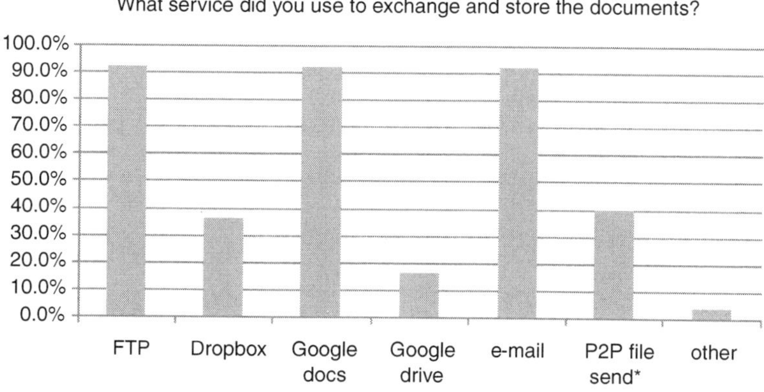

Fig. 1 The services, students used for document transfer. (* peer-to-peer file send using Skype, Google Talk, etc.)

transfer which works behind many cloud services as well. They also found another advantage in Google Docs, which is the possibility to concurrently work on the same document by several team members, while communicating over some synchronous communication channel.

The other advantage of online cloud service (e.g. Google docs, Dropbox, etc.) is also in accessibility to the documents through any web browser or special applications, which makes the documents and service independent on operating system and hardware platform and specifications. This fact together with the availability of highly portable IT devices enhances the possibilities to move the NPD out of the office if necessary.

When we check the services, which were used to communicate within the team (Fig. 2), we notice similar pattern as before. Namely, besides VC equipment, provided for formal meeting of the teams and coaches, all teams established their own communication channels. Here again, most of the students used e-mail as a tool of asynchronous communication, and Skype as a mean of synchronous communication. Surprisingly, the use of Skype conferencing was almost 100 % which is more than VC equipment, and there were no real alternative, despite of the availability. Google Talk was following with 63 %, while social networking communication channels like Facebook chat were almost unattended.

Here we need to mention also the use of the telephone as a communication tool. With the development of smartphones, a telephone is no longer a tool for the classical voice communication but can serve also as a platform for chat and VoIP talks. 55 % of the students said, that they used the internet access on their phones, which only confirms our statement. However, classical phone service was still used, mostly for the local research work and communication with local team members.

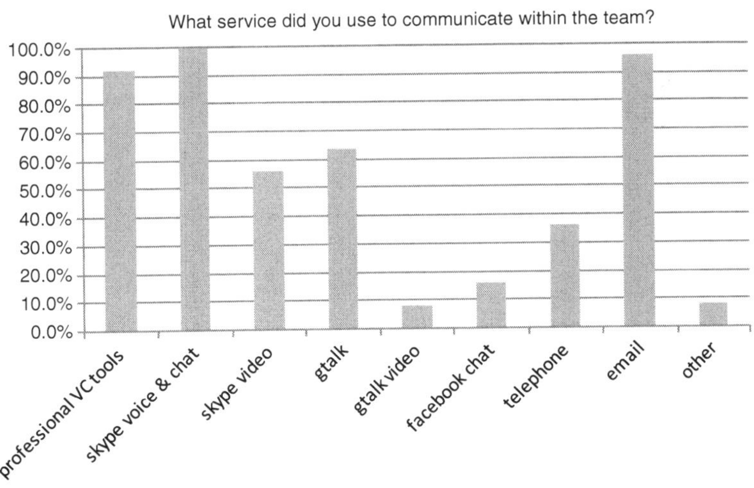

Fig. 2 Communication services used for the purposes of the EGPR 2012 course

If we return to the e-mail, we also observed the origin of e-mail accounts. Almost 90 % of the participants (students and even coaches) used their private e-mail addresses, created at some of the web e-mail providers, among which Google took again the biggest share of approximately 80 % (Fig. 3). Even in the previous years we observed the similar phenomena, when the students applied to the course with their faculty e-mails but in a few weeks they usually switched to their Google mail account, due to the services which Google provide and condition with the use of their e-mail: e.g., Google groups or Google documents.

Since, every third-party IT service always comes with its License agreement and Terms of use, we asked participants of EGPR course, if they normally read those conditions of use. More than 70 % of them answered negative (Fig. 4). That means that most of the course participants don't even know to whom they share their information and how this information will be treated by potential third-party persons.

As the course involves the cooperation of an industrial partner, the participants work also with some company's delicate information. In the year 2011 the company representatives therefore demanded not to use Google services for the course tasks, while the signing the NDA agreement prior the course start is already a standard procedure at the beginning of this course.

Therefore it will be necessary to establish an educational cloud service for the exchange of the information for the purposes of this course, and the service should allow concurrent work on the same document by several participants, as an alternative to free web-based services.

The results of this research show that it would be better if the course organizers prepare the IT protocols and standards which can be used by the students for this course, instead of teams doing this by themselves. This might demand of the

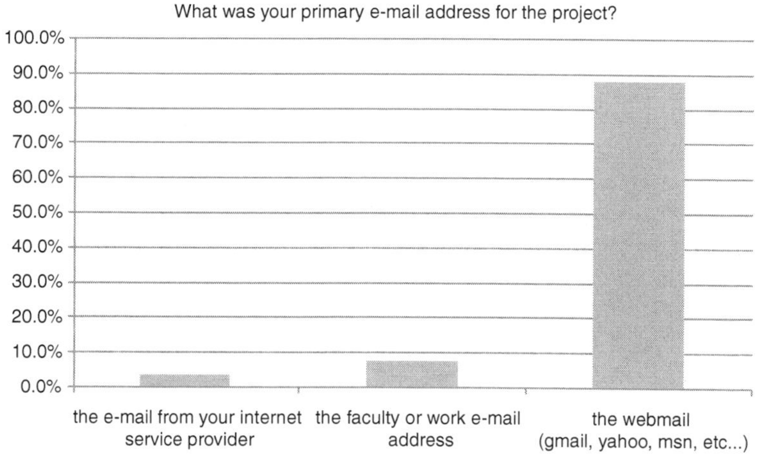

Fig. 3 E-mail provider for the primary e-mail addresses used during the course

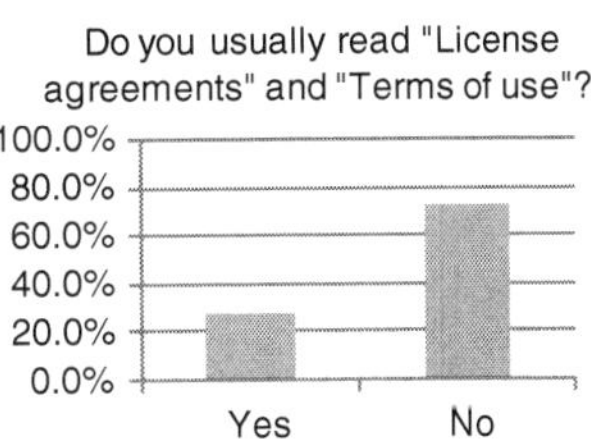

Fig. 4 Most of the participants don't read License agreements and terms of use, when using internet services

students some additional learning, but also more effort for the organizers to establish an IT system which would not limit the communication of the students, regardless to their location and electronic device they use.

3 Electronic Tools for Different NPD Activities

The other subject of our interest was what devices participants used for certain tasks of the NPD process. The tasks were divided into three major groups: *communication*, which included all flow of the information among team members: file and data exchange, synchronous and asynchronous communication regarding the project as well as informal communication which is essential for the team building and the level of trust among team members [3]; *Creative work*, which included all task connected directly to the development of the product: idea generation, concept selection, detailed design and prototype realization; *Research work*, which included all fuzzy-front-end process, SWOT analysis, and gathering the information, necessary for the creative phase. It included online surveys, internet research, on-site research and interviews with the companies and customers. The results are graphically represented in Figs. 5, 6 and 7. From our survey we noticed that participants used mostly their laptops, PCs and mobile phones, while they mostly didn't possess and use smaller laptops and tablet PC yet.

Figure 5 shows that participants frequently used for their communication tasks mostly their laptops (75 %), PCs (40 %) and also their mobile phones (together up to 45 %).

Figure 6 shows the frequency of use of different devices for the creative tasks. There was no need for the mobile phones, which seem to be inappropriate devices for serious creative work in the creative phases of the NPD. Even the use of laptops and PCs is slightly smaller as for the communication, which suggests that the creative process is not a process which could be much fostered by the use of electronic tools, since it is mostly the function of personal cognition and skills. The creative tasks depend on the results of communication and research work. And consist of three interfering and iterative operations: information filtering, contemplation and creation. Especially contemplation step is many times underestimated, but it is essential part of every creativity process, and it demands reduced information inflow to be successful. This is probably also one of the reasons for

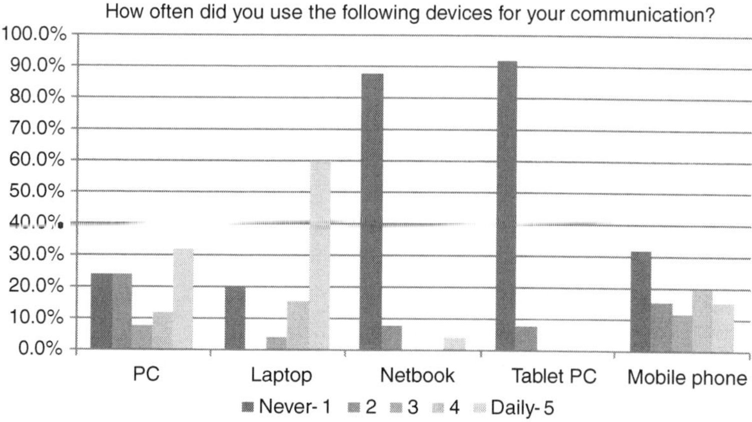

Fig. 5 Frequency of use of different electronic devices for communication purposes

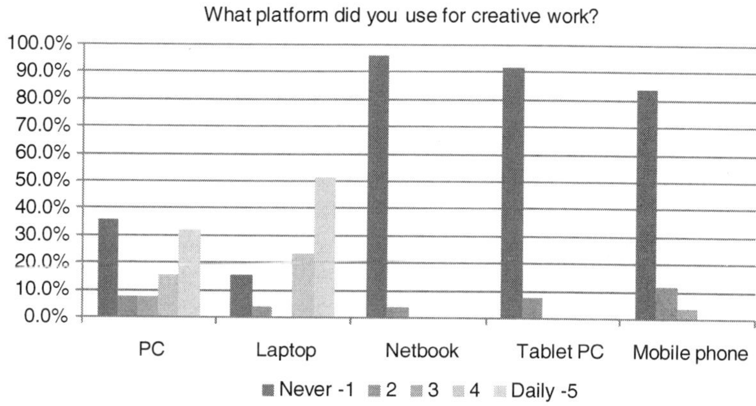

Fig. 6 Frequency of use of different electronic devices for creative activities

reduced use of all tools in the creative phase, but this would be good research topic for the future.

Again, slightly different results we got when we investigated the use of different devices during the research work as show on Fig. 7. The use of all electronic devices was even higher than compared to communication tasks, only there was slightly smaller use of mobile phones, which were mostly used for the interviews with the industrial partners and customers.

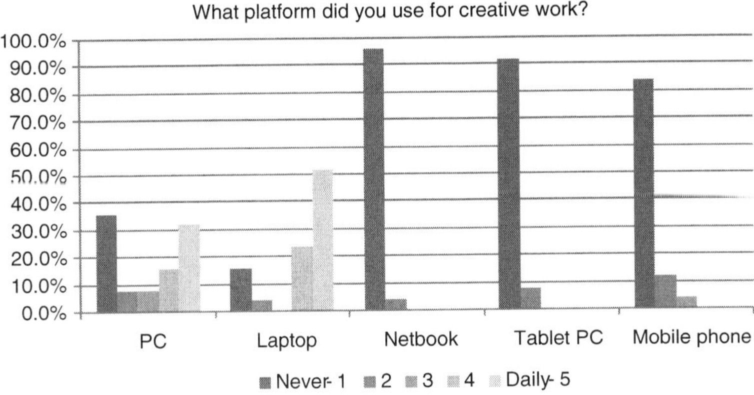

Fig. 7 Frequency of use of different electronic devices for research activities

4 Discussion

Hence, we can conclude that different devices do not have the same potential for different NPD tasks. Unfortunately, we didn't have good insight into the possibilities of the use of smaller versions of laptops and tablet PCs.

We believe that netbooks are not commonly used, since most of the participants already own the laptops, while netbooks and tablet PCs are usually bought as a second personal device. Since most of the participants have engineering background, and this was not their first NPD process, they chose their devices on previous working experience and needs.

Tablet PCs are a new market hit and still establishing their place in the engineering fields. Their use in the engineering future is supposed to increase but in the first phase mostly for the management, research and communication tasks. The use of tablets for the creative engineering work is not expected to rise in the near future for the same amount as for the other tasks for two reasons: The first reason is that the computer power and capacity of tablet PCs is still far behind performance of professional laptops and desktop workstations, while the second reason is the limited ergonomic performance of human-device communication interface: the screen size is limited by portability, while the input is done mainly through the touch display, which is unsuitable for an intensive long-time work.

From the research we can conclude, that most of the NPD work during all phases was still done within the office. However, some tasks, especially information digging and managing tasks are already moving out of the office for information to be available to other team members in the shortest possible time. Hereby, it is important, especially for the real industrial environments, to have established a proper network of communication and data storage and sharing protocols, which would prevent any leakage of delicate information to unauthorized parties, and allow users safe work regardless on device or platform they use. This will be priority also for the future of the EGPR course, since industrial partners play a crucial part of the process. The first step will include testing and implementation of different educational PDM systems. The selection of the best system will depend partly on the results of this research and the results of researches mentioned in this paper.

References

1. Fain N, Kline M, Duhovnik J (2011) Integrating R&D and marketing in new product development. J Mech Eng Stroj Vestnik 57(7/8):599–609
2. Žavbi R, Duhovnik J (2007) Experience based on six years of the E-GPR course, international conference on engineering design (ICED07)
3. Vukašinović N et al (2009) Education of NPD in virtual multi-x environments, international conference on research into design, ICoRD'09
4. Žavbi R, Tavčar J (2005) Preparing undergraduate students for work in virtual product development teams. Comput Educ 44:357–376
5. Fain N, Moes N, Duhovnik J (2010) The role of the user and the society in new product development. Vestnik 56(7/8):521–530
6. Darrel SFC et al (2012) Bringing employees closer: the effect of proximity on communication when teams function under time pressure. J Prod Innov Manag 29(2): 205–215
7. Montoya MM et al (2009) Can you hear me now? communication in virtual product development teams. J Prod Innov Manag 26:139–155
8. Leenders RTAJ, van Engelen JML, Kratzer J (2003) Virtuality communication, and new product team creativity: a social network perspective. J Eng Technol Manage 20:69–92
9. Durmusoglu SS, Barczak G (2011) The use of information technology tools in new product development phases: Analysis of effects on new product innovativeness, quality, and market performance. Ind Mark Manag 40:321–330
10. Wasiak J et al (2011) Managing by e-mail: what e-mail can do for engineering project management. IEEE Trans Eng Manag 58(3):445–456
11. Farris GF et al (2003) Web-enabled innovation in new product development. Res Technol Manag 46(6):24–35

Integrating the Kansei Engineering into the Design Golden Loop Development Process

Vanja Čok, Metoda Dodič Fikfak and Jože Duhovnik

Abstract One of the main tools for translating customer's psychological feelings and needs into product's design domain is Kansei engineering (KE). It is important that analogy of Kansei engineering enters into the product development process on time and fulfills the role of complementary methodology which gives an emotional value to the product. This paper describes how Kansei engineering works, what are its subgenres and options of integration in following design processes. We will describe standard procedure of Kansei engineering type 1, design process VDI 2221 and Global product or services development process called "Design Golden loop" (DGL). The detail description—research on possibilities of integration of Kansei engineering type 1 framework into the design Golden loop process will be described.

Keywords Kansei engineering · VDI 2221 · Design process · Customer · Global product or service development process

V. Čok (✉) · J. Duhovnik
Lecad Laboratory, Faculty of Mechanical Engineering, University of Ljubljana, Slovenia Askerceva 6 SI-1000 Ljubljana, Slovenia
e-mail: vanja.cok@lecad.fs.uni-lj.si

J. Duhovnik
e-mail: joze.duhovnik@lecad.fs.uni-lj.si

M. D. Fikfak
Clinical Institute of Occupational, Traffic and Sports Medicine, University Medical Center Ljubljana, Slovenia, Poljanski nasip 58 SI-1000 Ljubljana, Slovenia
e-mail: metoda.dodic-fikfak@guest.arnes.si

A. Chakrabarti and R. V. Prakash (eds.), *ICoRD'13*, Lecture Notes in Mechanical Engineering, DOI: 10.1007/978-81-322-1050-4_100, © Springer India 2013

1 Introduction

Today, product design for the global as well as for the local market is often very complex. The reason for that are contemporary industrial products which have to fulfill more and more user requirements. When a product reaches desired functionalities the decision of purchase is based on a customer impression, perception and his emotions about the product. This becomes very important when the customer has to choose from different products which fulfill the same functions. According to nowadays complexity of producing a competitive and user friendly product many research methods of integrated product development processes have been revealed. As mentioned before one of frequent used tool for customer's emotional impression detection and translation into the product features is Kansei engineering. While Kansei engineering covers an emotional aspect of the product other product development processes as VDI 2221 [1] and Design Golden Loop [2] focuses on holistic product design procedure.

1.1 Description of Kansei Engineering (Literature Review)

Kansei engineering was founded by Nagamachi at Hiroshima University about 40 years ago. It is defined as a tool for translating the customer's Kansei into the product design domain. The term Kansei used in Kansei engineering refers to an organized state of mind in which emotions and images are held in the mind toward physical objects such as products or the environment [3].

Kansei is an individual's subjective impression from a certain artifact, environment or situation using all the senses of sight, hearing, feeling, smell, taste and the sense of balance as well as recognition [3, 4]. The consumer Kansei is difficult to grasp and complex to measure. Nagasawa states that "autonomic nerve reflections are not the Kansei itself, but only correspond to the Kansei". This makes physiological measuring methods to an indirect measuring method [3]. The customer's Kansei has a diversity of expressions, from psychological to psychophysiological so both we can measure with different techniques and tools [5]. Lokman divided existing Methods of User's evaluation as Self-Reporting Techniques which are gathering data about feelings for product by asking people to report their feeling in adjectival words [5]. She also listed the following self-reporting techniques: Semantic Differential Methods, text completion, free following speech, Conjoint Analysis; Physiological and sub-conscious Method: video-recording, eye tracking cameras, Electro-Myo-Graphy (EMG), Electro-Encephalo-Gram (EEG), Electro-Cardio-Gram (ECG), Semantic Description of Environments (SMB), Quality Function Deployment(QFD), PrEmo, Kesoft [3], Kansei engineering (KE) [4], etc. There are few methods known in Kansei engineering for capturing consumer's internal sensation. It can be measured physiological or behavioral responses using electromyography (EMG), heart rate, electroencephalography (EEG), event-related potential (ERP), Functional magnetic resonance imaging (fMRI) or expressive by observing body or facial expression [5, 6]. Furthermore, psychological responses can be measured by semantic differential scales method, different personality tests or other questionnaires techniques [5]. There are six proved and tested types of Kansei engineering: Kansei engineering type I as category classification, type II as Kansei Engineering Computer System, Type III as Kansei Engineering Modeling, Type IV as Hybrid Kansei Engineering, type V as Virtual Kansei engineering and type VI as Collaborative Kansei engineering [5, 6].

1.1.1 Kansei Engineering Type 1 (Category Classification)

Our interest lays in Kansei engineering type I (KE1) which is described as "fundamental" technique using the process-ruled means [6]. This technique is also known by category classification or definition of targeted concept for new developing product to the design characteristic. We will pay attention on its basic procedure. The KE1 has his own process which is covered in ten steps. A following procedure is shown in Fig. 1.

At a beginning of the process engineer should pay an attention to R&D manager instructions and client company's CEO requirements. He must be informed with the company new product development strategy and understand each step. At this point it is important to gather and select a priority requirements and solving

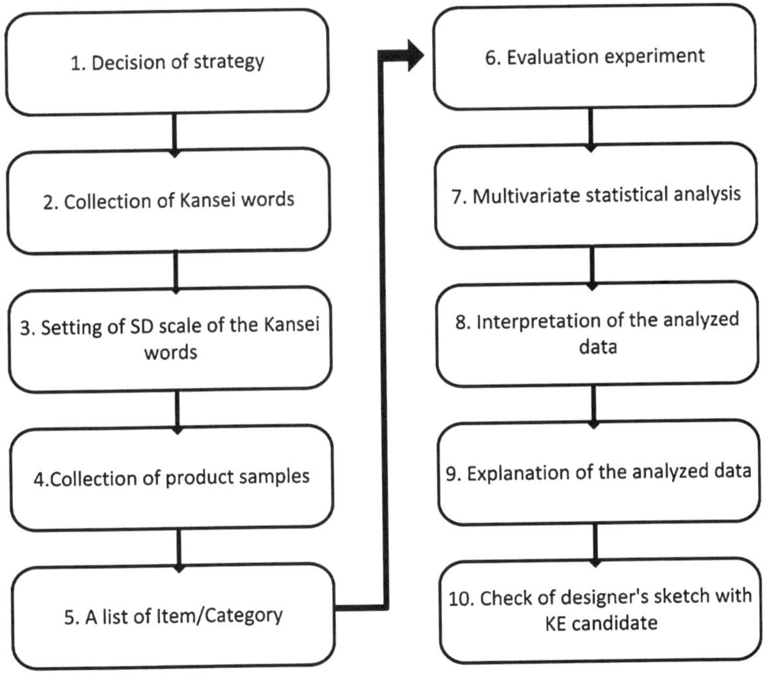

Fig. 1 The process of Kansei engineering Type I [13]

principles together. After clarifying those tasks an engineer collects Kansei words related to the product domain. Kansei words are adjectives, nouns, verb or even sentences. They are often collected from books, magazines, newspapers or other sources. First Kansei words must be collected and later to follow a selection of relevant and important ones. Later on a development of a psychological measurement scale devised by C.E. Osgood called a "Semantic differential scale" is followed. This method is used to make clear the psychological language structure [6]. Positive and negative words are arranged on both sides of a horizontal line. This kind of arrangement allows the respondent to value Kansei word that "feels" like a physical property of a product. There can be used different scales from 5, 7, 9 to 11-scale rates. Collection a number of 20–25 samples it's sufficient. There should be gathered samples of similar product as we plan to create. Later on the preparation of a list of items and categories related to the final design specifications is followed. For example, the evaluation experiment means, that a 7-point semantic differential scale is used for evaluation of each sample. From received data a multivariate statistical analysis can be done. Within it is used a Principal Component Analysis (PCA), Factor analysis and Quantification theory Type I (QTI). The next is the interpretation of the data and integration them into the product design properties. As the most important issue, Nagamachi exposed the interpretation of the data to designer. Designer must understand the final data interpretation correctly. His task is to create an emotional design based on those data. At the end of the process the engineer should do another check, to make sure that the product designer's interpretation was correctly interpreted and new design fits the customer's emotion. Otherwise a product designer should repeat the procedure and create a new proposal for product design [6]. KE1 is a standard procedure which aims to develop or improve products and services by translating customer's psychological feelings and needs into product's design domain [4]. However, there is a lack of information where occurs the integration of KE1 into the traditional product development process. Based on researches, many of them implement a general understanding, thinking processes, ontology, etc. [5, 7, 8]. However, research requirements regarding the general Kansei engineering procedure should be taken into consideration. Limited knowledge about combining different KE1 and DGL process conceals the new solutions and possibilities.

1.2 VDI 2221(Verein Deutscher Ingenieure)

VDI 2221 as a Product Development Process belongs to the history of design research in Germany carried out in this area. The name comes from Association of German Engineers. It is worth mentioning that all these research efforts resulted in VDI 2221 are a synthesis of various researches carried on the integrated product development. Guideline VDI 2221 is most often used by European schools in Central and Northern part of Europe, among others in German speaking countries. However, it is an approved standard product development process [9]. Its process structure is designed in a way that solves complex design tasks which are often complicated and unclear. This was solved as a systematic arrangement of the process which does split off on several development phases. This makes clearer review and evaluation of individual phases. We can easily return to the previous phases if a predicted model turns to be unsuccessful.

Methodological steps of construction work on the guidelines in construction process are:

- task definition
- defining the product conception through to a solution constituted,
- develop and result in the form of design solutions and
- detailed construction and solution of technical documentation.

Description of VDI 2221 Procedure

At the beginning of the design process search for solutions must be the task which is chosen to be solved a well-based. It is necessary to identify and meet the requirements that affect the solution. These tasks are categorized by their importance in the written document in which the task is broken down and described to the details. However, considering that the task meets the objectives of business and manufacturing opportunities, we approve the conceptual phase. Concept (draft) indicates a new, task-concept solution of the problem, which arises from the product, which is based on a new working principle. Even the conceptual phase is divided into several smaller working steps. Usually, we get a number of different solutions for each step, among which the best one is selected. Individual solutions are evaluated according to criteria contained on the list of design requests. If selected concept solution meets the criteria specified on the list of design requests, work can continue in the next work step, that is the design phase. However, if the solution does not correspond to the concept, a few preliminary steps are repeated to get better starting point for finding appropriate final solutions. The concept solution is only an approximate solution without design details which have to be solved during the design phase. Design phase represents therefore more or less a completion of a device or machine parts with the selected materials. Before that a plan procedure for production must be made. Design phase is divided into coarse and fine or refined design that can be economically evaluated. Based upon evaluation we select the best variant, and approve the functional decomposition phase. In the breakdown phase we make the product documentation for all components, after which the workshop produce a device or machine that gets to this final form.

Figure 2 shows the different phases of the overall workflow in the search for solutions till the choice of tasks to product documentation. In this respect, each stage of design decisions emphasis a choice of best solution. The advantage of finding new solutions by the methodical steps is that the solution is produced gradually, and anything of significance is not dropped. An experienced designer used in the construction of a new product numerous working steps unconsciously, often by combining some or even skipping. Described working steps are particularly suited to finding new solutions to new technical problems. Then we have the conceptual stage to find a new working principle is based on a new natural law.

VDI 2221 design process is based on the evaluation of each work step at a specific stage. If acquired solution does not meet the necessary criteria we should repeat the step, as shown in Fig. 2. Upgrade process is done in "Design Golden Loop—DGL" process [2], which is an iterative process or decision making procedure, where the decision for final product is made. An integrated process for developing a product or service is shown Fig. 3.

Description of "Design Golden loop" Process

DGL can be used in the Design Process (DP) by the Development and Design Process (D&DP). General paradigm of the DGL is based on the iterative steps in the design process [2]. A comprehensive D&DP model with the addition of DGL was first developed in 2001. There are three key parts of the whole D&DP. At first a goal definition is done, which covers a holistic process from abstraction to the goal definition. Second phase reveals key elements of process planning where an individual activities and team definitions are planned. Mentioned phases are parts of D&D process. Third phase represents construction process. Goal definition depends on activities analysis. Regarding on defining activities the design process

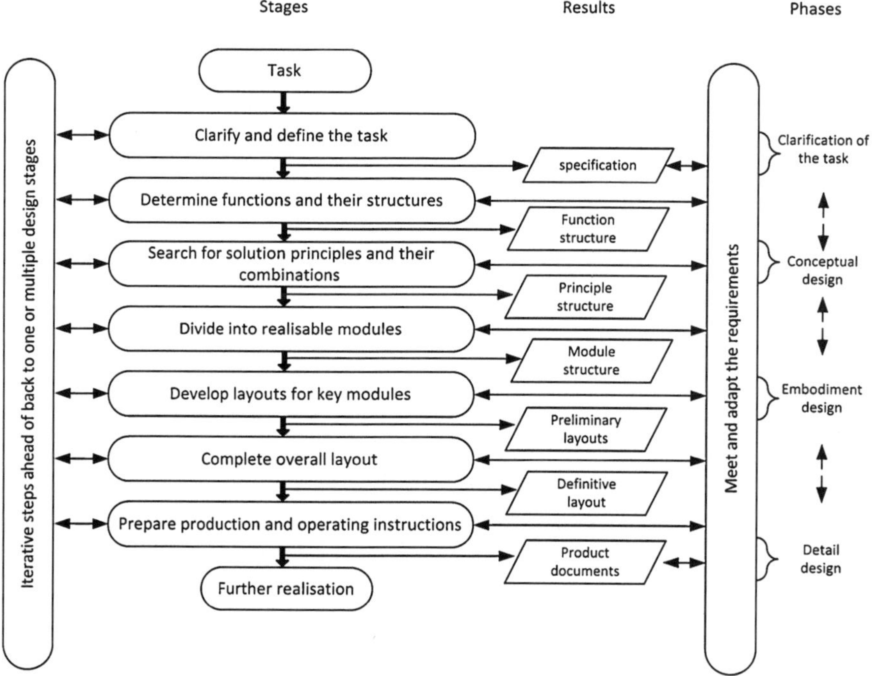

Fig. 2 Design process as guideline VDI 2221 (verein deutscher ingenieure)

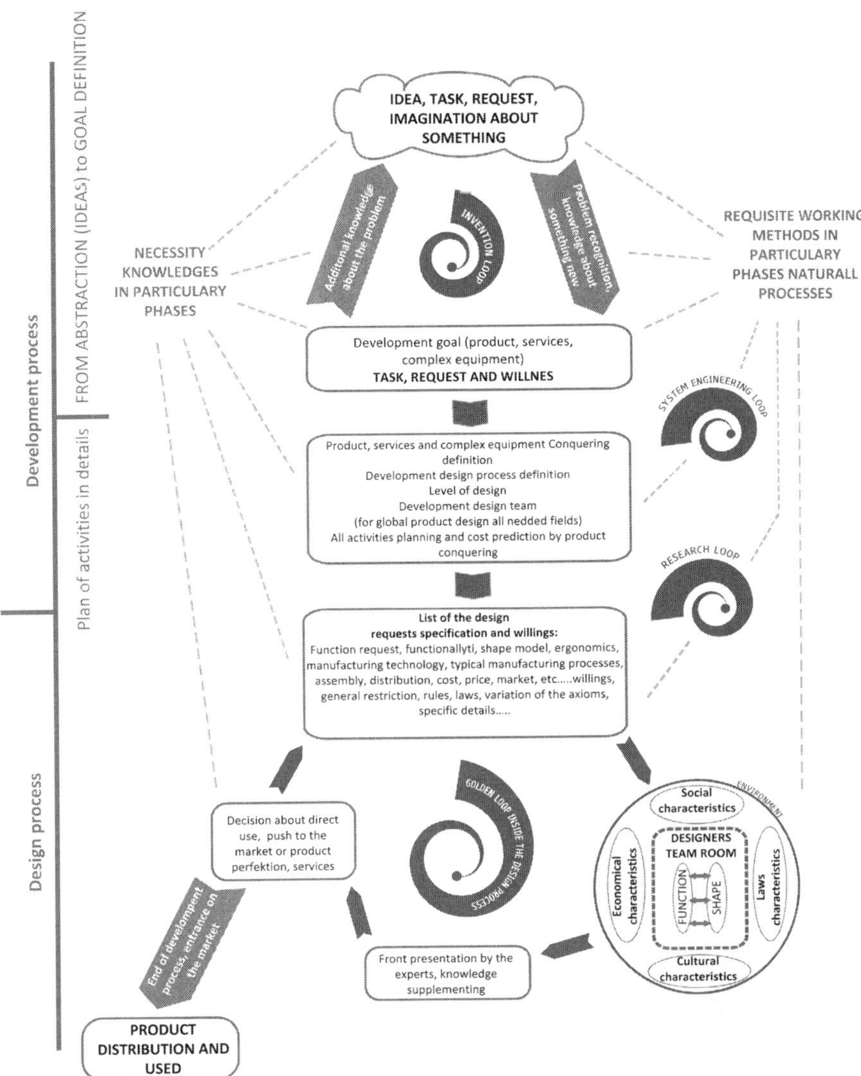

Fig. 3 Global product or services conquering process or "GDL" [2]

is more purified and gives us more clearly idea for development. The product development phase is divided into two tasks: the first one is goal definition and the second one is request, desires and tasks of the D&D process. Since the two jobs are closely connected, they are not performed sequentially or separately [2]. They take part of the iterative process of first development loop. Initial iterative steps are used when goals and tasks are defined. This is the second phase of the D&DP starts when the goal is very clear and the process must be defined more clearly and detailed. Cost analysis is also included in this part. However, it is of vital importance for a comprehensive monitoring of the economical side of a project. DGL was named since the shape of the product, resembling the end product, appears for the first time in front of the designer. Designers team room play an important role in defining the relationship between function and form, since both links. This way the difference between the aforementioned three loops in the development process and essential loop that brings a product into real life [2, 10–12].

2 Framework of Kansei Engineering Integration into Design Golden Loop Process

The starting point of a procedure shown at Fig. 4 is a methodological analogy of VDI 2221 design process. A previous description has given us more details on operations from task determination till final phase. Systematic structure of design process is based on evaluation or estimation of each work step, so to avoid mistakes the best solution at some stage should be chosen. In addition to VDI 2221 design process or guidelines aims to describe multiple design stages there are also mentioned the requirements which are meet and adapt during the procedure. This reference mostly refers to the product functionality requirements. There is a definition of requirements in VDI 2221 design process but they don't give us clear information on how customer needs and requirements should be met and adapted in product design. As already mentioned, improving products requires knowledge about how product properties affect customers. Anyway if the product fits to the customer's emotional needs there is more likely that he will decide to buy it. Product's appearance and added value through the customer esthetics preference within his emotional needs should be considered parallel with the development of features and product functionality. Therefore, it is necessary to integrate Kansei engineering methodology which aims to design and develop product/services that match customers' emotional, psychological feeling and needs [7] in traditional product development process as it is VDI 2221 or modified DGL procedure.

A model of Kansei engineering type 1 (KE1) integration into traditional product development process was made after a study of those three product development methodologies later on observation and comparison. Model of integration is shown on a Fig. 4. It represents steps where integration should be implemented. Nevertheless VDI 2221 design process is based on a strategic arrangement which guide us through iterative steps or multiple design stages till further realization in comparison with DGL process its final phase is more justified and upgraded. Therefore the integration should be done in DGL design process taking into account VDI 2221 procedure principles. According to the KE1 methodology which describe a basic procedure of KE1 the implementation in DGL design process should be done for a beginning in a first design stage where we define the task and product specification. Later on customer emotional needs and requirement

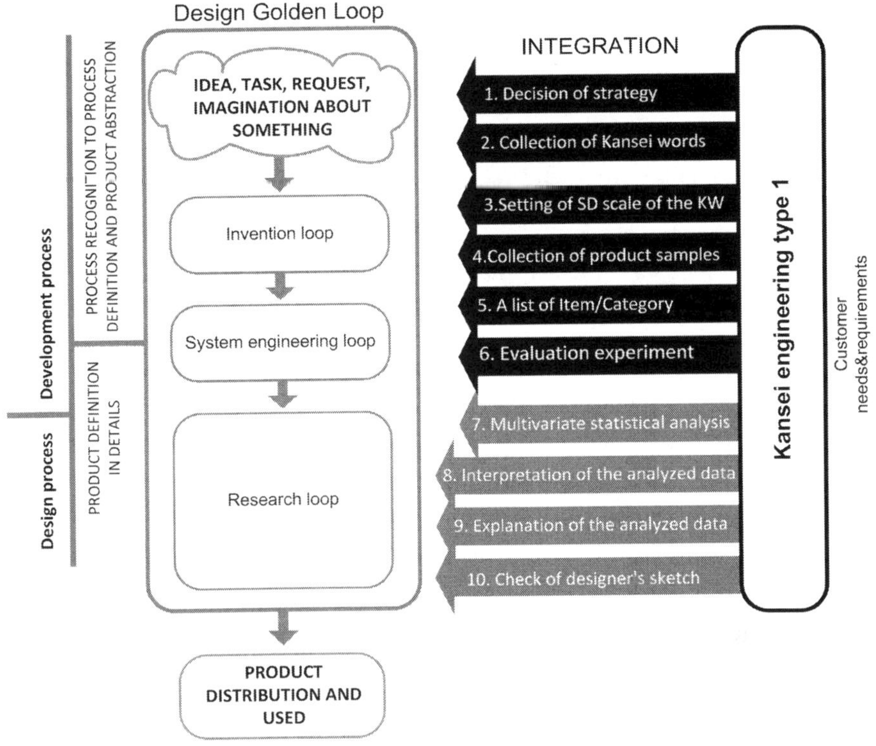

Fig. 4 A Framework of Kansei engineering integration into traditional design process. VDI 2221 design process was modified and supplemented by Duhovnik [2]. Description of Kansei engineering type 1 integration is revealed in a structural model

will be met already in the early phase of product development so the product design procedure will be adjusted according to this aspect. KE1 could be used as a tool which enables purifying the idea by means of requirements for clearer goal definition (in aspect of product visual appearance). As already mentioned KE1 procedure should be integrated from the first step of DGL design process where idea, task, request, imagination about something appears. Simultaneously when a decision for a product which will be developed is made a development strategy is followed. KE1 procedure should be considered when problem recognition appears and a new knowledge about something is needed. As soon as we have a clear goal of developing product or service, when task request and willings and development strategy are known we start collecting Kansei words. In this segment definition of product or service properties is made. The aim is to determine potential properties or design elements of the future product or service, which include the collection of existing product (20–25 samples), creation of new concepts, identification of potential customer and company images and priorities as well as the definition of properties, elements (i.e., attributes or characteristic) and design categories [13].

Therefore, a list of item/category must be made. The list is related to the final design specifications. Item (for example: color, shape, size, roundness) implies the design item of the sample product, and category (for example: yellow, green, red, blue) means the detail of the design item. After a selection of Kansei words is made we develop a semantic differential scale for evaluation experiment. The subjects receive an instruction and evaluate each sample with the 5, 7, 9 or 11-point SD scale of Kansei words. By this approach design assess patterns and design features we are looking for a product that will suit the user. The evaluated data are analyzed using a multivariate statistical analysis. Then we interpret the data using different statistical tools and integrate them into the product design properties. Afterward we explain interpretation data to a designer which is a member of a development design team in the nearly end research loop of DGL process. All the activities that are happening through the main three development phases in DGL process are intertwined with KE1 procedure. The development of function and functionality cannot be considered divided from form and visual characteristics which emotional value should be given inside the product development process like in our case DGL process.

3 Conclusion

This study showed a structured model of Kansei engineering integration into traditional design process as an proposal worth of future work. Therefore, the authors will continue investigating also other options of customer requirements and emotional needs integration in product development processes such as VDI 2221 and DGL process. As seen in previous descriptions KE1 procedure could not be an independent tool for new product development but is equivalent with DGL and other design processes. Integration of KE1 procedure is done simultaneously as other processes at different iterative stages because its focus is in customer's psychological feelings and needs and it's translation into product's design domain. It is hard to determine a clear border where KE1 should take a part in DGL process especially after a first design stage is carried out. Further research is required to develop a deeper understanding of a specific activities occurring in the different design stages and how integrated methodology influence a holistic traditional product development process.

References

1. VDI (1993) VDI 2221: systematic approach to the development and design of technical systems and products. Beuth Verlag
2. Duhovnik J, Balić S (2004) Detail functionality analysis using design golden loop. In: 4th International seminar and workshop, EDIProD

3. Schütte S (2005) Engineering emotional values in product design Kansei engineering in development. Dissertation 951, Linköping studies in science and technology
4. Nagamachi M (2002), Kansei engineering as a powerful consumer-oriented technology for product development. Applied Ergonomics, Elsevier, London, pp 33289–33294
5. Lokman AM (2009) Emotional user experience in web design: the Kansei engineering approach, Doctoral Thesis
6. Nagamachi M, Lokman AM (2010) Innovations of Kansei engineering. Taylor&Francis, UK
7. Dahlgaard JJ, Schütte S, Ayas E (2004) Kansei/affective engineering design: a methodology for profound affection and attractive quality creation. TQM J 20(4):
8. Levy P, Nakamori S, Yamanaka T (2008) Explaining Kansei design studies, conference of design and emotion. Kansei, Hong Kong
9. Tomiyama T, Gu P, Jin Y, Lutters D, Kind C, Kimura F (2009) Design methodologies: industrial and educational applications. CIRP Ann Manufacturing Technol 58:543–565
10. Zadnik Ž, Čok V, Karakašić M, Kljajin M, Duhovnik J (2011) Modularity solutions within a matrix of function and functionality (MFF). Tehnički Vjesn Tech Gazzete 18(4):471–478
11. Zadnik Ž, Karakašić M, Kljajin M, Duhovnik J (2009) Function and functionality in the conceptual design process. Strojniški Vestn J Mech Eng 55(7/8):455–471
12. Tavčar J, Benedičič J, Duhovnik J, Žavbi R (2004) Creativity and efficiency in virtual product development teams TMCE. In: Proceedings of the 5th international symposium on tools and methods of competitive engineering, pp 425–434 Millpress, Lausanne, Apr 13–17
13. Nagamachi M (2010) Kansei/affective engineering. Taylor&Francis, UK

Design and Development: The Made in BRIC Challenge

Luciana Pereira

Abstract The study examines the level of design and development in three multinational subsidiaries working on fluid power industry in Brazil. Some authors say that decentralization of R&D activities by MNCs is an important source of innovation for BRICS countries. Using a case study method this paper has found that the type of design and development activities performed by these companies are more related with redesign than original and innovative activities, which would demand more complex system approach. The answer for increasing innovation levels might be in domestic firms and not in multinational ones.

Keywords Design · Innovation · Fluid power · BRIC

1 Introduction

Recent literature on global research and development (R&D) location decisions by multinational firms (MNCs) has emphasized the importance of emerging economies like India, China, and Brazil as R&D center. The so called BRIC is an acronym for Brazil, Russia, India, China, plus South Africa, the leading emerging economies. Together, the five countries represent almost 3 billion people, with a combined nominal GDP of US$13.7 trillion [1].

Despite of all their cultural differences, Brazil, China, and India have followed a similar pattern of industrialization, which was based on foreign direct investment

L. Pereira (✉)
Center for Engineering, Modeling, and Social Applied Sciences, Federal University of ABC, Rua Santa Adélia, 166 A646-1, Santo André, SP, Brazil
e-mail: luciana.pereira@ufabc.edu.br

A. Chakrabarti and R. V. Prakash (eds.), *ICoRD'13*, Lecture Notes in Mechanical Engineering, DOI: 10.1007/978-81-322-1050-4_101, © Springer India 2013

attraction combined either with Import-Substitution Industrialization Model (India and Brazil) or Export-Oriented Industrialization Model (China) [2]. However, nowadays, instead of the traditional role of adapting products and services to local market conditions and supporting MNCs manufacturing operations, some claims that investments have being increasingly motivated by tapping into worldwide centers of knowledge as part of firms' strategies to source innovation globally [3].

In spite of this, relatively little attention has been given in this literature to the role of product design and product development in these emerging markets. There are a number of partial exceptions that have suggested that the new product development is becoming the fastest growing segment in India, Brazil, and China [4, 5]. Meanwhile, Manning et al. [6] suggest that this growth in innovation off shoring is driven by increased globalization of markets for technology and knowledge workers.

This paper is part of a large project that seeks to understand the quantity and quality of product design and development activities in some multinationals' subsidiaries. All over BRIC countries, governments have proposed policies to foster innovative activities inside their industrial park. This can be fully checked at Brazil Maior Plan, India National Innovation Council (NInC), and the Chinese 12th Five Years Plan.

The purpose of this paper is to gain a better understanding of MNCs regarding more complex activities. Are MNCs companies really relocating their R&D facilities to these countries? If this is the case, we may expect that design should be strategic because it has a key role both in product development and technological innovation. In order to reach this goal and learn what is going on inside the industry, we have conducted an exploratory case study in three MNCs located in Brazil working with fluid power systems technologies.

The importance of fluid power systems lies in the fact it provides key applications for industrial, mobile, and aerospace, all important sectors for the mentioned countries. For our data collection, we have used technical visits, and interviews. As a parameter we studied a Swedish group that work close with fluid power industry, so we can compare the type of activities they have performed with the Brazilian experience.

The case study has shown that more complex activities related with design or product development is far from happening. Different from what is found in the literature, there is no evidence that MNCs are moving R&D Centers because of knowledge pool. The reason seems to be related more with old strategies such as cutting costs and take advantage of local resources, including skilled and cheap workforce.

The contributions of this paper are twofold. First, it offers the opportunity to shift the innovation debate from the policy level to the position of those who actually are in charge of making innovation happen. In this case, designers and product developers solving problems at the firm level. Second, it brings together an interdisciplinary perspective in a multicultural context.

The paper is organized as follows: Section 2 presents a brief review of the literature concerning the *relationship* between design and innovation. Section 3 presents methodological tools and the triangulation data analysis as the research design. Section 4 discusses some important initial observations identified with the case. Section 5 delves into conclusion including directions for future research.

2 Design: The Materialization of Innovation

In this section we reflect upon the major themes we are interested in gaining more insight into design and development capabilities and the relationship between R&D decentralization. If it is true that MNCs have relocated its R&D centers in some emerging countries, we expect to see different design skills been demanded by them.

Design is defined as the core of innovation, the moment when a new object is imagined, devised, and shaped in prototype form. Design clearly plays an important role in the realization of the radical invention as an innovation [7]. In particular, systemic innovations need a great deal of design co-ordination in development and commercialization because systematic adjustments to other parts of the system have to be made [8].

But design is also important in the period of ‘swarming secondary innovation’ via competing designs, in product differentiation and reliability, and in price competition via the efficient use of materials and design for ease of manufacture. Thus design is important throughout the industry life cycle and at different stages of economic upturn and downturn, but plays a different role at each stage.

Over the years, BRIC countries have excelled in their manufacturing capabilities and today they are responsible for one fifth of the global manufacturing value added (MVA), or the relative share of value added to gross domestic product by manufacturing [9]. Although MVA indicates a country’s level of industrialization, it does not account for the technological structure of production neither its level of innovation. On the other hand, according to the same source, the design and production of high-technology manufacturing goods still been dominated by advanced countries.

In short, although manufacturing is a very important skill, its value depends on technological content of the products manufactured. In this paper we claim that innovation is close related to design skills. Therefore, existence of a competitive manufacturing sector alone does not, however, automatically lead to industrial upgrading [10]. We claim in this paper that innovation is close related to design. And even design has its own nuances according to the activity they carry out [11]:

1. Original design

   In this case the designer designs something that did not exist previously. Thus, it is also called new design or innovative design. For making original designs, a lot of research work, knowledge and creativity are essential. This type of design can take place when there is a new technology available or when there is enough market push.

2. Variant design

   This type of design demands design ability, in order to modify the existing designs into a new idea, by adopting a new material or a different method of manufacture.

3. Adaptative design

   In most design situations the designer's job is to make a slight modification of the existing design. This type of design needs no special knowledge or skill. Typically, products with higher technological content still have been designed, and, depending on the sector, even manufactured, in the headquarters. Therefore, if we want to check the quality of the innovation, we would expect to see more of the type (1) and (2) activities been performed in BRIC.

3 Fluid Power Case Study

This paper presents a study on the challenges for BRIC in product design and development activities given the aspiration of these countries to start playing a more direct role in fostering innovation. By product design and development we mean a multi-disciplinary problem-solving process which involves a series of complex skills that lead to innovative solutions.

In fluid power, the demand for new types of more efficient systems concepts has called for new design parameters, which has, consequently, affected the dynamic behavior of the whole system. Understanding the design rules is essential so that the problem can be controlled. Moreover, it is important to highlight that the design process should be performed at the system level rather than at the components level. In short, innovation in fluid power has been much more part of incremental, but constantly improvements in design, aiming systems performance optimization rather than technological advancements, or breakthrough innovation.

The work is based on an exploratory case study, which is a qualitative research method, in the fluid power technology. Fluid power is a designed system to transmit and control energy by means of a pressurized fluid, either liquid or gas. According to the US Fluid Power Association, fluid power technology has multiple applications in a wide variety of markets such as mobile, industrial, and aerospace. Table 1 gives an overview of the fluid power technology, divided by market, type of application, products, and end users.

The design of fluid power systems has generally focused on power and productivity giving little thought to the efficiency of the system. In recent years, however, new and stricter emissions regulations and increasing energy costs have caused the industry to look for more efficient system designs [12].

Therefore, the greatest challenge facing designers and product developer's teams working on fluid power systems are:

Table 1 Fluid power applications

Market	Fluid power used for	Product	End users
Mobile	Transporting	Backhoes	Construction
		Graders	Agriculture
	Excavating	Tractors	Marine
		Truck brakes	Defense
	Lifting	Suspensions	
	Controlling	Spreaders	
	Powering	Highway maintenance vehicles	
Industrial	Power transmission	Metalworking equipment	Machine tools
	Motion control	Controllers	Industrial automation
		Automated manipulators	
		Material handling	
		Assembly equipment	
Aerospace	Operating	Landing gear	Commercial aircraft
	Propelling	Brakes	Military aircraft
	Propulsion	Flight controls	Spacecraft
		Motor controls	Related support equipment
		Cargo loading equipment	

Source Elaborated by the author with data from US Fluid Power Association (NFPA)

1. Minimizing energy consumption
2. Decreasing power requirements
3. Achieving international emission requirements
4. Improving standard machine design
5. Identifying the technological barriers to achieving these goals

From a socio-economic point of view, all the sectors where fluid power can be applied play a key role in the process of economic development, employment creation, and income generation and redistribution in BRIC. At the same time, the emergence of new demand for innovative solutions creates opportunities for companies located in those countries.

In order to gather information on trends and changes in organization of design of the fluid power systems we have proposed a field research in sites such as universities and industries located in countries like Brazil, India, China, and Sweden. This paper is part of a larger project that has started in May 2011 called Collaborative Engineering for Sustainable Mobile Systems, COSMOS, aimed at promoting collaboration in research between Brazilian and Swedish universities and companies in fluid power engineering and system design.

3.1 Data Collection

A set of data collection techniques, such as key informant interviews, correspondence with experts, organization of workshops, and field observation has been used for conducting the research. However, as a starting point some literature

Table 2 Information on studied companies

Company	Headquarters	Products
A	United States	Tractors
B	United States	Components
C	Germany	Components

Source Elaborated by the author

review and research on Fluid Power Associations provided us with critical background information.

The next step was arranging meetings with academic experts in the field of our inquiry. In this sense, they are crucial tool to obtaining information from the state of the art of and the historic development of the field. These people are in a position to know the community as a whole and also in particular portion we are interested in.

It is important to remember that usually academic experts in fluid power have a close relationship with industry, which has fostered an environment of cooperation and mutual assistance. To sum up, the contact with academics help us selecting informants with different backgrounds and affiliation. Giving the geographic distribution of the project, another method to collect information and keep track of what was going on was correspondence with experts.

The first workshop on Innovative Engineering for Fluid Power and Vehicular Systems brought together Brazilian, Swedish and international industry and academia to promote collaboration in development of technologies, education, innovation management, and methods and tools for system development and design [13]. Organizing and attending the workshop was an opportunity for taking notes at lectures given by experts and to meet and talk with speakers and fellow attendees as well as to learn and practice the language of the field.

Following the workshop, technical visits to three multinational companies located in Brazil allowed us to observe the field. Table 2 provides the main characteristics of companies visited.

The visits have provided us with a wide variety of learning experiences. This time was possible to talk with product engineers on site, at the same time we saw how companies behave, and how they provide their products and services. Since this visit was made in group, later on a conversation allowed us to share our impressions differentiating between fact and opinion.

3.2 Data Analysis

In order to strengthen the case study findings and conclusions we propose the triangulation of the collected data as the analysis technique. Triangulation is broadly defined as synthesis and integration of data from multiple sources through collection, examination, comparison and interpretation. By first gathering and then

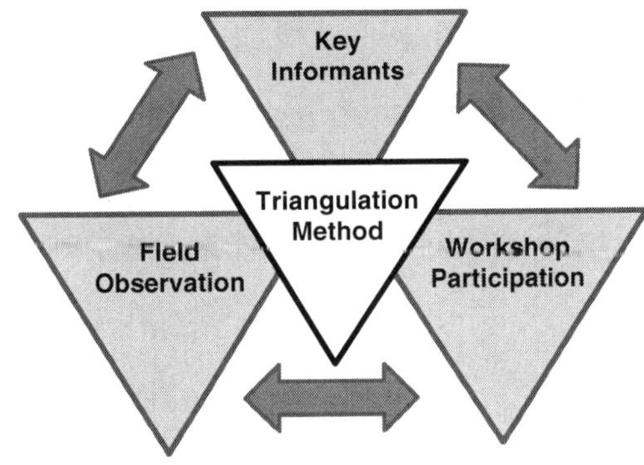

Fig. 1 Triangulation for fluid power case study. *Source* Elaborated by the author

comparing multiple datasets to each other, triangulation helps to counteract threats to validity in each approach [14]. Figure 1 illustrates the different methodological dimensions that were considered in this paper.

4 Initial Findings

Having the assumption that if MNC are relocating R&D activities, we would expect to find more design and product development activities taking place in their subsidiaries. Therefore, this section examines what type of design and development activities have been observed in selected Brazilian subsidiaries of MNCs working on fluid power technologies.

The question we asked was: How is R&D in these companies geographically distributed? First of all we found that all three sampled companies have R&D facilities in their headquarters. The German has R&D facilities in the US while the Americans have development centers in Europe. The American company working with tractors has recently opened a development center in Germany and Finland, but there is no mention about any emerging countries. The other American company working with hydraulics components has opened up new state-of-art manufacturing plants both in China and India, but there is no comments about expanding R&D centers to these countries.

This is also the company with the closest ties with Sweden. In the past thirty years this company has acquired some Swedish companies, which explain the collaboration with Linköping University. Although the ownership have changed, they were able to keep long term relationship with the local knowledge base supplying new employees and working to solve problems together.

The German company is the only one who is establishing a R&D center in China. However, as stated by them:

> The company hopes to take better advantage of the high-growth regions of Asia and South America with the continual expansion of its international development and production network. For example, products and solutions will be adapted to the specific requirements of each region based on platforms developed in Germany.

Looking at the Brazilian experience, none of the companies has even mentioned invest in R&D. They do have a Product Engineering Department, where design and product development are important functions. However, taking the stages of design framework into account, we would see that the activities performed by the Brazilian companies can be classified as a mix of adaptive and variant design.

In this case, the subsidiary works in the redesign of some components under the headquarters supervision. The goal is to adapt the original design to suit local conditions at various levels of the supply chain, both internal and external, such as technical and budget standardization, supplier development, and regulations. These design activities are closely related to achieve cost competitiveness while making the transition to manufacturing.

Not surprisingly, when the Brazilian experience is compared with the Swedish, we noticed that in Sweden, the company work close with the university to reach original design. In Linköping University's experience, companies interested in turning their fluid power systems more innovative have been partnering for decades with the division of Fluid and Mechatronic Systems (FLUMES). Together they have collaborated to maximizing performance in existing technologies results and investigating the dynamics of fluid power systems.

How much of the fact that subsidiaries engage in redesign can be put down to location and how much simply to the field in which they operate? Despite of the demand for new solutions, the problem solving is focused in some parts of the world. Not different from what was stated by the German company, there is the international development and the production network. Therefore, both redesign and location choices are close related with market size. It happens that MNCs do not have in their portfolio products that meet the real needs of these markets.

4.1 The BRIC Dilemma

Although some say research and development (R&D) location decisions by multinational firms are driven by increased globalization of markets for technology and knowledge workers, the truth is that the decentralization of R&D has been primarily motivated by the necessity of establishing a global presence with products that fits the local income and taste.

Given MNCs face even tougher competition; they have adopted strategies to meet consumer demands in markets previously considered irrelevant. The fact that there are high-skilled workers and public policies to attract these activities have been just a bonus, not the main motivator for companies to decentralize their development efforts. There are undeniable evidences of this phenomenon in BRIC automotive, consumer electronics, telecommunications equipment and, pharmaceuticals sectors.

However, in spite of a dramatic increase in the quantity of what MNC have called R&D centers located in BRIC countries—especially in China and India- the core of original design and development activities, which demands both structures and strategies for more complex problem solving are still been doing at the headquarters. The paragraph below shows how MNCs see the question:

> The design probably would not be acceptable in the U.S. or Europe where the business dynamics and customer needs are very different, and it eliminates the problem of giving them our premium technology to copy and sell at a much lower price.

The innovation performed by these companies is related with manufacturing, moving from Fordism mass production processes to the Toyotism lean mode. In this moment, there are some opportunities to redesign a component or two, always under the leadership of a more experienced project manager located overseas. When we look carefully at Sweden's experience, we noticed that a strong, long-term interaction between researchers and companies take place in the process of bring a new solution for a problem, most of them closely related with new design parameters.

Therefore R&D localization is mainly determined by the existence of economic incentives such as minimum project cost, which is possible due to a wage gap for local high-skilled workers. At the same time, the demand for skilled labor has raised manpower costs, putting pressures on companies to stay cost-competitive with other outsourcing destinations. In addition, governments' subsidies to attract foreign investments have been an incentive to relocate. In this sense, how much the name R&D is not just propaganda still remain to be proved with more detailed studies about the complexity of the activities performed.

On the other hand, BRIC's economic opportunity, leadership, and business development—and the ensuing distribution of new wealth—have led them to enter an era of new opportunities and challenges. Interesting enough is that the most innovative companies in these countries are the domestic ones such as the automaker Tata Motor's in India; the Brazilian aircraft maker Embraer; the telecom Huawei, in China, and the Russian jet engine maker NPO Saturn [15]. What do they have in common? They all have mastered the art of design and product development lifecycle, including complex technological systems that meet the particular needs of consumers and businesses.

5 Conclusion and Discussion

This study examined design and development associated with Brazilian efforts to upgrade the overall industrial structure with higher value added activities and skill—intensive labor. In order to get a better understand, we have done: (1) an assessment of product design and development process for three multinational subsidiaries in the field of fluid power technology; (2) an investigation of the academic and industrial collaboration in Linköping University.

The first part of this study indicates that MNCs have not totally decentralized their design and development activities, only allowing subsidiaries to work on less complex activities. Even in redesign, the subsidiary teams are always under the supervision of a headquarters project leader. Although these projects demand more skilled professionals, they have been done mainly to achieve cost and quality performance.

The second portion of this paper suggests that more innovative and complex design, many based on disruptive technologies, which have not seen before, are centralized in the headquarters with collaboration of traditional university partners. There appears to be no competition between them and BRIC countries. Even among BRIC there is some kind of natural selection when MNCs have to choose where to invest in a new project with Brazil lagging far behind the other BRIC mates in costs and availability of skilled workers. Despite of it, in order to increase innovation made in Brazil, government has proposed a policy that force some sectors to spend 1% of their revenue in R&D.

Domestic companies are a key part of turn BRIC countries more innovative. The challenge in design and development involves achieving a more proactive approach in complex problem solving. It has become increasingly clear that the dependence on MNCs R&D decentralization may be not enough to achieve the level of innovation BRIC expect to change their industrial structure. Hence, natural directions for future research include studying fluid power systems in other BRIC countries.

Acknowledgments This research was partially supported by São Paulo State Foundation Agency (FAPESP) grants #2010/12119-4 and #.2011/1475-2 Thanks to the academia, and industry interview participants, for sharing their experiences. My thanks to Petter Krus from Linköping University division of Fluid and Mechatronic Systems (FLUMES), and also, to Victor Juliano de Negri from the Laboratory of Hydraulic and Pneumatic Systems (LASHIP), Federal University of Santa Catarina for insightful discussions about fluid power. The usual disclaimer applies.

References

1. IMF 2012 April 2012 data. Retrieved 2012-04-21
2. Amsden A (2001) The rise of the rest: challenges to the West from late-industrializing economies. Oxford University Press, NY, p 405
3. Bruche G (2009) The emergence of China and India as new competitors in MNCs. Competition Change 13(3):267–288 22
4. Mundim A, SharmaM, Arora P (2012) McManus. Innovation in emerging markets: the new competitive reality. Accenture. http://www.accenture.com/us-en/Pages/insight-product-development-innovation-emerging-markets.aspx
5. Edwards D, Waite G (2011) Designing for emerging markets. Quarterly of the industrial designers society of America, Summer, pp 31–34
6. Manning S, Massini S, Lewin AY (2008) A dynamic perspective on next-generation off shoring: the global sourcing of science and engineering talent. Acad Manage 22(3):35–54
7. Walsh V, Roy R, Bruce M, Potter S (1992) Winning by design: technology, product design and international competitiveness. Blackwell, Oxford
8. Walsh V (1996) Design, innovation and the boundaries of the firm. Res Policy 25(4):509–529
9. UNIDO Report (2011) The international yearbook of industrial statistics 2011

10. Azadegan A, Wagner SM (2011) Industrial upgrading, exploitative innovations and explorative innovations. Int J Prod Econ 130:54–65
11. Pahl G, Beitz W, Schulz H-J, Jarecki U (2007) Engineering design. A systematic approach, 3rd edn. In: Wallace K, Blessing LTM (eds) Springer, Berlin
12. Center for compact and efficient fluid power. http://www.ccefp.org/index.php
13. First workshop on Innovative Engineering for Fluid Power and Vehicular Systems http://cisb.org.br/workshop-inovacao
14. Denzin NK (2006) Sociological methods: a sourcebook: Aldine transaction, p 590
15. The top 10 most innovative industry by countries in 2011 http://www.fastcompany.com/most-innovative-companies/2011

Stylistic Analysis of Space in Indian Folk Painting

Shatarupa Thakurta Roy and Amarendra Kumar Das

Abstract The paper aims at identifying the design principles of Indian folk painting by analyzing some paintings of master painters from Srikalahasti, Madhubani, and Raghurajpur. The authors further discuss the different initiatives to reinterpret the effectiveness of storytelling through graphic visuals among mass. The age-old *chitrakatha* tradition of narrative paintings with oration, as an intrinsic part of Hindu folk religion has played a significant role in the proliferation of the doctrines of Hindu epics and moral stories. The space division aims at an optimum clarity for impelling communication. The dimensions are often mandatory rather than arbitrary. Yet, often esthetic overpowers the theme. The authors as evaluators adopt the method of three folded analytical study of semiotic, iconic and thematic aspects. The tradition of narrative folk paintings although has emerged and grew in remote and isolated locations but is not confined to the temples and rituals any-more because of cross cultural exchanges. They are rich as artistic expression and potent enough to amalgamate and evolve with time. The paper, therefore briefly opines on the significance of folk art practice in the changing society.

Keywords Critical analysis · Design principles · Visual communication · Contextualization · Innovation

S. T. Roy (✉)
Department of Humanities and Social Sciences, Indian Institute of Technology Kanpur, Kanpur, India
e-mail: stroy@iitk.ac.in

A. K. Das
Department of Design, Indian Institute of Technology Guwahati, Guwahati, India
e-mail: dasak@iitg.ernet.in

A. Chakrabarti and R. V. Prakash (eds.), *ICoRD'13*, Lecture Notes in Mechanical Engineering, DOI: 10.1007/978-81-322-1050-4_102, © Springer India 2013

1 Introduction

The definition of folk art can be free flowing and ever changing as per the dynamism of its phenomena. The authors here make an attempt to indoctrinate the design principle of folk art in the Indian context. Folk art practice in India in spite of being vernacular and varicolored, shares a lot of commonality in them. The age-old traditions of narrative scroll paintings of regional India as well as the art practices intrinsically connected with the temple rituals have always played a vital role in the proliferation of Hindu epics and moral stories. Vatsyayan in her book titled "The Squire and the Circle of the Indian Arts" [1] writes, 'All ritual first establishes a point, a metaphorical center, around which lines are drawn to make triangles, squires and circles of great symbolic value. Notions of space and time are comprehended through it' (p. 163). The principle of folk art in spite of being rudimentary and down to earth also follows a somewhat similar concept of space organization.

2 Style and Narrative

The study involves a scrutiny of five such examples from the traditional folk paintings of India. They belong to five different states of India, namely Rajasthan, Bihar, Bengal, Odisha and Andhra Pradesh. It is to realize how the artworks hold similarity in their subject matters, expressions and purposes in a heterogeneous cultural and lingual periphery.

To encipher the orderly construct of space in traditional folk paintings the authors chose a wall art from Madhubani (Fig. 1), Bihar that is locally known as *Kohber*. They are drawn on the wall of the bridal chambers as popular ritualistic practice of the region. The conception of the space division is centralized and symmetrical to ensure harmony to the composition. The image is liner and radially balanced with a distinct point at the center, around which images that are symbolically related to fertility are drawn in plenty to adequately fill up the space in a geometric formation. The apparently intricate and semi-recognizable motifs are the derivatives of natural elements such as flowers, weeds, seeds, aquatic animals, amphibians and reptiles. The compositional formation is confined with the boundary of squires and circles with decorative and repetitive patterns to ensure an organized readability to the otherwise crowded visual.

The authors chose the next example from the traditional *Pichhvai* Paintings of Rajasthan to confirm this centralized formation of space that is based on several circles and squires. The word *pichhvai* means backdrop and the tradition is related to the worship of Shrinathji, at Nathadwara temple, Rajasthan. The philosophy behind the image of Shrinathji, a form of Lord Sri Krishna is vastly metaphorical. Lazaro, in his book Material, Methods and Symbolism in the Pichhvai Painting Tradition of Rajasthan [3], writes, "on-dualism or 'not-two' is that which

Fig. 1 Kohber, Madhubani

Pre-manifestation is. It is the complete, whole un-differentiated point, *bindu* (dot), into which all will ultimately withdraw". The image of Shrinathji in the paintings done in traditional Indian miniature technique may be decoded with the following diagram (Fig. 2). The length of the idol from the crown to the base of the feet can be primarily divided into four equal circles. A larger circle surrounds the entire body that touches the crown and feet. The geometric construction even if not so visible provides the composition with a desired balance and harmony.

Folk art is a countenance of traditional culture. The intrinsic association of art forms with ritualistic and religious purposes laid the foundation of folk art practice of India. It perpetually insured a continuity of the tradition but it is also the simplicity of mind that has always drawn the viewer toward it. Such traditions have also been generating highly individuated self-conscious artists. The above-discussed folk painting traditions of India are intrinsically associated to the customary rites. To trace their principle origin, one needs to understand its religious compulsion that operates the flow of the practice and is primarily responsible to shape up its stylistic identity. They are often budded in the vicinity of the Hindu temples, as sacred wall hangings, that harmonize the walls around the chambers and passageways of the temple. Roving minstrels (kathakas, story tellers) painted mythological figures and carried them to places, with the oration of the words of God. The painters over ages have been performing a role of a reformer by promoting the flare of rectitude in the minds of the common populaces not as a

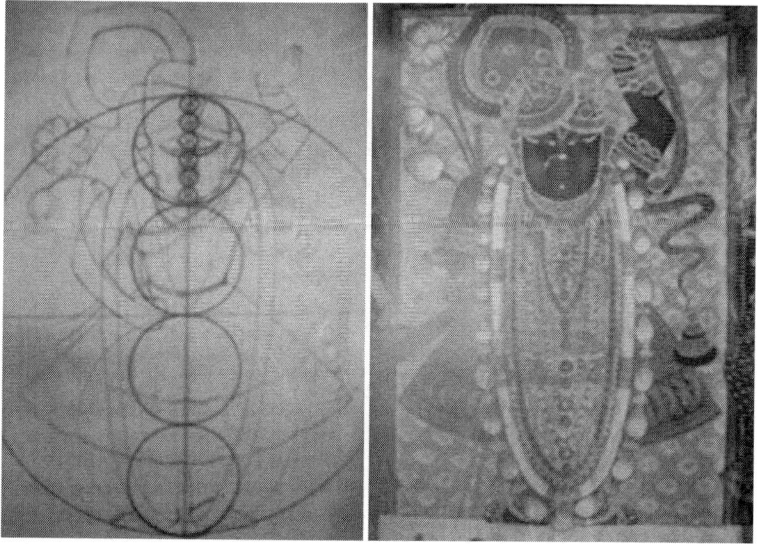

Fig. 2 Construct of Shrinathji at Nathadwara temple, Rajasthan

preacher but entertainer with their artistic ability of painting and singings. In the process of viewing and analyzing the art works to access the strength and weakness the research aims at achieving the right vision to determine the life of the cultural dynamism in the changing social milieu. "The meaning of art is similarly a mystic experience for which one needs *divya chakshu*, the inner perception. For understanding Indian traditional art, or for that matter, any art form, one needs Divine vision to understand the message of the artist. Beyond the outer image is the inner meaning, which can be perceived and shared both by the artist and the viewer. It is this vision, the shared experience that reveals the Soul of Art", Ramani [2].

The present study is an effort to indicate the notions of viewing and appreciating the artworks to conduct a comprehensive art critique. The art works although made by artists who are not educated in academic style in urban art schools, but inherit the skill and carry it through generations, may seem to contain elements in them that are ageless and primitive, and even be crude or amateurish. They are far beyond the representational norms of realism and naturalism and can be totally idealistic in nature. Folk art being a community practice the skill is often common in almost every inhabitant of a cluster. Not everyone qualifies to be a master, but only a handful of them. The rest of the people perform as followers to contribute to the process as skilled craftsmen. To understand and critically appreciate the artworks one must adopt the age-old three level method of critique. Thematic, iconic and semiotic levels of a thorough reading prove to do justice to the rich esthetic that the artworks contain. In the thematic level, an overview to the socio-cultural background is considered. Understanding the context is the most important part of it.

Can folk art be set apart from its historical context? The art, that takes birth and get nurtured to the serenity of the rural landscape and to the temples where the God is believed to dwell and thus being worshiped, when taken out and placed for the market display, juxtaposing numerous other items, may not always be successful in holding the similar identity. This contextualization and re-contextualization have had phenomenally influenced the viewers.

In the cultural center named Bharat Bhavan, Bhopal, J. Swaminathan initiated a unique coexistence of modern art and folk art in an exhibition in the early 1980s. Jayakar, as the director of Handicraft Board, Government of India, caused a paradigm shift for the female painters of Madhubani when she asked them to draw on paper instead of the mud walls. In 1966, during the drought and famine in that region Bhaskar Kulkarni represented Jayakar to offer them papers to work on so that they may generate some fund for them during the crisis. These incidents hold a lot of importance in the history of those traditions and cannot be kept unmentioned in the thematic critique. Realistic, representational and recognizable images are to be appreciated and criticized at the iconic level. In order to understand Indian Folk Painting, it is mandatory to be familiar with the iconic characters in the paintings that play the central role. In this level of critique the viewer mentally alter and reposition the figures within the frame according to their change in preference. The religious connection of folk painting seldom allows its practitioner to disturb the cosmological order and there by limit the scope of experiment in the execution. The use of color in most cases are mandatory over arbitrary. Vast use of hieratic scale contributes to the story telling process.

The following image (Fig. 3) of Ramayana inscribed on the body of Hanuman painted by Shilpaguru Ananta Maharana, Raghurajpur, Odisha, one can observe, as it has been pointed out to the authors by the artist's son Bibhu Maharana, that, all happy episodes are put on the face of Hanumana, while episodes related to Ravana, the antihero in the epic are drawn around the lower body. In the iconic level it is

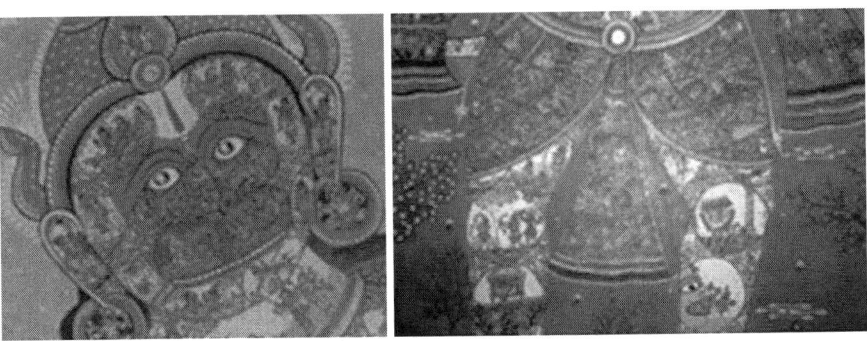

Fig. 3 Complete Ramayana on the body of Hanumana (detail), Raghurajpur

not only needed to recognize a figure but also to identify it with the prior knowledge of history and mythology that refers to the thematic level.

The layer of semiotic comprehension applies to a critical analysis free from any foreknown information. Non-representational and abstract images are best read applying this very level of critique. In the semiotic level the artwork can be viewed and analytically appreciated regardless of any prior knowledge of its history, background, method, medium or material. In this level each and every dot and pixels can be analyzed and thus be visually communicated. In the semiotic level the critic directly approaches to the elementary part of an artwork. It is read in terms of quality of line, form, shape, volume, color, texture etc. The elements here enable to establish a language for improved expression. The perception of the artist in this level makes an entry to the language of art. The vocabulary of visual art adds to the possibility of communication and connects it with the desired message irrespective of its existing cultural fabric. Indian painting tradition that spans over two millennia, in spite of having remarkable variety does not include folk art in the family of its major distinctive styles. Folk art practice has emerged and grew independently in regional locations there by holding the core essence and spirit of the soil. In the present study the authors primarily focus on the mandatory regulations that the practice prevail in them in order to assess how stringently they adhere to the rules. This is to scrutinize the tradition and change that keep up with the free spirit that the practice is desired to hold in them. The semiotic analysis of the space in the artworks hence will be conducted with a few relevant visual examples.

Ramayana is among the most popular religious texts of India. The central characters of this epic namely Rama, Lakshmana, Sita and Hanumana are worshiped as incarnations of God and God themselves in Hindus all over in the country. Episodes from Ramayana in single frame (ekachitra) as well as complete Ramayana both as narrative scrolls and single frames are to be found in almost all the regional folk clusters. The narrative that is orated with these pictures tells the stories in regional languages. In the past those narrations must have been the only source for the common people to learn the incidents and the paintings worked as illustration to add life to it. The paintings also have functioned as wall hangings in the temples. The story of Ramayana as all other epics have been told and retold for ages and never to lose its significance. But, regardless of the text the paintings hold a high quality esthetic value in them that can be viewed and appreciated independently as visuals irrespective of any clue to the story. The space is two-dimensional in general with no intension to create any illusion of depth. The use of perspective is aerial rather than linear. The compositions are dominant with linear qualities and all the characters and objects are confined in well-defined contour lines.

In Madhubani painting [4] of Bihar complete Ramayana is not so common. Episodes of Rama's wedding, Hanumana's faithfulness, Ravana's abduction strategies for Sita and finally Ravana's defeat to Rama are most popular as subjects. They use a technique known as *bharni* that literally means to fill. The painting styles defer in the major casts but the process in general is to draw the contours and fill them up with colors or decorative patterns. Empty spaces without any filling are rarely observed here, that enhances the decorative quality in these paintings. Angularities of all sorts including the limb joins are totally avoided in these works to ensure the harmonious lyricism. Visual abstraction achieves a representational dimension rich with naive esthetic excellence as the figures appcar in their most descriptive features with faces in profile with a frontal eye and torso, with hands and legs in their sidewise views. The lines do not show any rendering of light and shade in them. They follow an idealistic style with no sign of line variation in the style that may indicate and naturalistic quality anywhere in them.

Narrative scrolls (Fig. 4) depicting the story of Ramayana from Midnapur, Murshidabad and Birbhum, Bengal is not different as far as the principle of image making is concerned. The delineation with repetitive forms and decorative lines gives the images the required movement in the otherwise static space. In Bengal the narrative scrolls are known as *dighalpata* and *jaranopata*, which has related meanings. *Dighal* means long, *jorano* means rolled up and *pata* means canvas space. The dispersion of images is based on continuity. Continuity in composition is caused by repetitive forms and well-calculated proximity within the images. Even when the episodes are essentially separated and confined with a borderline the similarity in form causes the entire scroll to look united and harmonious with no visual break.

Complete Ramayana painting (Fig. 5) made by the master painter of Raghurajpur, Orissa still follow a regimented norm in the distribution of space in their composition. The skeleton structure if studied properly takes care of the marvelous balance with a preconceived and conniving repetition of circles, squires and rectangles.

Complete Ramayana in Kalamkari technique on cloth painted (Fig. 6) in Srikalahasti , Andhra Pradesh, incorporates typography in the horizontal borderlines to divide the space. The roundish regional scriptures blends as exactly and effectively with the drawing style that the scripts even if unreadable do not cause any hindrances to its smooth viewing. Painter M. Nagaraja Setty in his conversation with

Fig. 4 Narrative scroll paintings of Ramayana from Bengal

Fig. 5 Complete Ramayana, Raghurajpur

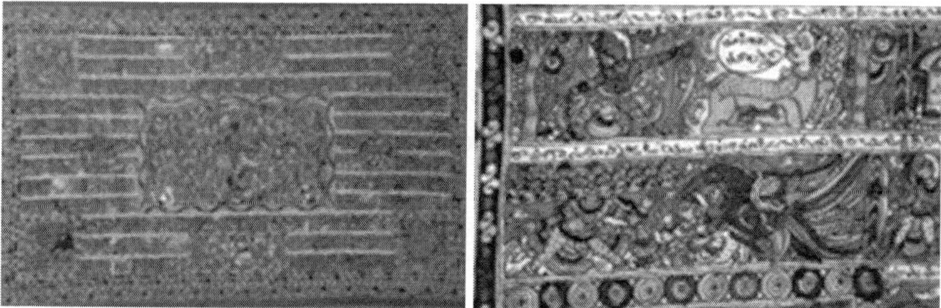

Fig. 6 Complete Ramayana, Srikalahasti

the authors put stress on that. It is indeed interesting to see how the Telugu script acted as a decorative pattern for the people who cannot read the language.

3 Conclusion

The expression of folk art is simplistic, but not as rudimentary as tribal art forms. In tribal art we see images that are related to votive cults that are untouched by the complexity of civil livelihood. Folk art belongs to the people who are exposed to many a cross cultural and cross media influences. They are literate and educated socialites who have complex religious foundations. Moreover they belong to the time that is dynamic and prone to change.

The ethnographic study of folk culture enables the researcher to understand the evolution of creative minds. Subramanyan writes, 'In present-day society like ours art does not have a decided role or patronage, and no decided thematic supports; hence the artist has either to devise his own supports, however small, or make more radical choices' [8].

The various government and non-government initiatives have caused a lot of innovation possibilities to the happening. The practice has lost its relevance in the society from time to time with the introduction of new ways of living. Comic books and television serials have replaced the practice of oration and painted illumination, with their big budget and glossy productions. The new modes of communication have left little to the folk painters to play any role that is vital and indispensable for a society. Nevertheless the condition like this, has offered newer openings to the folk art practitioners to create with no bindings and obligations. The folk art market of today is based on pure esthetics. The master artists who have always been extremely innovative and talented are now getting a platform to experiment and explore depending on various stipends and awards. The numerous trained artists from the community and form outside the community who are not involved into the innovation related to the image making are engaging themselves with the industries that deal with lifestyle products and earning their livelihood by trying their painting skill on designer saris, wall hangings, lampshades etc. The abrupt patronage that has helped the practice to sustain and come this long a way deserves a parallel study. The contribution of Kamaladevi Chattopadhyay and Guru Saday Dutta along needs an essential mentioning here. The folk painters are taking interest to explore the possibilities of new media. They are contributing and collaborating in animation filmmaking and book illustrations that are not away form their own principle purpose of image making. A few successful initiatives from recent time where the eminent folk painters of contemporary India took part that needs to be mentioned here are Moyna Chitrakar's Sita's Ramayana, Dulari Devi's Following My Paint Brush, Swarana Chitrakar's The Patua Pinocchio and Monkey Photo, T. Balaji's Mangoes and Bananas, Radhashyam Raut's The Circle of Faith Rambharas Jha's Waterlife, and many more published by Tara Books. Having been written in English language they have crossed the limitation of vernacular communication. However, it is the semiotic comprehension that proves to be the best way of communication outreach that has an appeal that is beyond all limitations and universal. The understanding is based on the fundamentals of human gesture that is rooted in the cosmos.

Tradition, religion, time, change and words like those if considered as the keywords for this article may sound as oxymoron in the present context, where the validity and significance of all these factors are primarily based on the uplifting of folk art as the thriving cultural industry for our country. The variety yet integrity in the folk styles hold the potential to establish a distinct cultural identity as well as can mirror the living visual culture of India. The process of semiotic viewing gives emphasis to its expressional prospective that is universally recognized and appreciated with due round of applause.

References

1. Vatsyayan K (1997) The squire and the circle of the Indian arts. Abhinav Publications, New Delhi
2. Ramani S (2007) Kalamkari and traditional design heritage of India. Wisdom Tree, New Delhi. ISBN ISBN 81-8328-082-X
3. Lazaro DP (2005) Materials methods and symbolism in the Pichhvai painting tradition of Rajasthan. Mapin Publishing Pvt Ltd, Ahmadabad
4. Anand MR (1984) Ministry of information and broadcasting: government of India. Madhubani Painting Publication Division, New Delhi
5. Archer M, William G (1974) The hill of flutes life, love and poetry in tribal India: a portrait of the santals. S. Chand Publications, New Delhi
6. Jayakar P (1975) Paintings of rural India. The Times of India Annual, Bombay, pp 53–62
7. Jayakar P (1980) The earthen drum: an introduction to the ritual arts of rural India. National Museum, New Delhi
8. Subramanyan KG (1987) The living tradition: perspectives on modern Indian art. Seagull Books, Calcutta. ISBN ISBN 81 7046 022 0

Classifying Shop Signs: Open Card Sorting of Bengaluru Shop Signs (India)

Nanki Nath and Ravi Poovaiah

Abstract Any Classification into categories aids in retrieving information. It develops a system for an object or phenomena. Hence, a classification of shop signs would provide an informed view about the system of elements that form the identity of a shop sign. The Philosophy of Classification as explained by Ereshefsky [1] brings to light three kinds of paradigms: Essentialism Sorts, Cluster Analysis and Historical classification. This study investigates the relevance of creating categories through cluster analysis. The analysis helps collate the pragmatic approach applied by the viewers of the shop signs. How people classify shop signboards mentally? What clues they use to attach qualities or concepts with a shop sign? Applying the method of Open Card Sorting [2] increased the analytical scope about the new values attached with the identities represented on these shop signs through text, images and materials. There is a paucity of published research in favour of the above statement. Therefore, this paper is a sincere attempt to substantiate the benefits of arriving at new categories via Open Card Sorting. This method provided the participants to design their own labels and classification structure for the given shop signs. A group of 30 participants (15 designers and rest 15 from other professions) underwent Open Card Sorting exercise. With formal instructions about card sort method, every participant was asked to 'think aloud' in order to resolve the 90 cards puzzle. Additionally, two standard questions regarding good and bad signs in the picture cards were asked. Around 20 new categories could be accumulated in the SPSS software. The viewers did not categorise total 10 cards of the 90 into any label(s). Cluster Analysis of this data gave rise to new classes/genres of these shop signs. It also clustered those cards that

N. Nath (✉) · R. Poovaiah
Industrial Design Centre (IDC), Indian Institute of Technology Bombay (IIT-B), Powai, Mumbai 400 076, India
e-mail: nanki.nath@iitb.ac.in; nankinath@gmail.com

R. Poovaiah
e-mail: ravi@iitb.ac.in

A. Chakrabarti and R. V. Prakash (eds.), *ICoRD'13*, Lecture Notes in Mechanical Engineering, DOI: 10.1007/978-81-322-1050-4_103, © Springer India 2013

were considered good and bad shop signs. It is a unique study to know how people view, read and form opinions about shop signs. Results of this study can be used to inform the designers about the new features/qualities of the content and form observed by viewers along with their opinions on good and bad signs. Therefore, these insights would be the essential parameters in terms of elements of design and related qualities that sign designers should apply in the design of shop signs.

Keywords Open Card Sorting · Generative Research Methodology · Classification · Structures · Taxonomy · Shop Signs · Identification · Signs · India

1 Introduction

There are different types of visual display that create the visible appearance of the shop signs in India. The perception and meaning of the elements and designs (say, posters in the streets, a magazine ad or a sign in a train station), must attract the attention of viewers [3]. The vast number of elements would make specific classification of every shop sign quite difficult. However, it is important to have an understanding of the basic elements to help us narrow down the research and selection of the most crucial elements of the shop signs. Designers and non-designers have a different way of viewing/reading/interpreting shop signs. Hence, for the experiment of Open Card Sorting here, aims to understand the overall rationale that people used to classify shop signs of Bengaluru:

1. To accumulate new nomenclature/terminology applied by people to name various sign classes, qualities/attributes assigned to these various classes to understand significant parts of the shop sign's language.
2. To identify the significant parameters that could be best used to arrive at a preliminary.
3. To accumulate all the parameters those were given preference against the left ones.

1.1 Basics of Classification

Classification is as an effective method to generate an order; that aims to formulate a system of understanding for objects, ideas, phenomena etc. Therefore, it is required that the approach to classification should be well articulated and should also project a logical correctness [4]. The following basics of classification explain the above requirements with simple visual examples:

1.1.1 Mutually Exclusive Classification

When an object is classified on the basis of many aspects, then each of these aspects have their own set of groups. For instance, shop signs can be classified on

Fig. 1 Genres: Indian commercial cinema. *Source* http://www.bollywoodwallpapersfree.com/

the basis of graphics, text, information and material display etc. Some people may notice further sub-aspects within these primary groups. For instance, 'hand-painted' and 'hand-crafted/3D fabrication' can be the sub-aspects of 'material display' for these shop signs. 'Material Display' is an exclusive aspect in itself, mutually aligned to others, but not a sub-class of others (others here being graphics, text and information).

To illustrate mutual exclusiveness approach, Indian Commercial Cinema has mutual and exclusive genres (see Fig. 1). In addition to mutual exclusiveness of respective categories, the respective categories display a *logical correctness* to them. For instance, 'Devotional' and 'Patriotic' movies express different themes. Both, in a way, are belief systems of the people of India, but their social meanings are different. 'Patriotic' genre is a special feeling for one's country (that is very different from devotional subject of 'Devotional films').

1.1.2 Collectively Exhaustive Classification

When an object can be easily classified under more than one category, the classification structure becomes collectively exhaustive. For instance, *faceted classifications* are generally collectively exhaustive. In such classifications, a facet includes "clearly defined, mutually exclusive, and collectively exhaustive aspects, properties or characteristics of a class or specific subject" [5]. For example, in case of shop signs, 'form' as the facet, 'shapes' (geometric, natural, oriental etc.), 'volume' (mass, contours etc.), 'proportion' (upright, uniform/non-uniform) etc. are the sub-categories within the facet.

In 1954, French typographer Maximilien Vox developed the Vox ATypI classification has nine classes (see Fig. 2). It elaborates on the evolutions of tools and techniques in respective time periods of specific typeface genres [6].

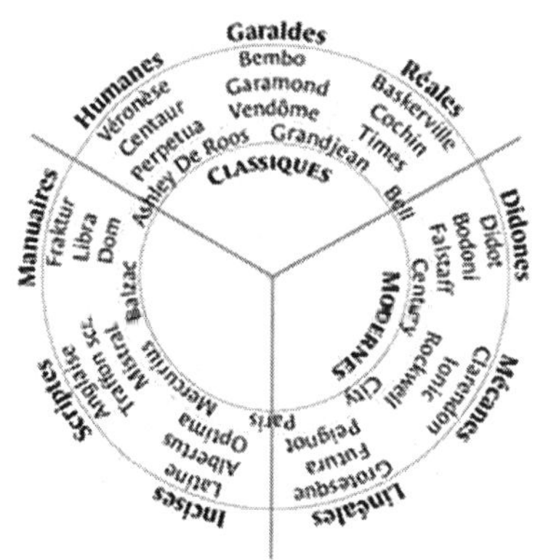

Classification VOX atypi

Humanes	Centaur, Golden Type, Hadriano.
Garaldes	Bembo, Garamond, Plantin, Sabon.
Réales	Baskerville, Perpetua.
Didones	Bodoni, Didot, Walbaum.
Mécanes	Clarendon, [illegible], Rockwell, Serifa.
Linéales	Futura, Gill Sans, Kabel, Univers.
Incises	Albertus, Optima.
Scriptes	Isadora, Shelley.
Manuaires	Banco, Libra, Ondine, Post Antiqua.
Fractures	Wilhelm Klingspor, Fette Fraktur.
Non latines	Ωλπ (Garamond grec), רטא (Hebraica).

Fig. 2 Vox ATypI classification map, from la Chose imprimée, chapter on type classification by the author, 1999. Classification Vox ATypI—additions of *Blackletter* and Non-Latin genres. *Sources* http://typofonderie.com/gazette/post/on-type-classifications/, http://illusive-pixel.com/art-general/typographie-histoire-et-pratiques/

1.1.3 Internally Homogenous Classification

When all the objects are grouped together under one category regardless of the common connection/likeability amidst them as a group, then this approach of classification is *Internally Homogenous.* For instance, in case of modern (commercial) shop signs, 80–90 % of the signs can be grouped under Digital Art group (see Fig. 3).

Fig. 3 Digital art, India. *Source* http://beclee.wordpress.com/gallery/signsweb/

Though, in the grouping, we would see various nuances of digital art in the text/graphics/material use and treatment/background visual effects or signboard in totality. But, these different nuances are of least concern being internally homogenous. There is advantage to this approach for commercial signs displayed here—the designer/maker of shop signs can find all the materials on a given topic in one area of shop signs.

2 Methodology

2.1 *Experiment Design and Process*

2.1.1 Experiment Setting and Instructions

A big space of four tables of regular height was selected for the open card sorting experiment. This space is part of a large working studio (Shenoy Innovation Studio, IDC, IIT Bombay). The lighting conditions were kept ambient. The experiment was conducted within the duration from 9 a.m. to 6 p.m.

2.1.2 Preparation of Stimulus

Each of the 90 shop sign picture cards, each of 6 by 3 inches, includes shop sign picture (in correct upright perspective). The sizes of the shop signs vary in sizes. The card sheet of 120 gsm (matte finish) was used for the prints. These 90 cards are from the city of Bengaluru, India. They have been collected from three different marketplaces of the city, namely Shivajinagar (oldest marketplace), Narasimha Raja Road (older one) and Avenue Road (new one). (Similarly, 30 signs have been collected from five more cities; in addition to Bengaluru. The basis has been devised to formulate a mathematical logic to quantify essential attributes to get changes (transitions) and co-relations; that would help discover new aspects of shop sign designs and the co-related factors that are instrumental in the process). This was not shared with the respondents. Rather, the whole bundle of 90 shop signs was handed over to every participant. But, at the back of each card, a linear code was written (for later information retrieval) in following way:

- [City—Marketplace—Time Period—Shop Kind—Serial number].
- [BA—MP1—30–44—A—01].
- [Bengaluru—Marketplace 1—Time period—Shop Kind—Serial number].

2.1.3 About Respondents and Instruction Session

Total 30 respondents participated in the experiment (15 designers and other 15 non-designers; range of age groups: 20–65 years.). In the designer gamut, design faculty, technical staff, one master student and research scholars from IDC, IIT Bombay participated. In the non-designer gamut, two B.Tech. students, M.Tech. students and Research Scholars from other departments of IIT Bombay participated. The following paragraph presents the instruction imparted to all the participants per card sort exercise:

The aim of this experiment is to know how you classify a group of 90 shop signs as picture cards here. Look at the group of cards and sort them into separate groups. All the groups that you make together would be a classification structure of 90 signs that you sort. Think Aloud and name or label all the groups you construct. You can put one particular card into as many groups as you like and include as many cards as part of a group. When you have finished sorting cards and labelling the groups, we would like you to tell us the approach that you followed to sort these cards.

2.1.4 Finalising Categories from 30 Card Sort Exercises

Every classification structure was observed for each respondent. When we started manually filtering the most common ideas/concepts/names as group labels; various aspects appeared to mix together that made labels very unclear. For instance, a group named 'Bilingual signs, with local language prominent and hand-painted signs'. With this multi-dimensional labelling, it was becoming difficult to lump all the selected shop signs by the participants under the label. Though, this being a major limitation of the labels we accumulated, we have tried to visually map all classification structures (by designers and non-designers) from most general to specific category (Table 1).

Table 1 List of main categories

Shop functions (SF)	Letterforms in signs (LF)
Information signs (IS)	Language in signs (LA)
Confused signs (CS)	Naming kinds in signs (NA)
Perfect signs (PS)	Price signs (PR)
Bad signs (BS)	Qualities in signs (QA)
Artistic signs (AR)	Representation in signs (RP)
Graphic signs (GR)	Layered signs (LA)
Graphic + typo-text signs (TTGR)	Typical signboards (BS)
Typo-text signs (TT)	Miscellaneous (MS)
Fonts in signs (FS)	Uncategorised signs (UnS)

2.2 Hierarchical Cluster Analysis

The method generates a system of clusters, collecting small clusters of very similar items to large clusters having only dissimilar items. This method helps create a sequential and logical order of segmentation by accumulating co-related clusters. It tells you the distance between two and more clusters, which clusters are important ones to consider. It presents the similarity and dissimilarity between clusters. The largest cluster in the resulting dendrogram from this study combines case/variable 3 (No. of Structures) with 32 (Photos)—see Fig. 4.

2.3 Data Feeding and Output

Putting the data in SPSS and then doing cluster analysis has given us following outcomes. Following is the analysis for each result:

2.3.1 Proximity Matrix (P)

Presents the co-relations that exist between cases/variables. How closely associated are some cases? With $n = 99$ (all variables), we are measuring here n-by-n matrix of similarities and dissimilarities (see Table 2 and footnote. Similarly, we have observed five more cases).

2.3.2 Agglomeration Schedule

An Agglomeration schedule (AS) table aids us to find relevant clusters to study as the final solution of cluster analysis. Also, significant cluster formations are those where we see sudden jumps in the coefficient values. This jump from one neighbour formation to the other indicates close cluster formations. Statistically it means, that for 0.031 coefficient value increase for clusters 12th and 28th at stage

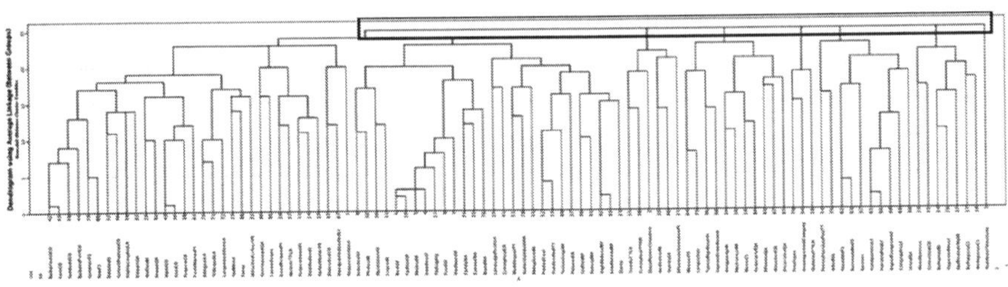

Fig. 4 The *black box* encloses the biggest cluster having most dissimilar cases together, with Case 3 and 32 being the most dissimilar, forming the extreme branches of the group

Table 2 Co-relation values of two adjacent cases/variables

FASF (Fashion shops)	PSF (Packaging shops)
−0.242	−0.365
−0.087	0.002
0	0
0.89	0.794
0.894**	0.769
0.946	0.893**

Source IBM 19.0 SPSS; output sheet of proximity matrix (*P*)
**Correlation is significant at the level of 0.01

68th, these clusters are more closer to the significance value in comparison to 23rd and 37th clusters with less significant coefficient value at stage 69th of the table (See Table 3).

2.3.3 Vertical Icicle (VI)

The presented parts of Vertical Icicle plots here represent how individual cases (variables) are combined into clusters per iteration of the analysis at the scale of 0–100 clusters formed. In the following two enlarged views of parts of the big Icicle that we got for 99 variables, we can see Fig. 5: Part a, we can see that all 'Letterform' (LF) cases are 60 at 60 clusters, 80 at 80 clusters and 100 at 100 clusters.

2.3.4 Dendrogram

Measure of average distances between two leaf notes (forming a close cluster). A cluster may have several leaf notes, but the average distance between the leaf notes may be more. This makes the group statistically not a cluster i.e., the group of variables are not that co-related as compared to others with lesser distance between the variable.

1. *Distribution into Groups:* The plot below shows that the cases are clustered into two main groups. Group 1 and 2 have further sub-clusters, demarcated by Gp.

Table 3 Highlighted part of the agglomeration table

Stage	Cluster 1	Cluster 2	Coefficients
38	65	69	0.486
39	55	56	0.470
68	12	28	**0.300**
69	23	37	**0.269**

Source IBM 19.0 SPSS; output sheet of agglomeration table

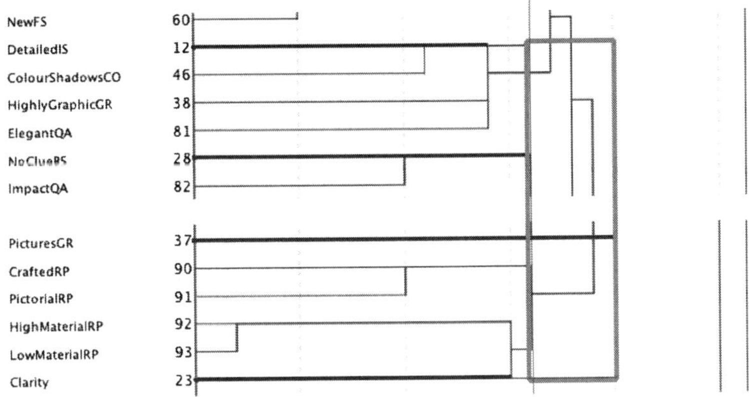

Fig. 5 *Vertical* icicle plot related to above discussion

1a (clustering 'Colour', 'Font', 'Graphics', 'Qualities', 'Information Signs' cases/variables in sequence) and Gp. 1b (clustering 'Bad Signs', 'Qualities', 'Colour' and 'Graphics' cases in sequence) and Gp. 2a (all 'Language' cases together) and Gp. 2b ('Quality' and 'Perfect Signs' in sequence) respectively (see Fig. 6).

2. *Analysis*

Total no. of *Variables* (categories under study): 99.

Total no. of Leaf notes (cases—indicated by dots in the Dendrogram (D)—see Fig. 7).

In the dendrogram, we could get 13 concrete and visible clusters. They are emerging from the branches anchored to the nodes of larger clusters. There are three large gamut' that encompass these three clusters. Of all the 13 clusters, cluster no. 1 has the maximum number of 16 leaf notes (see Fig. 8).

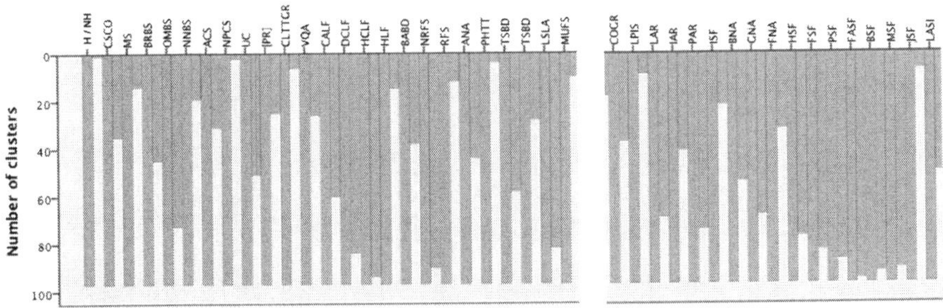

Fig. 6 Average distance between groups of clusters—an example

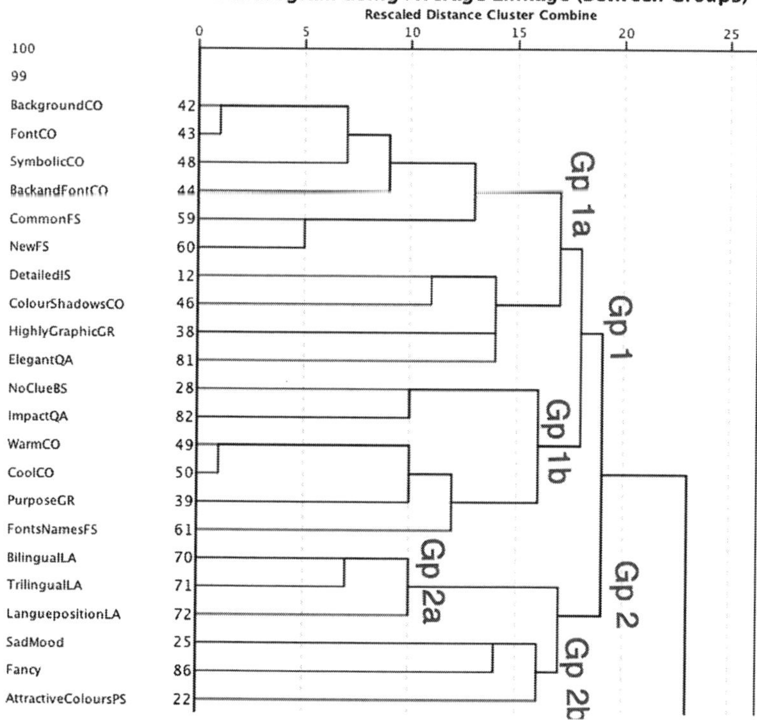

Fig. 7 The *grey box* highlights the jump from 0.269–0.300 for clusters 23–37 and 12–28 respectively. *Courtesy* Dendrogram representation of Hierarchical Cluster Analysis

2.4 Standard Questions: Qualitative Inquiry

Respondents were asked to give answers for following questions, keeping in mind a hypothetical scenario of imagining the 90 shop signboards erected on buildings and shop facades inside a street bazaar.

Q.1 a. Which signs are good/effective signs from the point of view of visual? communication?

b. Which signs are bad/ineffective signs from the point of view of visual communication?

Q.2 a. Which signs are good/effective signs from the point of view of typography/lettering styles used on these signs?

b. Which signs are bad/ineffective signs from the point of view of typography/lettering styles used on these signs?

Designer versus Non-Designer comments with respect to good and bad signs:

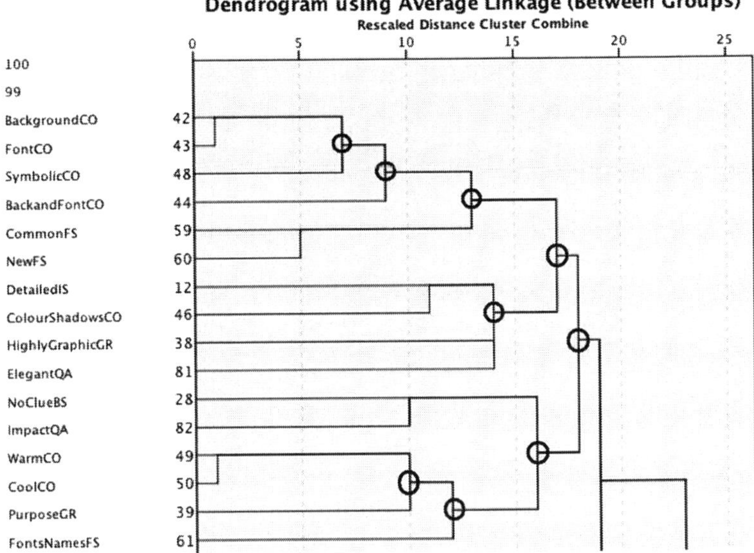

Fig. 8 Cluster 1—having maximum number of clusters. 9 nodes = 9 cluster formations

Designers made comments pertaining to design context, like likeability of font/lettering, visual representation, information flow. Non-designers looked into purpose, wordings, visual mood. Following views are summarised versions of six kinds of opinions given by designers as well as non-designers (Tables 4, 5):

Table 4 Designer versus Non-designer views about effective/good signs

Comments: good signs	Number of designer responses	Number of non-designer responses
Likeable (text + pictures/colours used)	10	14
Names/wording for names	8	12
Visual representations/styles/symbols	6	6
Likeable fonts	6	5
Clear/straightforward signs	5	4
Innovative/novel look/novel treatment to fonts	3	2

Table 5 Designer versus non-designer views about ineffective/bad signs

Comments: bad signs	Designer	Non-designer
Unlikeable fonts	15	8
Bad use of type and space	15	8
Cluttered/loud/messy/overflow/chaotic	12	20
Bad expressions/disconnect with meaning/message	8	15
Bad readability (size/effect/colour)	8	10
Over-persuasiveness like advertisements	5	4

Table 6 Working framework for Bengaluru shop signs

Elements (primary attributes)	Parameters (functional basis)	Values (selected units)	Context
CO	Colour contrast, colour perception	Foreground-background; warm-cool	Design
RP	Material culture	High-low quality	Design
TT	–	Painted-handcrafted	Design and content
FS	Font kind; readability	Regular-stylised; readable-non-readable	Design
LA	Multilingual identity	Bilingual-multilingual	Content
SF	Business strategy	General-specific products	Content

3 Conclusion

3.1 Working Framework

1. According to the dendrogram and related coefficient values in the agglomeration table, the closest primary variables in decreasing significance are: Colour (CO), Material Representation (RP), Text and Type (TT), Language format (LA), Fonts (FS) and lastly Shop Functions (SF). These are related to the principles that theoretically regulate the elements. In case of Text and Type group, it is a mixture of equal importance of the words used in shop names and other information, along with the visual identity of letters, their style and form (Table 6).
2. Maximum number of comments regarding the effective signs (why they function as 'good' signs) by the respondents good colours—their distribution in the layout and their visual mood and relative contrast, picture-text relation and visual styles. The colour dynamics and material pragmatics of shop sign are crucial principles that should be applied in the design of a shop's sign. The comments are almost equal for both CO and RP from both designers as well as non-designers. This is further supported by their close similarity cluster values in the dendrogram.

3.2 Advantages of Classifying Shop Signs

1. *Grouping*: With classification, we group things together. With grouping, we get a sense of features/characteristics that play a crucial role in defining these shop signs.
2. *Organisation*: This is required to make sense of the vast variety of elements of shop signs. This has been achieved with two different kinds of classification structures made by designers and non-designers in this study.

3. *Overlaps* (*Shared characteristics*): The Greek philosopher Aristotle first began classifying plants according to their shared characteristics. Groups of plant varieties helped biologists/scholars in the field to the details as well as the heritage associated with specific varieties. Similarly, application of the science of classification in the case would aid us in accumulating the details and the subjective basis of the shop signs.
4. *Hierarchy*: The relative importance of one group over the other would help us in understanding the thought process/approach of segmenting visual components/elements by designers versus non-designers.

Acknowledgments I express my utmost gratitude to my guide, Prof. Ravi Poovaiah for motivating discussions regarding the research aims and the methods used for the current study. My deepest gratitude goes to Prof. Uday A. Athavankar to share his insights as one of the participants and further aid in bringing more respondents for the study. This experimental study wouldn't have been possible without the support of Prof. B. K. Chakravarthy. He provided us with good space of IDC studio to conduct the experiment.

A Big Thank You to all the respondents from IDC and other departments of IIT Bombay for their keen interest to sort cards and solve the puzzle.

References

1. Ereshefsky M (2000) The poverty of the Linnaean hierarchy: A philosophical study of biological taxonomy. Cambridge University Press, Cambridge
2. Nielsen J (2004) Card sorting: how many users/to test. Alertbox: current issues in web usability. Available at:www.useit.com/alertbox/20040719.html
3. Nath N, Poovaiah R (2011) Understanding richness: a new typology of visual cues and meaning in Signs. Design, development and research conference, CPUT, Cape Town
4. Lazarsfeld PH, Barton AH, Bartholomew DJ (eds.) (2006) Qualitative measurement in social sciences: classification, typologies, and indices. Measurement 1:149–150 (SAGE Publications)
5. Maple A (1995) Faceted access: a review of literature. Music library association annual meeting
6. Dalvi Vinod G (2010) Conceptual Model for Devanāgarī Typefaces. Ph.D. Thesis, industrial design centre (IDC), Indian Institute of Technology—Bombay (IIT-B) 2010

PREMΛP: A Platform for the Realization of Engineered Materials and Products

B. P. Gautham, Amarendra K. Singh, Smita S. Ghaisas, Sreedhar S. Reddy and Farrokh Mistree

Abstract Integrated computational materials engineering (ICME), an integrated systems engineering approach is expected to (a) reduce the time and cost of discovery and development of materials and their manufacture, and (b) enable faster development of products assisted with richer material information. Development of a comprehensive IT platform that facilitates this through the integration of models, knowledge, and data for *designing both the material and the product* is a need of the day. In this paper, we introduce PREMΛP—Platform for Realization of Engineered Materials and Products conceptualized for this purpose. We also introduce two foundational problems that include (a) the development and production of steel mill products meeting stringent requirements of quality and cost and (b) the integrated design of gears and their manufacture, and how these are envisaged to be executed on PREMΛP. We envisage PREMΛP to be a platform for discovering new materials and concurrently designing materials, manufacturing processes and engineered components. The three associated papers in this series deal with application of the compromise Decision Support Problem construct for a manufacturing process

B. P. Gautham (✉) · A. K. Singh · S. S. Ghaisas · S. S. Reddy
Tata Consultancy Services, Pune 411 013, India
e-mail: bp.gautham@tcs.com

A. K. Singh
e-mail: amarendra.singh@tcs.com

S. S. Ghaisas
e-mail: smita.ghaisas@tcs.com

S. S. Reddy
e-mail: sreedhar.reddy@tcs.com

F. Mistree
School of Aerospace and Mechanical Engineering, University of Oklahoma, Norman OK 73019, USA
e-mail: farrokh.mistree@ou.edu

A. Chakrabarti and R. V. Prakash (eds.), *ICoRD'13*, Lecture Notes in Mechanical Engineering, DOI: 10.1007/978-81-322-1050-4_104, © Springer India 2013

design and component design problems, and perspectives on knowledge engineering application in the platform being proposed.

Keywords ICME · Integrated simulation · Materials design · Knowledge engineering · Robust design

1 Introduction

Traditionally, a component designer selects a material from a list of choices he/she has, to meet the design and performance requirements. For example, Ashby's charts [1] are widely used for preliminary material selection during the design of mechanical components. However, a designer uses these merely as look up database for short-listing of material options for further exploration. However, the actual performance of the material in the product often depends on parameters not addressed at this stage; for example the cleanliness of the steel or flow lines from a shaping operation. Having a closed-loop over the mechanical design of product, and the material and manufacturing process parameter selection will lead to better utilization of the material and better products. Such an integrated approach helps speed up the introduction of new materials and can support new material development in a target-oriented fashion.

Integrated Computational Materials Engineering (ICME) is conceived as a way future materials development will be done in close association with end product design [2, 3]. An integrated systems engineering approach envisaged in ICME is expected to reduce the time and cost of development of new materials and their manufacturing processes besides allowing for faster deployment of new materials in products and products designed with detailed material state information. Besides this, such an approach is expected to enable systematic development of materials as compared to the current dependence on heuristics or serendipity. The use of modeling and simulation tools to facilitate the exploration of the materials and product design space simultaneously is envisaged as the primary driver to reduce time consuming and expensive experimentation [3]. Modeling and simulation tools are expected to map the composition, processing, structure and properties of materials and ultimately link them to evaluate product performance. However, the current state of the art of materials science maps these only partially with significant gaps [2, 4]. Besides this, such models can, at best, only be used to answer questions on the impact of changes in composition and processing on material related performance through structure–property relationships. In view of the nature of these relationships, special methods are required to address inverse questions such as designing appropriate composition and processing parameters for desired end performance. To address the gaps in materials science, in addition to the use of physics-based models and established empirical models, one needs to look at other opportunities such as mining the vast amount of materials related data available for discovery of new phenomena or relations and the use of tacit or experiential knowledge [5]. All of these require multiple tools spanning a wide range of functionalities and complexity. Such a broad based study of material-manufacturing-component-performance problems must be facilitated by knowledge engineering systems to support human decision through the use of tacit knowledge and studies of past designs and systematically guided new experiments to reduce experimental efforts. Finally, such a platform will have many stakeholders such as scientists, manufacturing process engineers, product designers and managers, and may require considerable collaborative decision-making. Enabling all these possibilities can be achieved only through the use of appropriate IT systems [2].

There have been various attempts at building ICME systems for specific purposes. The AixViPMaP platform developed at RWTH Aachen University deals with linking models for a chain of manufacturing process related to steel processing [6]. A European consortium developed through-process-modeling tools for aluminum, the Vir* set of tools [7] which link processes from casting to sheet forming. Use of commercial tools such as iSightor ModeFrontierto integrate simulation tools [8] have also been reported. These integration tools facilitate communication between simulation tools, build surrogate models and carry out optimization. The PLM/PDM tools, can also help link simulation tools, however they need considerable customization for each application. None of these tools are domain specific and generally it is not easy to use them to set up knowledge support systems to assist guided design. A comprehensive materials engineering specific IT platform that addresses these problems with tools for modeling and simulation, materials informatics, knowledge engineering, decision support systems, collaboration, etc., is not available to the best of authors' knowledge.

In this set of four papers, we describe our work towards the development of a platform for ICME at Tata Consultancy Services.

- PREMΛP—A Platform for the Realization of Engineered Materials and Products.
- PREMΛP—Exploring the Design Space for Continuous Casting of Steel, [9]
- PREMΛP—Exploring the Design and Materials Space for Gears, [10]
- PREMΛP—Knowledge Driven Design of Materials and Engineering Processes, [11].

In this, the first paper, we describe our vision for an IT platform, PREMΛP, enriched with various tools for modeling and simulation and supported by informatics, knowledge engineering, collaboration, robust design and decision support tools along with appropriate databases and knowledge bases that will facilitate the simultaneous exploration of the material and product design spaces. We also elaborate on the way the platform is envisaged to be used by illustrating of two key problems concerning steel mill products and mechanical components made of steel. We illustrate the exploration of the material and the component design spaces using the compromise Decision Support Problem (cDSP) [12, 13] in two companion papers [9, 10]. In Ref. [9] we illustrate the exploration of materials space using the cDSP for a continuous casting operation, where operational parameters are to be evolved to meet target properties at maximum productivity. In

Ref. [10] we illustrate the exploration of the design space involving geometric parameters and material properties for a transmission gear where the total cost of the gear must be reduced while meeting performance requirements. In Ref. [11] the authors introduce the ontologies, knowledge engineering systems and the platform architecture of PREMΛP.

2 Purpose of the PREMΛP

2.1 Motivation

The advances in modeling and simulation tools for materials and manufacturing processes have significantly enhanced the capability of predicting evolution of properties of materials and also the performance of the material under service conditions. This facilitates addressing a number of material related decisions in industry to reduce the dependence on heuristics and individual knowledge. Some of the commonly encountered problems are listed below:

- New product design—material and manufacturing process selection that meets material performance requirements, low overall cost, etc.;
- Material and/or process substitution—to reduce cost, enhance performance, reduce weight, geographic constraints on cost and supply chain;
- New material development—to develop new material with enhanced performance properties or customize material composition and processing for given needs;
- Materials for special needs—to impart very specific properties through material and manufacturing process design;
- Develop and/or enhance specifications—to accommodate new needs or if components, which were designed and manufactured to specifications fail in service for no previously known or obvious causes.

All the above have to be carried out with a significant reduction in lead-time, minimum disruption to existing production and wider exploration of opportunity space for better material utilization in the future. PREMΛP is envisaged to be a platform to help address such problems in an industrial environment. In addition to its application in industry, the platform also will be useful for the research community.

2.2 Usage Scenarios

PREMΛP is a platform for concurrently designing materials, manufacturing processes and engineered components with increasingly exacting requirements through intelligent recommendations and decision support systems that engineers can use to make appropriate decisions in an industrial setting. With PREMΛP in place, in the long term, we envision using it to discover and improve materials, manufacturing processes, and product combinations. It is envisaged that the platform will be used by a variety of users, including experts involved in new materials or process development, process experts that can make standard templates for usage by non-experts in their day to day work, non-expert users who use the predefined templates and researchers testing their new models/tools for simulation. The potential utility of PREMΛP for various stakeholders is summarized below.

For expert users:

- Build simulation process steps for a chain of events that occur during material processing, testing and product design where simulation tools are available for a given need; knowledge assistance will be made available to facilitate this.
- Set the process chain in a decision support framework to make informed decisions about the design of the material, process parameters and product;
- Add models/simulation tools with appropriate connectors to the platform along with knowledge elements;
- Enable robust design in the presence of noise and uncertainty;
- Enable an engineer to explore material and component design spaces concurrently and to design the material and the component concurrently.
- Build simulation/design templates to make them available for non-expert users;

For non-expert end user:

- Use an existing template designed to solve specific problems (e.g., templates with a series of manufacturing operations to determine the process set points)
- Knowledge enablement for use with the process templates through guidance on inputs to be given or models to be selected (e.g., recommend an initial guess for temperature and time for a heat treatment process.)

For researcher:

- Ability to add/substitute models and verify them for improved performance
- Design and develop new materials and manufacturing processes
- Enrich the knowledge base through machine learning from past cases as well as systematic inputs from experts
- Statistical model development and knowledge enhancement through materials informatics/data mining
- Build low fidelity models from high-fidelity models for faster computation with desired accuracy.

In the next section we describe the key components of PREMΛP.

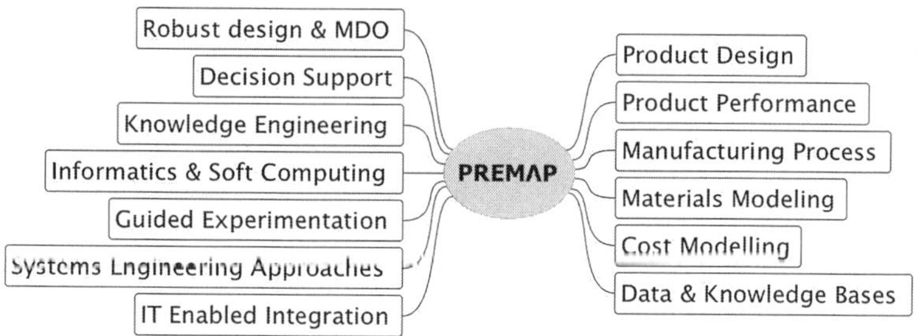

Fig. 1 Domain independent (*left*) and domain dependent (*right*) components of the platform

3 Components of PREMAP

In view of the vast numbers of diverse material systems and component/product application categories that such a platform has to deal with, we envisage both a set of domain dependent and a set of domain-independent components as shown in Fig. 1. On the right side of the figure are the components that are domain dependent and those on the left are domain independent. This categorization is done in order to build an extendable platform, which can be used for multiple domains. A domain may refer to a material category with associated manufacturing processes and/or a product category. Appropriate ontologies are being developed and applied to develop data and knowledge models of various components to make integration of various components and adaptation and extension to different domains easier, in addition to enabling drawing inferences based on ontologies.

PREMAP enables setting up design problems with a systems engineering approach. It offers support tools to capture requirements, constraints, goals etc., for a specified design problem. The key tool bases that are being developed include, (a) tools for robust design to facilitate the exploration of design spaces and determine ranged sets of specifications, and multidisciplinary optimization techniques (MDO), (b) decision support tools (e.g., the compromise decision support problem construct) driven by both formal mathematical techniques as well as heuristic techniques, (c) knowledge engineering tools based on case-based or rule-based reasoning, machine learning and data mining tools for capturing knowledge from the past cases and (d) materials informatics for the discovery of knowledge from data bases and (e) design of experiments and combinatorial experimentation tools to drive both simulation and experimental studies. Select tools for achieving the above are being added to the platform with facilities for adding more in the future. The platform also needs appropriate IT architecture that enables configurability of different tools and their communication in an easily adaptable and scalable way. Data and knowledge bases regarding materials, processes, products, use of simulation tools etc., are being captured in an ontological framework and a detailed description is given in the fourth paper of this series [11].

4 Illustration of PREMAP Usage

As an illustration, we look at an end-to-end problem involving a steel component starting from chemistry selection (secondary steel making). The life cycle of design and manufacture of a steel component, generally, involves two major stages; (a) production of an intermediate mill product form such as a rolled sheet/rod or an ingot carried out at the steel mill and (b) design and manufacture of the component carried out by the original equipment manufacturers (OEMs). The material related performance of the component is influenced by the design of the product as well as the processes involved both during the mill product production and manufacture to the final form from the mill product stock. Hence, the component design and material and manufacturing process selection needs a higher degree of coordination than that is in conventional practice today in order to achieve products with more stringent requirements (e.g., weight). The component considered for illustration is one of the critical components in an automotive industry, a gear. For gear design, information from the manufacturing process on the distributed properties of the materials such as strength, fatigue life and residual stresses from the initial bar stock from which it is manufactured can lead to a better design with reduction in total cost. Other aspects such as forming defects or final fatigue properties may depend on segregation and inclusions that come from mill product production processes, besides the material chemistry [14]. An integrated analysis allows for the interaction between the steel mill and component design and manufacturing in a more informed way.

We plan to design the steel and the gear in a closed loop involving design of the product, material and its manufacture, using physics based models, empirical models and knowledge engineering systems. The materials design problem and a discussion of the integration of the two problems are described next.

4.1 Development of Steel Mill Product

A steel mill produces a variety of intermediate mill products such as slabs, billets, and blooms and finished flat and long products such as sheets, bars, etc. These intermediate and finished products are used for component manufacturing with or without further processing. We consider two such applications from automotive domain.

The first one is for the production of steel sheet, which is used as a raw material for the manufacture of car body and other sheet metal components. The production of a steel sheet in a typical flat product production line is illustrated in Fig. 2. Starting from the chemistry and cleanliness of the molten steel, casting, rolling, heat treatment, forming and other processes influence the properties and performance of the steel sheet produced in the mill. The performance of a formed sheet

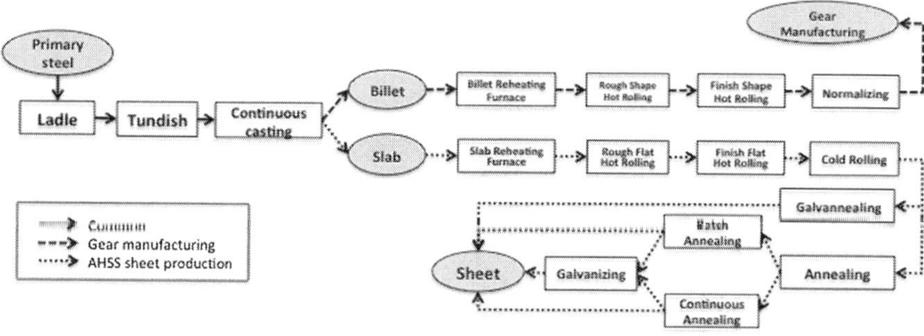

Fig. 2 Flow diagram for steel mill products

component in an automobile is influenced by the entire life cycle the material has gone through in its production including the mill processes.

The second mill product, a steel bar, is produced in a long product line as shown in Fig. 2, and is used as raw material for the production of a transmission gear. Again the evolving properties of the bar are dictated by the chemistry and cleanliness of molten steel, the micro and macro structure of the cast products and the subsequent deformation and thermal processes.

Tracking the evolution of the properties as well as determining the influence of evolving structure on various properties is a challenging task. Apart from developing models to track the evolution of properties, an enabling platform is required to integrate these interconnected models across multiple-scale as well as multiple unit operations with knowledge and data to obtain the optimized design set points of various unit operations of the entire production path. This will assist develop materials and processes to meet the ever-increasing demands of lighter sheet materials or bar with better properties. PREMΛP, the enabling technology described in this work, is aimed at addressing such complex problems.

Here PREMΛP links various physics based models (e.g., fluid flow, heat and mass transport, diffusion and deformation) along with empirical correlations (e.g., for property prediction) as appropriate, for various processes illustrated in the picture. The models, after successful validation, will be subsequently put in a decision support framework to assist design of the chemistry and process conditions to achieve desired end properties.

Among the processes given in Fig. 2, continuous casting operation is a crucial step involved in production of a variety of steel products such as slabs and billets from liquid steel owing to the complex nature of physics involved (multi-phase flow, phase transformations from liquid to multiple solid phases, chemical segregation, etc.) and its effect on further downstream operations. Productivity and yield are two key parameters indicative of caster performance and both are required to be on the higher side while meeting quality requirements from an industrial perspective.

In a companion paper [9], we explore the design space consisting of process and equipment parameters for this important intermediate operation of slab casting, which is used for sheet manufacturing. The same methodology is valid for exploring the design space for casting a billet, which is used for gear manufacturing. An integrated design framework comprised of a cDSP using metamodels for continuous casting has been developed to explore its design space. Our focus is on demonstrating the potential of integrating mathematical modeling and multi-objective robust design formulations in an integrated design framework specifically for the materials and design communities.

4.2 Integrated Design of Steel Gears

One of the largest consumers of steel gears is the automotive industry. Automotive transmission gears are generally helical or bevel gears and are made of case carburized steels. While fatigue life and reliability are the most important driving parameters, reducing transmission weight and noise are key drivers for successful gear design. Some of the challenges articulated in the Gear Vision document [16] that are affected by materials include the need to (a) enhance power density, (b) achieve 100 % reliability for 20 years for large gears, (c) increase overall achievable accuracy during manufacture at the same perceived cost, and (d) reduce time to market for gear components by 25 % in 5 years and 50 % in 10 years. The document further states that achieving these require innovative gear design and development with predictive tools (virtual testing). It also anticipates the use of steels that heat treat to RC70+ and higher with cleaner steel to enable greater power density; new steels to reduce/eliminate intergranular oxidation; shot peening and other innovative manufacturing processes for improved performance; and high quality and stable production processes.

This vision can be achieved by concurrent design of products and materials and manufacturing process with greater information exchange between the design and manufacturing teams to address the interconnected problems. In Fig. 3, the "manufacturing" block illustrates the choices of manufacturing processes to generate the shape and tailor the material properties to the desired product performance. The upper block illustrates the design requirements. Currently, a product designer uses this information along with the limited amount of information about material properties available from material and manufacturing departments. ICME paradigm enables tightly coupled materials and manufacturing processes development and product design.

If simulation assisted detailed material state information is available, the designer can take advantage of the actual variation of properties of the material (e.g., hardness distribution in place of two parameters—case and core hardness; impact of residual stresses), on the gear fatigue life to make a better design. Often the impact of processes such as shot peening are not fed back into the design

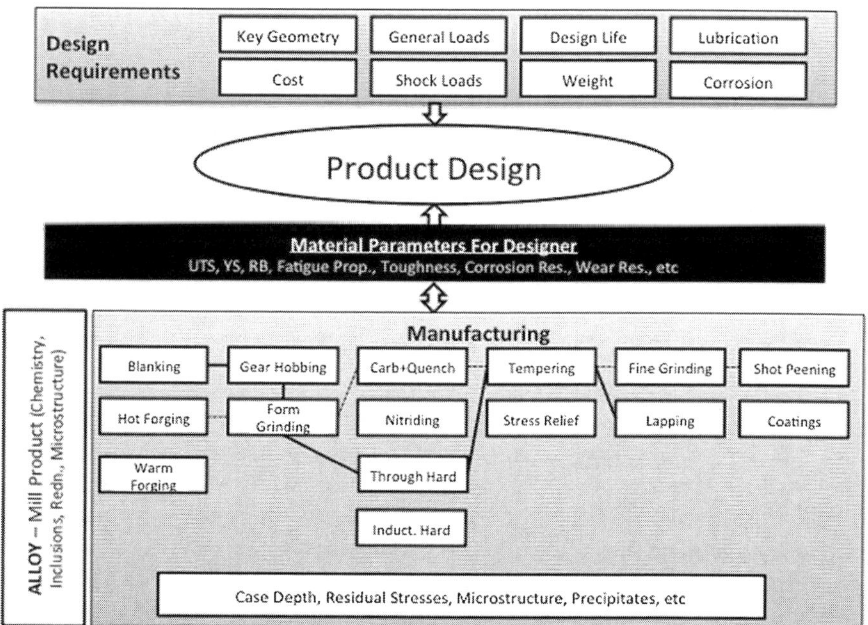

Fig. 3 Interaction between design and manufacturing for gears

process due to lack of appropriate tools. An ICME platform such as PREMΛP is expected to facilitate this.

As illustrated in Fig. 4, a number of alternative paths are possible to achieve the final desired properties and a systematic evaluation of some of these can lead to a more cost effective design. Besides this advantage, such an enablement can reduce the dependence on empirical correction factors of AGMA (American Gear Manufacturers Association) to lead to better designs, effectively bringing in a paradigm shift in gear design. When the simulation chain also incorporates "virtual testing" (e.g., simulation of operating conditions incorporating fracture phenomena), it may lead to discovery of new failure modes, not addressed in standard design procedures, when designs are significantly modified from the range of standards (e.g. slender teeth to reduce noise may lead to a new failure mode called tooth interior fatigue failure [13]).

Tools for addressing a subset of the above problems to establish and demonstrate such an approach for gear design and manufacture are being developed on PREMΛP. Here we concentrate on one major manufacturing path with two key manufacturing processes, carburizing and tempering. These are linked to AGMA based material design and fatigue analysis for design and virtual testing segments, as illustrated in Fig. 4. The purpose of this exercise is to carry out geometric design, material selection and manufacturing process parameters in a closed loop for overall cost reduction. This will be put in a formal decision support framework through use of compromise decision support problem (cDSP) framework for

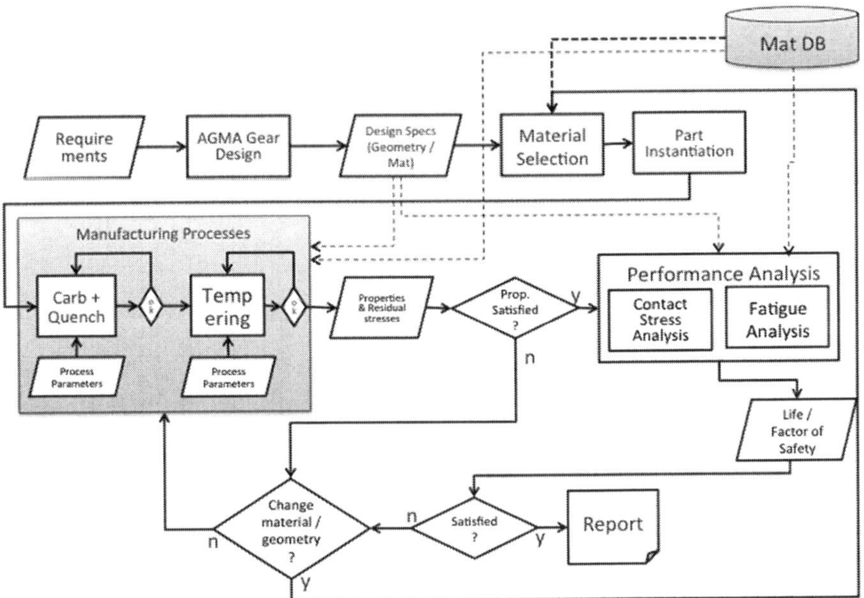

Fig. 4 Flow chart for the integrated gear design

obtaining good engineering solutions to meet the required targets. A companion paper, third in this series [10] illustrates the application of cDSP for a subset problem consisting of geometric design and material selection.

In Fig. 4, the key functional and performance requirements, constraints and other information are captured in the requirements block. The AGMA gear design block allows a geometric design and specification of material requirements. Subsequently, material selection will be carried out from the database. With the information on geometry and material, a gear part will be instantiated in appropriate CAD format and FEA mesh in PREMΛP for further analysis. We simulate the carburization-quenching and tempering operations using thermo-mechanical-metallurgical models to predict the residual stresses, hardness and other material properties with point to point mapping. The output of these simulations will subsequently be used in stress and fatigue analysis. Here we use standard empirical relations to link the hardness to fatigue strength and also incorporate the effect of residual stresses. This linkage allows the designer to check the feasibility of various combinations of manufacturing process parameters and make informed decisions and further use in virtual testing to verify design in an iterative mode.

At various stages, knowledge engineering tools guide selection of appropriate parameters. For illustration, the user would get guidance on appropriate temperature and time for carburization through simple formulae, which are input for detailed simulation so that even if the user is not very familiar with the process, he/she will start with a reasonable combination. Armed with this information, a designer can either modify geometry or material along with process parameters to improve the design (weight, cost, etc.). In future we will add more upstream processes such as gear blanking, machining etc., and the impact of the mill product state from the previous study will be incorporated.

5 Closing Remarks

In this paper we introduced PREMΛP a platform, being developed by Tata Consultancy Services, for concurrently discovering and designing new materials, manufacturing processes and engineered components. With the advent of modeling and simulation tools and other enabling technologies from the information technology field such as knowledge engineering, the future of materials, manufacturing processes and product design is set to make a paradigm change. PREMΛP is being developed to fulfill this vision. Various aspects of PREMΛP such as usage scenarios, its segments and its architecture are briefly discussed. In the next two papers in this sequence we introduce two problems of relevance to industry that are being developed as test examples for PREMΛP and outlined how we propose to use a combination of modeling and simulation tools along with knowledge engineering and decision support tools in these cases. This paper also sets the context for three related papers on using cDSP and knowledge engineering framework in PREMΛP.

Acknowledgments The authors acknowledge the management of TCS for encouraging this research. The authors acknowledge various colleagues who have contributed to this endeavor through discussions and suggestions. Farrokh Mistree acknowledges financial support from the L.A. Comp Chair at the University of Oklahoma.

References

1. Ashby MF (2005) Materials selection in mechanical design. Butterworth-Heinmann. ISBN: 9780750661683
2. Pollock TM, Allison J (2008) Integrated computational materials engineering: a transformational discipline for improved competitiveness and national security. The National Academies Press, National Research Council, Washington, D.C
3. McDowell D, Panchal J, Choi H-J, Seepersad C, Allen J, Mistree F (2010) Integrated design of multi-scale, multifunctional materials and products. Elsevier Inc, New York. ISBN 9781856176620
4. Fullwood DT, Niezgoda SR, Adams BL, Kalidindi SR (2010) Microstructure sensitive design for performance optimization. Prog Mater Sci 55:477–562
5. Rodgers JR, Cebon D (2006) Materials informatics. MRS Bull 31:975–980
6. Schmitz GJ, Prahl U (2009) Toward a virtual platform for materials processing. J Metals 61(5):19–23
7. Hirsch J (ed) (2005) Virtual fabrication of aluminium alloys. Wiley-VCH, Weinheim

8. Backman DG, Wei DY, Whitis DD, Buczek MB, Finnigan PM, Gao D (2006) ICME at GE: accelerating the insertion of new materials and processes. J Metals 58(11):36–41
9. Kumar P, Goyal S, Singh AK, Allen JK, Panchal JH, Mistree F (2013) PREMΛP—exploring the design space for continuous casting of steel. In: Proceedings of the 4th international conference on research into design, IIT Madras, Chennai, India, Paper no. 65
10. Kulkarni N, Zagade PR, Gautham BP, Panchal JH, Allen JK, Mistree F (2013) PREMΛP—exploring the design and materials space for gears. In: Proceedings of the 4th international conference on research into design, IIT Madras, Chennai, India. Paper no. 64
11. Bhat M, Das P, Kumar P, Kulkarni N, Ghaisas SS, Reddy SS (2013) PREMΛP—knowledge driven design of materials and engineering processes. In: Proceedings of the 4th international conference on research into design, IIT Madras, Chennai, India, Paper no. 66
12. Mistree F, Hughes OF, Bras BA (1993) The compromise decision support problem and the adaptive linear programming algorithm. In: Kamat MP (ed) Structural optimization: status and promise, Washington, D.C, pp 247–286 (AIAA)
13. Chen W, Allen JK, Mavris D, Mistree F (1996) A concept exploration method for determining robust top-level specifications. Eng Optimiz 6:137–158
14. Şimşir C, Hunkel M, Lütjens J, Rentsch R (2012) Process-chain simulation for prediction of the distortion of case-hardened gear blanks. Materialwiss Werkstofftech 43:163–170
15. Mackadlner M (2001) Tooth interior fatigue fracture & robustness of gears. PhD Thesis, KTH Sweden
16. "Gear Industry Vision: A vision for the gear industry in 2025." (http://agma.server294.com/images/uploads/gearvision.pdf). Accesed on 27 May 2012

PREMΛP: Knowledge Driven Design of Materials and Engineering Process

Manoj Bhat, Sapan Shah, Prasenjit Das, Prabash Kumar, Nagesh Kulkarni, Smita S. Ghaisas and Sreedhar S. Reddy

Abstract In this paper, we present the knowledge engineering aspects of an IT infrastructure for a Platform for Realization of Engineered Materials and Products (PREMΛP). PREMΛP enables harnessing available knowledge, learning emerging knowledge and continually creating new knowledge. It consists of an ontology-based, knowledge-assisted method and platform to capture, structure, configure and reuse knowledge for designing materials and engineering systems. The PREMΛP ontology provides extensible representation of data and knowledge. The semantic mappings of concepts in the ontology are used to draw inferences and provide pro-active guidance while designing manufacturing processes and selecting parameters to meet the product specification that addresses a given engineering problem. We show how the ontological models can help in (a) automating process design starting from a requirements statement to selecting a suitable design process and (b) creating a parameterized instance of the selected

M. Bhat · S. Shah · P. Das · P. Kumar · N. Kulkarni · S. S. Ghaisas · S. S. Reddy (✉)
Tata Consultancy Services, Pune 411013, India
e-mail: sreedhar.reddy@tcs.com

M. Bhat
e-mail: manoj.bhat@tcs.com

S. Shah
e-mail: sapan.hs@tcs.com

P. Das
e-mail: prasenjit.d@tcs.com

P. Kumar
e-mail: kumar.prabhash@tcs.com

N. Kulkarni
e-mail: nagesh.kulkarni@tcs.com

S. S. Ghaisas
e-mail: smita.ghaisas@tcs.com

A. Chakrabarti and R. V. Prakash (eds.), *ICoRD'13*, Lecture Notes in Mechanical Engineering, DOI: 10.1007/978-81-322-1050-4_105, © Springer India 2013

process. In the context of two engineering examples, namely, the simultaneous design of a gear and the steel from which it is to be made, we illustrate the salient features of knowledge driven design.

Keywords Material design · Manufacturing processes · Ontology and knowledge engineering

1 Introduction

Knowledge of structure–property relationships is at the heart of material science. Properties such as strength, hardness, toughness are the consequence of the material microstructure. Materials engineers experiment with compositions and processes in order to arrive at the right microstructure and thus the desired properties. But the structure–property relationships are not well understood;—especially across multiple length scales. While physics-based models do exist for some materials and phenomena, for many others there are no such models and one has to rely on experiential knowledge, past cases and rules of thumb. A lot of research has gone into this area over the years, a lot of results have been published and experimental data has been generated. A wealth of useful knowledge can be gleaned from these results, and insights can be drawn if this data is systematically collected and mined. PREMΛP is an attempt to accomplish this [1].

Knowledge guided process design is one of the core activities within PREMΛP. Different processes may be used for different purposes. Also, models of different levels of accuracy may be used within a given process—an approximate model for quick answers, and a detailed simulation model for more precise answers. The ability to set up such process chains, and carry out simulations is an integral part of the platform.

The vision for PREMΛP is to build a comprehensive knowledge base of materials, components and manufacturing processes, and use this knowledge to (1) significantly accelerate the design and development of materials and manufacturing processes, (2) significantly improve component design practice by integrating material information into the design process. Knowledge of different kinds needs to be captured: material composition, microstructure and properties; manufacturing processes and their effects on material structure and properties; component models, their design requirements and so on. This knowledge exists in a variety of forms: factual data, design guidelines, rules of thumb, process templates, simulation models, empirical models, case histories, and so on. This knowledge has to be captured and structured so that it can be effectively utilized, bringing the correct knowledge to bear on the right problems. It should effectively guide different stakeholders in their respective activities, such as capturing product specifications, setting up the manufacturing process chain, selecting input materials, executing and optimizing the process chain, and so on. A high quality ontology is a critical foundation to this kind of a platform.

So also is a high quality infrastructure for data extraction and integration. Data comes from a variety of sources such as online databases, manufacturer's catalogs and publications, in a variety of forms such as tables, graphs, diagrams and so on. It has to be extracted from all these sources and integrated. Also, data is produced in a variety of contexts, such as an experimental context and application context, etc. It is critical to identify these contexts, because data from one context may not be applicable in another context. We expect the PREMΛP platform itself will produce a lot of data as a part of process design simulation and lab experiments. This data also has to be captured and integrated with rest of the data.

In order to support all these things, the platform should have a highly flexible and open architecture. It should be possible to add new processes, models, rules, procedures, and so on without modifying the code of the platform. It should also be possible to add new simulation tools without affecting the rest of the infrastructure. Similarly it should be possible to add new data sources without changing the data architecture. Ideally, all these elements should be first class artifacts in the knowledge base, so that one can simply pick them up and use them as required, without having to hardwire them into the platform.

The rest of the paper is organized as follows. In Sect. 2 we present two materials engineering problems, namely, development of steel mill products and gear design, and discuss their knowledge requirements [2, 3]. This sets the context for subsequent sections. Section 3 presents the ontology for the materials engineering domain. In Sect. 4, we discuss the knowledge engineering required for PREMΛP. Section 5 presents the platform architecture and discusses its components. Section 6 discusses data extraction and integration. In Sect. 7 we discuss related work, and Sect. 8 concludes the paper.

2 Sample Design Scenarios

2.1 Gear Design

Figure 1 gives an overview of the gear design process.

This process template is specific to gear design. Similarly other engineering components have their own specific templates. The knowledge base should have the flexibility to store such component specific process templates. It should be possible to easily add new components and corresponding process templates, thus extending the capabilities of the platform.

The steps in Fig. 1 are as follows:

1. Requirements specification: The requirements are specified in terms of parameters such as rated torque, rated speed, reduction ratio, weight, cost and smoothness. Performance requirements such as fatigue life are also specified.

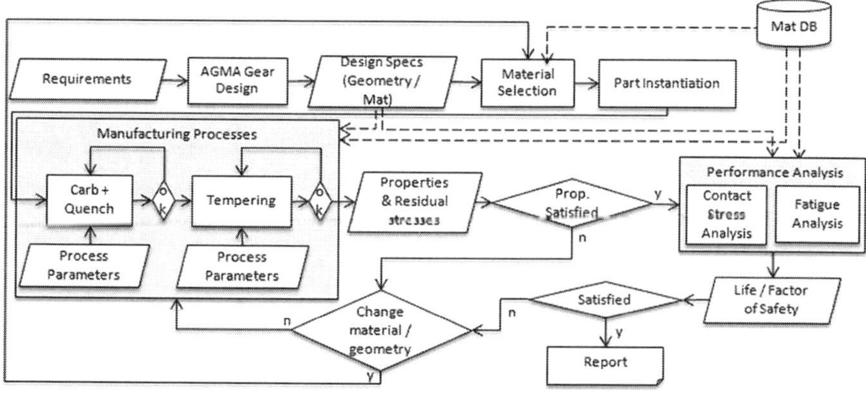

Fig. 1 Gear design process [3]

These are gear specific requirements. Similarly other components will have their own specific requirements. The knowledge base should provide the flexibility to capture component-specific requirements; so that while designing a component we know which requirements to select. There should be an ontology for capturing component specific requirements.

2. Gear design: To compute design parameters such as module, number of teeth, face width, addendum, dedendum, strength and hardness required, American Gear Manufacturers Association (AGMA) [4] guidelines are used. These guidelines are available as a set of rules. Knowledge base should store these rules, and these rules should be invoked when we reach this step in the process. Alternatively one might use an instantiated compromise Decision Support Problem (cDSP) [5] which internally invokes AGMA rules in order to determine the parameters. In [3] we have described this step in detail. We have explored gear geometry and material space using the cDSP construct. Different practical gear design scenarios are created (in cDSP) and solved using DSIDES.
3. Materials selection: Candidate materials for the given strength and hardness are selected. This information should be available in the database.
4. Carburization and quenching: We need to select design parameters such as carburization potential, carburization time, carburization temperature, diffusion time and diffusion temperature. There may be a set of rules that can determine these parameters based on property requirements. Once these parameters are selected, we can simulate the carburization and quenching process using an appropriate simulation model. Therefore, associated with the process step we need several elements in the knowledge base: process design parameters, rules to compute these parameters, an approximate model to compute them, and a simulation model to simulate the process. The knowledge base should be extensible enough to accommodate new processes, design parameters, rules and models. At the end of a simulation run, if we find that the requirements are not met, we may want to repeat the process by changing the design parameters.

There are a set of rules in the knowledge base that determine how the parameters should be changed.

Similar knowledge requirements exist for the subsequent process steps as well.

Another important requirement is an ontology for describing microstructure. The output of several of the simulation steps is a description of the microstructure. For instance, *carburizing and quenching* produces an output microstructure that goes as input to the tempering process. Knowledge about the microstructure consists of elements such as crystal structure, grain size, grain boundaries, vacancies, dislocations, segregations, inclusions and so on. This structure has to be available at the right level of detail so that one can reason (manually or with automation) about it. These indicate the need for an ontology that is specifically designed for modeling the microstructure.

We do not discuss the rest of the process due to space considerations, but the requirements on the knowledge base and platform should be clear from the description above. The design of gear using PREMΛP is discussed in detail in Ref. [3].

2.2 Development of Steel Mill Products

We want to be able to support three different use cases for development of steel mill products. These cases are discussed briefly here.

Case 1: For a given steel grade and requirements, find out a suitable process route, operating parameters and design modifications if any in an existing plant configuration.
Case 2: For a given requirement, find out suitable grades, process route, operating parameters and design modifications in an existing plant.
Case 3: Incorporation of new technology and unit operations for stretching existing plant capacities and suggesting new configuration.

Figure 2 highlights the information flow for these use cases. These three use cases need different information flows, and therefore need three variants of a process template. Hence the knowledge base should support storage of different use cases and use case specific process templates.

The process chain has several processing steps such as ladle, tundish, casting, annealing, rolling, etc. which are discussed in Ref. [1]. The knowledge requirements of these processes are similar to those for carburizing and quenching above. Suffice it to say that for each of these processes we need to capture design parameters, rules/models to set these parameters and simulation models. A part of process chain for development of steel mill products, exploration of design space for slab casting has been discussed in detail in Ref. [2]. We use cDSP to explore the space of design parameters such as super heat, casting speed, mold oscillation frequency, cooling conditions and so on to meet productivity and quality requirements.

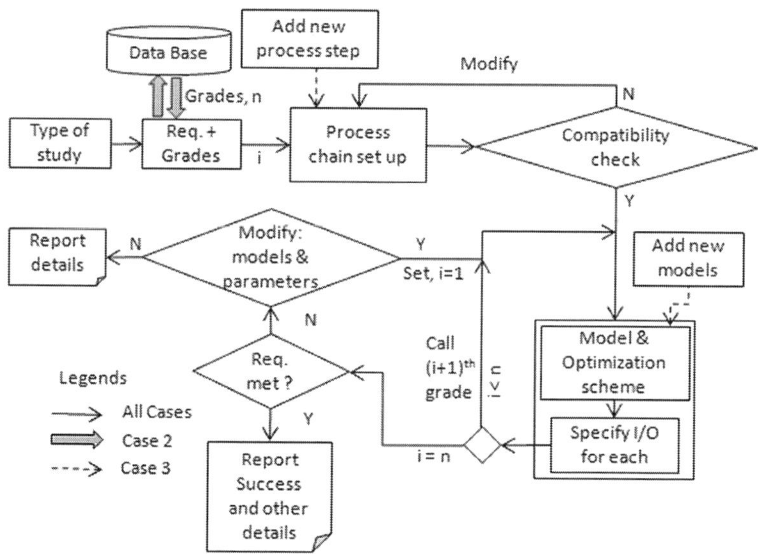

Fig. 2 Development of steel mill products

3 Ontology

The examples above emphasize that PREMΛP needs to extract and store different kinds of knowledge such as material compositions, their properties, material structure, manufacturing processes, their design parameters, rules, models of different kinds (approximate models, simulation models), engineering components, their specifications, component design process templates and use case specific process templates. The foundation to facilitate this can be provided by formalizing the information in terms of an ontology. The ontology can then serve as the basis for building different kinds of knowledge elements and associated knowledge services.

Ontologies are extensively used to capture the shared understanding of a domain by defining concepts, properties and relationships between them. Ontologies are machine-interpretable and enable reasoning. They enable reuse of domain knowledge and provide semantic interoperability between resources, services, databases, and tools [6]. PREMΛP ontology is categorized into Material, Process and Product ontology.

While evolving the ontologies we considered the following view points:

- The engineering goals of the platform, namely integrated design of materials, components and processes
- Views of stakeholders such as researchers, designers and platform builders.
- Physical phenomena that govern the engineering processes.
- Evolutionary nature of structure and properties across multiple length scales.
- Modeling of the above at different levels of precision.

We considered different design scenarios of the two example problems and held extensive discussions with subject matter experts to arrive at the structure and organization of different parts of the ontologies and their relationships.

We are in the process of validating the correctness and completeness of these ontologies by building proof of concept implementations.

3.1 Material Ontology

The Material ontology captures information related to materials, material structure and material properties along with various associations including structure–property relationships. The partial model of the Material ontology is represented using UML notation in Fig. 3.

The *iMaterial* represents the material in its bulk state. It is associated with the *form* (bar, sheet, billet, etc.) and *state* (solid, liquid and gas). The evolution of microstructures and properties of the materials are captured at each *representative unit*. The *representative unit* is classified as *point* and *representative volume element* with specific shape. A *material* and a *phase* within a *material* have specific *composition* which is classified as *element* and *compound composition*. The *composition* of the material is expressed as *element composition*, *compound composition* or both. For instance, the chemical composition of AISI 8620 steel

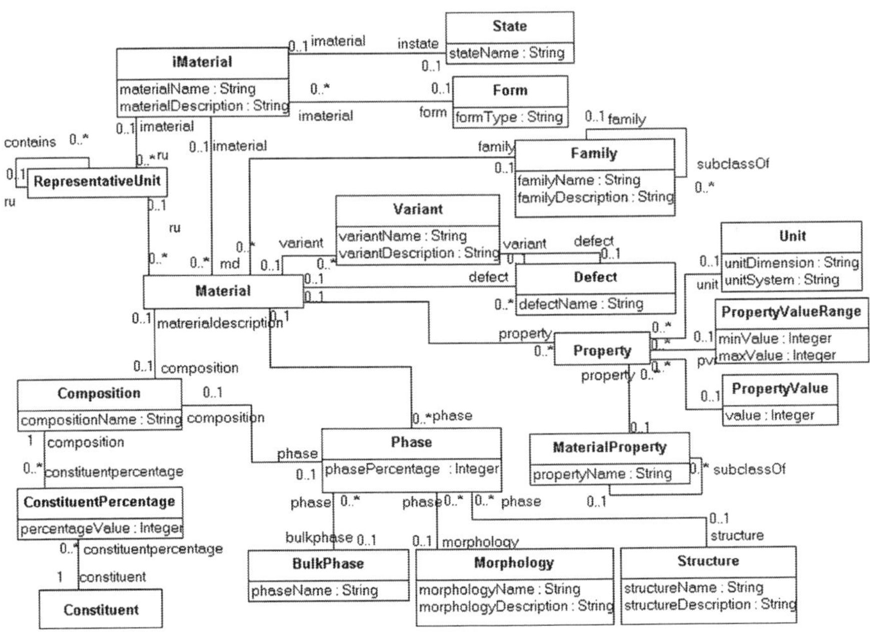

Fig. 3 Material ontology

could be *element*: Carbon (typical value: 0.2 %, range: 0.18–0.23 %) and *element*: Chromium (typical value: 0.5 %, range: 0.4–0.6 %). The constituent percentage of oxides of Aluminum, Manganese, Ferro-chromium or Ferro-manganese is captured. The microstructure of the material consists of *bulk phases* (Ferrite, Martensite, Austenite, Cementite, Pearlite and Bainite), *inclusions* and *precipitates*. Each of these *phases* has attributes such as *phase percentage* and *phase distribution* and is associated with *morphology* (lath, plate), *structure* (crystalline or amorphous) and *composition*. A *material* has specific *properties* which are broadly classified into *mechanical*, *physical*, *thermal*, *chemical*, *electrical*, *biological*, etc. which are further sub-classified. Each property has measurable value or range of values and associated units.

3.2 Process Ontology

The Process ontology explicates material and manufacturing processes along with property-process and process-performance relationships at different length scales (ranging from nanometer onwards to macro). The partial model of the Process ontology is represented using UML notation in Fig. 4.

Process ontology provides placeholders to capture the *processes* and *sub-processes* along with their *inputs* and *outputs*. For instance, *ladle* has *degassing* and *desulphurization* as *sub-processes*. The input and output associated with a *process* is an aggregation of properties and characteristics of one or more *work-material*, *non-work material*, *equipment*, *process* and *simulation tool*. The concepts are further detailed in the ontology to meet the framework's requirements. For instance, *simulation control* is classified as *mesh control parameter*, *time step*, *contact tolerance* and *convergence tolerance*.

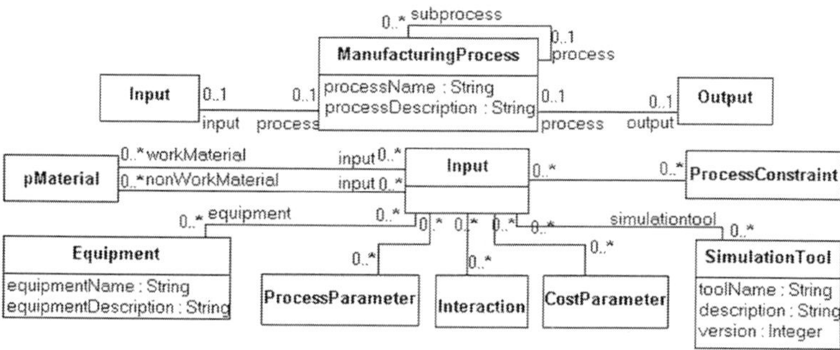

Fig. 4 Process ontology

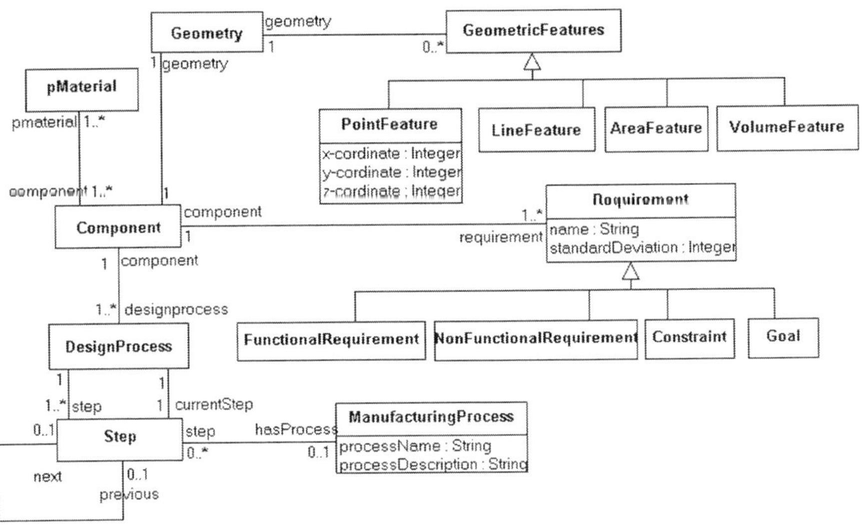

Fig. 5 Product ontology

3.3 Product Ontology

The Product ontology captures the problem description and the product specifications. The focus of the Product ontology is to facilitate product designers to capture the end *requirements* of the components. *Requirements* are classified as *functional requirements*, *non-functional requirements*, *goals* and *constraints*. For example *gear* has *geometry* which has one or more *geometric features* such as *pitch points*, *tip points*, *root points*, *line features*, *area features* and *volume features*. The partial model of the Product ontology is depicted in Fig. 5.

4 Knowledge Services

Knowledge services provide a mechanism to interact with the knowledge base of the PREMAP framework. These are published as web services so that any authorized external entity can use them to access the knowledge base. Knowledge needs of different stakeholders are discussed in paper 1 [1]. A knowledge service may need to reason across multiple knowledge elements such as rules, cases, equations, models and so on, to satisfy a given knowledge requirement. We have a knowledge service modeling mechanism (Fig. 6) that allows us to compose services that use knowledge elements spanning across multiple representation formalisms. A knowledge service is specified as a procedure with a number of steps, each of which is implemented by a knowledge element. Different knowledge representation mechanisms such as database, rule base, case base, equations and

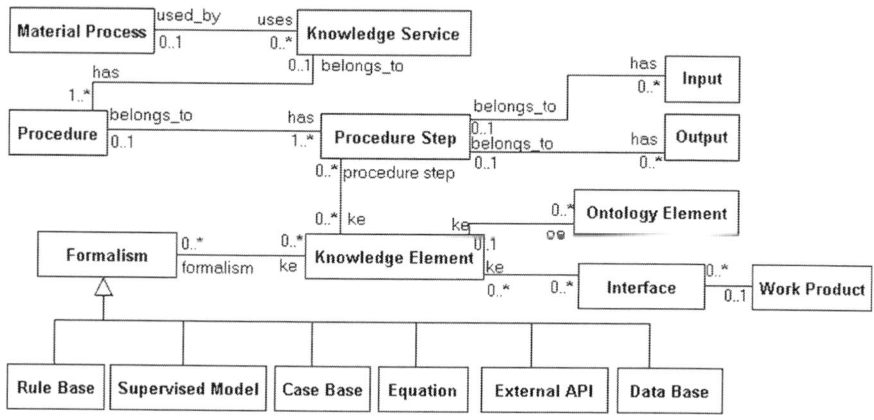

Fig. 6 Knowledge service

models need to be supported. Ontology provides the common vocabulary for expressing knowledge in all these forms.

To illustrate knowledge services, let us consider an example from the gear design problem. The first step in the gear design process is to select gear type such as helical, spur, bevel, etc. depending on requirements for power transfer capacity and direction. The selection is done by using a set of rules. This requires a rule engine to be invoked. We also use a knowledge service to determine design parameters such as billet geometry, austenization temperature, forging temperature, etc. This is achieved by solving a set of equations using an equation solver.

5 Platform Architecture

The PREMΛP platform is designed as a set of components plugged into the central integration bus. Each component offers specific functionality. The integration bus acts as a communication channel between the components. The platform is extensible and enables plug-in of new components to the integration bus to satisfy custom requirements. The newly added components will be able to interact with other components. Figure 7 shows the high level view of the PREMΛP platform architecture.

5.1 Knowledge Base

The knowledge base is the central information repository. It provides knowledge services to access the knowledge. The knowledge pertaining to materials, processes and products are captured in PREMΛP ontology which is classified into Material, Process and Product ontology.

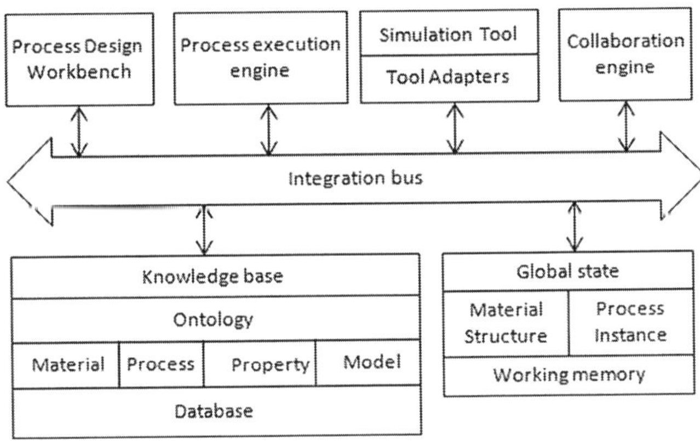

Fig. 7 PREMAP platform architecture

5.2 Process Design Workbench

The process design workbench enables definition of new material engineering processes. Existing process templates can be suitably customized and reused or entirely new processes can be defined. The workbench captures the material engineering processes in a notation similar to business process model notation (BPMN) [7]. Suitable services are bound to the activities in BPMN to enable execution of simulation tools, etc. These processes are executed by the process execution engine.

5.3 Process Execution Engine

This component uses a general purpose business process execution engine. The capabilities of this component are exposed as platform services. There are services to start execution of a defined material engineering process, to abort the execution and to check the status of process execution.

5.4 Global State

Global state is used as a working memory for tracing the evolution of material state during process execution. A process instance reads from and writes to the global state. If the desired state is not achieved one can go back and modify the design parameters and rerun the simulation. At the end of execution, the global state holds the final material state generated by the material engineering process. This information can be verified by an expert and saved to the knowledge base if deemed suitable.

5.5 Tool Adapters

Tool adapters allow the integration of external simulation tools with the PREMΛP platform. The capabilities of the platform can be extended by plugging in new tools through tool adapters. Once the new adapter is registered in the platform, the new tool is available to the platform and the same can be used during process definition.

5.6 Collaboration Engine

The collaboration engine allows various stakeholders such as material scientists, process and product designers to work collaboratively. The collaboration engine addresses the gaps that exist between material scientists and product and process designers by providing an integrated collaborative environment.

6 Data Extraction and Integration

Data comes from a variety of sources such as online databases, manufacturers' catalogues and publications. It is produced in a variety of contexts such as research context, experimental context, application context, and so on. It has to be extracted from all these sources and integrated into the PREMΛP database. Figure 8 shows the architecture of the data extraction and integration component. The component uses ontology mappings as a conceptual framework for integration. PREMΛP ontology provides a unified conceptual view of the data that exists in the database. Each source has a context specific ontology that provides a conceptual view of the data content available at that source. We map these source ontologies to the PREMΛP ontology to specify how their data should be integrated into the PREMΛP database. A source specific wrapper extracts data from the source and presents it as an instance of the source ontology. The architecture provides a flexible means for adding new data sources into the platform.

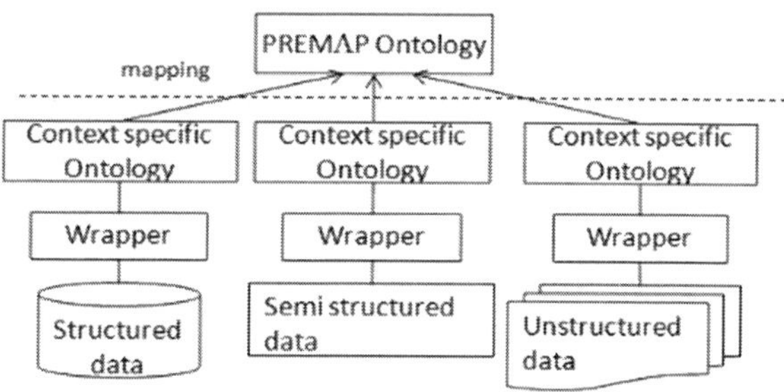

Fig. 8 Data extraction and integration architecture

7 Related Work

Work has been reported on (a) capturing and structuring knowledge for semantic interoperability in materials domain (b) providing an integrated platform for simulation and optimization of materials processing. MATML [8] is a well-known materials markup language proposed for information exchange. MATONTO [9] is based on DOLCE upper ontology [10] and builds on MATML and the works of Ashino [11] and Tanaka [12]. It classifies the knowledge of materials domain in an ontological framework. Materials Ontology [13] captures various concepts of materials structure, property, environment and processes. These models provide a unified representation to integrate heterogeneous data sources. However, they fail to address several important aspects such as microstructure of materials and process-structure–property relationships. Additionally, they do not address aspects relevant to engineering of components. They also do not provide constructs for capturing knowledge elements such as rules, cases, approximate models and so on. PREMΛP ontology addresses these.

Integrated Computational Materials Engineering (ICME) is identified as a strategic approach for future competitiveness [14]. It aims to integrate computational materials science tools into a holistic system that can accelerate materials development and unify design and manufacturing processes. The benefits of ICME have not yet been realized to its full potential in terms of significant reduction in time and cost involved in material and process development [15]. The ICME platform discussed in [16] lacks the product designers' perspectives. It also lacks the capability of knowledge guided process design, and knowledge engineering to learn from data and experimental results. The PREMΛP platform provides these capabilities.

8 Closing Remarks

In this paper, we have outlined the realization of the vision [1] for an integrated materials engineering platform–PREMΛP. To the best of our knowledge this is the first ever attempt that comprehensively addresses knowledge-critical core processes in the materials engineering value chain. In this context, we make three specific contributions to the materials engineering domain at large: (1) Foundational ontologies to formalize and organize domain knowledge, (2) Mechanisms for context-specific data retrieval and presentation and (3) A platform to seamlessly integrate process, data and simulation models.

The foundational ontologies provide a robust and extensible formalism to structure knowledge. The data extraction component maps varied sources of data onto this formalized structure and provides a context-specific view of relevant data to stakeholders. Finally, the platform architecture provides for mapping of target requirements onto appropriate processes, relevant materials knowledge and data needs and thus facilitates exploring of design spaces to arrive at solutions acceptable both in terms of their quality and economy.

The two illustrative examples bring out the need for a flexible and extensible design of a platform such as PREMΛP so that it can be employed to achieve target properties in materials and components in a variety of design scenarios. We have also discussed how PREMΛP can be used to iteratively (1) design engineering processes (2) experiment with process parameters (3) execute processes and verify outcomes.

With realization of PREMΛP, we hope the materials engineering domain can benefit from harnessing available knowledge, learning emerging knowledge and continually creating new knowledge.

Acknowledgments We thank Professor Farrokh Mistree, Professor Janet K. Allen and Dr. Jitesh Panchal for their valuable suggestions and comments.

References

1. Gautham BP, Singh AK, Ghaisas SS, Reddy SS, Mistree F (2013) PREMΛP—a platform for the realization of engineered materials and products. In: Proceedings of the 4th international conference on research into design, IIT Madras, Chennai, India. Paper number: 63
2. Kumar P, Goyal S, Singh AK, Allen JK, Panchal JH, Mistree F (2013) PREMΛP—exploring the design space for continuous casting of steel. In: Proceedings of the 4th international conference on research into design, IIT Madras, Chennai, India. Paper number: 65
3. Kulkarni N, Zagade PR, Gautham BP, Panchal JH, Allen JK, Mistree F (2013) PREMΛP—Exploring the design and materials space for gears. In: Proceedings of the 4th international conference on research into design, IIT Madras, Chennai, India. Paper number: 64
4. "Design Guide for Vehicle Spur and Helical Gears", ANSI/AGMA 6002-B93
5. Rolander N, Rambo J, Joshi Y, Allen JK, Mistree F (2006) An approach to robust design of turbulent convective systems. J Mech Des 128:844–855
6. Chandrasekaran B, Josephson J, Benjamins V (1999) What are ontologies, and why do we need them? IEEE Intell Syst 14:20–26
7. White SA (2004) Introduction to BPMN. BPTrends, July
8. Begley EF (2003) MatML version 3.0 schema. NIST 6939, National Institute of Standards and Technology Report, USA, Jan 2003
9. Cheung K, Drennan J, Hunter J (2008) Towards an ontology for data-driven discovery of new materials. AAAI workshop on semantic scientific knowledge integration, Stanford University, pp 26–28
10. Gangemi A, Guarino N, Masolo C, Oltramariand A, Schneider L (2002) Sweetening ontologies with DOLCE. In: Proceedings of the 13th international conference on knowledge engineering and knowledge management. Ontologies and the Semantic Web. Springer-Verlag
11. Ashino T, Fujita M (2006) Definition of a web ontology for design-oriented material selection. Data Sci J 5:52–63
12. Tanaka M (2005) "Toward a proposed ontology for nano-science", CAIS/ACSI: data, information, and knowledge in a networked world. The University of Western Ontario, London
13. Ashino T (2010) Materials ontology: an infrastructure for exchange materials information and knowledge. Data Sci J 7:54–61

14. Committee on Integrated Computational Materials Engineering, National Research Council (2008) Integrated computational materials engineering: a transformational discipline for improved competitiveness and national security. National Academic Press, Washington, DC ISBN:0-309-12000-4
15. Bradford AC, Backman D (2010) Advancement and implementation of integrated computational materials engineering (ICME) for aerospace applications. Interim Report, DTIC Document, March 2010
16. Schmitz GJ, Benke S, Laschet G, Apel M, Prahl U, Fayek P, Konovalov S, Rudnizki J, Quade H, Freyberger S, Henke T, Bambach M, Rossiter EA, Jansen U, Eppelt U (2009) Towards integrative computational materials engineering of steel components. JOM 61(5):19–23

Bridging the Gap: From Open Innovation to an Open Product-Life-Cycle by Using Open-X Methodologies

Matthias R. Gürtler, Andreas Kain and Udo Lindemann

Abstract Open-X methodologies describe the application of Open Innovation to different stages of the Product-Life-Cycle (PLC). Open Innovation deals with involving external players in a company's innovation process. Those can provide ideas from any stage of the PLC, such as lead users in the development stage or product-users in the utilization stage. These ideas themselves can initiate innovations in any PLC stage. However, this typically affects the development of new products. This means that ideas collected are incorporated into early PLC stages only. There is significant potential in using ideas not only for early stages but also for later stages, which means for existing products. Open-Utilization, as one form of the Open-X methodologies, can create innovations in the form of upgrades or new services for a product. Because respective PLC stages are not equally suitable for Open-X methodologies, this paper presents an assessment concept for evaluating each PLC stage regarding their Open-X capabilities and possible constraints. To illustrate the utility of the assessment concept, this paper identifies two PLC stages which demonstrate exemplary capacity for Open-X methodologies.

Keywords Open-X methodology · Open innovation · Product life cycle

---

M. R. Gürtler (✉) · A. Kain · U. Lindemann
Institute of Product Development, Technische Universität München, Boltzmannstraße 15, Garching D-85748, Germany
e-mail: guertler@pe.mw.tum.de

A. Kain
e-mail: kain@pe.mw.tum.de

U. Lindemann
e-mail: lindemann@pe.mw.tum.de

A. Chakrabarti and R. V. Prakash (eds.), *ICoRD'13*, Lecture Notes in Mechanical Engineering, DOI: 10.1007/978-81-322-1050-4_106, © Springer India 2013

1 Introduction

Open Innovation opens a company's innovation process to its environment [1]. This means new innovations are no longer solely created in isolated R&D departments but with the support of actively integrated customers, suppliers, other companies or even competitors. We can distinguish between two possible kinds of innovations according to the flow of information: (1) the outside in innovation, which uses external knowledge for the development of new or improved products, and (2) the inside-out innovation, which specifically gives information to the environment to enable external innovations [2].

Till this point, the research in this area has mainly been driven by economists such as Henry Chesbrough, Eric von Hippel, Ralf Reichwald, Frank Piller and others. These economists normally stopped short of considering the transition from the economic concepts to the technical realization. Engineering research started filling this gap by operationalizing and adapting those concepts in practical use [3, 4]. Currently the main research focus is on the early stages of the Product-Life-Cycle (PLC) such as the Open-Product-Development.

Thus far, examinations of the entire PLC from the perspective of engineering science have been rudimentary, at best. Howard mentions the relevance of considering the entire PLC from the view of the design stage by, for example, collecting information concerning later stages such as disposal [3]. Though the information relates to all stages of the PLC and can be gained from each stage, it is mainly used for the development of new products—which means an application in the early stages like concept or development. An application to existing products—in later PLC stages—is widely neglected, as illustrated in Fig. 1.

In the eyes of the authors this is a huge deficiency: although the early stages are of great importance in determining the costs of later stages, as shown in Fig. 2, these costs primarily appear in later stages. Concepts like Systems Engineering already consider the whole PLC but also only from the point of development.

Mistakes made in early stages might lead to high (changing) costs in later stages. Also, varying customer needs, changing markets or new technologies can

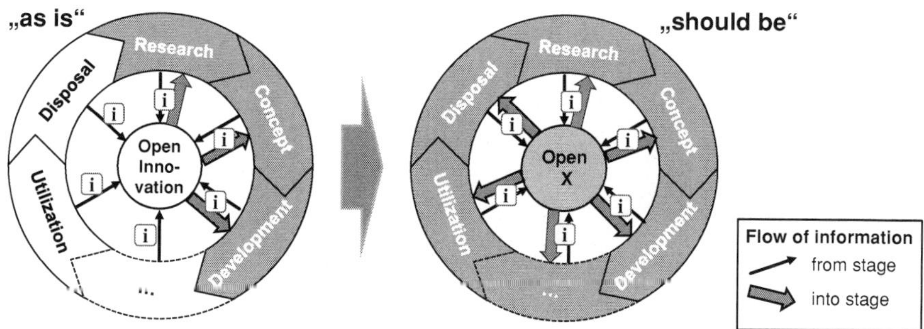

Fig. 1 Application of gained information in every stage of the product-life-cycle

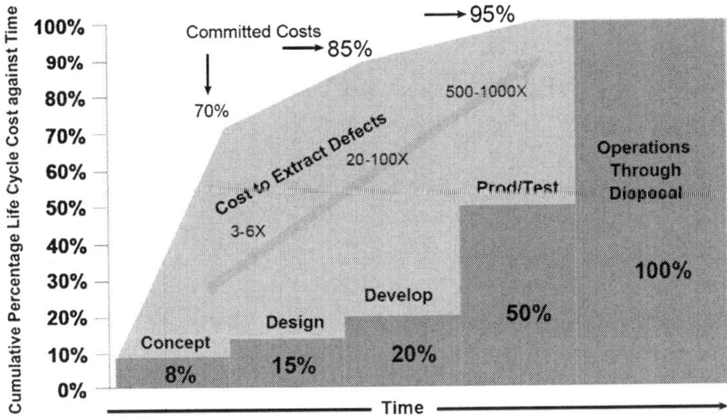

Fig. 2 Committed life cycle cost against time [6, Figs. 2–3]

cause product adoptions. With this, Open-X methodologies bear the potential for near-term improvements in later stages: for example Open-Utilization can help creating product updates or new services. As shown in Fig. 1, information which is gained all over the PLC should be used not only in early stages but across the entire PLC.

However, each stage has its own characteristics and requirements to Open Innovation methods: the range from the internal information being published, the retaining of intellectual property, supporting processes inside the company, and so on. Braun stresses the importance of defining in advance which stages of the innovation process are possible, advisable or necessary to be opened [5]. Due to the complexity that arises when all characteristics are considered, this is not a trivial task. Hence, there exists the need for a method to determine those aspects. This paper presents an evaluation sheet as part of an assessment concept to examine the characteristics of each stage of the PLC and analyze possible points of application for Open-X methodologies.

The following chapter introduces the relevant stages of the Product-Life-Cycle considered in the Open-X approach.

2 The Product Life Cycle

This section briefly presents the single stages of the Product-Life-Cycle (PLC). To design the Open-X approach on a workable foundation it is based on the Systems Engineering concept and its PLC considerations [6]. The Systems Engineering approach is widely accepted and used in research as well as in industry. Thus, the following examination is oriented towards the stages considered in Systems Engineering (shown below).

In this context a product also includes Product-Service Systems which consists of a technical product part and an intangible service part [7], as we expect a broad understanding of a product to support our PLC analysis. Figure 3 illustrates the enhancement of Chesbroughs' innovation process funnel [1] towards an entire Product-Life-Cycle view. The inner cycle symbolizes the company with its PLC stages and superordinate units such as organization, (long term) strategy and structure. The outer cycle illustrates the company's environment. Through the permeable borders innovations can be exchanged.

The following section gives a rough overview of the PLC. Due to spatial constraints the overview is restricted to half of the stages. For more detailed consideration please see the literature referenced in this paper.

2.1 (Product) Development

The purpose of the Development stage is to design a product that meets customers' requirements and is possible within the constraints of the company (e.g. available technologies, special production machines). In a classical innovation process the customer requirements are surveyed by the marketing department, preprocessed and transferred to the development department. This stage also constitutes strategies for integration, verification and validation of the designed product [6].

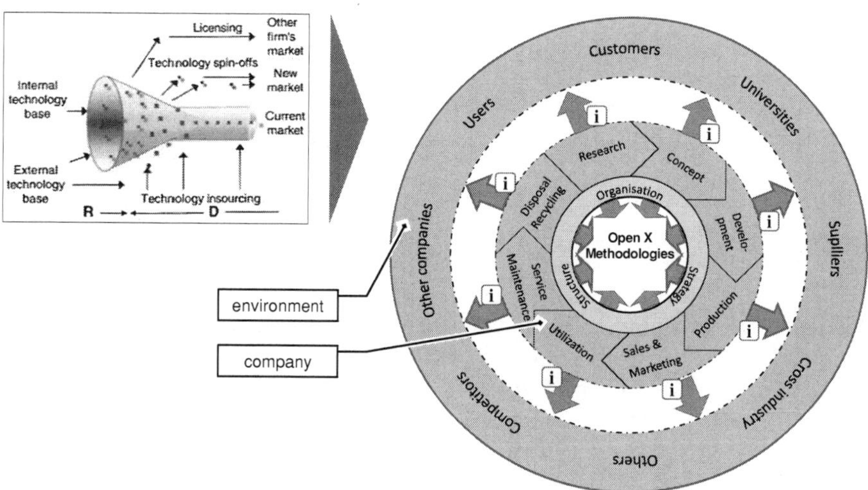

Fig. 3 Enhancement of open innovation towards the entire PLC (left figure after [1, Fig. 1.2])

2.2 Production

In the Production stage the previously designed product is manufactured, single components are assembled and the final product is tested. In some cases it might be necessary to modify the product to resolve production problems or enhance the product's capabilities [6].

2.3 Utilization

"The Utilization Stage is executed to operate the product, to deliver services within intended environments and to ensure continued operational effectiveness." [6]. To enhance the capability of a product upgrades might be designed and applied

2.4 Disposal and Recycling

This stage disposes of old products in an economic and ecofriendly way and provides supporting services. Normally planning for disposal is already part of the concept stage [6]. However, in the case of old products manufactured decades ago the disposal process can be challenging and expensive.

3 The Open-X Evaluation Sheet

As shown above, each stage has a different task in the innovation process, including different players, processes and resources. Thus, methods which are applicable for one stage might not fit to another due to specific characteristics and constraints. It is therefore not possible to give a blanket statement about the applicability of Open-X methods, demonstrating the need for a systematic analysis and assessment of each PLC stage.

The following section describes the structure and setup of the developed Open-X evaluation sheet as part of an assessment concept. The primary goal is supporting research to find possible points of application for Open-X methodologies and potential research areas. In a second step the evaluation sheet can also be used in industry for analyzing the potential benefits and threads of opening a specific PLC stage.

The sheet consists of five main categories which were determined by literature review and projects' experience. They are: actors, classification of innovation, effects, issue and provided information, and risks. Each main category comprises several first- and second-level subcategories, which are explained in more detail below. In this paper we focus on the outside-in innovation process due to its strong awareness in research and industries [8]. Our future considerations will deal with the inside-out innovation process as well.

3.1 Actors

This section describes: who can participate, how they can participate, how a company can access them and if there is a critical number of participants.

3.1.1 Participants

This category classifies the kind of external partner participating in the Open Innovation process in this stage of the PLC. It can be distinguished between random persons or experts [9]. It further differentiates between private persons and institutional participants as well as between direct or indirect customers or suppliers of the regarded company [10].

3.1.2 Accessibility to Participants

Closely linked to the kind of participants is the question of their accessibility. Is it sufficient to involve a random group of people or is it necessary to invite special groups with specific expertise, like Lead Users who might be determined in advance? [11].

3.1.3 Required Number of Participants

Depending on the issue and the stage of the PLC, either a small or a large number of participants might be suitable, which also directly affects cost and complexity.

3.1.4 Type of Participation

Depending on the issue, participants can support the innovation process with ideas/ information, services or even products.

3.2 Classification of Innovation

This section characterizes the types of innovation the opening of a specific stage can enable.

3.2.1 Target of Innovation

The innovation project can aim at a product (e.g. a new or improved product), a process (e.g. an improved distribution process) or an entire business case (e.g. the decision to enter a new market) [12].

3.2.2 Value Gain/Main Objectives

This section describes potential objectives for Open Innovation in this PLC stage. The corresponding Open-X project could target market share (e.g. by modifying the product portfolio), production (e.g. by lowering the costs, more flexibility), better insights in products' application and social and environmental improvements (e.g. less emission).

3.2.3 Type of Innovation

Innovation can either be incremental or radical. Incremental innovations are related to small improvements of existing products or processes, whereas radical or breakthrough innovations go along with fundamental changes [13].

3.3 *Effects*

Opening one stage can also affect other stages and can influence the whole company itself. These effects are discussed below.

3.3.1 Effects on Other PLC Stages

This category classifies the effect on other PLC stages resulting from changes in the stage being opened. Normally changes in early stages will affect all following stages. But changes in later stages might affect preceding stages, e.g. changes in the Production stage might lead to modified product architecture.

3.3.2 Possible Side Effects

The effects can be further distinguished as internal effects (e.g. on company departments), external effects (e.g. a better PR) and the impact on the company's network (e.g. on the supply chain).

3.3.3 Financial Effects

Each PLC stage should be analyzed as to whether extra value can be added to the product in this stage and if it is possible to gain direct revenue from this stage [adopted by 3].

3.4 Issue and Provided Information

In order to gain potential innovations by Open-X, the company first needs to provide some information by itself. The kind of information and its preparation is classified in the following.

3.4.1 Effort for Issue Definition

This section describes the effort and difficulty to define a suitable issue for an Open–X project and the amount of corresponding information needed to be published to ensure a sufficient result from a project. It also determines the kind of information, e.g. just text, photos, special data or a combination of them.

3.4.2 Complexity of Issue/Task

The complexity of an Open-X issue determines the kind of participants (e.g. a random crowd or a group of experts). Thus, it is essential to analyze the complexity of potential issues in each PLC stage.

3.4.3 Effort for Evaluating Gained Information

During the Open-X project a large amount of ideas and information will be collected. To operationalize them they need to be analyzed and categorized, and have useless items filtered out. Depending on the issue the evaluation can be carried out by a "random" employee, special experts, by a jury, or even by the participants themselves.

3.4.4 Transparency and Accessibility

In each PLC stage the quantity and quality of the information provided to the participants can differ along with the accessibility of information amongst the participants. For example, when considering critical parts or processes it might be expedient to publish little information and filter information gained by participants before giving them to the other participants.

3.4.5 Interaction/Feedback

This section deals with the question of whether interaction and feedback between the participants themselves as well as with the company or special experts might improve the outcome of the Open Innovation process [14].

3.5 *Risks*

This section describes potential risks that can occur by opening a PLC stage, which should be considered in advance.

3.5.1 Data Security/Knowledge Drain

This category defines the expected amount of information needed to be published to get a sufficient result. It also determines the degree of necessary confidential information.

3.5.2 Replicability of Accessible Information

This aspect deals with the question of whether participants or competitors can use the information provided by the company and the participants for their own business, or even for rebuilding the regarded product or system [3].

3.5.3 Strategic Risks

Here, possible strategic risks are determined. Risks include the drain of knowledge (company's knowhow as well as information from the Open-X project), the partial loss of system's control to participants or competitors [10], uncertainty of gained information, among others.

3.5.4 Possible Operative Barriers

To ensure the success of an Open-X project it is necessary to identify and classify possible interferences with the daily business and its impact on the company's processes and structures.

4 Examples: Open-Product-Development and Open-Disposal

To illustrate the concept and demonstrate its practical use, the Open-X evaluation sheet is applied to two examples from different PLC stages. The first one is located in the Development stage and is based on the experience from industrial idea contests. Here, two companies had developed new pre-products and were looking for innovative fields of application. For this they published a description and photos of their pre-product on an idea contest platform on the internet with the proposition: "What would you make out of…?" Here everyone was able to register and post their ideas.

The second example is generic and located in the Disposal stage. It considers the disposal of an old cargo ship designed and manufactured decades ago without caring about retirement or recycling issues. In this case, the fictitious shipping company looks for efficient, economic and ecofriendly ways to retire the old ship.

The choice of an early and a late stage will serve to emphasize the different characteristics of each stage. Due to space constraints just a part of the categories can be presented in detail (Table 1).

5 Discussion

As we can see, the Open-X methodologies bear great potential. For Open-Development this potential has already been verified by practical utilization in several industrial projects. Open-Development methods support the innovation process not only by gaining ideas for new fields of application, as shown in the prior example, but also with ideas for future products based on real stakeholder needs or solution ideas for challenges during the development process. In the case of retirement, Open-Disposal can contribute to gather both ideas to dispose and recycle an old product (e.g. a cargo ship) in an efficient and ecofriendly way, and maybe also support in becoming acquainted with specialized disposal-companies which offer corresponding services for the whole product or single subsystems.

Additionally, the evaluation sheet exposes similarities but also differences between the two Open-X stages. While in the case of Open-Development a large group of participants without much expertise in a specific field can participate, Open-Disposal requires a group of experts which on the other hand might be smaller. The kind of the expected innovations also differs: for Open-Development, the focus is on new products or business cases. Here, innovations are mainly gained in the form of ideas, drawings or first technical concepts. In contrast, Open-Disposal aims for process innovations in the form of ideas/information for recycling/disposing old existing products, or the consideration of services offered by participants. The type and amount of information provided by the company is also different in each case: when collecting ideas for new applications of a building

Table 1 Comparison of open-development and open-disposal

Category	Open-development	Open-disposal
Actors		
Participants	Due to the low complexity of the task no special expertise is required. Thus, amateurs as well as experts can participate	The complexity of the task requires special expert knowhow which usually only a minority of amateurs possesses
Accessibility to participants	Since the task can be performed by a random crowd the accessibility is relatively easy. This can become more difficult with an increasing complexity of the issue (e.g. solutions for technical problems)	Due to the required expertise, the number of potential participants is limited, which also complicates the accessibility
Type of participation	Primary participants provide ideas for (in this case) possible applications in the form of text, drawings, etc. They might also act as potential partners for realizing their ideas	In this stage, it is likely that participants not only provide suggestions for disposal steps but also for services (e.g. "If you prepare component A and B in a specific way, my company can dispose of it and would even pay you money for these components")
Issue and *provided information*		
Effort for evaluating gained information	Due to the low complexity of the task, the contextual evaluation of the received ideas is relatively easy. Challenges can arise out of an usually high number of ideas and diversity of content and formulations. In some cases, an evaluation by the participants themselves might also be possible	Though the amount of information is smaller, the more complex content increases the evaluation effort required. Normally, an internal expert group needs to perform detailed analysis and calculations to determine whether suggestions are applicable and economic
Risks		
Data security and knowledge drain	The amount of provided information is medium: though the most important properties of the pre-product needs to be published, these might be in a rough level of detail. Additionally no information regarding the manufacturing process is required. This also limits the risk of knowledge drain to a medium level: other companies might adapt some ideas to similar products of their own but they do not gain insights into the production process itself	The disposal of an old and complex product requires a high amount of information and high level of detail, e.g. technical drawings, photos, visits, etc. Due to the age of the product and the contained technology the risk of knowledge drain usually is relatively low. Exceptions might be "top-secret" products like military systems

(*Source* own data)

material or other products, rough information about properties and perhaps some photos are sufficient. For disposing of a product more and detailed data is necessary, such as detailed descriptions, photos and technical drawings. From these examples it is clear that the Open-X methodologies as part of Open Innovation, containing method- and tool-sets, need to be adapted to the characteristics of each stage. At this, the Open-X assessment makes a great contribution to determine these characteristics and the potential and potential barriers of an Open-X project.

6 Conclusion and Outlook

As described in the discussion section, Open-X methodologies bear great potential for a sustainable innovation process. The presented Open-X evaluation helps research to analyze the characteristics of each PLC stage and determine potential research areas and constraints for new methods and tools. In a second step, industry can use it to improve their innovations processes. With this, two levels of focus are possible: (1) a company focus: which PLC stages can be opened, which benefit can be expected doing this and which constraints need to be considered? and (2) a product focus: at which stages of the PLC can the product gain which input?

The illustrated verification of the Open-X evaluation on two PLC stages showed the challenge on finding the right combination of categories and a convenient level of detail. However, despite the first successful results the long-term add value will manifest after multiple applications in practice.

Hence, this paper lays the foundation of a holistic approach to utilize the potential of Open Innovation for the entire Product-Life-Cycle. In the next step we will apply the evaluation sheet to the remaining PLC stages and analyze their potential for Open Innovation. We will also refine and further improve the evaluation sheet and enhance it to inside-out innovations. To ensure applicability, the Open-X evaluation will be verified in further industrial projects. Finally, based on the evaluation results we will develop efficient Open-X methods for each PLC stage.

References

1. Chesbrough H (2006) Open innovation. Researching a new paradigm. Oxford University Press
2. Gassmann O, Enkel E (2004) Towards a theory of open innovation: three core process archetypes. R&D Management Conference (RADMA), pp 1–18
3. Howard TJ, Achiche S, Özkil A, McAloone TC (2012) Open design and crowd sourcing: maturity, methodology and business models. International design conference (Design 2012), pp 181–190

4. Kain A, Kirscher R, Lindemann U (2012) Utilization of outside-in innovation input for product development. International design conference (Design 2012), pp 191–200
5. Braun A (2012) Open innovation—Einführung in ein Forschungsparadigma. In: Braun A et al. (ed) Open innovation in life science. Springer Link
6. Haskins C (ed) (2006) INCOSE systems engineering handbook, version 3
7. Tukker A (2004) Eight types of product-service system: eight ways to sustainability? experiences from suspronet. Bus Strat Environ 13:246–260
8. Gassmann O, Enkel E, Chesbrough H (2010) The future of open innovation. R&D Manage 40(3):213–221
9. Sloane P (2011) A guide to open innovation and crowd sourcing. Kogan Page
10. Enkel E (2009) Chancen und Risiken von Open Innovation. In: Zerfaß A, Möslein KM (eds) Kommunikation als Erfolgsfaktor im Innovationsmanagement. Gabler
11. Von Hippel E, Franke N, Prügl R (2008) Pyramiding: efficient identification of rare subjects. MIT sloan school of management working paper 4719-08, Oct 2008
12. OECD/Eurostat (1997) OSLO manual, version 2
13. Inauen M, Schenker-Wicki A (2012) Fostering radical innovations with open innovation. Eur J Innov Manage 15(2):212–231
14. Ebner W, Leimeister JM, Krcmar H (2009) Community engineering for innovations: the ideas competition as a method to nurture a virtual community for innovations. R&D Manage 39:342–356
15. Thomke S, von Hippel E (2002) Customers as innovators: a new way to create value. Harvard Business Review

Researching Creativity Within Design Students at University of Botswana

Chinandu Mwendapole and Zoran Markovic

Abstract Creativity is considered the mental and social process of generating ideas, concepts, and associations. According to Bingalli (2007) concept generation is critical to the design process because it provides the designer with the necessary tools to picture the qualities of the desired design through the use of words or images. The paper discusses research done at University of Botswana, with 35 students from all years involved. Similar researches were done at several places in past (e.g. Brazil), but we were trying not only to establish students' creativity level, but also to improve our learning environment and curriculum. The Paper addresses our intentions, research hypothesys, basic principles, methodology, preliminary and final results, and conclusions.

Keywords Creativity · Innovation · Concepts · Ideas

1 Introduction

As part of the process of exploring how to create a learning environment that promotes and nurtures creativity we undertook a preliminary study to explore how design students in the Department of Industrial Design and Technology (IDT) understand and experience the concept of creativity. Data collection methods

C. Mwendapole (✉) · Z. Markovic
Department of Industrial Design and Technology, University of Botswana,
Gaborone, Botswana
e-mail: mwendapolec@mopipi.ub.bw

Z. Markovic
e-mail: markovicz@mopipi.ub.bw

A. Chakrabarti and R. V. Prakash (eds.), *ICoRD'13*, Lecture Notes in Mechanical Engineering, DOI: 10.1007/978-81-322-1050-4_107, © Springer India 2013

included the use of observation, practical test and a self administered questionnaire. Findings from the study helped begin the process of mapping or assessing the Department of Industrial Design and Technology as an environment that supports the nurturing of creativity among students.

35 students were involved in the preliminary research. They showed not only their level of the creativity, but also views on the existing learning environment and its ability to nurture creativity. Similar researches were done at several places in past (e.g. Brazil and Estonia), but we are trying not only to establish students' creativity level, but also to improve our environment, physically and as a teaching-learning process. The Paper addresses conceptual frame work, intentions, research hypothesis, basic principles, methodology, preliminary and final results, and conclusions.

2 Conceptual Framework

For the purposes of this research, creativity is defined as the process by which ideas are generated, connected and transformed into things that are of value [2]. In other words, creativity is the production of novel, appropriate ideas in any realm of human activity which includes the sciences, education, the arts, business and everyday life [3]. All though creative people can have certain straits which distinguish them from others, such characteristics can be developed with dedication and practice. Educational research on creativity provides evidence that creativity training can enhance an individual's level of creativity [4].

The term '*innovation*' has its roots from the Latin '*novus*', which means '*new*' and in the broadest context to innovate is "to begin or introduce (something new) for the first time" [5]. Innovation in companies can assume many forms, including incremental improvements to existing products, application of new technology to new markets, and the use of new technology to serve existing markets. We now live in what is viewed as weightless or knowledge economies, where knowledge and information have become important economic resources, and creativity is considered a major asset in the innovation [6].

Since creativity is both an individual and social construct, and design covers a wide range of activities which all include the creative visualization of concepts, plans and ideas, the production of those ideas. They has been a growing number of studies that have began to address issues surrounding creativity processes, thinking skills and environments within design education [7–10]. In keeping with current trends, the authors elected to explore the creative thinking skills of the Department of Industrial Design and Technology (IDT) students and their opinions on creativity and their environment.

3 Methodology

In this study the research was divided into two phases. First, in order to gauge the students creativity a 30 min practical test inspired by the Panamericana School of Art and Design test was given to the 3rd, 4th and 5th year students in the of the Industrial Design and Technology Department. The test was then followed by a self administered questionnaire.

3.1 Practical Test

Creativity thinking skills are defined as a novel way of seeing or doing things, that is characterized by four components: fluency (generation of ideas), flexibility (shifting of perspective, originality (conceiving something new), and elaboration (building on other ideas) [11]. Creative thinking skills include both output and process: creative output is end product of a creative process, whereas the creative process is the methods applied to develop ideas to solve a particular problem [12].

The study made use of a creativity test originally developed by the Panamericana School of Art and Design in Brazil, a simplified version of the Torrence Tests of Creative Thinking—Figural which originally consisted of three activities: picture construction, picture completion and repeated figures of lines and circles [13, 14]. The main function of the original Torrence Test—Figural was to encourage respondents to view the tests as series of fun activities, thereby reducing test anxiety [13]. One of the main aims of the test was to use the results, as a means of gauging the level of the students' creative thinking skills such as fluency, flexibility, originality and elaboration of ideas. In order to assist us in mapping or planning a way forward for the development of a learning experience that enhances the students' creative abilities.

The Panamericana School of Art and Design test focuses primarily on the use of repeated figures of lines and circles. The test requires 30 min of work time, so speed is important and artistic quality is not a relevant factor. The test was developed to understand and nurture the students' capacity for creativity. The practical test consisted of three A4 sheets of papers that included three basic shapes (circle, lines).

The students were then asked to continue the basic shapes drawings with as many different symbols as possible within the allocated time.

3.2 Self-Administered Questionnaire

After the visual test the students were asked to fill in a self-administered questionnaire in order to gauge the students understanding of creativity and their opinions on how their learning experience and environment contributes to their creativity.

The objective of the questionnaire was to suggest how the Department of Design and Technology could create a nurturing teaching-learning environment for the students. The students were therefore asked to self administer a questionnaire in which they had to put a general ranking regarding creativity in design, as per their understanding. Questions were organized in a manner considered to be appropriate and familiar to all students (study year 3–5). Contrary to the Practical Test, here the students had to show their understanding of creativity and to recommend, as per their understanding, how the teaching-learning environment could be developed to nurture in creative learning. The questionnaire was divided into two sections, two ranking comments and part open for suggestions.

The first section comprises eleven situation/approaches/intentions and students were asked to rank the level of creativity in each of them. From "Problem identification" to "Ability to communicate new ideas to others", students had to recognize creativity and its level in each of them.

The second section was connected to factors that had a bearing on the Practical Test, which was given to them previously. Students had to choose between different areas of knowledge, creative thinking skills and motivation, and to rank them. In order to not only show their understanding of the whole research, but also to indicate factors that had an impact on their perceptions of creativity in the design and learning process.

The third part content only two questions and students have to rank (from 1 to 10) their impression of the test, but also influence and impact of the learning experience and environment in promoting their creativity. This was the main point of the research.

Students were intentionally given unlimited time for questionnaire. The idea was to create environment without limitation, including time-concern, so students would fully feel free to express themselves.

4 Findings

Immersion into data allows the researcher to identify patterns, possibly surprising phenomena and also to become sensitive to inconsistencies such as the divergent views offered by different groups of individuals [15]. Since this study was primarily about the students abilities and experiences of creativity. Grounded theory was therefore considered the most appropriate choice for analyzing data from the study. This was because it provided a flexible framework to sort out the ideas, issues and themes emerging from the raw data for analysis and interpretation.

4.1 Practical Test

Creativity requires a certain baseline of intelligence, it also requires domain relevant skills… Domain relevant skills include a minimum level of factual knowledge and technical proficiency [16]. Since creativity is considered the ability to imagine, explore, synthesize, connect, discover, invent and adapt information.

The analysis of the drawings from the practical test was divided into three categories:

1. The level of fluency—the number of relevant ideas produced by the students
2. The level of originality—the ability to produce uncommon or unique responses
3. The level of elaboration—the students ability to develop and elaborate on ideas.

The practical test showed that the students could draw a wide range of images from one basic shape. Observing the images the researchers found that with regard to fluency or the number of relevant ideas produced by the students the 3rd year students rated much higher than the 4th and 5th year students. It was interesting to note that they all made an effort to draw images using all three basic forms (circles and lines). Examples of drawings included eyes, stop signs, faces and bicycles (see Figs. 1, 2).

The ability to produce uncommon or unique responses as an indicator of originality was high in the drawings by the 3rd and 5th year students and medium for the 4th year students. Examples of uncommon or unique responses drawings included: cherries, cats air pump and the use of words (see Figs. 3, 4).

The ability to elaborate and build on ideas was consistently high in the 3rd and 4th year students and medium for the 5th years. Examples of drawings that included the elaboration of the basic shapes into larger images included: a chair, man with a hat, digital clock and house (see Figs. 5, 6). Overall the practical test seemed to indicate that the 3rd year students were consistently high in all three categories, while the 4th were high in the elaboration of ideas and 5th year students were high in the ability to produce uncommon or unique responses (Figs. 7, 8, 9).

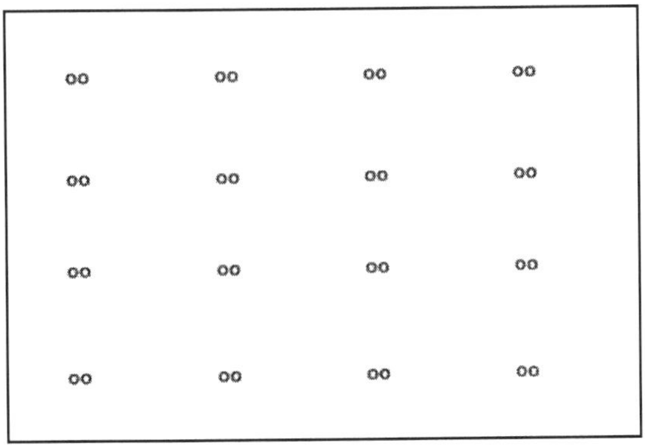

Fig. 1 Empty test "00". *Source* research team

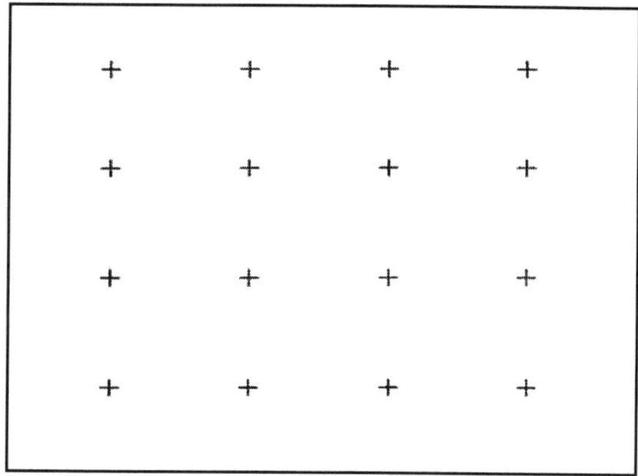

Fig. 2 Empty test "+". *Source* research team

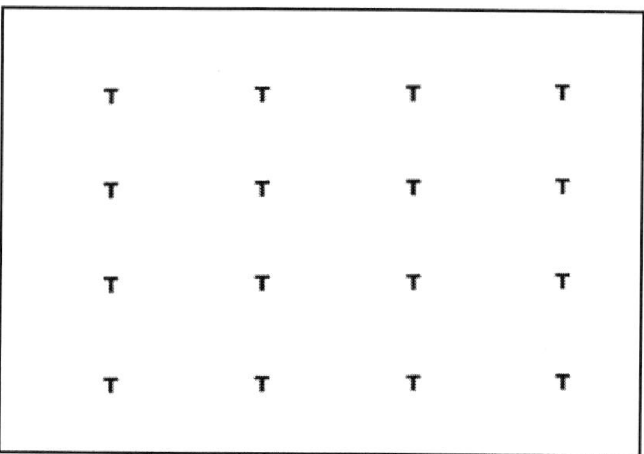

Fig. 3 Empty test "T". *Source* research team

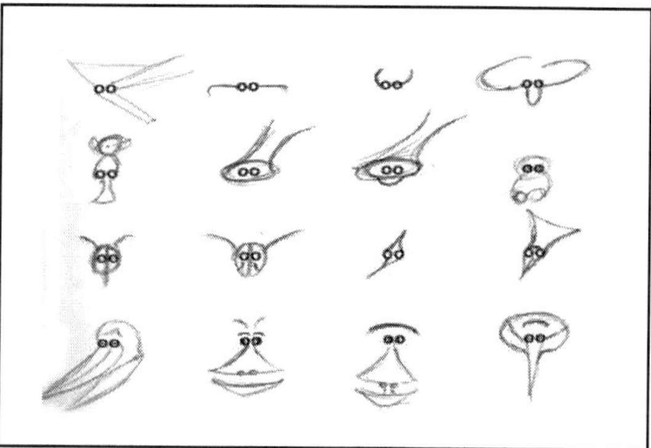

Fig. 4 Test showing fluency. *Source* research team

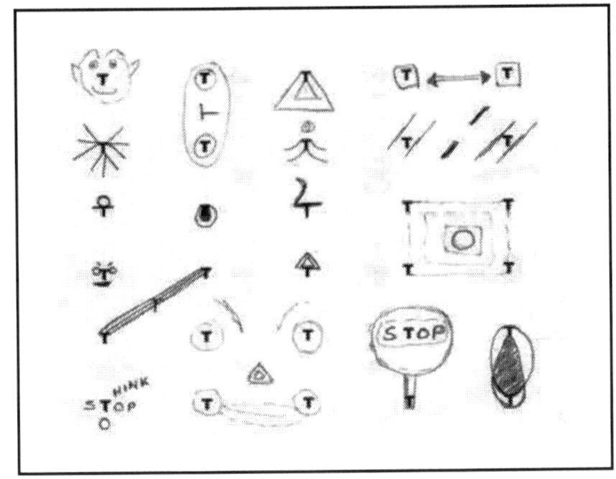

Fig. 5 Test showing fluency. *Source* research team

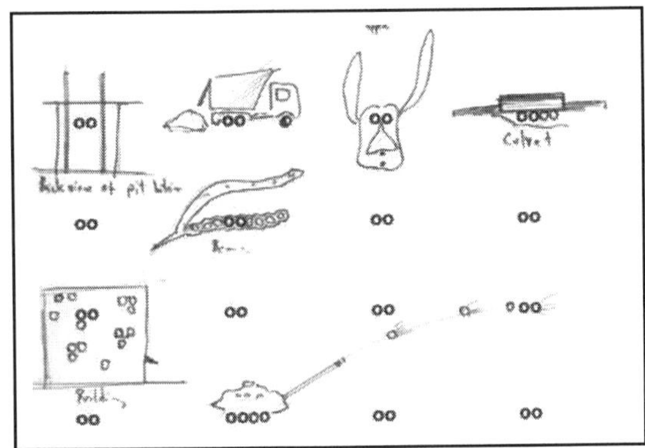

Fig. 6 Test showing originality. *Source* research team

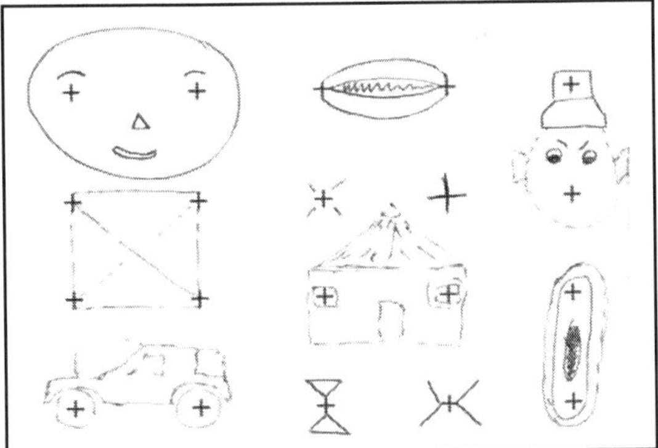

Fig. 7 Test showing originality. *Source* research team

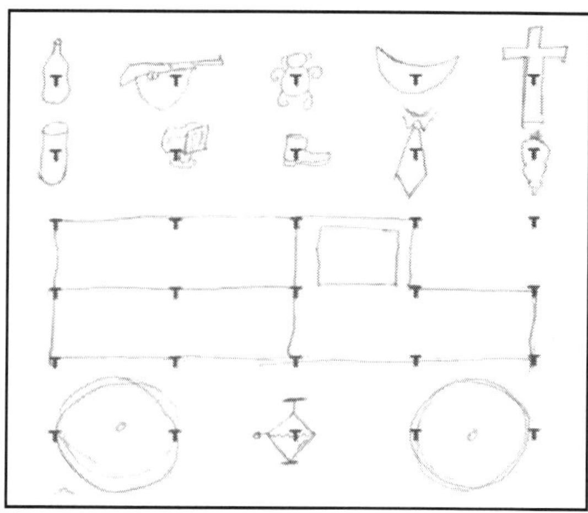

Fig. 8 Test showing elaboration. *Source* research team

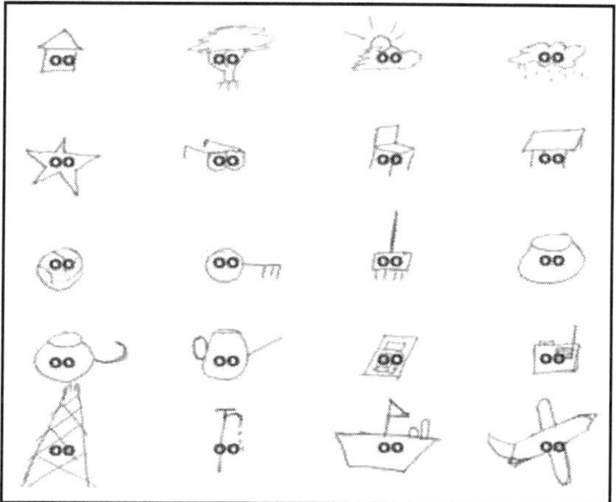

Fig. 9 Test showing elaboration. *Source* research team

4.2 Self-Administered Questionnaire

The questionnaire was administrated amongst students from the 3rd, 4th and 5th year of study, in the Department of industrial Design and Technology. Design students were chosen due to their developed interest and knowledge regarding creativity. The questionnaire was distributed separately to different groups, during a regular class, in classrooms and without time-limit. So students felt free and answers were more objective. However, a large majority of the students finalized the questionnaire within 20 min.

Results were analyzed separately per years of the study, and a final comprehensive schedule was prepared. It is very interesting that results per groups show very little differences.

Table 1 Final result for all groups (year 3, 4 and 5), summary

University of Botswana
Department of Industrial Design and Technology

Creativity: The purpose of this questionnaire is to collect information about the students understanding of creativity. The results of this survey will be published by the department of industrial design and technology, university of botswana and they will be availed to participants upon request

Total number of the participants: 35 (3rd year: 7; year 4th: 14; year 5th: 14) Results: all groups

A	In your view which of the following best demonstrates creativity	Strongly agree		Agree		Neutral		Disagree		Strongly disagree		Average
1	Problem identification or articulation	12	34,3 %	5	14,3 %	8	22,9 %	5	14,3 %	5	14,3 %	2,60
2	Ability to identify new patterns of behaviour or new combination of actions	10	28,6 %	6	17,1 %	7	20,0 %	6	17,1 %	6	17,1 %	3,06
3	Integration of knowledge across different disciplines	10	28,6 %	10	28,6 %	6	17,1 %	3	8,6 %	6	17,1 %	2,57
4	Ability to originate new ideas	13	37,1 %	7	200,0 %	3	8,6 %	7	20,0 %	5	14,3 %	2,54
5	Comfort with the notion of no right answer	7	20,0 %	8	22,9 %	13	37,1 %	5	14,3 %	2	5,7 %	2,63
6	Fundamental curiosity	7	20,0 %	9	25,7 %	6	17,1 %	8	22,9 %	5	14,3 %	2,86
7	Originality and inventiveness of work	13	37,1 %	5	14,3 %	8	22,9 %	3	8,6 %	6	17,1 %	2,54
8	Problem solving	12	34,3 %	6	17,1 %	4	11,4 %	6	17,1 %	7	20,0 %	2,71
9	Ability to take risks	12	34,3 %	5	14,3 %	6	17,1 %	8	22,9 %	4	11,4 %	2,62
10	Tolerance of ambiguity	4	11,4 %	11	31,4 %	13	37,1 %	6	17,1 %	1	2,8 %	2,69
11	Ability to communicate new ideas to others	10	28,6 %	9	25,7 %	3	8,6 %	7	20,0 %	6	17,1 %	2,71
	Which of the following factors do you think contributed to your responses to the panamericana creativity test	Frequently		Often		Sometimes		Seldom		Never		Average
	Knowledge											
1	Procedural–cultural factors	3	8,6 %	14	40,0 %	9	25,7 %	4	11,4 %	5	14,3 %	2,82
2	Intellectual-thinking skills	14	40,0 %	7	200,0 %	6	17,1 %	5	14,3 %	3	8,6 %	2,31
3	Technical-drawing skills	6	17,1 %	7	200,0 %	11	31,4 %	7	20,0 %	4	11,4 %	2,89
	Creative thinking skills											
1	Intuition	12	34,3 %	7	20,0 %	6	17,1 %	6	17,1 %	4	11,4 %	2,51

(continued)

Table 1 (continued)

University of BotswanaDepartment of Industrial Design and Technology												
2	Imagination	12	34,3 %	11	31,4 %	2	5,7 %	5	14,3 %	5	14,3 %	2,43
	Motivation											
1	Intrinsic-inner passion	15	42,9 %	7	20,0 %	7	20,0 %	3	8,6 %	3	8,6 %	2,20
2	Extrinsic–monetary reward	2	57 %	7	20,0 %	10	28,6 %	5	14,3 %	11	31,4 %	3,46
B. Overall impression of the creativity test (tick only one)												
	How would you rate the Panamericana test on creativity, on a scale of 1–10 (1 being poor, and 10 being excellent)	1	2	3	4	5	6	7	8	9	10	Average
		1			1	2	5	10	9	7		7,14
		2,86 %	0 %	0 %	2,86 %	5,71 %	14,21 %	28,57 %	25,71 %	20,00 %	0 %	100 %
C. How would you rate the role of your learning experience and environment in promoting your creativity												
	How would you rate your current learning experience and environment in promoting creativity on a scale of 1–10 (tick only one)(1 poor and 10 excellent)	1	2	3	4	5	6	7	8	9	10	Average
				1	2	2	4	8	8	8	2	7,34
		0 %	0 %	2,86 %	5,71 %	5,71 %	11,43 %	22,86 %	22,86 %	22,86 %	5,71 %	100 %
D. Suggest one thing that you think could contribute to a more creative learning experience and environment for students in the department of design and technology												

Source research team findings

General findings were that majority of the students had similar impressions regarding the questionnaire. Younger students (3rd year) already have a strong opinion regarding creativity, the learning environment and process. Majority of the latter years of study students followed the same pattern. The final matrix is shown under Table 1.

The results we split in two groups—Part one addressed the students' understanding of creativity. Part two the students' opinions of the existing teaching-learning environment.

Originating new ideas, inventiveness of work, problem identification and articulation, problem solving and ability to take a risk were on the top when students had to indicate areas which demonstrate creativity. Interesting is that students ranked tolerance of ambiguity very low. Also, students preferred creative thinking skills to knowledge, and when talking about motivation inner passion was rated higher than monetary rewards.

Regarding existing teaching-learning environment, students have a positive opinion, with an average ranking of 7, 34 (from 100 %). However, they recommended several improvements, for the teaching process (more design oriented courses and less theory) to physical environment (better organized design studios). This shows students' understanding of multifaceted approach of creativity. To develop students' creativity and nurture a creative learning and teaching environment, the Department of Design and Technology has to improve their curriculum, consider a more creative delivery of information and invest in a learning environment that encourages creativity.

5 Conclusions and Recommendations

Creativity and innovation are generally recognized as vital to commercial success in the 21st century. In order for Botswana to achieve its vision and strategic position of moving more toward economic growth and diversification, the design and technology curriculum must engender creativity and innovation. Because new products, new services and new manufacturing processes, no less than artistic works or scientific advances, have an idea as their origin.

The study on students in the Department of Industrial Design and Technology (IDT) offers us the opportunity to understand and experience the concept of creativity through the experience of the students. The students' responses to both the practical test and self administered questionnaire indicated the students' creative strengths as well as their personal opinions on their understanding of creativity and how the current learning experience and environment nurtures that creativity.

Since the most difficult skills for a design student to learn is the ability to develop a concept for a design, because concepts require a leap from written data or needs to communication of a design [1]. The study offers an opportunity to develop a plan that will help the Department of Design and Technology to nurture and harness the creativity of students in the 3rd, 4th and 5th year. There is a need for a curriculum audit that addresses the contribution of the different strands to the innovation process possibly with the end result of developing strategies that engenders creativity, innovation and technical proficiency.

The recommendations of the researchers are as follows:

- Create a learning environment that includes creative spaces and curriculum innovation
- Review the current curriculums teaching of creativity form the 3rd–5th year
- Encourage creative behavior amongst the students and lecturers.

References

1. Binggeli C (2007) A survey of interior design. John Wiley and Sons
2. United Nations conference on trade and development (UNCTAD) 2010 Creative Industries, Creative Economy Report
3. Amabile TM (1997) Motivating creativity in organizations: on doing what you love and loving what you do (Creativity in Management) 40(1):39–58
4. Evans JR (1991) Creative thinking in the decision and management sciences. OH: Southwestern Publishing Co, Cincinnati
5. Mutlu B, Alpay E (2003) Design innovation: historical and theoretical perspectives on product innovation by design. Barcelona Conference paper
6. Smith K (2000) What is the knowledge economy? Knowledge-intensive industries ad distributed knowledge base. Paper prepared as part of the project innovation policy in a knowledge based economy commissioned by the European Commission, Oslo
7. Wylant B (2008) Design thinking and the experience of innovation. Massachusetts Inst Technol Des Issues 24(2):3–14
8. Magano C (2009) Explaining the creative mind. Int J Res Rev 3(1):10–19
9. Glaveanu VP (2011) Creating creativity: reflections from fieldwork. Integr Psychol Behav Sci 45:110–115
10. Runco MA (2007) Creativity, theories and themes: research development and practice. Elsevier Academic Press, New York
11. Alvino J (1990) A glossary of thinking-skills terms. Learning 16(6):50
12. Csikszentmihalyi M (1996) Creativity: flow and the psychology of discovery and invention. Harper Collins, New York
13. Kim HK (2006) Can we trust creativity tests? A review of the Torrence test of creativity thinking (TTCT). Creativity Res J 18:3–14
14. http://www.toxel.com/inspiration/2009/05/06/school-of-art-and-design-creativity-test
15. Nonaka I, Teece D (2001) Managing industrial knowledge: creation, transfer and utilization. Sage Publications
16. Hammersley M, Atkinson P (1995) Ethnography: principles in practice. Routledge

Role of Traditional Wisdom in Design Education for Global Product Development

Ar Geetanjali S. Patil and Ar Suruchi A. Ranadive

Abstract Tradition is the codified research over centuries—with tremendous feedback. It is important to realise that tradition is a result of research with feedback over the centuries, Joseph Allen Stein. For Indians the concept of globalization is not new. India has been trading partner to various nations since ages. What is new is the speed and the borderlessness of the transactions characterizing the present cache of globalization, made possible in large measure by the networking technologies, and in part by the object of such transactions, viz., information and knowledge, and the ensuing new economy. India being an old civilization with kaleidoscopic variety and rich cultural heritage, faith in the idea of growth and change remains the driving force of modern India. The material evidence of art crafts and architecture inevitably becomes the principal source for understanding the historical and cultural context. Our culture's ability to sustain uniform and repetitive means of production and reproduction, and implicit in this uniform repeatability its high level of technical coordination. Design education needs to integrate these existing knowledge systems with the emerging knowledge economy, especially standing as we are at the threshold of forces of a globalization, and the opportunities that this could throw up for the traditional sectors. Design education has to redefine itself as the response to fast changing scenarios on the global front. Collaborative cultures form a invariable aspect of design studio. The interdisciplinary dialogues and debate nourish the atmosphere in the design studio. This paper intends to understand the ability of traditional societies of assimilation and reflection of various cultural and technological changes which result in the plurality of Indian architecture, culture and craft, which forms one of

A. G. S. Patil (✉) · A. S. A. Ranadive
MVP'S College of Architecture and Centre for Design, Nashik, India
e-mail: geetashri@gmail.com

A. S. A. Ranadive
e-mail: suruchiranadive@gmail.com

A. Chakrabarti and R. V. Prakash (eds.), *ICoRD'13*, Lecture Notes in Mechanical Engineering, DOI: 10.1007/978-81-322-1050-4_108, © Springer India 2013

the major aspects of the global product development. Various examples from the different design fields like architecture and product design are taken for presenting the connect between the traditional wisdom and global product development..

Keywords Traditional wisdom · Environmental coherence · Design education · Global product development

1 Introduction

Design like all creative disciplines, involves an explorative process. A problem is posed or need is identified, then through a complex series of questions, thinking and actions, a solution or answer is identified and realized. Where once industrial designers focused primarily upon form and function, materials and manufacturing, today's issues are far more complex and challenging. New skills are required, especially for such areas as interaction, experience, and service design.

India is a multi-cultural society existing in different time zones. It is one of those countries where many centuries are telescoped into one and which has a multicultural society united by a deeply shared experience. Sustainable development is the core emerging issue in the field of design.

Since traditional environments are created in a field of tension between reason, emotion and intuition, design education should be viewed as training toward the manifestation of the ability to conceptualize, coordinate, and execute the idea of building/product rooted in tradition.

The profession and its education today face severe challenges that threaten this traditional role of the designer. This paper has attempted to understand and analyze the role of traditional craft and structure, and it is done with case studies of the two design oriented disciplines via architecture and the product design.

1.1 Understanding Traditional Wisdom

When we look into the architectural heritage of India, we find an incredibly rich reservoir of mythic images and beliefs all co existing in an easy and natural pluralism. Centuries of contemplation and synthesis have gone into traditional architecture and design to maintain its environmental coherence. The surrounding and the built form, products are both attuned to each other. Traditional environments are those that enhance, celebrate, and support human activities, those that reflect behavioural and cultural norms defined by society, those that ultimately integrate economy, ecology, and society into those everyday environments.

Traditional wisdom is born out of the assimilation of the various experiments and experiences of the individuals and the group over centuries. Tradition and modernity are not opposed to each other but part of continuous process. Modernity functions as an economic and social tool to achieve some wealth, flexibility and innovation for individuals and groups.

Therefore study of the traditional craft design should form an integral part of the designer's education. One should try to understand the character of the traditional environment where the design of artifacts and structures form a integral part of each other. As design educators seek to develop new curricula and adopt new teaching methodologies that transcend the regional barriers, we need to emphasize the relevance of well-established design philosophies, regional traditions, and cultural sensitivities.

1.2 Tradition Verses Modernity

Modernity functions as an economic and social tool to achieve some wealth, flexibility and innovation for individuals and groups; Tradition functions, partly and at times largely, as a mythological state which produces the sensation of larger connectedness and stability in the face of shockingly massive social change over last half century. One might also say that modernity is an economic force with social, cultural, and political correlatives; Tradition is a cultural force with social, economic, and political correlatives.

1.3 Role of Traditional Wisdom

Many changes we now experience in our environments, communities, identities, and requirements are an impact of globalization. Although driven by economic practices, globalization is to a significant extent enabled by design. While the debate of what the overall benefits and drawbacks of globalization are continues, we are faced with designed changes that challenge and celebrate our understandings of place and identity.

The powerful product semantics in India governs the use of objects not only in religious rituals but also in daily life, not just in the forgotten past but also in the living present. The material products of a culture can never be regarded as user independent in function or separately understandable entities. They acquire meanings in use, become integrated in everyone's whole life experiences, and interact with the mythology from which they derive their symbolic strength. They collectively participate in and carry forward the message of what that culture is about.

Traditions are not only part of history which far from being a factual representation of the past but also is a man made cultural construct for the process of selection.

2 Environmental Coherence

Traditional settlements are excellent examples of the man and nature relationship in terms of planning and crafts. The centuries of assimilations of the knowledge and skill is actively used in the traditional settlements where sustainability is the inherent character of the lifestyle.

Rural houses, in contrast to urban houses are built on lines evolved over thousand years of aesthetic traditions, indigenous techniques and judicious use of local material such as mud, grass, bamboo and cane. The Kutch desert houses, for example are built with thick walls and small openings, and narrow passages between the houses to overcome the gale of strong winds, the dust and heat. Vernacular building traditions all over the world display remarkably mature thermal adaptation. Early builders consistently used forms and materials that effectively moderated prevailing climatic conditions. By 400 BC, Persian engineers had mastered the technique of storing ice in the middle of summer in the desert. The ice was brought in during the winters from nearby mountains in bulk amounts, and stored in a Yakhchal, or ice-pit. These ancient refrigerators were used primarily to store ice for use in the summer, as well as for food storage, in the hot, dry desert climate of Iran.

3 Design Pedagogy

Tagore's education marked a novel blending of the ideas of the East and West According to Tagore, teaching should be practical and real but not artificial and theoretical. As a naturalist out and out, Tagore laid emphasis on the practicality of education. That will definitely increase the creative skill within a learner. As one of the earliest educators to think in terms of the global village, Rabindranath Tagore's educational model has a unique sensitivity and aptness for education within multi-racial, multi-lingual and multi-cultural situations, amidst conditions of acknowledged economic discrepancy and political imbalance.

If design education is going to assert that designers should have an impact on the development and performance of the design involved with today's markets and communities, then we must consider how designers can resolve the increasing tension between the global and local that is experienced in a growing number of communities around the world.

The way design performs determines much of our environment, experience, behaviour, emotions, motives, desires, understandings, identities, etc. This makes it necessary for design curricula to provide interdisciplinary and multidisciplinary exposure to gear up students to tackle design problems with a holistic approach and to make things work within practical constraints.

Today most of the design schools are found in urban areas. Urban cities show standardization in many aspects where as the rural areas carry forth with age old traditions in place. It becomes then necessary for design school to accommodate the Indian context and concern for the integration of built form and landscape. To achieve this, the interface between students and the actual site is essential. The material evidence of architecture and crafts inevitably becomes the principal source for understanding the historical and cultural context of these settlements.

4 Global Product Development

Globalization is associated with new dynamics of re-localisation. It is about the achievement of a global–local nexus, about new and intricate relations between global space and local space. Globalisation is like putting together a jigsaw puzzle: it is a matter of inserting a multiplicity of localities into the overall picture of a new global system (Morley and Robins 1995, p. 116).

Rapid advancement in the field of information technology has led to the emergence of the knowledge economy as the new power house. In this context, it is only natural that design education should address the new and emerging socio cultural and economic aspects.

A designer like all other professionals is first a human being and then a professional. His skills must stand firmly on this "human base" lest he becomes a "human robot." The human base is becoming increasingly necessary because the technological advances of the present are such that the skill part of the human activity is being rapidly replaced by mechanical and electronic gadgets ever more efficiently than before. What is now required more and more is not a skilled designer (by skill I mean knowledge and aesthetic sense included) but a broad based, socially well integrated, humane designer with a broad global vision.

5 Case Studies

The case studies which we have considered provide us with an example of transformation of the product in the global and the local scenario. The Case studies of the Ceramic Water Filter and Stone Grinder was result of a study conducted for indigenous technologies used for product design in the college. The Turkish Teapot is selected for its transformation in terms of form and technology used for brewing. The architecture case study talks about the basic necessity of need to create with the reference with the local traditions, was conducted as a part of settlement study in architecture studio.

Fig. 1 Evolution of the Turkish teapot from local to the global level

5.1 Turkish Tea Pot

The interesting point about teapot designs are, as tea has become a tradition, or is thought to be a tradition in Turkey, teapot might be said to be a traditional or archaic device, in the sense of the ways of brewing and preparing has been an outcome of years experience. The teapot could be said to have gone through certain transformations due to the changing technology, ways of life, new needs, products and functions. What is kept stable is the basic functional structure of the object that is water is boiled in the pot that is underneath the one in which brewing is done. The top part, whether of porcelain or metal is heated by the boiling steam applied from under.

Sameness is transferred into difference by the new setup. While it could be seen as divergence in the global market, as the marketing of a different product; it is convergent in the local market by turning the already different, culture-specific teapot set into a combination of the Western typologies of objects on top of a tray (Fig. 1).

5.2 Stone Grinders For Dosa, Idlis

Stone Grinders viz sil batta, ragda etc. have formed the integral part of Indian kitchen. The food processor and the mixer grinders where supposed to be standard

Fig. 2 Transformation from ragda to stone grinder

answers for the same. The revival of the practice of using ragda for preparing wet batter was done by industries like Ultra Grind. The tradition of the using stone grinders is revived through the mechanical designed ragdas (Fig. 2).

5.3 *Ceramic Pot Filters*

Study about indigenous techniques for water purification is gaining importance all over the world and is being promoted by WHO for creating user friendly products using the same.

Purpose of using the traditional techniques is because they are; Sustainable, local, natural availability of material, suitable for climatic condition, made by local people or craftsman, developed by experimentation n logics and need, cost effective, ecofriendly, gaining knowledge from traditional wisdom, reviving techniques, Regional influence, character.

The ceramic water pot (CWP) was identified as a product that provides an excellent potential for improvement through optimizing materials and processes as well as an opportunity for creating an updated design with greater aspirational appeal for consumers. Locally produced ceramic pot-style filters have the advantages of being relatively inexpensive, low-maintenance, portable, effective, and easy to use. The filters remove microorganisms from water by gravity filtration through porous ceramics, with typical flow rates of 2–3 liters per hour 6.

The design for the portable water filter has to fulfill easy manufacturing process, economic and have the aesthetic component which is not alien in the user

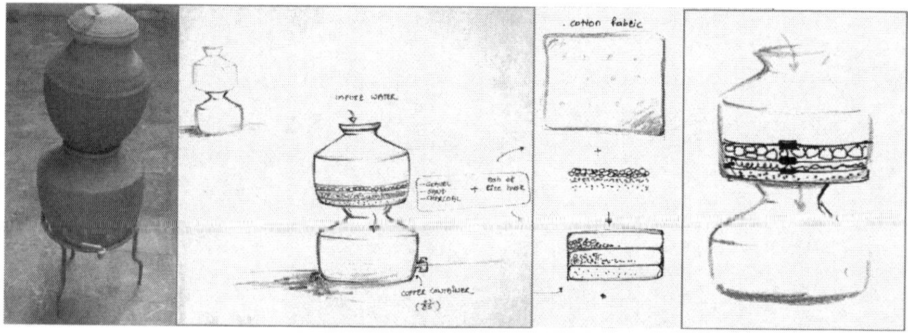

Fig. 3 Explorations done in the college using the local variation

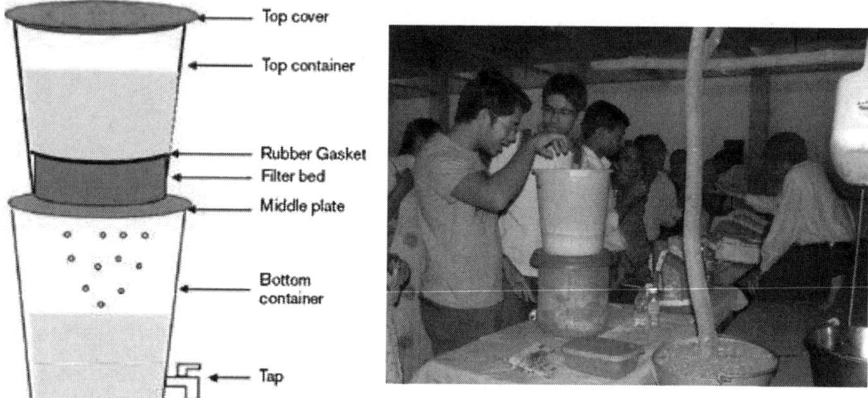

Fig. 4 Arogya water filter, vardha

environment. We present a few explorations done in the college using the local variation of mixing rice husk and crushed laterite for the formation of the filter bed (Figs. 3, 4).

5.4 Settlement Study: Pragpur

Architecture practices have often voiced concerns that schools of architecture do not provide students with the right set of skills needed in practice. The students need to connect to the environment prevalent not only in urban India but also rural India. Urban cities show standardization in many aspects where as the rural areas carry forth with age old traditions in place.

This is also an opportunity to establish links that provide enduring benefits by mobilising students, faculty, and neighbourhood organizations to work together to find new solutions to problems, while borrowing from the old.

With a view of sensitizing the urban students with the rural set up, its needs and understanding of sustainable dwellings and environment responsive architecture a design problem is set every year which is based on the study of a settlement carried out by the students on site. The settlement chosen is at least 200 years old. The settlement has to support a traditional craft as a means of livelihood for its residents. This can be weaving, pottery, toy making, handicrafts, leather goods etc. which are traditional arts and a part of the fabric of its residents. These are traditional skills which are handed down from one generation to another. The various aspects of the settlement are then studied by plotting and mapping the growth of the settlement over the years. The sections into which the settlement is divided is on basis of caste, work patterns etc. The students are made to understand the cultural context, imagery, roots, the health and educational facilities which are currently available. Various settlements have been studied by the students of CANS Nashik. The documentation and analysis of Pragpur shown herewith is a representation of similar towns spread over the country.

Pragpur, a small village near Dehra sub division of Kangra distt. of Himachal Pradesh is India's only officially declared heritage hamlet. Pragpur is located in the lush green Kangra valley in Himachal Pradesh surrounded by snow covered Dhauladhar mountain range.

Pragpur was declared as the country's first "Heritage Village" in December 1997, credit for which goes to villagers who have preserved their rich history and heritage with such a dedication and determination Pragpur village still have cobbled stone streets, mud-plastered walls and slate-roofed houses (Fig. 5).

Since the settlement of Pragpur is built in the hilly region, the streets are sloping. The buildings are a judicious mix of single and double storied dwellings. The cobbled streets are interspaced with interactive spaces which lead the visitor on an interesting journey.

Taal and its surroundings. The water level of Pragpur was extremely low before the 1800s. Nahar community directed the overflow of the nearby village of Lag-Belgana. They constructed a canal system using bamboo and supplied the excess water to Pragpur. The Taal was constructed to accommodate this overflow. The

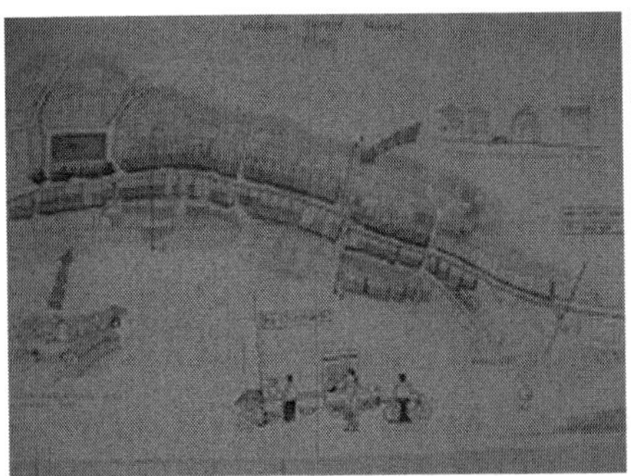

Fig. 5 *Left* Taal and surrounding area. *Right* The Taal

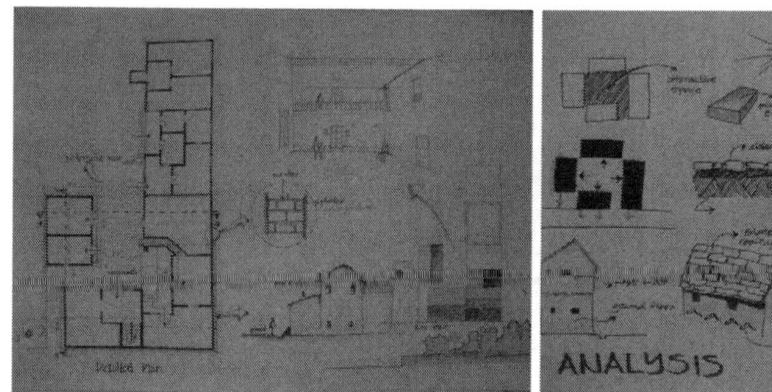

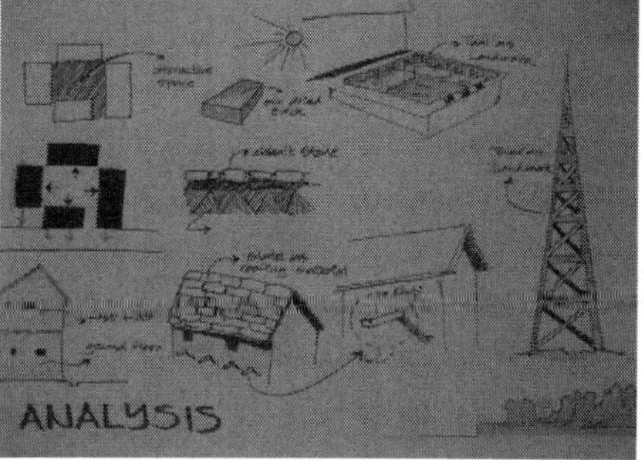

Fig. 6 *Left* Typical cluster. *Right* Analysis

surrounding area of the Taal developed over the years and is today an important public space (Fig. 6).

Typical dwelling Unit. Typical houses in Pragpur have the superstructure made of sun dried bricks. Foundations use local stone for stability. The walls are plastered either with mud or a mixture of cow dung and husk. Small windows towards the external walls restrict wind during the cold winters and also allow diffused sunlight into the interiors.

The houses are clustered around a central court which while acting as an interactive space also serves to act as an wind breaker.

Thatched roofs are used which are tied with a rope to a wooden member projecting from the wall.

Analysis. The various striking elements of the settlement were analysed. The play of light and shadow was emphasized due to the varying building heights. The cluster, the Taal and the various environmentally cohesive structures were the basis of analysis.

6 Conclusion

Nature, culture and society are endless resources for designers' inspirations, in the past, today and in future. Cultural variety is of equal importance as biological diversity. Global reflection and knowledge can create more global awareness. It will provide us with an opportunity to deepen the understanding on the values of the cultural diversity and the natural environment, and to learn to better "live together." Helmut Langer.

Any design with a carefully considered performance potential can be a powerful and sustainable influence within particular local places and identities. With a developed understanding of design as a performance in context, designers can design for local context with a respect for its values, use of its resources, and knowledge of its various dynamics and patterns of change and stability. This protection from the arbitrary influences of globalization comes from contextual design's ability to perpetuate what is essential within a culture while adapting what is new into a relevant form that is beneficial for the design with global sensitivity.

Acknowledgments 1. MVP s College of Architecture. Centre for Design Nashik, S Y BArch (2010-11).

2. Pranjal Duberkar (2010–2011)

References

1. Stein JA (1998) Building in the garden. In: White S (ed)
2. Norman D Why design education must change. http://www.core77.com/blog/
3. Balaram. Thinking design
4. Michael C, Salama AM () Design education: explorations and prospect for a better built environment
5. Yatin P () Concepts of space in traditional Indian architecture. Majun Publishing
6. Kenn F (2003) A critical pedagogy of space. Doctoral Dissertation, Flinders University
7. Timur Ş (2001) Reading material culture: an analysis of design as cultural form March
8. Standardized manufacturing practices–path. www.path.org/projects/safe-water-standardized-manufacturing.php
9. Langer H (2005) The world challenge for a global designers generation, design education, tradition and modernity, DETM 05

Color Consideration for Waiting Areas in Hospitals

Parisa Zraati

Abstract Color is one the most important factors in the nature that can have some affects on human behavior. Many years ago, it was proven that using color in public place can have some affect on the users. Depend of the darkness and lightness; it can be vary from positive to negative. The research will mainly focus on the color and psychological influences and physical factors. The statement of problem in this research is what is impact of color usually applied to waiting area? The overall aim of the study is to explore the visual environment of hospitals and to manage the color psychological effect of the hospital users in the waiting area by creating a comfortable, pleasant and cozy environment for users while spend their time in waiting areas. The analysis concentrate on satisfaction and their interesting regarding applied color in two private hospital waiting area in Malaysia.

Keywords Hospital environment · Human psychology · Color · Waiting areas

1 Introduction

This research will be the application of color and how to apply to public areas in hospitals should take account of the emotional and psychological factors which can affect their well-being at waiting room. The skillful use of color can help to overcome the sensory deprivation caused by lack of visual stimuli associated with

P. Zraati (✉)
Malaysia University of Science and Technology (MUST), No.17, Jalan SS 7/26, Kelana Jaya, Petaling Jaya 47301 Selangor, Malaysia
e-mail: p.zraati@gmail.com

A. Chakrabarti and R. V. Prakash (eds.), *ICoRD'13*, Lecture Notes in Mechanical Engineering, DOI: 10.1007/978-81-322-1050-4_109, © Springer India 2013

drab or monotonous environments. Hospital has a wide range of users with different requirements, from the elderly to the very young. A well-designed visual environment can be particularly helpful to visitor with partial sight [1].

Color can play a major role in creating accessible environments. Color contrast can identify obstacles and hardware that might prove difficult to negotiate. It is important to address the subjective needs and preferences of the users especially in buildings of anthropogenic character and service oriented building such as hospitals. Viewing environment as requires us to address the issues between living and lifeless spaces, life renewing nourishing and life sapping spaces [2].

This research will mainly focus on the color and psychological influences and physical factors. Research in environmental psychology has demonstrated that different environmental stimulus can affect on both mood and behavior [3, 4]. The effect of the physical environment can be part of importance in healthcare setting, where people experience a relatively high degree of uncertainty, fear and stress. Possible effects of the physical healthcare environment on the healing process of patients have received some attention [5–7].

The primary objectives of this research are:

1. To study the color consideration for waiting areas in hospitals.
2. To determine whether color consideration could influence on mood and behavior of users in waiting areas.
3. To determine what are the most suitable color scheme to consider for waiting areas in hospitals.

In this research, researcher was able to study relationship between color and emotional on waiting areas in hospitals and also study is to establish current color application in the design of hospitals, revealed a wide range of literature presenting mixed evidence on this aspect of color as well as a diversity of strategies for color usage in interior design of the public area.

2 Color

Color is the visual perceptual property corresponding in humans to the categories called red, green, blue and others. Color is considered informative and a way to interpret and understand meaning of designed environments [8].

Color has a strong impact on our emotions and feelings [9, 10]. The relationship between color and emotion is closely tied to color preferences, i.e. whether the color elicits a positive or negative emotion. Some color may be associated with several different emotions [11], and some emotions may be associated with more than one color [12]. Emotions can be divided into moods and feelings. A mood is a state of mind, an attitude, or a disposition that may take into account memory, language, context and physiological state. A feeling is an emotional state that is the result of sensation, a more immediate perceptual Response.

2.1 Colors in Healthcare

The color of our surroundings can both create stress and ease the stressed in life. Many of the effects of color on our moods may be the result of social and psychological associations with a particular color.

Color as property of designed environments may not have intrinsic meaning. Much research has demonstrated that healthcare occupant-patients, users and staff experience considerable stress, and one of the major stresses is produced by poorly designed physical environments [13]. Research has shown that certain colors directly affect human emotions, human feelings and human behaviors. To better explore this belief, researchers developed a theory called the psychology of color which is concerned with the effects that specific colors have on individuals' moods, emotions and behaviors as they perceive colors.

As a fundamental element f the physical environment, color in healthcare setting is increasingly considered as an environmental factor that can impact users' and staffs' stress, safety, fatigue and way finding. On the contrary, color palettes have also been found to positively affect people's healing processes as well as increase the work efficiency of healthcare staff [14, 15].

2.2 Color Design for Waiting Areas

Color design for waiting area covers all materials and surfaces. Furniture, color and lighting can do much to alleviate stress and enhance those areas. Good design can provide a visually calming environment. Comfortable seating with flexible configurations of small group arrangements could provide a friendly, welcoming atmosphere. Daylight and a view out, particularly of planting, make a waiting area much more pleasant and should be provided wherever possible [1].

Image of nature, shown in number of studies to distract patients, reduce stress and alleviate pain, can be used to great advantage [16]. The color of walls should be soft earth tones, yellows, greens or blues, which all promote a calming effect.

In waiting areas, this device can be interesting and engaging. However, care should be taken with over-enthusiastic flooring designs as people may tire of two dominant a design and find these motifs unfashionable or even annoying after some time.

3 Methodology

This part explains the methodology, which was used in achieving the research objectives.

Main methods of investigation comprised of:

- Interviews with patients and staff
- Conducting literature reviews
- Observation and gathering information from site.

Observations of the physical attributes of the waiting areas were taken. Photographs and note related to emotional and psychological of using color in this area were documented.

Each of waiting area of the hospitals was interviewed based on their willingness to participate.

3.1 Conducting the Study

This study has been conducted on two private hospitals' waiting areas in Malaysia, Subang jaya (A) and Bandar Sunway (B) cities. Table 1 shows the characteristics of the two waiting areas for each hospital (Figs. 1, 2).

Table 1 Characteristics of waiting area

Hospitals	Waiting areas	Design characteristics
Hospital A	Public waiting area	The study witnessed an attempt to create a comfortable waiting area with the play of soft and pastel colors for the walls, seats, floor, screen, view to outside and lack of light
	Outpatient waiting area	To create more comfortable waiting area compare with public waiting area. Mini artificial garden corner, though the attempt is admirable but it creates a lifeless environment due to the artificial plant used and its non strategic location
Hospital B	Public waiting area	There was an attempt to design the interior space but still lacks aesthetic input and coziness to the feel. However, comparatively more comfortable that the outpatient waiting area
	Outpatient waiting area	Enclosed outpatient waiting area with seats arranged linearly in rows which does not encourage social interaction among users. The interior was purely functional and lack aesthetic and coziness

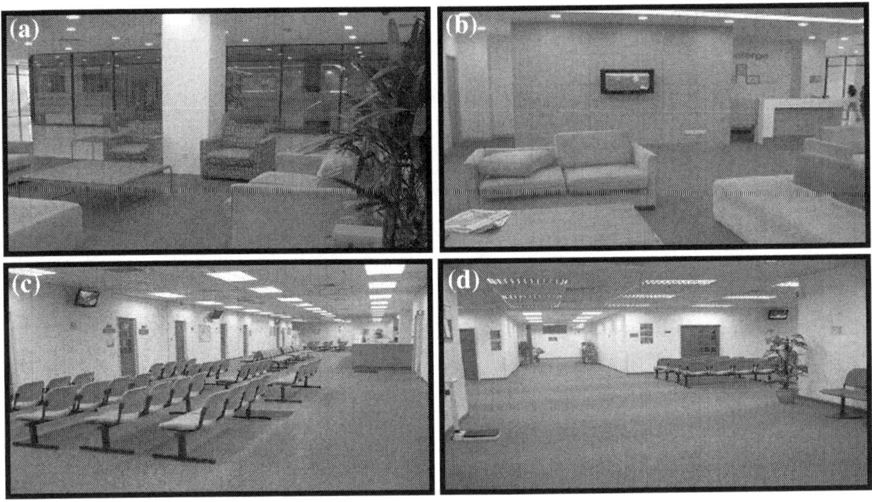

Fig. 1 **a**, **b** Sunway Hospital: In General waiting area has been used more of *orange* and *light brown* color also sofa and furnished are in same color, as a harmonic of colors. There was an attempt to design the interior space with artificial plant, so it created a lifeless environment. Applying of television and windows to view outside; make more welcome feeling to users (**c**, **d**) Sunway Hospital: Outpatient waiting area. Enclosed waiting area with row of seats in narrow corridor which does not encourage social interaction. The interior was purely and can see lack of aesthetic. Color of furnish is not harmony with area

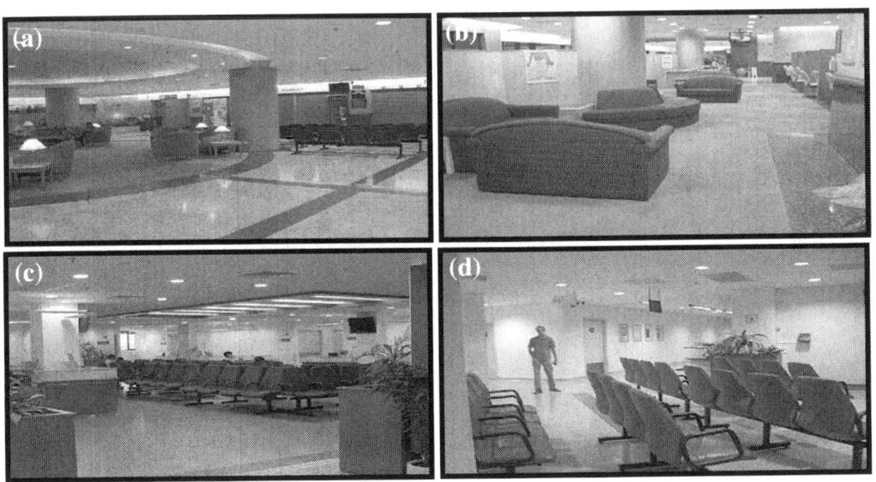

Fig. 2 **a**, **b** Subang Hospital: General waiting area. Use more artificial garden corner, though the attempt is an admirable but it created a lifeless environment due to the artificial plant. Cold color used for furniture and hot color used for painting wall. Make area bright, friendly, and cozy and relax (**c**, **d**) Subang Hospital: outpatient waiting area. Used warm and cold color for design seating area and wall painting. Design square shape with most artificial nature design for decoration, but use of weak range of light that make this place not bright

3.2 Data Collection

The data sources were classified into prime sources and secondary sources. Primary research was carried out to enable the collection of data that fits the exact purpose of this research and to increase the reliability of the information. The methods used to collecting data form face-to-face interview and observation of two waiting areas

Interviews were conducted only with those who were willing to participate. The secondary sources are journals, books and previous study, articles and paper which has been published or available online. The study recorded the responses by taking note from in view of the relatively number of respondents (n = 20) selected for each hospitals. The users Age ranged are 20–55 years old. The users were from various religious and gender and races. Table 2 presents the main interview questions.

Base on question 2 of interview, we indicate users emotional responses to five principle hues(i.e., red, yellow, green, blue, purple) and five intermediate hues (i.e., yellow-red, green-yellow, blue-green, purple-blue, and red-purple), and three achromatic colors (white, gray, black). Table 3 shown the Munsell notations.

4 Data Analyze

4.1 Color Consideration for General Waiting Areas in Both Hospitals

4.1.1 Overall Layout

As discussed in literature review chapter, color design for waiting area covers all materials and surfaces. Furniture, color and lighting can do much to reduce stress in waiting area. Good design can provide a visually calming environment. Based on observation, most warm color has been used for sofa, floor, wall and ceiling lighting in Sunway hospital. Yellow color which has been applied for walls make self-confidence and encourages optimism to area and feelings of fear. Brown color that used for furniture and carpet will bring feelings of stability and security.

Table 2 Interviews questions used in the study

Interview questions	Research objectives
1. Which color do you prefer for applying in waiting area? Why?	1
2. How do you feel in the waiting area? What emotional response do you feel with this color?	2
3. Are you satisfied with the color scheme that used in waiting area? What's your suggestion?	3

Table 3 Munsell notations for color samples

Color	Hue	Value/Chroma
Red	5R	5/14
Yellow	7.5Y	9/10
Green	2.5G	5/10
Blue	10B	6/10
Purple	5P	5/10
Yellow--red	5YR	7/12
Green-yellow	2.5GY	8/10
Blue-green	5BG	7/8
Purple-blue	7.5 PB	5/12
Red--purple	10RP	4/12
White, Gray, Black	N/9, N/5, N/1	

In Subang hospital, most used light color especially pink and blue for sofa, floor, wall, ceiling lighting for waiting area thus we feel relaxed and calmed and light blue that used for floor make patients and staff feels quite and away from the rush of the day. In Subang hospital consideration on interior design and make place more welcoming for visitors and patients comparing with Sunway hospital.

4.2 Color Consideration for Outpatient Waiting Areas in Both Hospitals

4.2.1 Overall Layout

Color which has been observed in waiting areas is; green, gray and white color combines for color of sofa, floor, wall, and ceiling in Sunway hospital. Furthermore used suitable lighting make area bright but in Sime Darby Hospital mostly apply green color for furniture and type of lighting that used, make waiting area darker. Both hospitals used white color painting for wall.

As discussed in literature review, the color of walls should be soft tone, like yellows, greens or blue, which all promote a calming effect. Too much white color can give feelings of separation and can be cold and isolation.

In Sunway hospital has been used gray color for carpet. Plastic and ceramic finishing for floor combined variety range of colors such as; dark and light blue, green, light yellow in Subang hospital. As discussed before, well-maintained flooring could be light in tone and preferably warm in color.

Nature elements and artwork using in Subang hospital, are providing for more positive energy. Both nature and artwork help users having a greater "sense of well-being" where spaces lend themselves toward contemplation and feeding the senses.

4.3 Color Emotional Analysis

Base on the color has been used in both hospital The finding outcome from the users' of the waiting areas, shows the green color had the maximum number of positive emotions (85 %), comprising the feelings of relaxation, comfort, hope, peace, and happiness. The second color which had more positive emotional was yellow (80 %) including clear thinking, memory, and self confidence and encourages optimism. The third number of positive emotions was given for the color blue (70 %); Blue was associated with the sky and so comprising relaxing and calming effect and make us feel quite and away from the rush of the day. The color purple attainted 50 % more than color red positive emotion (45 %). Purples have been used in the care of mental of nervous disorders because it help balance the mind and transform obsessions and fears. Red has been associated with love and blood, but result showed that red was not a favorite color.

For the achromatic colors, white had a highest amount of positive responses (62, 5 %), compared with color gray 30 % and black respectively. Freedom, uncluttered openness, peace and hope have been associated with color white. The gray color tend to independence and self-reliance also feeling of sadness, depression, although usually through of as a negative color. The color Black was also negative emotions color such as darkness, fear.

4.4 Staff and Visitors Perception

One of the most important things in design color scheme in waiting area is whether visitors, patients and staffs satisfy. This research gathers some statistic information where users were interviewed at random on their satisfaction in 2 waiting areas in both hospitals. Figures 3 and 4; shows staff and visitors satisfaction in both waiting areas.

5 Conclusion

The mainly purpose of this research was to determine factors that must be taken into consideration colors in waiting areas in two private hospitals in Malaysia from users satisfaction perspective, to identify which impact of color most commonly applied in waiting areas and to determine whether color consideration could influence on mood and behavior of users in waiting areas in hospitals.

Comparison that done for two mention hospitals in waiting areas specified that general waiting area in Sunway Hospital applied modern design with harmony of color for each parts. Therefore result is leading to relax and pleasant environment combining with calming, cozy, quiet, nice, home comfort but the outpatient

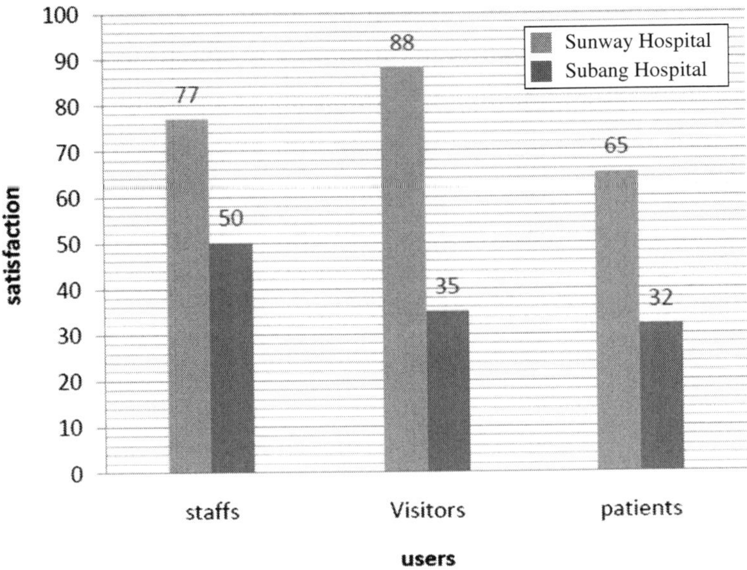

Fig. 3 Users satisfaction in general waiting areas in both hospitals

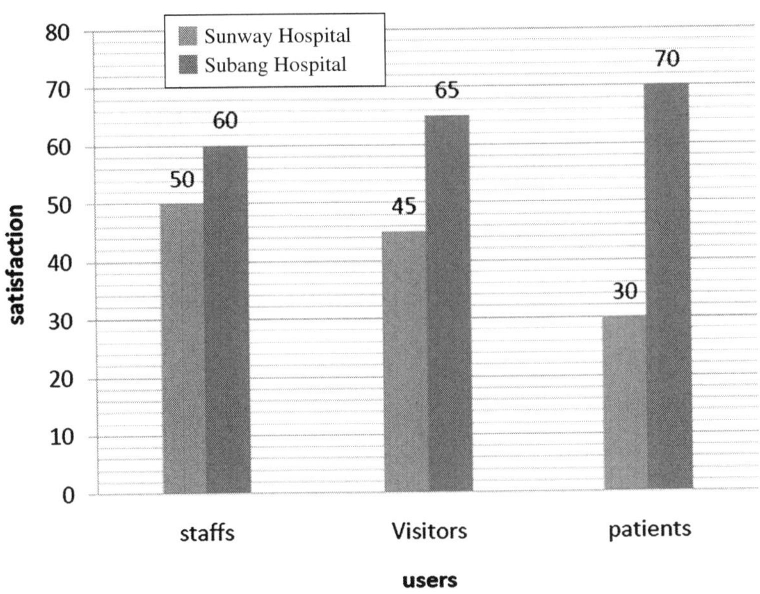

Fig. 4 Users satisfaction in outpatient waiting areas in both hospitals

waiting area is so simple, lake of design, narrow and quite boring area. Both waiting areas in Subang Hospital used lighter color for floor, wall and ceiling. In waiting areas of both hospitals can see artificial garden that it created a lifeless environment due to the artificial plant. After observation can say waiting areas in Subang Hospital are inviting, cheering, fresh that get arousing affective quality.

Color in architecture has multiple applications. Warm color can be used to reduce scale of area and cold colors visually enlarge a space, making it less confining. It can make appearance of space boring or pleasant. That's why color and emotion relationship is closely tied to color preferences.

It is recommend that, the color use, both in the interior and lighting must be an important factor in the waiting area. The color of the interior is makes an area pleasant and welcoming. It is also one of the things that remained people of being at home.

According to interview, users prefer warmer colors more positively rather than colder colors. Colors of a warmer temperature, such as yellow-brown, make the room feel warmer. Strong colors like red-orange are used to decorate the waiting room increase the patient's patience level which lead them to be restless while they wait in the room.

Floor color is better use a lighter shade of color like; white, cream and light gray tone. Light colors were consistently preferred or all objects such as ceiling, wall, floor, furniture. It should be emphasized that even white was a desirable color for ceiling.

For future study, we can develop this article in: direct communicators, the arrangement of the furniture, color use and messages that communicate the waiting room is a nice place to be.

Acknowledgments I would mainly like to thank my thesis supervisor, for this invaluable time spent to providing me guidance, suggestions, comments and encouragement. Also, I wish to express my appreciation to the staff and management of both hospitals for having guided and assistance me at searching and collecting data in the hospital.

References

1. Dalke H, Littlefair PJ, Loe DL (2004) Lighting and color for hospital design. London South Bank University, pp 351–356
2. Mariah W, Harun W, Ibrahim F (2009) Human—environment relationship study of waiting areas in hospitals. In: Journal of international association of societies of design research, proceedings of IASDR 2009, Seoul, pp 691–700
3. Kenz I () Effects of color of light on nonviable psychological processes. J Environ Psychol, pp 201-2-8
4. Leather P, Beale D, Santos A, Watts J, Lee L (2003) Outcomes of environmental appraisal of different hospital waiting areas. Environ Behav 35:842–869
5. Schweitzer M, Gilpin L, Frampton S (2004) Healing spaces; elements of environmental design that make an impact on health. J Altern Complem Med 10(Supplement 1):S71–S83
6. Ulrich RS (1992) How design impacts wellness. Healthcare Forum J, pp 35, 20–25, 1992

7. Arneill AB, Devlin AS (2002) Perceived quality of care: "the influence of the waiting room environment". J Environ Psychol 22:345–360
8. Mahnke F (1996) Color, environment, human response. Van Nostrand Reinhold, New York
9. Boyatzis CJ (1994) Children's emotional associations with colors. J Genet Psychol 155:75–85
10. Hemphill M (1996) A note an adults' color emotion associations. J Genetic Psychol, pp. 157, 275–281
11. Wexner LB (1954) The degrees to which colors are associated with mood-tones. J Appl Psychol 38:432–435
12. Linton H (1999) Color in architecture: design methods for buildings, interiors and urban spaces. McGraw Hill, New York
13. Clark PA, Maline M (2006) What patients want: designing and delivering health services that respect personhood. In: Marberry SO (ed) Improving healthcare with better building design. Health Administration Press, Chicago, pp 15–35
14. Malkin J () Medical and dental space planning: a comprehensive guide to design, equipment, and clinical procedures, 3rd ed. Wiley, New York
15. Ulrich RS (1991) Effect of interior design on wellness: theory and recent scientific research. J Health Care Interior Des 3(1):97–109
16. Ulrich RS, Zimring C et al (2008–2009) A review of the research literature on evidence based healthcare design. Health Environ Res Des J 2(2):61–125

Hybrid ANP: QFD—ZOGP Approach for Styling Aspects Determination of an Automotive Component

K. Jayakrishna, S. Vinodh and D. Senthil Kumar

Abstract Styling of automotive products is a vital issue and it need to be imbibed with increased customer expectations during every stage of product design. In order to achieve effective styling, it is necessary to apply quality function deployment (QFD) approach which is an effective product and system development tool. This study presents a decision framework where analytic network process (ANP) integrated with QFD and zero–one goal programming (ZOGP) models are used in order to resolve the design requirements which are more efficient in achieving aesthetic design. The first phase of the QFD is the house of quality (HOQ) which transforms customer requirements into product design prerequisites. In this study, after determining the sustainable requirements named voice of the customers (VOCs) and Engineering metrics (EMs) of an automotive component, ANP has been employed to determine the importance levels in the HOQ considering the interrelationships among EMs and VOCs. Additionally ZOGP approach is used to take into account different goals of the problem. A case study was presented to exemplify the approach.

Keywords Analytic network process · Quality function deployment · House of quality · Zero–one goal programming

K. Jayakrishna (✉) · S. Vinodh · D. Senthil Kumar
Department of Production Engineering, National Institute of Technology, Tiruchirappalli 620015 Tamil Nadu, India
e-mail: mail2jaikrish@gmail.com

S. Vinodh
e-mail: vinosh_sekar82@yahoo.com

D. Senthil Kumar
e-mail: dvsenthilkumar.mit@gmail.com

A. Chakrabarti and R. V. Prakash (eds.), *ICoRD'13*, Lecture Notes in Mechanical Engineering, DOI: 10.1007/978-81-322-1050-4_110, © Springer India 2013

Nomenclature

QFD	Quality Function Deployment
ANP	Analytic Network Process
ZOGP	Zero-One Goal Programming
HOQ	House of Quality
VOCs	Voice of the customers
EMs	Engineering Metrics
CRs	Customer Requirements
ECs	Engineering Characteristics
EPE	Environmental Performance Evaluation
DEMATEL	Decision Making Trial and Evaluation Laboratory
MCDM	Multiple Criteria Decision-Making
TOPSIS	Technique for Order Preference by Resemblance to Ideal Solution
LOC	Location of the component
U	Uniqueness
R	Reliability
EF	Enhanced functionalities
E	Ergonomics
I	Illumination
A	Aesthetics
D	Durability
W	Super matrix representation of the QFD model
W2	Matrix that denotes the influence of the VOCs on each EMs
W3	Matrix that represents the inner dependence of the VOCs
W4	Matrix that represents the inner dependence of EMs
WVOCs	Interdependent priorities of the VOCs
WEMs	Interdependent priorities of the EMs
WANP	Matrix that represents the overall priorities of EMs
wES	Weight vector of environmental sustainability
wM	Weight vector of manufacturability
wE’	Adjusted weight vector of environmental sustainability
wM’	Adjusted weight vector of manufacturability
ω	Matrix that represents the relative importance weights of the goals
C	Unit cost of EMs
UC	Adjusted unit cost vector

1 Introduction

Designing and manufacturing aesthetic products is a major, high-profile challenge to the industry as it involves highly complex, interdisciplinary approaches and solutions. Most research and applications so far, however, have not heavily focused on styling of aesthetic products. By designing a product with styling parameters in mind, companies can increase profit by reducing the cost incurred for their brand publicity, advertisements etc. Capturing and translating VOCs is a tough and tedious job. In modern challenging environment, QFD act as a key strategic tool to aid companies in developing stylish products that can satisfy customer needs. The fundamental concept of QFD is to interpret the requirements of customers i.e. VOCs into EMs and consequently into parts features, process strategies and production needs. HOQ is a type of intangible record that offers the means for efficient planning and communication [1]. In this case study, a methodology for determining the influential EMs that are to be considered in styling of a automotive product by integrating analytic network process (ANP) and zero–one goal programming (ZOGP) decision making techniques. ANP helps to bring up the interdependences among the VOCs and EMs. ANP was used for calculating the final relative importance of the EMs with maximum probable consideration of the EMs in the design phase as goal. Consideration of other metrics, such as cost budget, environmental sustainability and manufacturability of EMs marks the uniqueness of this study. Environmental sustainability quantifies the improvement of one EM over the other in styling improvement of the product. Manufacturability measures the degree of possibility of manufacturing the product with respect to one EM over the other. The relative importance weights of these goals are determined by pair wise comparisons. The ZOGP model is solved to determine the EMs, which will be considered in the design phase, in order to minimize the chances of divergences from the prioritized goals. The proposed framework adds quantitative precision to the judgmental decision process of product styling of the case assembly.

2 Literature Review

The literature was reviewed from the perspectives of applications of QFD, ANP and ZOGP

2.1 Review on Applications of QFD

An extensive practice in industry to cope with global competition is the adoption of Quality Function Deployment (QFD), which is well known as a customer-driven product development method originated in Japan in the late 1960s [2]. QFD is a planning and problem solving tool that is gaining growing acceptance for translating customer requirements (CRs) into engineering characteristics (ECs) of a product [3]. The QFD technique is a systematic procedure for defining customer needs and interpreting them in terms of product features and process characteristics. The systematic analysis helps developers avoid rushed decisions that fail to take the entire product and all the customer needs into account [4]. It is a process that involves constructing one or a set of interlinked matrices, known as "quality tables." The first of these matrices is called the "House of Quality" (HOQ). The Quality Function Deployment (QFD) is a well-known and a systematic method which is based on the idea of adapting technology to people [5]. QFD analysis method can be compared to 'participatory ergonomics' where the end-users are involved in developing and implementing the technology [6]. QFD has been used successfully by many Japanese firms, most notably Toyota [7]. Toyota halved their design costs and reduced product development time by a third after they started to use QFD [8]. Yang et al. [9] proposed a framework based on QFD for Environmental Performance Evaluation (EPE) to improve the environmental management system. QFD helps to identify the key performance indicators by transforming the environmental requirements to quantitative indicators. Rough set theory is integrated to analyze the incomplete and vague information. They remarked that QFD provide a systematic way to determine the performance indicators.

2.2 Review on Applications of ANP

ANP can produce priorities of technologies with consideration of their direct and indirect impact and was utilized in a systemic approach towards identification of core technologies from the perspective of technological cross-impacts [10]. The advancement of pair wise comparison matrices, employment of interdependencies among decision levels and expansion of more consistent results are considered as the unique features of ANP. Wu this paper proposed a combined ANP and DEMATEL approach to evaluate and select KM strategies. KM strategy selection is a multiple criteria decision-making (MCDM) problem. ANP deal with all kinds of interactions systematically and the Decision Making Trial and Evaluation Laboratory (DEMATEL) convert the relations between cause and effect of criteria into a visual structural model helps to handle the inner dependences within a set of criteria. The application of the proposed method is explained through empirical study. ANP approach can be used for choosing, evaluating, prioritizing and ranking etc., if mutually dependent relationships have considerable impacts on the decision model and have been used for developing an expert selection system to select ideal cities for medical service ventures jointly with technique for order preference by resemblance to ideal solution (TOPSIS) [11].

2.3 Review on Applications of ZOGP

Badri et al. [12] developed a project selection model for health service institution which incorporates all of these factors such as decision maker priorities and preference, benefits, cost, project risk. The model is formulated as mixed 0–1 goal programming and validated by using a real world Information System (IS) project selection data. Lee and Kim [13] suggested an improved Information System project selection methodology which reflects interdependencies among evaluation criteria and candidate projects using analytic network process (ANP) within a zero one goal programming (ZOGP) model. The authors concluded that exploiting project interdependencies is one way of saving IS costs and frugality resources.

3 Case Study

The case study has been carried out in an automotive plastic component manufacturing organization located in Bangalore, India (hereafter designated as XYZ) with the help of a cross functional team involving participants from the organization and customers. XYZ was facing the problem of generating new stylish components for their customers including their valuable ideas into product design and development.

Step 1 *Identification of VOCs and EMs*

The cross functional team revealed the most important VOCs of the product under study with respect to styling as, Location of the component (LOC), Uniqueness (U), Reliability (R), Enhanced functionalities (EF) and EMs as, Ergonomics (E), Illumination (I), Aesthetics (A), Durability (D)

Step 2 *Consideration of interdependencies among VOCs and EMs within HOQ and resolving the overall precedence of the EMs by ANP approach*

The super matrix representation of the QFD model proposed by [14] (Eq. 1) is used in this case study to imbibe the dependencies inherent in QFD process into analysis.

$$W = \begin{matrix} Goal\,(G) \\ Voice\,of\,Customers\,(VOCs) \\ Engineering\,Metrics\,(EMs) \end{matrix} \begin{pmatrix} 0 & 0 & 0 \\ w_1 & W_3 & 0 \\ 0 & W_2 & W_4 \end{pmatrix} \qquad (1)$$

where, $\mathbf{w_1}$ is a vector that represents the influence of the goal, namely improving the styling aspect of the product based on VOCs. $\mathbf{W_2}$ is a matrix that denotes the influence of the VOCs on each EMs, $\mathbf{W_3}$ and $\mathbf{W_4}$ are the matrices that represent the inner dependence of the VOCs and EMs respectively. The interdependent priorities of the VOCs ($\mathbf{W_{VOCs}}$) and the interdependent priorities of the EMs ($\mathbf{W_{EMs}}$) are computed by using Eqs. 2 and 3.

$$\mathrm{W_{VOCs}} = \mathrm{W_3} \times \mathrm{w_1} \qquad (2)$$

$$W_{EMs} = W_4 \times W_2 \qquad (3)$$

The overall priorities of the EMs ($\mathbf{W^{ANP}}$) which imitate the interrelationships within the HOQ are computed by using Eq. 4, as

$$W^{ANP} = W_{EMs} \times W_{VOCs} \tag{4}$$

$$W^{ANP} = \begin{pmatrix} 0.550 \\ 0.269 \\ 0.104 \\ 0.079 \end{pmatrix}$$

The ANP analysis results indicate that the most important design feature is the ergonomics (E) with a relative importance value of 0.550.

Step 3 *Identification of the metrics*

Cost budget, Environmental sustainability, manufacturability of the case product were selected as the metrics by the cross functional team. The weight vector of environmental sustainability ($\mathbf{w}^{\mathbf{ES}}$) and the weight vector of manufacturability ($\mathbf{w}^{\mathbf{M}}$) are calculated by pairwise comparisons.

Step 4 *Determination of preference ratings of the EMs with respect to metrics by pair wise comparisons*

The cost budgets of all the EMs are limited to Indian National Rupee (INR). 90 in this case study by the cross functional team. The unit cost (**C**) of EMs were estimated based upon the actual possible cost that will be incurred to implement the respective EMs. Table 1 presents the **C** of the EMs.

Step 5 *Computation of weight vectors of EMs with respect to the metrics for dependencies encountered in the HOQ*

To infuse the adjustment for dependencies, the adjusted unit cost vector (**UC**) is calculated by multiplying the inner dependence matrix of the EMs ($\mathbf{W_4}$) with the unit cost vector of the EMs (**C**) as shown in Eq. 5.

$$UC = W_4 \times C \tag{5}$$

Correspondingly the adjusted weight vectors of the environmental sustainability and manufacturability are computed using Eqs. 6 and 7 respectively, as

$$w^{E'} = W_4 \times w^{ES} \tag{6}$$

$$w^{M'} = W_4 \times w^{M} \tag{7}$$

Table 1 Unit cost of the EMs

EMs	Cost in INR (C)
E	25
I	15
A	40
D	10

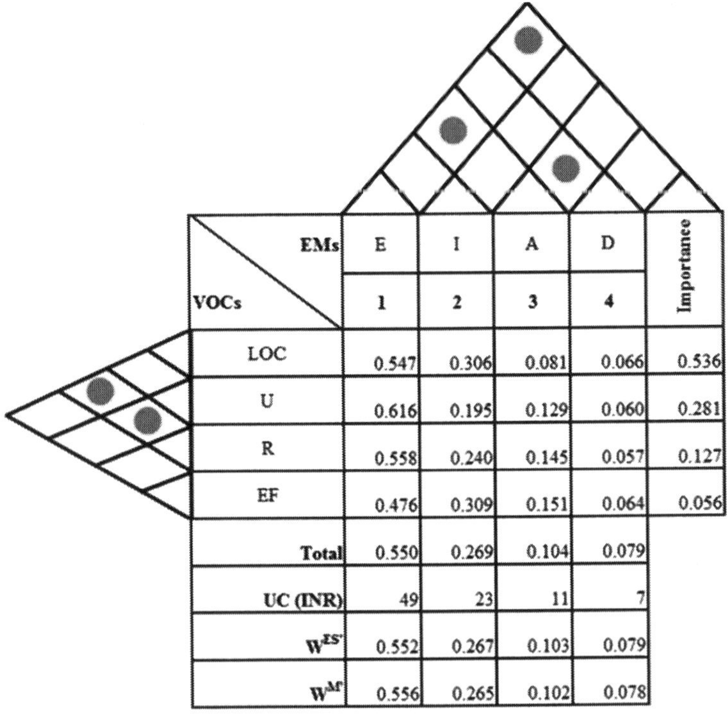

Fig. 1 HOQ presenting the dependencies of EMs with metrics

$$w^{E'} = \begin{pmatrix} 0.552 \\ 0.267 \\ 0.103 \\ 0.079 \end{pmatrix}$$

$$w^{M'} = \begin{pmatrix} 0.556 \\ 0.265 \\ 0.102 \\ 0.078 \end{pmatrix}$$

The HOQ attained using all the above data from the earlier steps would resemble as illustrated in Fig. 1.

Step 6 *Calculation of relative importance weights of the goals using pairwise comparisons*

In this step, the relative importance weights (ω) of the goals are considered in enhancing the product styling of the case product which are computed by pairwise comparison matrix as shown in Table 2.

Table 2 Computation of ω for goals

Goals	WANP	U	WES	WM	Relative importance weights (ω)
WANP	1.000	3.000	5.000	5.000	0.537
U	0.333	1.000	3.000	7.000	0.289
WES	0.200	0.333	1.000	3.000	0.117
WMFG	0.200	0.143	0.333	1.000	0.058

Table 3 Group of EMs selected by integrated hybrid approach

EMs	ZOGP solution
E	x1 = 1
I	x2 = 0
A	x3 = 1
D	x4 = 0

Step 7 *Development of ZOGP model and solving to determine the EMs to be consider in the process of product styling*

The generic form of ZOGP model as proposed by [15] is used in this case study. The ZOGP model developed using Eq. 8, based on the results computed is shown below, The ZOGP model developed with the data acquired from the previous steps was solved using LINDO software

$$\min 0.537 d_1^- + \left(\frac{0.289}{800}\right) d_2^+ + 0117 d_3^- + 0058 d_4^- \tag{8}$$

Subjected to,

$$0.550 x_1 + 0.2698 x_2 + 0.104 x_3 + 0.079 x_4 + d_1^- - d_1^+ = 1$$

$$49 x_1 + 23 x_2 + 11 x_3 + 7 x_4 + d_2^- - d_2^+ = 90$$

$$0.552 x_1 + 0.267 x_2 + 0.103 x_3 + 0.079 x_4 + d_3^- - d_3^+ = 1$$

$$0.556 x_1 + 0.265 x_2 + 0.102 x_3 + 0.078 x_4 + d_4^- - d_4^+ = 1$$

$$x_j \in (0, 1), j = 1, 2, \ldots\ldots, 4 d_i^-, d_i^+ \geq 0, \quad i = 1, 2, 3, 4.$$

Table 3 shows the group of EMs selected by integrated ANP and ZOGP approach include Ergonomics (E) and Aesthetics (A).

4 Conclusions

The ever increasing market dynamism, forces modern organizations to practice integrated approach as they remain the need of the hour to survey. The integrated approach illustrated in this case study aims to quantify the interdependencies and multiple objectives inherent in the styling design problem in a systematic way based on expert opinions, considering goals, resource limitations and metrics, appears as an effective solution support. The study also revealed that ergonomics and aesthetics as the major EMs to be more concentrated in developing the design of the case automotive product. The integrated approach presented in this paper can be extended for real time applications by considering goals, additional resource limitations and design metrics.

Acknowledgments The authors are thankful to Department of Science and Technology (DST), New Delhi, India for sanctioning the fund towards the execution of project titled "Development of a model for ensuring sustainable product design in automotive organizations" (Ref. No.SR/S3/MERC-0102/2009). This research study forms a part of this major research project.

References

1. Hauser JR, Clausing D (1988) The house of quality. Harvard Bus Rev 66:63–73
2. Cristiano JJ, Liker JK, White CC (2000) Customer-driven product development through quality function deployment in the US and Japan. J Prod Innov Manage 17:286–308
3. Akao Y (1990) An introduction to quality function deployment. Quality function deployment (QFD): integrating customer requirements into product design
4. Cohen L. (1995) Quality function deployment: how to make QFD work for you. Addison-Wesley, USA
5. Andersson R (1991) QFD—ett system for effektivare produktframtagning. Studentlineratur, Lund, Sverige. (In Swedish.)
6. Imada AS, Noro K (1991) Participatory ergonomics. Taylor and Francis, London
7. Lochner RH, Matar JE (1990) Design for quality. Quality resources, New York and ASQC Quality Press, Wisconsin
8. Bergman B, Klevsj B (1994) Quality from customer needs to customer satisfaction. McGraw-Hill, London
9. Yang YN, Parsaei HR, Leep HR, Chuengsatiansup K (2000) Evaluating robotic safety using quality function deployment. Int J Manuf Technol Manag 1(2/3):241–256
10. Kim CH, Lee H, Seol C (2011) Identifying core technologies based on technological cross-impacts: an association rule mining (ARM) and analytic network process (ANP) approach. Expert Syst Appl. doi:10.1016/j.eswa.2011.04.042
11. Ayag Z, Ozdemir RG (2009) A hybrid approach to concept selection through fuzzy analytic network process. Comput Ind Eng 56:368–379
12. Badri MA, Davis D, Davis D (2001) A comprehensive 0–1 goal programming model for project selection. Int J Project Manage 19:243–252
13. Lee W, Kim SH (2000) Using analytic network process and goal programming for interdependent information system project selection. Comput Oper Res 27:367–382
14. Saaty TL, Takizawa M (1986) Dependence and independence: from linear hierarchies to nonlinear networks. Eur J Oper Res 26:229–237
15. El-Gayar OF, Leung PS (2001) A multiple criteria decision making framework for regional aquaculture development. Eur J Oper Res 133:462–482

Kalpana: A Dome Based Learning Installation for Indian Schools

Ishneet Grover

Abstract For children in India, the day of visit to a science center brings lot of fun and curiosity about the observed installations. But they visit this learning playground only once or twice a year. This kind of learning experience is missing in schools. In this project we attempted to bring the experience of science centers to schools. We propose "Kalpana"; an interactive dome based learning installation. It empowered students to visualize the phenomenon that sun changes its trajectory in the sky with change in location or time of the year. Students could interact with the physical model and corresponding changes were observed on dome supported with contextual audio feedback. This installation based teaching method was evaluated against conventional paper based and software based applications. It found to be significantly better and entertaining for students.

Keywords Learning installation · Dome based learning · Indian school education · Astronomy for kids · Interactive installations · Science museum

1 Introduction

There are 13,62,324 (provisional as on 30th Sept 2011) number of Primary Schools in India as stated by National University of Educational Planning and Administration under District Information System for Education (DISE) [1]. In contrast to the number of schools there are only 27 science centers throughout India (under National Council of Science Museum) [2]. Science Center, an excellent place for effective experiential learning is either out of reach of students

I. Grover (✉)
Interaction Design, Industrial Design Center, IIT, Bombay, India
e-mail: ishneetgrover@gmail.com

A. Chakrabarti and R. V. Prakash (eds.), *ICoRD'13*, Lecture Notes in Mechanical Engineering, DOI: 10.1007/978-81-322-1050-4_111, © Springer India 2013

or they have insufficient time during their visit. In India, students visit a science center, once or twice a year. There also, they just move around in a hurry to complete the visit and do not have sufficient time to experience the phenomenon or to understand the rationale. Personal experience from a visit to a science center motivated us to answer the question: "How can we bring experience of Science Center learning to schools?" and hence we started exploring in this area.

2 User Studies

The project started with reviews from five experts in the field of design, teaching, cognitive psychology and learning. The group introduced us to various methods of teaching, education system in schools and installations in science centers. Following this, a detailed study was conducted in two stages:

2.1 Contextual Inquiry at Science Centers

User studies were carried out at four different science centers in Bangalore, Mumbai and Pune. The purpose of these studies was to understand the effectiveness of learning environment that a science center provides through physical and digital installations. *Observational study* was carried to figure out characteristics of installation that could attract students the most. This was followed by *informal discussions* with students and teachers to gather insights on aspects of installation they like and don't like. *Contextual Interviews* were carried out with 12 students, 4 accompanying teachers, and 2 science center managers. A typical interview lasted for about 15 min. 10 out of 14 interviews were good and interpreted. The interpretation was done on the same day.

In science centers students were in a learning environment where they could interact with the content and discover new things (Fig. 1). This encouraged students to actively use their intuition, imagination and creativity [3]. Students could relate better to the topics already covered in schools. As quoted by one of the teachers "Children respond better after their visit to science center" [4]. A lot f installations at science center required teamwork, which encouraged students to work together (Fig. 1). The use of audio and visual feedback in installations helped students to understand the content being explained. It also allowed students to compare different results on the go.

During visits a short *survey* was conducted and it was found that 34 out of 40 students were visiting this place after a year. Sometime, there were instances when students felt need of assistance to understand the installation and so they preferred company of an elder, teacher or friends rather than visiting alone. A few sections in the science center were underutilized (Fig. 1). According to the Science Center Manager, these installations explained topics that are taught only in higher

Fig. 1 (*left* to *right*) Student interacting with installations at science centers, students playing together with the installation in a group, least visited computer section in science centers, teacher explaining using static models

secondary e.g. logic gates. Students could connect very fast to the installations if it shows familiar content.

2.2 Contextual Inquiry with Students and Teachers in Schools

This study was conducted at six different schools that include government, private school and international schools from Mumbai, Mohali and Bangalore. A total of twenty *contextual interviews* were conducted with over ten students (of class vi–vii), eight teachers (teaching science, social science and math) and two school principals. The interviews were conducted like an informal discussion. The purpose of the study was to understand the actual learning environment, problems faced by teachers while teaching and problems faced by students to understand abstract concepts.

Master apprentice model of interviewing was adopted for interviews with two teachers. Teachers were asked to teach us "What to teach and How to teach the class in their absence". It was found that teachers were not able to communicate and explain clearly certain topics using just textbook and blackboards; static models helped little, but confusion still remained (Fig. 2). Teachers tried to take help of external resources like YouTube videos and Internet based applications. These videos and applications were helpful, but there was difference between actual topic that was to be taught and the focus with which the video/application was made. Hence, the topic was not efficiently explained with required more emphasis. Students faced lot of difficulties especially when they had to imagine and visualize 3-dimensional information. Some of the private schools used e-Beam technology and interactive board equipped classrooms. These type of technological interventions enhanced teaching but 3-dimensional abstract concepts mentioned above still remained unclear.

3 Available Learning Modules

Parallel product survey was done to understand what kind of products are available that are used for teaching abstract 3-dimensional concepts (Fig. 2). Below is the analysis (Table 1) explaining the advantages and disadvantages of few products.

Fig. 2 (*left* to *right*) Do it yourself kit, electric bell from Iken, mechanical model, Stellarium (software) and Taramandal installation

Table 1 Evaluation of existing products, which teach scientific concepts

Product	Advantages	Disadvantages
Do it yourself kits astrolabe and planetary probe	Teaches interesting concepts using simple day-to-day objects	Product quality is crude, does not provide feedback, lacks fun element
Mechanical model	Simulates the actual process of rotation mechanically	Limited content; requires assistance, no instruction set available, labels on the installations were not correct
Online applications Stellarium [5] and Celestia [6]	Very informative and flexible to use. Gives complete control in the hands of the user, interesting and exciting to use	Too much information and options, difficult to grasp due to panoramic views that are confusing for students
Physical installations Taramandal and mini planetarium	Very good immersive learning medium with fun element. Mini-planetarium is portable and thus can be carried around	Purely observational in nature, lacks interactivity, manually operated by staff, information is hard coded and thus limited information

4 Content Selection

Rotation, revolution and the sun's trajectory are factors that form the basis for few important concepts. Seasons on earth, Time zones, day/night, solar/lunar eclipse, etc. are some of the phenomena affected by this. Through our preliminary survey we realized that these concepts were not clear to many students. Important for the selection of this phenomena as the subject to be taught were reasons for this were:

- It's 3-dimensional in nature,
- It's abstract since it can't be observed and
- It takes long duration of time say years/months to observe these changes in reality.

More precisely the exact subject that was finally chosen as the subject matter was: "Sun changes its trajectory in the sky with change in location or time of the year". Such topics are taught in schools and appear in NCERT books of class vi, Social Studies subject Chapters 2 and 3 of *The Earth is Our Habitat*. Below are some of the images from the book [7] (Fig. 3).

In order to explain this concept effectively to students, the content was restricted and following points formed the basis for designing content:

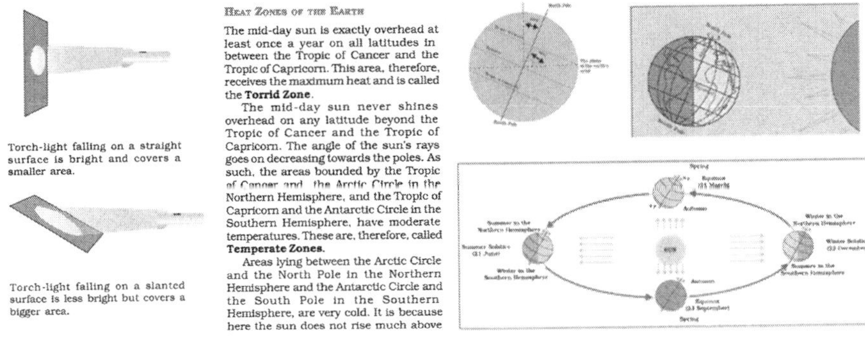
Torch-light falling on a straight surface is bright and covers a smaller area.

Torch-light falling on a slanted surface is less bright but covers a bigger area.

HEAT ZONES OF THE EARTH

The mid-day sun is exactly overhead at least once a year on all latitudes in between the Tropic of Cancer and the Tropic of Capricorn. This area, therefore, receives the maximum heat and is called the **Torrid Zone**.

The mid-day sun never shines overhead on any latitude beyond the Tropic of Cancer and the Tropic of Capricorn. The angle of the sun's rays goes on decreasing towards the poles. As such, the areas bounded by the Tropic of Cancer and the Arctic Circle in the Northern Hemisphere, and the Tropic of Capricorn and the Antarctic Circle in the Southern Hemisphere, have moderate temperatures. These are, therefore, called **Temperate Zones**.

Areas lying between the Arctic Circle and the North Pole in the Northern Hemisphere and the Antarctic Circle and the South Pole in the Southern Hemisphere, are very cold. It is because here the sun does not rise much above

Fig. 3 (*left* to *right*) Image of NCERT book explaining heat zones formed, rotation, revolution and seasons [7]

- These concepts should be taught with reference to the "Observer's location on earth" in addition to "Space" as the viewpoint.
- The concept should be taught by comparing sun trajectories of two locations or two different times of the year.

Following tables explain the content redesign in detail: A set of basic principles was derived which were later used in the final design (Table 2).

Table 2 Basic principles and comparisons of sun's trajectory at two locations

Phenomena	Reason	Examples	Diagrammatic representation
Two different locations have same seasons and same sun's trajectory	Two places have same latitude and are in same hemisphere	New York and Kashmir	New York / Kashmir
Two different locations have opposite seasons and mirror image sun's trajectories	Two places have same latitude values but are in opposite hemispheres and in opposite time of the year	Red sea and Madagascar	Red Sea / Madagascar
Two nearby locations have similar seasons but prominent changes in sun's trajectory	Two places are near same poles but little distance apart	North Pole and Greenland	North Pole / Greenland
Sun's trajectory is perpendicular to horizon	The place is on the equator	Singapore	Singapore
Sun's trajectory is parallel to horizon	The place is on any of the poles	North Pole	North Pole

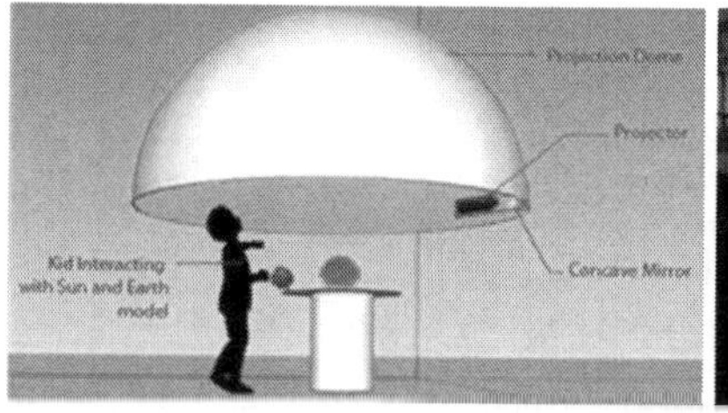

Fig. 4 (*left* to *right*) Diagrammatic representation of Kalpana, actual life size prototype

5 Proposed Solution

Based on the insights from user study and principles derived from the content; an immersive installation of size (2 × 2) was proposed. Hence, we designed *Kalpana*, an immersive dome based, learning installation for schools that allows students to interact using tangible objects (Fig. 4). *Kalpana* is made using one projector, a spherical mirror, a hemispherical projection surface and replica models of earth and sun. Students interact with the physical model to select location on the earth and change time of the year. This in turn shows them the sun's trajectory, which is projected on the surface of dome. For the purpose of evaluation a high fidelity prototype was developed [8].

The system consist of:

1. *Physical Model*, for input
2. *Visual Response* and
3. *Audio Response*

Figure 5 shows the basic workflow of the installation.

5.1 Physical Model (Interactive Input System)

A physical model was constructed using known objects like a sun (represented by glowing ball) and a globe. It was made interactive using simple buttons. At any time during interaction student could give two types of input.

5.1.1 Location Input

Student can select the location by pressing the button on the globe (Fig. 6).

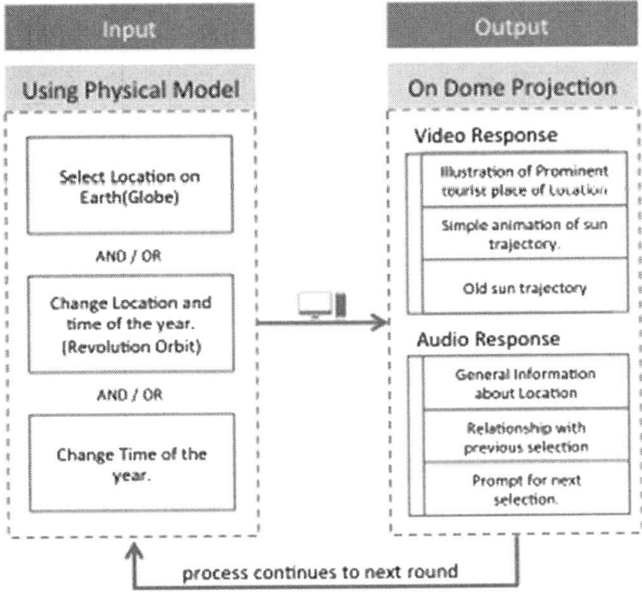

Fig. 5 Interaction flow diagram of Kalpana

5.1.2 Time Input

Student could select the time of the year by revolving the globe around the sun. It could be revolved only in one direction replicating the revolution of Earth. The base of the model was labeled with four prominent months of the year. [Summer Solstice (21st June), Autumnal Equinox (22nd September), Winter Solstice (21st December), Spring Equinox (21st March)]

5.2 Visual Response

The visual response projected on dome showed three things:

- Illustration: Location and a visual showing the prominent tourist place/monument of that location e.g. Statue of Liberty for city New York (Fig. 6).
- New Trajectory: The trajectory of the sun path with simple line animation (Fig. 6).
- Old Trajectory: The trajectory of the sun for the previous selection to show comparison (only if relationship exists).

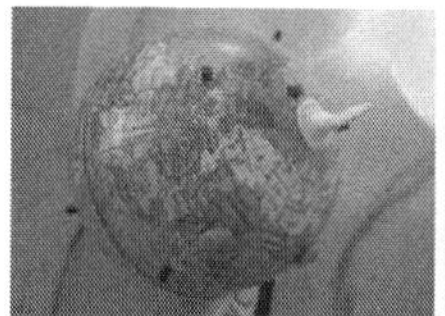 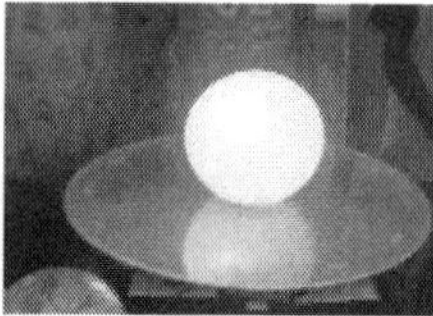 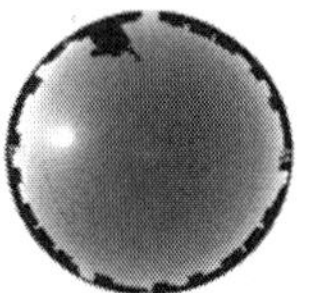 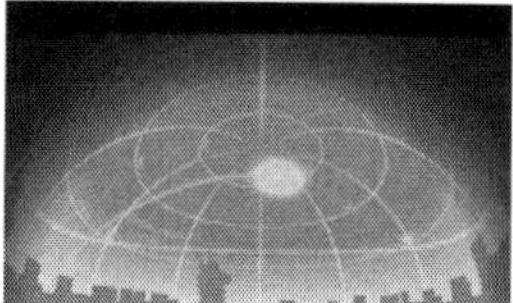

Fig. 6 (*left* to *right*) Globe with buttons to select location, model that allows student to revolve the earth around the sun to select time of the year, Output visual of New York as seen on the dome (polar grid), video screenshot of New York with sun's trajectory warped for projection on dome

> *"We are watching the March sky in North Pole. It is the northern most ti p of the earth located at 90 degrees N latitude. In the month of March the sun encircles on the horizon in clockwise direction.*
>
> *Did you notice? The sun did not set, even at the end of the day. This is because at Poles, sun rises only once in 12 months. Which means they experience 6 months of long days. However At North Pole, when the sun is about to rise, it is about to set at SP. At NP the sun will set only in the month of September. [pause]*
>
> *At poles sun is always parallel to the horizon. As we move towards the equator the path of the sun starts becoming angular and we see the sun higher in the sky. Tr y and observe this in Greenland, which is located south of NP at 72 deg ree North latitude.*

Fig. 7 Example of an audio script for North Pole in month of June

5.3 Audio Response

An audio feedback was given to explain the reasons of the observed sun's trajectory (Fig. 7). Depending on the last two inputs given by the user, the audio response for any input will have all or some of the following audios:

- Information about the Location.
- Information about the sun's path at that location.
- Comparison of sun's new trajectory with previous path (only if relationship exists).
- Triggers to prompt the student for the next Input.

6 Usability Evaluation

Kalpana, Paper based learning method and Stellarium (screen based learning method) were evaluated with students. Students of class VI scoring average marks in Social studies were shortlisted and distributed into three groups. The goal of the evaluation was to compare paper based learning method, Stellarium and Kalpana

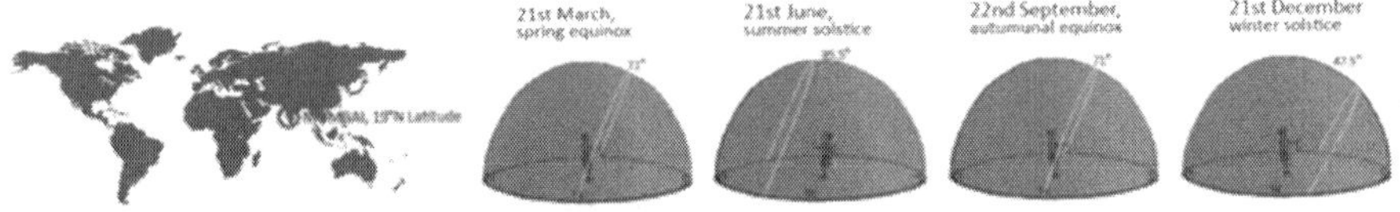

Fig. 8 Clip from the chapter prepared for paper based module

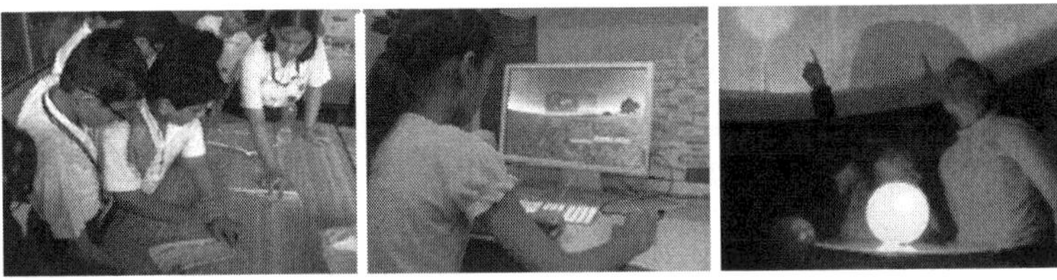

Fig. 9 (*left* to *right*) Evaluation of paper based, Stellarium, Kalpana

with respect to understanding of the concept (Fig. 9). For the evaluation of Paper based method an improved version of book chapter, a well-labeled world map and worksheets (to indicate their answers) were used (Fig. 8).

In all eighteen students participated in the study, with six students per learning method. Initially a 10 min orientation of the concept was given to each student for the method being evaluated. After this, an informal question and answer session (pre-test) was conducted with each student to evaluate understanding of the concept. In all, total of six parameters were defined to evaluate. The pre-test included the evaluation concepts such as: *Latitude Change*, *Longitude Change*, *both Latitude and Longitude*, *Month Change*, *Extreme cases* (E.g. North Pole, South Pole, Equator, Tropic of Cancer, Tropic of Capricorn). The sequence in which the parameters were evaluated was the same but the questions were framed as per the user to make the student feel comfortable and not to make them feel as if it is an exam.

6.1 Findings

Table 3 summarizes the success rate of the students for each method. Kalpana scored the maximum score per student (92 %). The difference between Kalpana and Paper based method was significant beyond $\alpha = \mathbf{0.01}$ (one tailed). The difference between Paper based and Screen based was not significant with $\alpha = \mathbf{0.17}$ (one tailed).

The installation based learning environment empowered students to visualize easily. Physical model helped students to see the over all picture and visual projection on dome to give location specific view. It helped student to relate the content to their day-to-day life and thus easier to remember. The support of audio

Table 3 Success rate of students for each method

	Paper based module	Stellarium	Kalpana
Correct answers (average out of six)	2.16 (S.D 0.4)	3.17 (S.D 1.6)	5.5 (S.D 0.8)
Percentage (%)	36 %	53 %	92 %

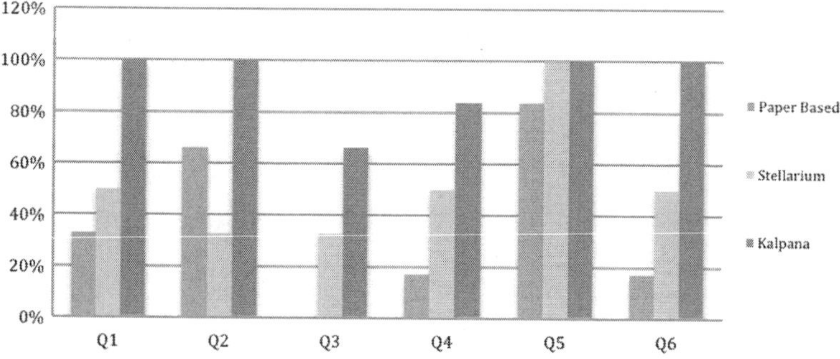

Fig. 10 Graph showing the comparison of the three methods w.r.t each evaluation parameter. (*x* evaluation parameter, *y* percentage of questions answered) (n = 6)

feedback helped students to learn more about the location. Prompts were useful to guide the student for the next input. When the data is reviewed as per the evaluation parameters, the average performance of students using Kalpana increased drastically when compared to average performance of students using Paper based method (Fig. 10).

7 Conclusion

'Kalpana'—meaning imagination, made students fantasize about traveling to different places on earth and observe changing trajectories of sun. It helped them to learn about earth's tilt, the changing sun's trajectory and how this change affects their city; with an immersive medium. Kalpana proved to be significantly better teaching method when compared to other paper and screen based methods. This was result of physical models of globe and sun, which gave a bird's eye view in combination with dome projection that gave actual view of how it would look in real sky. It helped student to relate the content to their day-to-day life and thus it became easier for them to remember and explain the concept. The support of audio feedback helped students to learn more about the location and ensure message is conveyed clearly. Students were actually enjoying the installation while learning was automatically happening.

However, a set of evaluation and comprehensive feedback session with teachers who would be actually using this installation at schools is yet to be done. The initial set of feedback from school stakeholders threw light on more interesting challenges such as; need of installation to be modular in nature, incorporate different kind of contents/concepts which can taught using similar setup, etc. The cost of the setup was one of the major challenges and with the availability of low cost projectors; we are trying to make it affordable for every school.

8 Future Work

The project can be developed on two front product design and other applications:

8.1 Product Design

The dome in 'Kalpana' prototype was made of FRP (Fiber Reinforced Plastic). When this prototype was tested it was found that there is a need to make the dome portable and foldable with minimal complications to setup. Also, there is need to make the whole installation visually attractive and robust for children.

8.2 Other Applications

Kalpana can also be used to teach other concepts that can be understood better by children through interactive immersive installations. For Example:

- Earth-Sun relationship: Shadows, weather.
- Astronomy: direction of starts, constellations.
- Concepts of electromagnetism: directions of lines of force, and many more.

All of these concepts would require a lot of study and expertise to develop.

Acknowledgments We thank Prof. Anirudha Joshi for his constant guidance throughout the project that helped a lot. We would also like to express sincere appreciation for Prof. B. K Chakravarthy, Prof. Sudesh Balan and Prof. Mandar Rane for their help to design the prototype and Prof. U. Athavankar to enable me to think at a broader perspective. We also thank the teachers and students at Kendriya Vidyala, IIT Bombay, all participants, users who provided feedback. We would also like to thank Shant Sagar, Chitra Chandrashekhar, Nikhil Autade, and Kumar Ahir for their help.

References

1. Mehta AC (2010–2011) Progress towards UEE—Elementary education in India. District information system for education, India, Analytical Reports
2. National council for science museums (2011) http://www.ncsm.gov.in
3. Brune J (1996) The culture of education. Harvard University Press, Cambridge
4. Grover I (2010) Interview with teacher at Nehru Science Centre, Mumbai
5. Stellarium (2011) http://www.stellarium.org/
6. Gregorio F (2011) http://www.shatters.net/celestia/index.html
7. NCERT (2010) Social science: earth our habitat, 10 July
8. Brouke PD (2005) Using a spherical mirror for projection into immersive environments. Graphite (ACM Siggraph), Dunedin Nov/Dec 2005

Design and Development of Hypothermia Prevention Jacket for Military Purpose

S. Mohamed Yacin, Sanchit Chirania and Yashwanth Nandakumar

Abstract This article discusses the design of a jacket that will help the military personnel stationed at the extreme cold climatic conditions to prevent going into hypothermia, a medical condition where in the core body temperature drops below 35 °C (95 °F), which causes shivering, mental confusion and ultimately leads to death. Our aim is to develop an external heating system in the form of heating pads that will aid the body to avoid going into hypothermia state. The set of three heating pads are controlled through a programmable microcontroller and temperature sensors. The pads are allowed to get heated only within the set range i.e. 37–45 °C. All these components are finally embedded in a polar fleece jacket which is an ideal substitute for wool and is much lighter than its predecessor. An inner layer of styrofoam and copper sheet lining helps in further distribution of heat evenly to the rest of the body surface. Thus, a virtual environment (around 37 °C) is created around the human body and does not allow it to lose heat further. This reduces the hypothermic effect on the body by preventing the loss of heat from the body to the cold surroundings, and hence makes the individual feel much better in such harsh climatic conditions and prevents any adverse effects from happening.

Keywords Hypothermia · Heating pads · Microcontroller · Sensor

S. Mohamed Yacin (✉) · S. Chirania · Y. Nandakumar
Biomedical Engineering, Rajalakshmi Engineering College, Thandalam, Chennai, India
e-mail: s_yacin@yahoo.co.in

S. Chirania
e-mail: sanchit.chirania.2008.bme@rajalakshmi.edu.in

Y. Nandakumar
e-mail: yashwanth.n.2008.bme@rajalakshmi.edu.in

A. Chakrabarti and R. V. Prakash (eds.), *ICoRD'13*, Lecture Notes in Mechanical Engineering, DOI: 10.1007/978-81-322-1050-4_112, © Springer India 2013

1 Introduction

Environmental temperature and humidity play an important role in the maintenance of body temperature [1]. Humans are Endothermic species of animals and maintain their body temperature within a range so that it will assist in carrying out the metabolic activities of the body. Mainly the temperature maintenance of the body is controlled by the preoptic area of the anterior hypothalamus of the brain, which is responsible for recognizing alterations in the body temperature and responding appropriately [2]. Hypothermia is very common condition that occurs in the cold climatic conditions where the outside temperature is so low that the body begins to lose its internal heat to the surroundings in the form of radiation as well as convection. Despite all this, the human body tries its best to adapt to the situation and does numerous activities listed below to avoid any fatal condition.

1. Sweat stops being produced.
2. The minute muscles under the surface of the skin called erector pili muscles contract, lifting the hair follicle upright. This makes the hairs stand on end which acts as an insulating layer, trapping heat.
3. Arterioles carrying blood to superficial capillaries under the surface of the skin can shrink, thereby rerouting blood away from the skin and towards the warmer core of the body. This prevents blood from losing heat to the surroundings and also prevents the core temperature dropping further.
4. Shivering of muscles increases heat production as respiration is an exothermic reaction in muscle cells. This means that less heat is lost to the environment via convection.
5. Mitochondria can convert fat directly into heat energy, increasing the temperature of all cells in the body.

Despite all the efforts put by the body, sometimes it will not be able to generate the necessary heat or loose that extra bit of heat in order to keep the stable normal internal temperature. This is when hypothermia sets in and causes fatal issues, which needs to be attended immediately. Although many traditional methods are there to help the people revive back from hypothermia but very few methods to prevent it.

1.1 Existing Technologies

The existing technologies include the equipments that are used to bring back the person who has already been affected by hypothermia and helps the victim to return to normal condition. The following subsections will discuss in detail about all the equipments used today for reviving the victim from hypothermic state.

1.1.1 Blizzard Survival Blanket

The blizzard blanket consist of a extra large blanket made up of Reflexcell™ materials that is used to wrap up the casualty and it reduces the risk of heat loss from the patient. This is another way to cover up the casualty completely from head to toe thereby preventing any more loss in body temperature. It provides the total warmth and weather protection with the help of Reflex cell™ material, thereby reducing shock and the risk of hypothermia in all conditions [3].

1.1.2 Res-Q-Air

This technique involves administering warm air to the person who has gone into hypothermia; by which we can increase the body temperature internally. The system functions exactly like that of a ventilator, the only difference being the temperature at which the oxygen is administered is warmer in this case [1]. The inlet oxygen is heated and humidified to a temperature of 42 °C. Thus, it requires an oxygen cylinder.

1.1.3 Hot IV Fluid Bag

In this method hot IV fluid bags are placed in the weak points of the body, i.e. the neck, arm pits and the groin. These are the points where maximum heat loss takes place. Thus tapping the heat would result in overall increase in the body temperature [1]. This method can only be used in the case if the person is suffering from mild hypothermia.

All the three above methods may provide some benefit; however, for massive transfusion therapy counter-current heat devices seem more effective maintaining the temperature of infused fluids and blood to higher than 33 °C. Thus preventing a person to fall in hypothermia in cold climatic conditions is a challenging task and requires a specially designed system which should maintain the normal body temperature and metabolism.

1.2 Challenges and Difficulties in Designing a Preventive Jacket for Hypothermia

Treating hypothermia is a multi step procedure and is usually performed by trained doctors or nurses using some assist devices, which is cumbersome and takes the more time to do. So preventing the person from going into hypothermia is better than reviving a person after he/she has gone into a hypothermic condition. The main objective of our work is to design a system which prevents the interaction of human body with external cold condition without affecting the normal phycological functions of the body. This will actually minimize the heat loss from the body to the outside cold surroundings.

The first attempt of developing a Hypothermia Prevention jacket was by using Infrared radiation. Infrared has a unique property that it can pass through the human tissues causing minimum side effects. During this process, it tends to increase the blood supply in the region it passes through. The increased blood supply results in increased temperature of the body, thus serving the purpose. This seemed to be the best possible solution to the problem of providing a controlled external heating. When we started designing, the problem that surfaced was the infrared source. The infrared LED's that is available in the market is of the range below 1 μm (NIR). But to have the desired effect quickly, the infrared source should be in the range of 1.5–5 μm (FIR) and none of the sources matched the above specifications were small enough to be fitted inside the jacket. Using such a small LED's to pass infrared radiation for providing external heat to the body became very difficult.

The next thing that could possibly be used for heating up any object is a heating pad (for e.g. Sauna belt). Usually all the heating pads run on direct AC supply which hinders its portability. Thinking on similar lines we got the idea of having Nichrome as our heating material which has the capability of heating up to 100 °C that too when the source of current is D.C. Because of this, Nichrome became the perfect material to be used for our heating purpose.

The biggest hurdles here were the cost and the availability of a heating pad made up of Nichrome that could work in the required temperature ranges desired. The sizes of the pads were also an issue as increasing the pad size would increase battery consumption. DC run heating pads were used and with the introduction of a programmable microcontroller coupled with a MOSFET drive, the heating pads were made to work in the required temperatures. To reduce the pad size, a policy of retaining the maximum temperature was employed. Special fabric setup was designed constituting components like Polar fleece, Styrofoam and copper sheets lining.

2 Proposed Design for Hypothermia Prevention Jacket

The aim of hypothermia prevention jacket is to protect the user from the cold surroundings and at the same time prevent him from going into hypothermia. The making of the jacket can be easily divided into fabric technology and electronic circuit's part. Both are explained in detail in the following subsections.

2.1 Electronic Circuit

This section discusses the electronic circuits and components used in the making of the control unit for the jacket. This includes microcontroller (PIC 16F877A), temperature Sensor (LM35), metal oxide semiconductor field effect transistor (MOSFET—FQPF 5N60C), voltage regulator chip (7805) and heating Pads.

The microcontroller clock is generated by an external 10 MHz crystal. It produces a single instruction cycle time of 0.4 μs. This times all the circuit operations of the PIC IC effectively. A stable 5 V operating voltage is supplied to the controlling circuit using a voltage regulator (7805) followed by rectifiers, help avoid any further peaks that tend to be produced by a Lithium battery. MOSFET's are used as an electronic switch that's used to switch ON or OFF the heating pads. Also, heat sinks are used with the MOSFETs to remove the excessive heat generated from them. The temperature sensor (LM35) are placed in four different body locations three of which are placed on the heating pads to monitor their temperature and the fourth on the free armpit (Left) to read the body temperature. All the LM 35 are serially connected to the control unit. The control unit will house the microcontroller and its connections, LCD to display the temperatures of pads, body temperature and outside temperature, an On/Off switch, and a mode selector to switch between automatic and manual modes. Dimensions of the box being 17 × 10 cm made of iron and coated with rust proof paint. A specially developed algorithm is used by the programmable microcontroller [4] that controls the operation of the circuit and hence maintains the virtual environment in between the body and jacket preventing hypothermia.

2.2 Theoretical Calculations

This section discusses the theoretical calculations on heat generation and transfer of that through the body surface [3]. The amount of heat generated must be capable of maintaining the required outside temperature (in between jacket and body) and allow the body to perform its physiological functions. The temperature at a point is related to the heat density at the point as follows:

$$\text{Heat density} = \text{Mass density} \times \text{Specific heat} \times \text{Temperature} \tag{1}$$

Differentiating with respect to time:

Rate of heat density change = Mass density × Specific heat × Rate of temperature change

$$Q = M \times Cp \times dT \tag{2}$$

where, M—Mass of the air/body (approx. 70 kg), C_p—Specific heat of air/body (C_p for human body = 3,470 J/kg.K), dT—Change in temperature acquired (here, 1 °C).

This equation gives the amount of heat in Joules required to raise the temperature of the object. Now substituting for the various values in the equation,

$$Q = 242,900\,\text{J} = 242.9\,\text{kJ} \tag{3}$$

Convection is the transfer of thermal energy from one place to another by the movement of fluids or gases. Convection is usually the dominant form of heat transfer in liquids and gases [5–7]. Although often discussed as a distinct method of heat transfer, convection describes the combined effects of conduction and fluid flow or mass exchange. The amount of heat required is describes as,

$$Q = U \times A \times dT \tag{4}$$

where, U—heat transfer coefficient, A—Area of contact, ΔT—Temperature gradient, Q—Amount of heat required in watts.

This equation gives us the exact area of contact required to increase the body temperature. First step towards this is to calculate the temperature gradient and is given by,

$$\Delta T = \Delta T_1 - \Delta T_2 / \ln(\Delta T_1 / \Delta T_2) \tag{5}$$

In this case, the value were taken from the heat transfer coefficient for transfer of heat from heated car seats to driver or passengers and it was assumed that the same rate of heat flow to take place ($\Delta T = 3.47$ K). Q here has to be taken in watts and hence we have to convert the Q got in the above equation from calories to watts. The point here to be noted is that for conversion we need a time factor. The cold water survival of humans is more than an hour. To be on a safer side, we take the time for the exposure of heat to be 15 min. Then the Q becomes, 269.88 W. After substituting all the values in the equations, A becomes 0.539 m^2. Now this is the area of contact required for raising the temperature of the body by 1 °C.

Various calculations have been published to arrive at the body surface area (BSA without direct measurement. The formula that was used for thc calculation by Mosteller formula:

$$\text{BSA}\left(\text{m}^2\right) = ([\text{Height(cm)} \times \text{Weight(kg)}]/3{,}600)^{1/2} \tag{6}$$

Therefore, assuming weight = 70 kg and height = 180 cm. We have, Surface area = $70^{0.5} \times 180^{0.5}/60 = 1.87$ m^2. Thus, by the above calculation the area required for the heat transfer to raise the body temperature by 1 °C is 0.539 m^2. As per the design, the Surface area of the pad is 0.0735 m^2. The difference in values can be compensated by other means, i.e. by using copper sheets, etc. to increase the re- radiating heat. The above calculations are in ideal conditions, whereas practically considering the body is covered with the clothing and there is re-radiating heat that compensates for the above.

2.3 Jacket Fabrication

The jacket fabrication consists of keeping several layers of different fabrics together to get the desired result. The different materials used for fabrication of

Fig. 1 Jacket measurement diagram

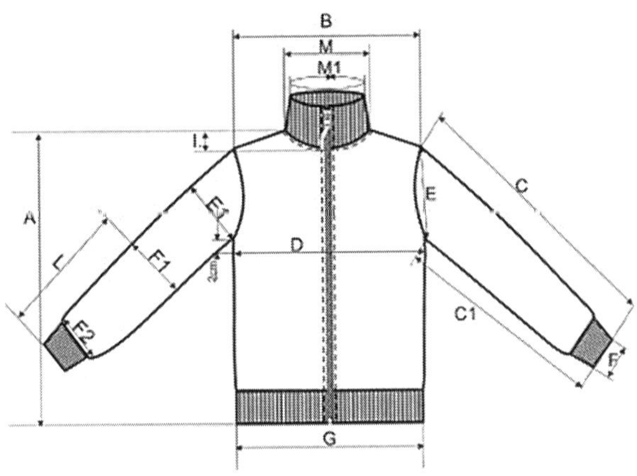

jacket were: (a) polar fleece which is a soft napped insulating synthetic fabric made from polyethylene terephthalate (PET). It is soft, lightweight and warm in nature. It is hydrophobic and retains very less water, making it moisture free and thus the best material that can be used in winter times. The fabric weighs 300 g/m^2 and is 300 g/m^2 thick. (b) Single jersey which is a type of knit textile made from cotton or a synthetic blend. It is made from a light weight yarn but it still maintains a very stretchy outlook because of it single knitting, flat at one side and piled on another. In this jacket, 140 g/m^2 thick single jersey materials were used.

The outermost layer of the jacket was kept as the polar fleece material. Its excellent insulation capability helps in trapping the heat generated inside and prevents it from escaping to the surroundings. The second layer was kept as the pads that are going to generate the heat necessary for maintaining the temperature of the body. The last and the final layer will be that of the copper sheet, whose main purpose is to distribute the heat effectively and uniformly throughout the jacket thus increasing the surface area of heat transfer by tapping the heat on the other side of the pad. The Fig. 1 shows the drawing of the standard size of the jacket selected for the work. In this study, the size of large (L) was considered for stitching process.

The fabrication was done using a skilled tailor and materials were supplied as per the design and requirement. The various stages of the fabrication are shown below (Figs. 2, 3).

2.4 *Testing*

The fabricated jacket was tested in an environment where a constant temperature of 18 °C was maintained. Faculty and student volunteers in the age group of 19–

Fig. 2 Copper strips used for heat transfer and placed in the fabric

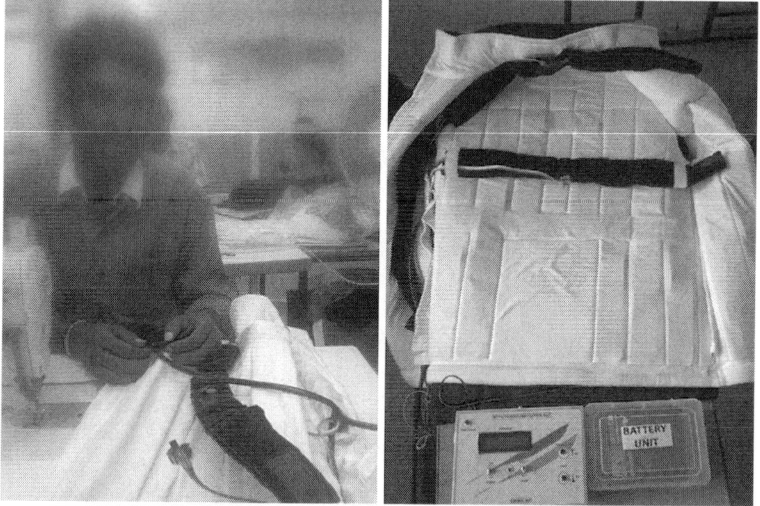

Fig. 3 Placement of the pads in the fabric and the completed one with the control unit

35 years were considered for this study. The set point of 42 °C was kept in the control unit. Standard thermometers were used for temperature measurements. During testing, the volunteers were asked to lie in supine position for nearly 1 h without doing any physical activity. The readings were recorded one from the inside of the jacket (T_1) and the other from the body (T_2, thermometer kept under the tongue) at two instances of time with the gap of 30 min. The following table shows the readings taken from the volunteers (Table 1).

Table 1 Temperatures recorded inside the controlled room

Volunteers	T_1	T_2	T_{12}	T_{22}
1	39.50	34	39.52	34
2	38.5	34	39	34.2
3	39	33	39.4	33.2
4	38.4	33.6	39	33.8
5	38	34	39	34.2
6	39	34	39.2	34
7	38.2	33.2	38.8	33.6
8	38.4	33.6	38.8	33.8
9	38.6	34	39	34.4
10	38.8	33.8	39.2	34.2

Temperatures are recorded as °C, T_{12} and T_{22} are recorded after 30 min after first reading

3 Conclusions

In this article, a hypothermia prevention jacket for military purpose was proposed, fabricated and was tested to meet the desired parameters. The findings show that the virtual environment created by the proposed jacket maintains the normal body temperature without any discomfort to the subjects. The challenges we faced in this design, was the life of the battery. The battery quite often drains due to the long durations of the circuit operation. If the life of the battery preferred was long, then the size of the battery becomes big and heavy to carry. It is very difficult to recharge the battery in the battle fields or any cold environments where the soldiers are stationed for a long period. Hence it is necessary to think of other alternate ways for recharging the battery.

4 Future Works

Probable use of lightweight solar panels for recharging the battery. The size of the battery and the controlling unit may also be reduced, to bring down the overall weight of the jacket. Also, embedding the Nichrome material in the inner part of the jacket will help in avoiding the numerous wires used inside. The embedding of the nichrome will also include proper contact with the copper strips which extend to both the upper and lower limbs thereby increasing the area of heat generated and in turn increase the efficiency of heat retaining capability. Commercially, the jacket can also be modified for use by civilian people in the northern parts of India during heavy winter, where many deaths are reported every year due to the harsh cold climate.

Acknowledgments The authors sincerely thank the management of Rajalakshmi Engineering College, Chennai for their support and encouragement while carrying out this work. We also thank Mr. G. Loganathan, Managing Director, The Imperial Pvt. Ltd, Tiruppur, and his colleagues for their technical support in the fabrication of the jacket.

References

1. Sessler D (1997) Mild perioperative hypothermia. N Engl J Med 336:1730–1737
2. Morrison SF, Nakamura K (2011) Central neural pathways for thermoregulation. Front Biosci 16:74–104
3. http://www.blizzardsurvival.com/product.php/105/blizzard-survival-blanket as on July 29
4. Predko M (2007) Programming and customizing the PIC microcontroller. McGraw-Hill, New York
5. de Dear RJ, Arens EA, Zhang H, Oguro M (1997) Convective and radiative heat transfer coefficients for individual human body segments. Int J Biometeorol 40:141–156
6. Thome JR (2004–2010) Wolverine tube heat transfer data book, pp 21–24
7. Boutelier C, Bougues L, Timbal J (1977) Experimental study of convective heat transfer coefficient for the human body in water. J Appl Physiol 42(1):93–100

Decoding Design: A Study of Aesthetic Preferences

Geetika Kambli

Abstract When implementing a global product development approach, it becomes important to note that local aesthetic preferences and cultural semiotics can play a leading influence on brand preference and purchase behaviour. While much has been done to gain a broad understanding on culture and its impact on design, efforts to specifically map local certain aesthetic preferences has been limited. This paper seeks to create such a mapping for the Indian consumer. Armed with this understanding, global design efforts may gain empowerment to create appealing aesthetics that provide the right messages to Indian consumers. Thereby addressing issues related to local implementation, which is an important aspect of global product development. Moreover, this paper uses a mixed methods approach, (though rooted in strong established theory), which encourages the construction of new models to understand and map creative practice.

Keywords Semiotics · Design research practice · Global product development · Aesthetic preferences · Culture · India · Industrial design

1 Introduction

Much work has been conducted to address semiotics and industrial design.

Griffin [1] speaks of products as carriers of meaning. He proceeds to explain that the process of understanding unfamiliar products involves a knowledge-based and an emotion-based reaction. It is clear that thoughts, beliefs and values form an important part of this reaction. Therefore, it can be understood that beliefs and

G. Kambli (✉)
Future Factory LLP, Mumbai, India
e-mail: geetika@futurefactory.in

A. Chakrabarti and R. V. Prakash (eds.), *ICoRD'13*, Lecture Notes in Mechanical Engineering, DOI: 10.1007/978-81-322-1050-4_113, © Springer India 2013

values form a cultural perspective, and therefore help give meaning to products by virtue of the cues that their designs emanate. The study and application of such an intersection of culture, meaning and design will be useful to study and map. As this has direct implication on product design for marketing and commercial purpose.

This paper extends such an implication to map these meanings (semantics) and their relation to certain product cues (through product design). It therefore aims to use established theories to put together the two faculties of semiotic research and industrial design, with specific relevance to commercial implementation in Indian markets.

1.1 Objectives

The study aims to assist global product development in its design efforts by:

(a) Mapping specific aesthetic preferences; especially mapping form, shape color and finishes, to reflect preferences of the Indian consumer
(b) Understanding the semantics of sub-cultures with specific reference to certain groups that may emerge
(c) Arriving at a set of design principles that can assist in local customization to drive consumer preference.

1.2 Challenges

The study of semiotics has largely remained a marginal and academic field with limited acceptance in commercial research work. It remains limited to the area of social studies and societal impact, with most influence on the community of academia bad least on industry and global commerce.

This seems like a missed opportunity, as the faculty of semiotics offers significant opportunity to supplement tradition communication and marketing communities which today hold a strong influence on industry. By exhibiting a collaboration between the faculties of design & social sciences, this study works towards a wider acceptance of these faculties and tools.

2 Methodology

The study was rooted in a strong theoretical framework and soundness of research practice, but with a certain experimental approach to the interviewing methods. Specifically, it uses a mixed methods approach of integrating ethnographic practices and simulation of shopping behavior.

Reference [2] have been made to the practice of ethnography and its application to industrial design and on how the practice of ethnographic research for design is different from that used in academia by anthropologists. The objective of working on commercial applications and the strong need to arrive at results that address real market requirements brings a shift in the approach used by industrial designers when practicing ethnographical research. Therefore, adaptations of a strong theoretical methodology such as ethnography is usually accepted in the practice of industrial design. Here, in this study, the method of ethnography has been adapted for such a purpose.

While maintaining its identity as an observation based tool, this methodology allows the observer freedom to map his findings by integrating another widely practiced method—the simulation of shopping behavior [3]. Modeling shopping behavior often uses techniques to measure consumer choices and evaluate buying behavior. By using instrumented technologies in virtual supermarkets, researchers record customer actions and mine data to map reactions to products on shelves. While such technological methodologies map reactions, they often don't yield reasons or motivations for such reactions. A qualitative adaptation of evaluation of shopping behavior may be modeled as a "product exposure test" where consumers are exposed to a set of pre-decided products and their responses and choices are measured and evaluated. In this study, such a test has been integrated with the process of ethnography to arrive at a methodology that is rooted in strong social science theory but is generative and exploratory in its application to industrial design.

The study therefore operated with mixed methods interspersing detailed ethnographic interviews with several product exposure tests modeled on shopping basket experiences. Such a combined methodology allowed for real insight generation to work alongside a large sample base for application to a wider scale. The large sample was specifically important because it allowed a closer examination of specific sub-cultures and smaller geographies.

2.1 Selection of Stimuli

The stimuli were chosen by a team which consisted of industrial designers, graphic artists and design researchers. The stimuli reflected a wide variation on choices of, color, finish etc. and was pre-selected to allow designers to help analyze winning products on specific design aspects. Given that the study aimed towards understanding aesthetic preferences for implementation of successful product design for Indian markets, there was a strong need to consider artifacts and their relation to semiotics. According to Krippendorff [4], we are "perceptually guided" by the visual appearance of the object in order to fetch the right cognitive model for its use from long term memory. This may cause conflict sometimes. For instance when the product aesthetics cue both known and unknown functions at the same time. The study selected a set of stimuli while acknowledging the possibility of such instances to allow a complete set of stimuli and make analyses more reliable.

2.2 Interviewing Techniques

The interview guides were designed around cognitive models that would allow for insight generation and yet allow collection of useful data around groups that emerged during the study.

2.3 Methods of Analysis

Methods for analysis were mostly means-end analysis [5] using value-cues research [6] typically used in communication applications. Inherently, such analysis has been used towards arriving at key motivations that drive consumer choices. They are qualitative and generative in nature, mapping all intermediate motivations and exposing the cues that lead to the messages of product communication.

For the purpose of this study, such a methodology was adapted to product cues (generated by design) instead of product communication. It was then used to generatively map the motivations and messages leading to the product cues; a method often used by paint companies when modeling color forecasts and trends. Such analysis was mostly qualitative with scores assigned for progressing the trend towards the conclusion. Also for the ethnographic work, metaphors were analyzed towards generating insights that could help frame hypotheses for the exposure tests. Designers worked with the analysis team to evaluate the progression of the trend in terms of specific design learnings.

3 Findings

Results proved that semiotics play an important roles in defining aesthetic preferences and that different subcultures in India have distinct aesthetics choices. What is of real relevance is how aesthetic preferences differed across groups of Indian consumers and what defined these groups to be different.

3.1 General Perception

There is a general perception of design preferences in India markets, which is assumed by managers in marketing positions and sometimes even by Indian designers. There are largely 2 broad tenets to this general perception:

(a) Common perception of design in India is that style needs to be accentuated and that which is bold and overstated is appreciated.

(b) It is also commonly believed that quiet sophistication and mild curvatures are easily overlooked, as it is believed that Indian sensitivity to design is weak.

The study found that such a generalized assumption is largely incorrect and mostly insufficient.

For example,

When the study examined results for even Tier-2 towns, it showed that consumers were sensitive to the slightest change in texture, or finish. These differences were obvious in consumer choices across products in the exposure tests, and this result disproved the notion that for Indians—style needs to be accentuated and overstated designs are appreciated.

Moreover, the generalized assumptions are insufficient for the following reasons.

3.2 Geographical Clusters

The diversity of India causes distinct geographies to behave as sub-cultures. This is because settlements have mostly grown around singularly large cultural groups. However, it may be insufficient to state that each geographical cluster seems to exhibits a specific style preference. It is more important, to examine the elements of such a style preference: (specifically—form, shape, color, finishes) and explain how they differ across geographies.

One clear finding that emerged was that specific geographies prioritizing form over graphic content. This has specific relevance to design in commercial practice. In places where form gains a clear priority over two dimensional aspects such as graphic content and color, design calls for more investments in tooling to deliver superior form. In other places, global product investments are more efficiently disbursed by compromising tooling for visible, striking graphic and color design effort. Understanding which geographies need form, and assessing their contribution to the business, will help decide whether to tool up, or to invest instead in a focused graphic design effort.

Another clear trend is that within specific geographies, similar cultural values prevailed. Therefore, the use of a specific color (in it's semantic sense) may gain high scores in some geographies and may completely fail in another. The commercial application of such a result, is that the application of color needs a fair consideration for cultural implication during the process of design. Specifically, this has strong relevance as during the study, most designers were found to use their own semantic construct when applying color. Which might have been very distinct from the culture in which the product would eventually find use.

However, some patterns were applicable across geographies.

For example, one finding clearly showed how Indian consumers were drawn towards gloss and sheen. Products with matt finishes were often overlooked and not considered "attractive". As a design sensibility this emerged pan-India for the category of small appliance products.

3.3 Product Clusters

More importantly, by applying the same aesthetic to different categories of products, the consideration changed and the means-end analysis threw up a different assignment.

For example, across most markets in India, clear transparency in plastic won a high preference score when associated with technology products as the aesthetic was associated with a value-cue related to technological advancement. However, when associated with shelving in refrigerators plastic transparency implied fragility and possible breakage. Therefore, the same aesthetic elicited different metaphors when associated with different objects (read: categories).

Such findings seem to validate the semiotic premise by Kippendorff [7] and Wickstrom [8] where meaning in design is defined as an object's features being connected to the context of use or where the design is self-instructing. So the semantic meaning is obvious in its design implication. By means of the cognitive model.

3.4 Demographic Clusters

India's progressive economy has caused large shifts in consumer demographics. Such shifts are surprising and contrary to popular belief. For instance: consumer segments grouped by similar income showed huge variance within the group on important semiotic aspects of belief and behavior.

It is possibly for these reasons, that the hypotheses on aesthetic preferences that succeeded in geographic clusters, failed when the geography held diverse demographic groups. This demanded that grouping aesthetic preferences be regarded independently, across consumer segments as well.

While such a detailed examination might be too exacting, a preliminary analysis yielded some broad findings. It emerged that within India, one specific consumer demographic groups behaved like global consumers reflecting global trends in their aesthetic preferences. Interestingly, this trend was noted across categories of products giving strength to the premise that global consumers have lower semiotic attachment and cultural ties are weaker.

4 Success and Critique

The study finds strong use in implementing localized styling efforts to globally developed products for successful launch in India. Specifically it has vast data on geographic, demographic, and category-wise preferences in form/shape/color/finishes. Which when applied can help effectively allocate design and tooling investments for global companies looking to leverage investments across countries.

The study was also successful in implementing a mixed methods approach, which combined theoretical frameworks with experimental adaptations. Also, the methodology was true to the profession of design, incorporating a strong ethnographic approach to the interview technique despite an extraordinary high sample size.

However, the lack of a specific quantitative tools might be seen as a lost opportunity especially given the large sample size. The advantage in using statistical significance or variance, may prove more reliable when examining micro-trends, or those that have a serious implication on design application.

This study is expected to be repeated for recency and relevance to new context, during which a quantitative tool may be considered.

Acknowledgments The author wishes to express her appreciation to: Designers at Future Factory for their countless, patient discussions on design research methods. The support and co-operation of the team at Futuring, where she floated the first version of this study. Prof. Navalkar, who encouraged the basis of semiotic study in her early research years.

References

1. Griffin T (1999) Semantic communication through products
2. Jonathan V (2011) The industrial designer as a mediator and semiotic translator: mediating clients' and users' needs. Hebrew University of Jerusalem, Bezalel Academy of Arts & Design, Israel, pp 75–76
3. Schwenke C (2010) Simulation and analysis of buying behaviour in supermarkets, (ETFA 2010). IEEE conference department of technical information system, Dresden University of Technology, Dresden, Germany
4. Kippendorf K (1989) On the essential contexts of artifacts. In: Margolin V, Buchanen R (eds) The idea of design, vol 1995, pp 156–184
5. Reynolds TJ, Gutman J (1988) Laddering theory, method, analysis, and interpretation. J Advert Res,
6. Zaltman G (2003) How customers think–essential insights into the minds of the market
7. Kippendorff K (1992) Transcending semiotics: towards understanding design for understanding. In: Vihma S (ed) Publication series of the University of Industrial Arts, Helsinki
8. Wikström L (1996) Methods for evaluation of products' semantics. PhD Thesis, Chalmers University of Technology, Sweden

Earthenware Water Filter: A Double Edged Sustainable Design Concept for India

M. Aravind Shanmuga Sundaram and Bishakh Bhattacharya

Abstract Approximately 1.5 million children die annually in India, before they attain 5 years in age. The deaths are mainly attributed to pneumonia and water-borne diseases like dysentery and cholera. Unavailability of clean drinking water is the root cause of the problem. Apart from sources of water being equally available to the masses, unavailability of extensive filtration systems, either at the point of distribution (POD) or at the point of use (POU), is a reason for consumption of impure water. An affordable water filter (POU) that can effectively kill all the disease causing pathogens will help in stemming the infant deaths. In India, earthenware pottery dates back to 1500 BC. Currently, potters are almost facing extinction because of changes in society. This paper, explores the possibility of using the skills of the Indian potter in making water filters out of earthenware with cheap filter substrate and provisions for germicidal UV based disinfection system powered through energy harvesting from transducers.

Keywords Water filter · Potter · Earthenware · Sustainability · Affordability

1 Introduction

The number of live births in India is estimated to be around 12 million per annum which almost accounts to 20 % of the world's birth [1, 2]. However, India also contributes to more than 20 % of child deaths in the world. India loses

M. Aravind Shanmuga Sundaram (✉) · B. Bhattacharya
Design Programme, Indian Institute of Technology Kanpur, UP, India
e-mail: aravind@iitk.ac.in

B. Bhattacharya
e-mail: bishakh@iitk.ac.in

A. Chakrabarti and R. V. Prakash (eds.), *ICoRD'13*, Lecture Notes in Mechanical Engineering, DOI: 10.1007/978-81-322-1050-4_114, © Springer India 2013

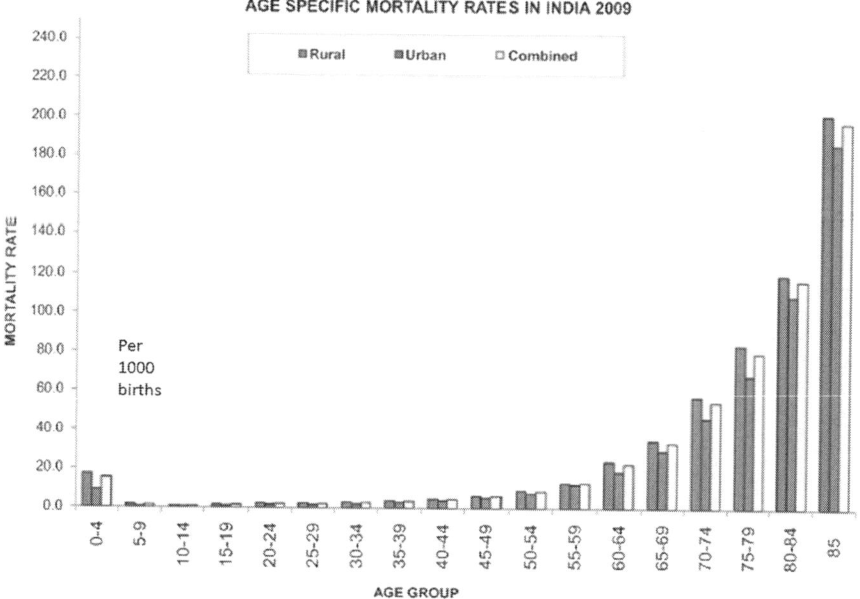

Fig. 1 Latest mortality census [2]

approximately about 1.83 million children before they reach 5 years of life [1]. A majority of the deaths are due to preterm births [2]. Even though preterm birth is a global nemesis, South Asia and Africa contribute to 60 % of the deaths that happen and India's contribution to this is 15 % [1–3]. The infant mortality rate (number of children dead per 1,000 live born) in India is 50 [2]. The under 5 mortality rate (U5MR) in India is around 65 [2]. 88 % of the deaths are due to preventable communicable diseases like Pneumonia and Diarrhoea [1]. Other diseases like Cholera, Typhoid and Hepatitis A contribute to deaths as well. Most of these diseases are water borne. Majority of these deaths occur in rural and urban poor, where the living conditions are very conducive to easy spreading of infections (Fig. 1) [2].

Lack of clean portable drinking water is a major reason for infections to spread. Infected water sources are often shared, ensuring widespread infection in the community. State controlled water supplies are usually guarded against these infections by having effective filtration systems at the PODs. The reach of these supply lines are not usually extensive and does not reach to every corner of the populated topography. Urban population has been on the rise and metro cities have their own pockets of high density areas where the supply is close to zero. In such places, people are forced to use water from uncontrolled to unmonitored water sources like private suppliers or local sumps which are open to contamination because of drainage or industrial waste. In rural areas, the water sources are mostly open wells, rivers or canals which are again easily polluted. Use of effective

Type of need	Quantity	Comments
Survival (drinking and food)	2.5 to 3 lpd	Depends on climate and individual physiology.
Basic hygiene practices	2 to 6 lpd	Depends on social and cultural norms
Basic cooking needs	3 to 6 lpd	Depends on food type, social and cultural norms
Total	7.5 to 15 lpd	Lpd : Liters per day

Fig. 2 Water needs per person per day: WHO [5]

filtration methods will help in stemming the progress of infections in these societies. The prevailing problem is lack of awareness and that of affordable filtration apparatus. Most of these households cannot afford the contemporary water filters that are being sold in India. The average water filter is far beyond the reach of the majority of Indians. This leaves majority of Indian population vulnerable to preventable diseases. This paper explores the ways of making the water filter an affordable commodity. Contemporary water filter manufacturers in India are aiming for cost of ten paisa per litre [4] on the basis of one single unit output.

The average water needed for a person to survive a day as per WHO standards is a minimum 7.5 L (Fig. 2) [5]. The minimum quantity of water to be consumed per person is at 2.5 L per day [5]. WHO India too recommends the same based on Indian conditions.

For Rural India, the water need per capita is as given below (Fig. 3) [6]. The rural average is more than the recommended average of 2.5 L for drinking water. 3 L per day is the stipulated need. The average size of Indian household is 5.3 [2] (approximated to 6). Based on this, the average need for drinking water per family in India would be 18 L per day.

Fig. 3 Average water needs per person per day in India

Type of need	Quantity (lpd)
Drinking	3
Cooking	5
Bathing	15
Washing utensils	7
House ablution	10
Total	40

2 Sustainability: The Need of the Hour

Many factors like changes in environmental conditions resulting in skewed monsoon behaviour, increased temperatures, failure of crops and dwindling of agricultural prospects and economic factors like globalisation, open market economy has led to a marked exodus of people from rural areas to urban cities. This has created an imbalance in the demographic dynamics of India. The divide between the affected and the not so affected is widening. The affected people form the majority and are poor. The market driven economic practises mostly concentrate on to the needs of the affordable masses and price points are fixed based on their purchasing power. The ballooning urban scene has given rise to unforeseen contention for resources like water, energy and space. The rural areas have become further alienated. The need of the hour is to attain sustainability in all possible fronts, whether be it demographic shifts, energy consumption, or environmental impacts. From a machinist view, all the events are causal and hence this intertwined complexity can be solved or controlled, only by adapting little sustainable parts that sum up to a bigger sustainable whole. This model is portable and can be replicated in an identical situation, but a radically different practise. For example, similarities could be drawn between the traditional mud pots giving way to metal utensils and the dhobi's services being overtaken by washing machines that consume and contaminate more water per cloth than the traditional way. A sustainable solution that could meet the demand of the changing scene as well as a sustainable livelihood of the famished potter or the dhobi is the need of the hour.

2.1 The Case of the Indian Potter

Pottery has existed in India ever since civilisation could be historically accounted in the subcontinent. Earthenware pottery has been in existence since 1500 BC in India. They had mostly started from being storage instruments to other artifacts like ornaments, idols and toys. Pottery in India has been influenced and improved by many influences from Persia to Mongolia. Some of the artisans who belonged to the invading armies from these parts of the world made India their home and have enriched its pottery practise [7]. India has wide variety of clay soil which had a big influence in the evolution of pottery. The government of India has also played its role in improving the condition of pottery by industrialising it [7]. It was the first big deviation that was attempted to improve the prospects. Second World War created new demands and also brought in western influences resulting in mushrooming of lot of ceramic industries. These industries were able to employ a certain amount of skilled and unskilled labourers. However, the plight of the Indian potter took a hit with changing life styles. The traditional and ubiquitous earthen pot gave way to utensils made of metals like aluminium and stainless steel wares. The potter has been relegated to cater the needs of traditional occasions like

Sl. No.	Category of Pots	No. of families	No. of pots produced / annum	Cost of clay in Rupees	Cost of burning	Total Cost	Average cost in Rupees.	Average Revenue (Price) in Rupees.	Average No. of pots produced per family per annum	Average Income per day in Rupees of a family from the occupation
1	Up to 1000	72	50050	25550	35900	61450	1.22	9.36	695.13	50.48
2	1000 - 1500	23	29500	15100	18800	33900	1.14	9.55	1282.60	78.11
3	Above 1500	5	9300	5100	7700	12800	1.37	10.43	1860.00	93.26

Fig. 4 Income of potters in Warangal, AP [8]

festivals and other obeisance. The potter was also a vital cog in the wheel of agricultural setup of India, making storage utensils and also huge terracotta idols which the local farmer revered a lot. The deterioration of agriculture had its effect on the potter too. Many have migrated to cities in search of other menial labour destroying his family fabric. The plight of the potters of Warangal, Andhra Pradesh is a good sample to elicit the tough times the community is going through [8]. The average salary for a potter's family per day is around 50 Rs (Fig. 4). This goes up based on the number of pots produced. But their per capita income with 5.3 as the household average size [2], are very much below the poverty line by any standards. This has mostly resulted in the Indian potter moving to urbanised societies in search of employment.

With minor design interventions, the potters could be made to regain their lost ground. A different product line by the potter meeting to the current demand is needed. Similar exercises where rural communities can stay in their rural setup yet get connected to the dynamics of changing societies through a new product relationship [9] have been explored in many places.

3 Current Point of Use Filtration Systems in Use

There are a multitude of POU filtration systems that have been put forth in the market. Most of them use reverse osmosis as a means of filtering assuring effective filtration. These setups are augmented with an Ultra Violet chamber in which most, if not all pathogens are killed. Some of them have additional chemical systems to soften the water and to improve the taste as well. These are state of the art systems and would cost on an average, 10,000 Rs (Fig. 5) based on the quantity of water through put. From a sustainability point of view, the water wasted and the power consumed to produce one litre of pure water is phenomenal. The capital and operational cost of these systems are not easily affordable by all.

Other forms of filtration like boiling, chlorination are also done. These affect the taste of the water but are effective against pathogens. Old filtration systems use ceramic based filter units that provide filtration of suspended impurities only. Pathogens are not effectively removed. The ceramic candle filtrate is expensive.

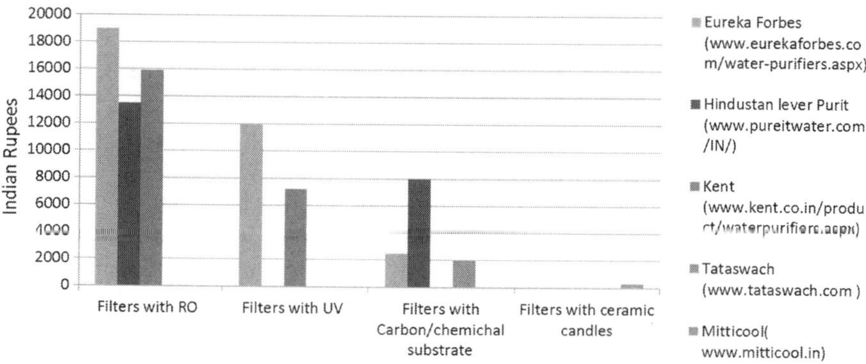

Fig. 5 Comparison of water filters in Indian market

Sand bed filters like fast sand filters and slow sand filters have been used extensively in POD centres. Fast sand filters are effective in filtering suspended impurities. Since fast flow of water disturbs the growth of bacteria that filter, they are not effective against pathogens. Slow sand filters ensure effective filtration of pathogens and chemical impurities. The process is very slow. Most of the water filtration plants adapt this method of water filtration since they are very effective and can treat large amount of water. Slow sand filters [10] are gaining ground amongst the impoverished societies of the world as a household filter (Fig. 6). The filtrate is acquired in the POU.

4 Proposed Solution

Design in the context of alleviation of the poor has not received much attention or has been pursued earnestly by design practitioners both at the industry as well as in academic levels [11]. States and voluntary organizations have attempted to find a market for the artifacts produced by the poor by organising selling points in the form of national and international fairs. They have not proved to be sustainable. Huge communities cannot be effectively included in this practise. The artifacts produced have strong cultural identities, since most of them are elements like ornaments, cloths like scarfs, toys and idols. These have a very sporadic demand. If consumables made by the artisans, that meet a steady demand, it will be sustainable. Design interventions will have to be made through out the supply chain [9, 10]. An earthen ware water filter could be made this way (Fig. 7).

The proposed water filter (Fig. 9) will have earthenware shell which could be easily made by the potter. The filtrate shall be slow sand bio filter that shall be made up of washed sand at an average of ten micron in size to create a dense filtrate medium. This shall be rested on a thick and thinner brick chip bed. In addition, the water chamber shall have a germicidal UV powered by poly-

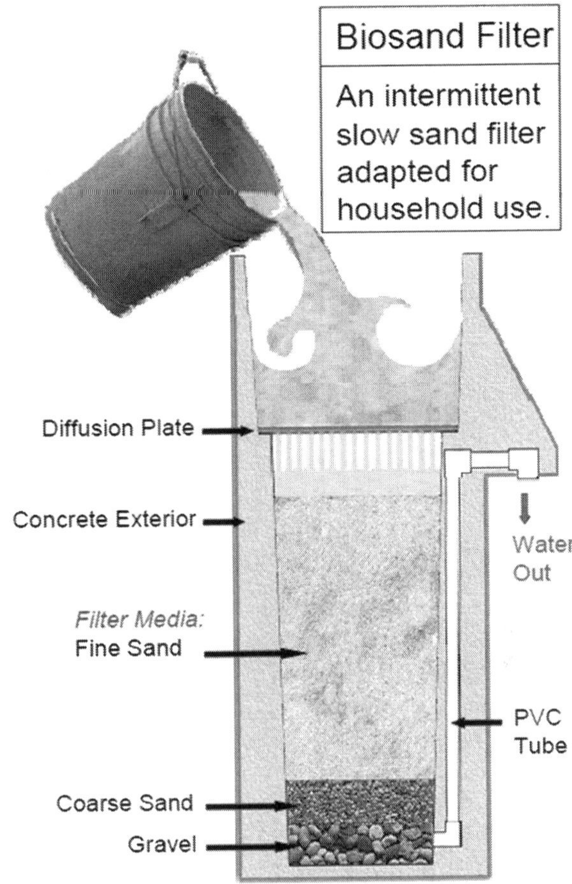

Fig. 6 Bio sand filter for home use (www.cawst.org)

crystalline solar cell array. Effective, disinfection can be ensured by using germicidal UV. The germicidal UV has two single watt lamps accounting to four watts of power need. These are provided by a solar panel made of four polycrystalline solar cells. Poly crystalline solar cells are far cheaper as compared to monocrystalline solar wafers (Fig. 8). The water has to be maintained in the UV reservoir for a maximum of 30 s for all the pathogens to be killed effectively. The water chamber is a 20 L container. A participatory design process needs to be adhered here to enable the chambers being made as per design and assembly of the electronic elements to be incorporated in the body. Once the process is matured, the potter would be able to replicate the design.

The filter substrate comprises of sand available at the point of use that could be washed and used. Initially the filter needs to be curated for beneficial bacteria for a week. The water would be usable after a week. The sand should not be disturbed or changed frequently for substantial results.

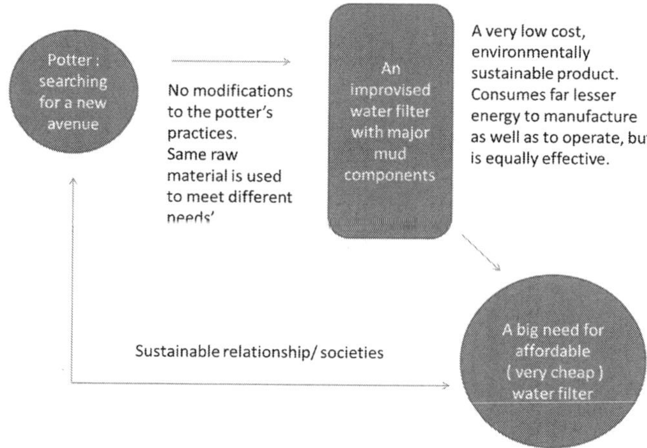

Fig. 7 Sustainable supply chain: earthenware water filter

Element	Property
Solar panel	7.8 V polycrystalline solar module
Battery	6.6 V
UV lamp	Wavelength 240-255 nm, 3V (2 numbers)

Fig. 8 Preferred configuration for the components of the filter

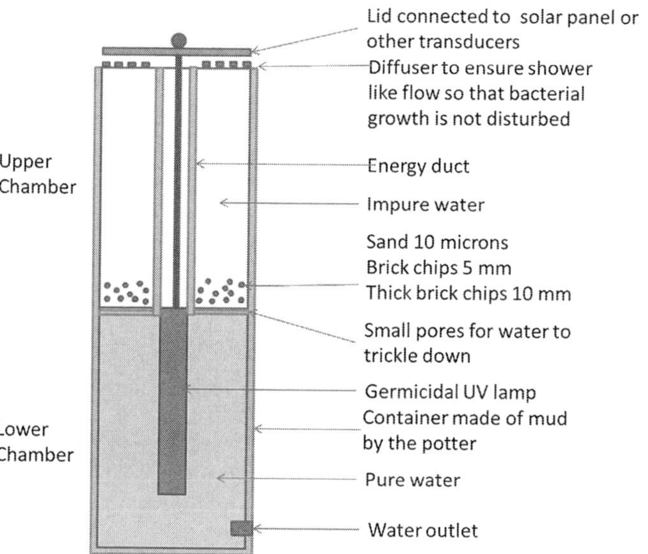

Fig. 9 Earthenware water filter

Element	Approximate cost (In rupees)
Mud Chambers	50
Battery	80
Solar panel	80
Germicidal UV lamps	80
Total	290

Fig. 10 Approximate cost of the proposed water filter

The approximate cost of production for the water filter will be 290 Rs (Fig. 10). The cost per litre for the buyer would come to very low as 0.0004 RS/L of water at the rate of 20 L production per day.

$$\begin{aligned}\text{Cost/litre} &= \text{total annual cost/total annual litres of water produced.}\\ &= 0.00047\,\text{RS/L.}\end{aligned}$$

4.1 Sustainability Parameters

Low-fire clay needing a heating temperature of around 1–4 cones accounting to 1,120 C. There are no new skills for the potter to be learnt. As compared to a polycarbonate body of state of the art filters or the stainless steel bodies of yesteryear filters, the clay body consumes lot less energy to make [9]. Midrange clay is to be used, since, if they are not recycled, the easily disintegrate into natural elements.

4.1.1 Potter's Perspective

The average income of a successful potter family is around 1,500 Rs/month [6]. They have peak sales during festivals and runs close to dry on other months. The damages that are caused to his inventory during storage as well as transport are also borne by them. The new cost of the filter could be around 350 Rs.

$$\begin{aligned}\text{Revenue for the potter from one sale} &= \text{cost of filter} - \text{cost of production}\\ &= 60\ \text{Rs.}\end{aligned}$$

60 Rs/filter are a sizeable amount that a potter could make as compared to the old revenues they get from selling a pot.

The loss of chambers during storage and transport could be lessened by adapting a hexagonal cross section for the chambers (Fig. 11). This helps in packing them like a honey comb which ensures better shock resistance.

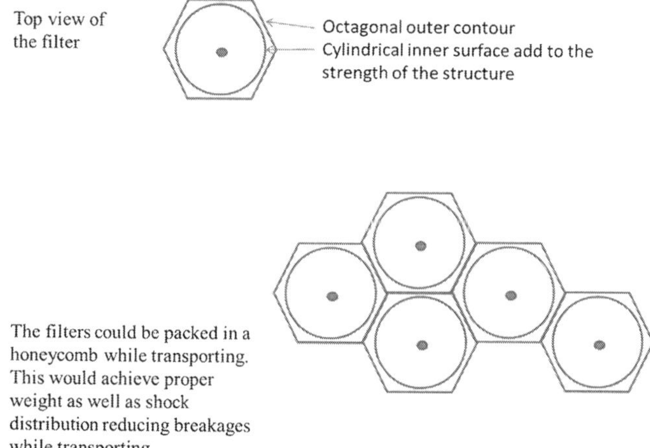

Fig. 11 Earthenware water filter—top view

4.1.2 Designer's Perspective

Such practises are achievable through participatory design process where in addition to user study to identify the needs, the process of manufacturing needs to be keenly noted to stem any inadvertent process related losses. The same model and learning could be exported to other systems. When working with a community, designers should be able to identify all the other peripheral activities as well. Many of them could be made to come under the sustainability dragnet easily. Designers should also pay attention to the importance of localisation, for example, the filter substrate could be tweaked or altered based on the impurities that are present locally.

5 Conclusion

A complete supply chain perspective for water filter design is presented here which will help identifying a sustainable ecosystem benefiting both the producer and the consumer very effectively. Traditional practises have been mastered over generations and their outcomes are all that we may need as solutions for complex problems of today, whether it is health, education or food. The task of the designer is to identify these pockets of important practises and tradition and find ways of retrofitting them with the current challenges to achieve equanimity and sustainability.

References

1. The situation of Child in India—A profile: UNICEF May 2011
2. National health profile India (2011) Central bureau of health investigation, Ministry of health and family welfare, India
3. Maternal and child mortality and total fertility rates (2011) SRS, Office of the registrar general, India
4. Delivering safety: Corporate reports, Business India, June 14, 2009
5. Technical notes on drinking water, sanitation and hygiene in emergencies: WHO
6. Rajiv Gandhi national drinking water mission. National rural drinking water programme: Framework for implementation Ministry of rural health and development, Government of India (http://www.indiawaterportal.org/sites/indiawaterportal.org/files/National%20Rural%20Drinking%20Water%20Programme_MoRD_2010.pdf)
7. Barkin D and Barón L (2005) Constructing alternatives to globalization: strengthening tradition through innovation. Development Pract 15(2):175–185 (Published by: Taylor & Francis, Ltd. on behalf of Oxfam GB Stable URL: http://www.jstor.org/stable/4030078)
8. Suresh LB Rural artisans—Indigenous technology: an empirical study on village potters in Warangal, Department of Economics, Kakatiya University, Warangal (http://community.eldis.org/lalbsuresh/.59e05148/.59e8c693)
9. Project Bravo (2006) Field study in Haiti, a summary of results, June 2006. Centre for affordable water and sanitation technology (www.cawst.org)
10. Thomas A (2006) Design, poverty, and sustainable development. Des Issues 22(4):54–65 (Published by: The MIT Press Stable URL: http://www.jstor.org/stable/25224076) (Autumn, 2006)
11. WHO (2012) Born too Soon: The global action report on preterm birth

Designer's Capability to Design and its Impact on User's Capabilities

Pramod Ratnakar Khadilkar and Monto Mani

Abstract Extending the dictionary meaning of 'capability' as individual skills or abilities, the current paper adopts a Capability Approach (CA) based definition. Accordingly *capabilities* are effective (operational) options available to an individual to be and do as aspired, in leading a life of value upon reflection. This concept primarily evolved in development economics, and has remained confined to the Bottom/Base of the Pyramid, wherein the *capabilities* of the poor are envisaged to be leveraged upon in alleviating poverty. Design for the BoP aims at designing appropriate products to serve as means to realize or augment the *capabilities* of the poor. However, the designer's role as an individual with *capabilities* is taken for granted. *capabilities* to design are the effective options/ resources available for the designer to design effectively. The current paper extends the concept of CA to the designer, and evaluates the related *capabilities* for its impact on designing products aimed at the BOP to alleviate poverty.

Keywords Design · BoP · Capability approach · Capability space

1 Introduction

The definition of Design as used in this paper is, *'activities that actually generate and develop a product from a need, product idea or technology to the full documentation needed to realize the product and to fulfill the perceived needs of the*

P. R. Khadilkar · M. Mani (✉)
Center for Product Design and Manufacturing, Indian Institute of Science, Bangalore, India
e-mail: monto@astra.iisc.ernet.in

P. R. Khadilkar
e-mail: pramod@cpdm.iisc.ernet.in

A. Chakrabarti and R. V. Prakash (eds.), *ICoRD'13*, Lecture Notes in Mechanical Engineering, DOI: 10.1007/978-81-322-1050-4_115, © Springer India 2013

user and other stakeholders' [1]. Throughout the design process a designer (as a design team or firm) requires access to extensive information starting from the end user/beneficiary needs and peculiarities, affordability and pricing, market size, and product life-cycle, viz., material and manufacturing data, service life and disposal. A successful product essentially is determined by the market response, and in the case of the poor, in terms of the products ability to serve as a means in alleviating poverty. Besides customers, various stakeholders are involved, viz., designer, manufacturer, transport, sales-network, service and disposal. Product design essentially connects the designer to the consumer, with the role of each stakeholder being characteristic in determining product success, i.e. gratified beneficiary with a profitable product [2].

To reiterate, the definition of *capabilities (italicized henceforth)* in CA are the options available to an individual to be and do, as aspired, in leading a life of value [3]. *Functionings* are realized *capabilities*, e.g., for an individual aspiring to become literate, going to a tutor or self-learning using Tablet-PC's are *capabilities*. Eventually opting for a Tablet-PC would become a functioning in the individual's aspiration to become literate. A designers role in designing a tablet-PC conducive to self-learning is determined by the *capability* options available during the design process, viz., conversancy in touch-screen technology, dynamic team, freedom to innovate, etc. In the case of products designed for the Bottom of the Pyramid (BOP), Capability Approach essentially values the *capabilities* of an impoverished individual, which through product means, can provide opportunities to come out of poverty. However, in designing for the BOP, the CA perspective has only been applied to the beneficiary. However, it is crucial to understand that CA applies to an individual, and every stakeholder comprises individuals, whose attitude essentially determines its role and effectiveness [4]. Thus extending the CA to evaluate the *capabilities* of other stakeholders would be more appropriate in determining the success of a product. Since Designers and Users (BOP) are the two critical stakeholders, the current study focuses on evaluating CA through the *capabilities* of the individuals characterizing these two stakeholders (see Table 1 for the Tablet-PC example).

The following section illustrates the distinctive features of CA. The designers *capabilities* (to design) is discussed in detail in Sect. 3, followed by an appreciation of how the users capabilities are impacted by that of the designers' capabilities (embodied in the designed product) in Sect. 4.

1.1 Research Motive, Questions and Contributions

Prahalad [2] in his influential work proposed that businesses (products and services), targeted at the BOP can infuse development amongst the poor while ensuring profit in business, i.e. eradicating poverty through profits. However, the impact of this approach remained inert as it lacked clarity on what *development for the poor* really meant. Products/services can potentially leverage on the existing

Table 1 Designer's capability to design and its effects on capabilities of users/beneficiaries through appropriate product design

Stakeholders		Users/beneficiary (poor/BOP)		Designer (formally trained)		Manufacturer (OEM), others, ...
Aspirations		• To be literate • To enhance social stature, ...		• To design trendsetting products • To improve sales/profit, ...		...
Configuration of available Capabilities	Internal abilities	• Go for tuitions or • Buy a Tablet-PC and self-learn	• Basic visual ability • Conversancy with electronic gadgets/PCs	• Company value position to design & allocate resources for the BOP	• Ability to sketch, visualize • Creativity, ...	...
	External resources /means	• Income or affordability	• Tablet PC • Self-teaching apps • Electricity, Antivirus, ...		• CAD and prototyping support • Collaboration with GUI team	...
Functioning		• Being literate by self-learning using a Tablet PC,		Design of simple user-friendly Tablet PC to promote literacy for the BOP		...

skills and strengths of the poor in supporting them to achieve what they want to do and be in life. Understanding the needs of poor is a challenge, as a product acceptance (by the poor) depends on factors more than a product's functionality alone. It deals with complex socio-cultural value judgments that the poor perceive, and does not vibe with simple 'return on investment logic' [5].Well intended products designed for the poor have failed due to inadequate understanding of the users' needs, inadequate infrastructural setup [2, 20] and for being too product/ technology focused [6] clearly traceable to inadequacies in the analysis phase of product design. The analysis phase is attributable to the designers *capabilities*, wherein the authors strongly believe can be addressed through the CA normative framework. This would require the designer to understand 'why' a product is designed beyond product functionality and 'what' product is to be designed keeping in mind the beneficiaries aspirations and *capabilities*. CA provides a lens to appreciate an individual's value judgments in terms of capabilities, i.e. effective options to 'be' and 'do' something of value. Embodying a product with functionalities attributed to users' (beneficiary) *capabilities* provides a broader perspective for the designer to accommodate and respond to. The objective of the current study is to provide a CA based design methodology (for designers' *capabilities*) to effectively design for the poor. This would place the designers' *capabilities* to be mapped from the conventional requirements of design-experience, sketch-ability, creativity [7] to mechanisms to permit interaction with remote BOP users [8]. To evaluate a successful design (product) intervention it would be useful for the designer to foresee the likely impact on the lives and *capabilities* of the poor. A model to permit this assessment would provide crucial foresight *capability* to the designer. The current paper is an attempt in this direction towards developing a '*capability* space for design' based model that would permit the delineation of various stakeholder (Design for BOP) *capability* spaces.

2 Capability Approach

2.1 Introduction to Capability Approach

Capability Approach was proposed by Nobel Laureate Amartya Sen as a normative framework to broadly evaluate individual wellbeing and social arrangements. It fundamentally relies on the freedom of individuals operating at two levels, viz., firstly in the effective individual freedom to decide what to be and do, i.e. *capabilities*, based on ones perception of a valuable life, and secondly in the freedom to achieve and/or realize from the list of capabilities, referred as *functionings*. Sen [9] articulates '*functioning*' as achievements and 'capability' as the options to achieve. As discussed earlier, an appropriately designed product can potentially expand the *capabilities* of the user/beneficiary. Thus, a design process converts a want/need into user *capabilities*. The effective adoption of a product is an achieved functioning (see example in Table 1).

2.2 Distinctive Features of CA

Understanding CA requires an appreciation of its distinctive features. Three distinctive features [10] include (1) Distinction between means and ends—instruments, such as mobile-phones, can be perceived as means to facilitate a sense of individual achievement as in 'improved social stature', or for 'business networking', etc. This distinction is important, as (intrinsically) just owning a mobile-phone can be mistakenly evaluated as an end. (2) Distinction between means and capabilities—means are distinct from *capabilities* in that just access/possession to a means does not translate to *capability*. Further means would require other essential means for its effectiveness, e.g., For the *capability* of business networking, an individual should be conversant with using the means of a mobile phone, the successful operation of which depends on other essential means such as mobile connectivity, power, etc. (3) Importance in understanding relation among means—multiple interdependent means, such as with mobile phones, connectivity, talk-value, etc., can be in '*and*' or '*or*' dependency. *And* dependency indicated concurrent availability, e.g., Owning a mobile phone *and* available connectivity *and* talk-value would support the *capability* of business networking. *Or* on the other hand, a person could be engaged for business networking. While this and/or relation between means might seem trivial, it is in reality fundamental [11, 12] to how a product can succeed in a market, for e.g., access and availability of basic infrastructure was mistakenly taken for granted in initial ICT projects in developing countries, resulting in failures [13]. A fourth distinction of CA [11] is in its qualitative vs quantitative dimension, for e.g., an achieved functioning of adequate food (to satisfy hunger for wellbeing and productivity), must include adequate quantity *and* adequate quality (nutrition value). Though this might seem apparently obvious or trivial, it has not been explicitly dealt with regards *capabilities* and carries the risk of being overlooked.

3 Capability to Design

3.1 A Capability Approach Perspective

According to CA, a designers' (as an individual and/or a design firm) capability are the available (internal and external) effective options/resources to design (a product) as intended. This would include, besides intuitive and trained ability to design, available design tools, Computer Aided Design (CAD), prototyping facilities, set design-values (guidelines), information support, etc. However, dictionary based meaning of capability would imply an individual's 'quality of being capable—physically, intellectually or legally' [14] to design in the current context. The discussions in this paper are targeted at products designed for mass consumption in free market, as in Sen [3].

3.1.1 Capability to Design: Distinction Between Means and Ends

A design statement specifies the exact motives and requirements of the product/service being designed. However, generally this statement fails to permeate to the level of questioning the fundamental purpose 'why' of designing (a design individual and/or a firm/organization). Designing can be a means of achieving something intrinsically important, for e.g., *designing* for purpose of profit would differ from *designing* for a more nobler cause such as charity/service, and both would vary in terms of *ends* achieved. Further, companies running for profit, could also base their design on other intrinsic values such as sustainability, abolishing child-labor, abolishing poached product-use, etc. A company's mission statement would reveal the underlying values guiding the means of *design*. Design is responsible for the resources consumed throughout the lifecycle of the product. Thus for the same design statement (challenge), firms with different underlying values would differ both in the means adopted and ends achieved. Design firms working with strong sustainability values will adopt appropriate capabilities to design appropriate products (ends). For e.g., 3M a global company known for its

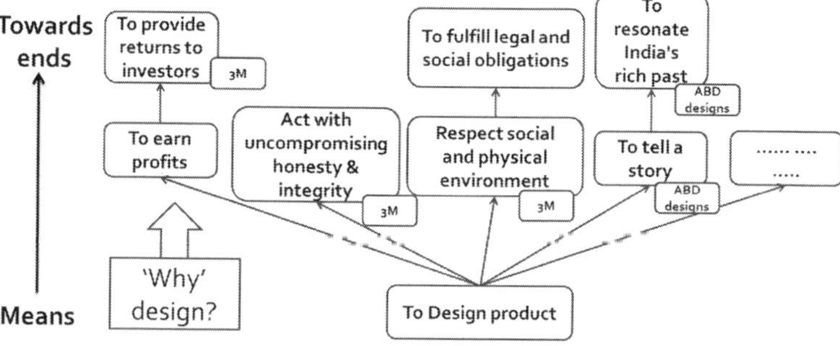

Fig. 1 Distinction between means and ends

innovative products states the values that defines (see Fig. 1) the ends of what they do, such as [15]:

> Act with uncompromising honesty and integrity in everything we do
> Satisfy our customers with innovative technology and superior quality, value and service
> Provide our investors an attractive return through sustainable, global growth.
> Respect our social and physical environment around the world
> Value and develop our employees' diverse talents, initiative and leadership
> Earn the admiration of all those associated with 3M worldwide

An Indian design firm, Studio ABD, a recipient of Red dot award 2010 states: '*At Studio ABD we design from our heart. Emotion underlines our products, giving them poetic and inspired meaning. They connect deeply with the user by* ***telling vivid stories****, by overlaying the familiar with the new and surprising ..... propelled by humor, craft, rituals, people, situations and Indian heritage, we create products that speak a unique language—an Indian design vocabulary.*'[16]

3.1.2 Capability to Design: Distinction Between Means and Capabilities

Designers need skills linked with imagination to help them in generating (innovative) artifacts that are yet non-existent. Understanding design methodology, basic engineering knowledge, ergonomics, rendering skills and CAD tools are some core skills linked with the design. The designer's experience plays a critical role. Further, there is convincing proof in the design literature [17, 18] and in practice that design requires significant external support. Top management support positively impacts design quality/effectiveness [19]. Designs targeting BOP customers pose different demands, a different methodological approach [20].

Thus, the list of resources required to support the designer is quite varied and difficult to comprehend without a tool/aid. Kleine [21] provides a 'choice framework' to aid the listing of resources (interchangeably referred as 'means' in the

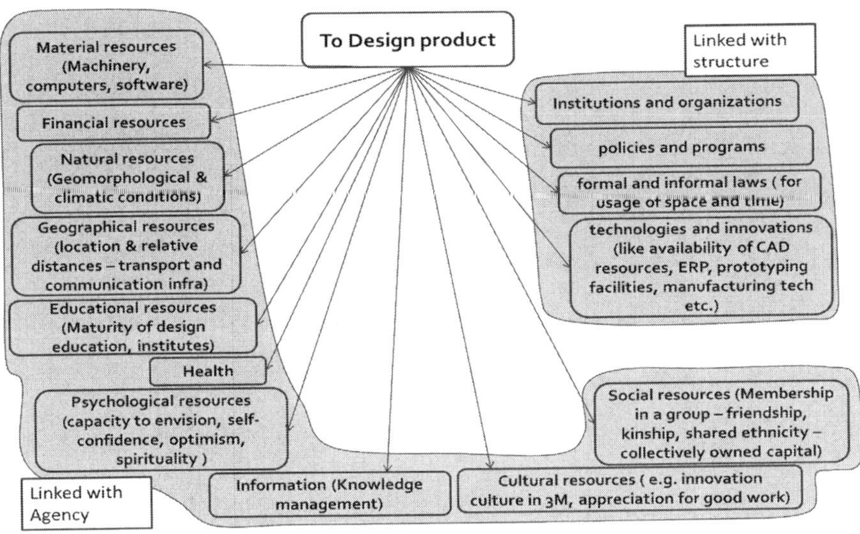

Fig. 2 Distinction between means and capabilities. *Source* Adapted based on choice framework [21]

current paper) required to fulfill a given *capability*. Figure 2 illustrates the list for the paper's current theme 'to design a product'. This framework differentiates *agency* linked resources (resources that directly interface with an individual to help in taking charge of their destiny) from *structural* resources available to an individual.

3.1.3 Capability to Design: Importance Given to Understanding Relation Among Means

Relation ('*and*'/'*or*') between resources would depend on the case being considered and requires to be discerned explicitly with regards *capability* to design. The design firm's policies, cultural and social resources, which significantly influence the psychological resources of an individual, could be considered as being in '*and*' relation working as a set. Absence of any one of the resources (in '*and*' relation) would make the set ineffective. Natural resources in realizing a product (e.g., biodegradable fabric), could be in 'or' relation, viz., cotton '*or*' jute.

3.1.4 Quality and Quantity Dimension of Capability to Design

The mere availability of resources as illustrated in Fig. 2 in adequate *quantity* would not ensure a fulfilled *capability* to design for the designer (individual/firm).

The *quality* of material resources is as important as *quantity* and are in *'and'* relation as inadequacy in either would result in a design inadequacy. *Quality* reinforced through design-firm's policy, influences the *capability* to design. If the *quality* of a PC's hardware-software configuration is unsuited for a desired function then its physical presence (*quantity*) is ineffective.

4 Effects of Design on User's *Capabilities*

4.1 Capability Space

Technologies/products, appropriately designed, can serve as the potential means for *capability* expansion [22]. An intention is definitely in-built the product attributed with the purposeful nature of the design [23]. Each product is designed to satisfy customer's needs. That satisfied need could be considered in terms of CA as the achieved beings and doings (*capabilities)*, thus *functionings* of the user. A given product may work as instrument to achieve some high level need, which in turn will fulfill other higher level need, till the intrinsic end is achieved (refer Sect. 3.1.1), for e.g., with access to a sewing machine, a beneficiary's health and ability to use the machine effectively translates as a *capability* to sew. This, through proper market mechanisms, provides a vocation to earn income (level 1) which beyond providing adequate nutrition (alleviating hunger) can support the education of children (level 2) through which there's a perceived improvement in social stature. The beneficiaries, as parents, can now achieve an intrinsic end of being proud parents (an achieved functioning). Thus a single product like a sewing machine, along with maze of other products/services/resources, renders an individual with different *capabilities*. Thus many products can converge to serve a single purpose. Likewise, in a market, a product can satisfy multiple purposes, e.g., a mobile phone can serve for business communication as well as for esteem/stature.

The concept of *capability* space is illustrated through Fig. 3 to discern the effect of design on *capabilities*. *Capability 1* is possible by adopting either or a combination of Products A, B and C. The space bound by these three products and linked resources can be outlined as Capability 1 space (thick black line in Fig. 3). The three product alternatives may vary in terms of linked resource needs, e.g., a solar LED lantern vs. a rechargeable CFL lantern vs. a user-replaced battery light have varying resource requirements. The first permits self-reliance; the second requires a charging-point, while the third requires periodic battery replacement. The choice of either of these would determine the linked *capabilities* (and profit), e.g., for a poor street-vendor requiring illumination to better his sales to make a living. Some resources might be shared amongst product alternatives while some could be unique.

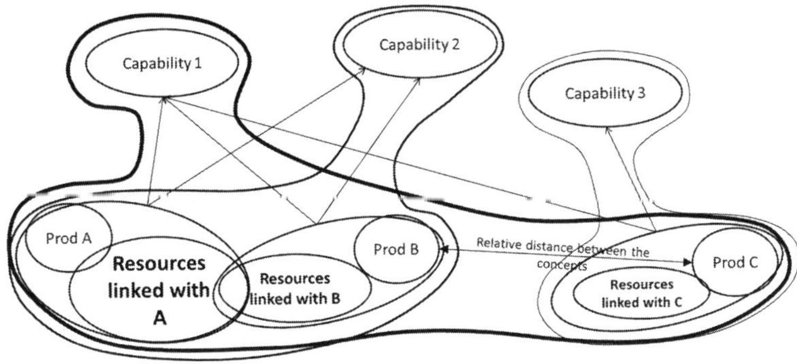

Fig. 3 Capability space

4.2 *Effects of Design on Capabilities*

Design affects the capabilities of the users/beneficiaries in following three ways:

4.2.1 Design Operating in a for-Profit Market Results in a Restrained Capability Space

Products 'for' profit tend to monopolize to maximize market share (volume of sales). The designer has to ensure that the products meet functional requirements, is appealing and still meets production cost targets to maintain competitive edge. With more than one product, with varying quality (and pricing) to serve the same purpose, increases the potential *capability* space for the user. However, for a company, aiming to monopolize sales, maintaining many products for the same purpose may not be profitable, whereby they tend to retain only few such products. Consequently customers' *capability* space and available options are restricted. Development by definition can be number of options available to people to lead life [3]. A variety of traditional hand-made toys are vanishing (along with the associated skills and local material use) from the market with the dominance of standard mass-produced products using cheap (and often toxic) plastics. Thus, design for profit may not necessarily be amenable to sustainability of *capabilities*.

4.2.2 Design Decisions Directly Affect the Capability Space

Product design determines effectiveness of achieving product functionality. Besides material resources for manufacturing, design dictates the resources (energy, accessories, training, etc.) required for its successful usage by the user/beneficiary. A product designed to utilize resources commonly available to people (consumers) further increases their *capability* space, e.g., water cans (to fetch drinking water) that can be fitted on bicycles that are ubiquitous.

Variety in design is crucial, and is defined as the distance of a product concept from the center of the concept space [24]. A product design that is farther away from the center can potentially expand the users' *capability* space (see Fig. 3). For e.g., for communication and socializing, video conference provides an opportunity which is far varied from actual face to face meeting and significantly expands the *capability* space, such as in tele-medicine where the deprived and impoverished from Africa and Asia have access to the best of medical diagnosis. However, disruptive innovations can bring in too much variety that can kill existing products, and severely constrict the *capability* space. In some cases certain section of society may not have the new resources required to benefit from the new product, and old products are no longer available. This usually leaves a void of helplessness in the *capability* space. Products designed and developed with the prime motive of profit can cause such voids amongst the lower economic strata or BOP. The prevalent evidence of the BOP having little or no access to basic healthcare contrasted with thriving multispecialty hospitals within the same landscape is one example [25]. Software's developed for a particular Operating System is a good example. Emergence of microfinance is an example of filling such a void in the *capability* space amongst the poor's *capability* to access capital [26]. As discussed in the earlier section, the product/service quality dimension is crucial as access to inferior quality products would represent fulfilled capabilities, but with limitations.

4.2.3 Design by Virtue Cannot Shy Away from Being Paternalistic

Emphasis on freedom of choice (for the BOP) places enormous importance to participatory design to include customers' preferences and aspirations. This infuses customer's voice in Capability Approach integrated design. Though designers' tend to give due respect to customer aspirations (social, cultural, esteemed, etc.), they often are compelled to take unilateral decisions in interest of the beneficiary. Given the number of stakeholders and the consequent plurality of the considerations involved (as discussed in earlier sections), the designer's role becomes inadvertently paternalistic. The designer is thus the architect of the users/beneficiaries' *capability* space. This is valid even if the design has been approved by the users, as the designer takes a paternalistic stand from the instance he proposes a set of design/product alternatives for the users/beneficiaries' to choose from. As in the case of disruptive innovations, users might not be able to fully comprehend the entire benefits resulting from the envisaged product [27].

5 Summary

This paper provides an overview of Capability Approach and its critical role in designing for the BOP. It moves beyond conventional definition of capability, to include effective (operational) options available to an individual to be and do as aspired, in leading a life of value upon reflection. The paper views various stakeholders involved in a successful design (product/service) as individuals with *capabilities*. The current paper evaluates the *capabilities* of the designer (individual or a firm) for its impact on designing products aimed at the BOP. A concept of *capability* space has been adopted to provide a vantage for the designer to evaluate the effect of design on the *capabilities* of users (beneficiaries). Such studies would permit the evaluation of successes and/or failures associated with products/services aimed at expanding/fulfilling *capabilities* of the users. Mapping the *capability* space before and after introduction of a product can help in further discerning the multi-stakeholder system and underlying influences.

Acknowledgments This research has been made possible with a research grant from the Netherlands Organization for Scientific Research (NWO) for the project 'Technology and Human Development—A Capability Approach'

References

1. Blessing L, Chakrabarti A (2009) DRM, a design research methodology. Springer, London
2. Prahalad CK (2005) The fortune at the bottom of the pyramid: eradicating poverty through profits. Wharton School Publishing, Singapore
3. Sen A (2000) Development as freedom. Oxford University Press, New York
4. Mani M, Ganesh LS, Varghese K (2005) Sustainability and human settlements. Sage Publications, New Delhi
5. Banerjee A, Duflo E (2011) Poor economics. Randon house publishers, Noida
6. Margolin V, Margolin S (2002) A "social model" of design: issues of practice and research. Des Issues 18(4):24–30, MIT, Cambridge
7. Ahmed S, Blessing L, Wallace K (1999) The relationships between data, information and knowledge based on a preliminary study of engineering designers. ASME design theory and methodology, DETC/DTM-8754, Lag Vegas, Nevada
8. Diehl JC (2009) The first learning experiences of design for the BoP. In: Kandachar P, Jongh I, Diehl JC (eds) Designing for emerging markets: design of products and services. Delft University of Technology, Delft
9. Sen A (1983) Economics and the family. Asian Dev Rev I
10. Robeyns I (2005) The capability approach: a theoretical survey. J Hum Dev 6(1):93–114
11. Khadilkar P, Monto M (2012a) A methodological framework to investigate the connect between capability approach and livelihood enhancement using ICT. In: Dahmani A, Ledjou J-M (eds) The information and communication technologies (ICTs) in southern countries: from promises to the socio-economic reality. Karthala Publications, Paris
12. Khadilkar P, Monto M (2012b) Assessment of technology in the view of sustainability and capability approach. In: Mukhopadhyay C, Akhilesh KB, Srinivasan R, Gurtoo A, Ramachandran P, Iyer PP, Mathirajan M, BalaSubrahmanya MH (eds) Driving the

economy through innovation and entrepreneurship: emerging agenda for technology management Springer, India
13. Gigler B (2004) Including the excluded—can ICTs empower poor communities? Towards an alternative evaluation framework based on the capability approach, international conference on the capability approach, University of Pavia, Italy
14. Word web software (2010) Version 6.3. Princeton University. Accessed in April 2012
15. 3M India official website. http://solutions.3mindia.co.in/wps/portal/3M/en_IN/About3/3M/. Accessed in July 2012
16. Studio ABD official website. http://www studioabd.in/about.html. Accessed in July 2012
17. Cross N (2000) Engineering design methods strategies for product design. Wiley, West Sussex
18. Ulrich KT, Eppinger SD (2004) Product design and development. Tata McGraw-Hill, New Delhi
19. Swink M (2000) Technological innovativeness as a moderator of new product design integration and top management support. J Prod Innov Manag 17:208–220 Elsevier
20. Ramani SV (2011) On the diffusion of toilets as bottom of the pyramid innovation: lessons from sanitation entrepreneurs. Tech Forecasting Soc Change. Elsevier Inc 79(2012):676–687
21. Kleine D (2010) Ict4what?—using the choice framework to operationalise the capability approach to development. J Int Dev 22:674–692
22. Oosterlaken I (2009) Design for development: a capability approach. Des Issues 25(4):91–102. MIT, Cambridge
23. Simon H (1982) The sciences of the artificial. MIT Press, Cambridge
24. Shah J, Vargas-Hernandez N (2003) Metrics for measuring ideation effectiveness. Des Stud 24:111–134 Elsevier
25. UN (2012) The millennium development goals report 2012. New York
26. Yunus M (2007) Banker to the poor. Penguin books, New Delhi
27. Leonard D, Rayport J (1997) Spark innovation through empathic design. Harvard Bus Rev. Harvard business school publishing, 102–113

Author Index

A. Chakrabarti and R. V. Prakash (eds.), *ICoRD'13*, Lecture Notes in Mechanical Engineering, DOI: 10.1007/978-81-322-1050-4, © Springer India 2013

V (*cont.*)

W

Y

Z

Printed by Publishers' Graphics LLC